THE MINES AND MINERALS OF BERKS COUNTY, PENNSYLVANIA

Ronald A. Sloto

COPYRIGHT 2016

By Ronald A. Sloto

Second Printing

Cover Photograph: The Jones Mine, Caernarvon Township, 1910

TABLE OF CONTENTS

	Page
Introduction	1
Format of this book	1
Mineral Products of Berks County	3
Minerals of Berks County	4
Natural Science Organizations	5
The Reading Cabinet	5
Reading Society of Natural Sciences	5
Reading Chapter of the Agassiz Association	6
Spencer F. Baird Association	7
Mt. Penn Association of Sciences	7
Berks County Mineralogical Society	8
Reading Public Museum	8
Acknowledgments	9
Albany Township	11
Lynnport Group Slate Quarries	11
Quaker City Slate Company Quarries	11
Hemerly Quarry	12
Mammoth Quarry	12
Oswold Quarry	12
Centennial Quarry	12
Pittsburg Quarry	13
Unnamed Slate Quarry 1	13
Unnamed Slate Quarry 2	13
Greenawald Group Slate Quarries	13
J. Wibur Company Quarry	13
Atlas Mineral Products Albany Quarry	14
B. Frank Ruth and Company Quarry	15
Focht Quarry	17
Unnamed Slate Mine	17
Other Quarries	17
Greenawalt Quarry	17
Kunkle Quarry	17
Gilt Flagstone Quarry	17
Legler Flagstone Quarry	17
Berg's Farm Locality	18
Brobst (Union) Furnace	18
Alsace Township	19
Temple Quarry	19
Trexler Mica Mine	19
Gottschall's Mine	20
Gottschall's Farm Locality	20
Valentine Hartman's Mine	20
Salem Church Ocher Pit Locality	21
Weist School Locality	21
McKnights Gap Locality	21
Mount Laurel (Alsace) Furnace	21
Amity Township	22
J. Rhodes Sandstone Quarry	22
Bechtelsville Borough	22
Miller Quarry	22
Bechtelsville Furnace	23
Bern Township	26
Essig Quarry (Ohnmacht Cave)	26
Rickenbach Area Quarries	27
Epler Quarries	27
Reading Railroad Quarry	28
Red Bridge Quartz Crystal Locality	28
Bernville Borough	28
Bernville Area	28
Bethel Township	28
Frystown Fetid Barite Nodule Localities	28
Huber Farm Fetid Barite Occurrence	29
Miller Farm Fetid Barite Occurrence	29
Sadler Farm Fetid Barite Occurrence	29
Kline Umber Prospect	29
Birdsboro Borough	29
A Very Brief History of the E & G Brooke Iron Co	29
Boyertown Borough	36
Early History	36
Geology	40
Eckert Open Cut and Slope	41
California Mine (Phoenix Mine Lower Slope)	41
Phoenix Mine Upper and Middle Slopes	43
The Great Trespass	45
A Trip into the Boyertown Mines	46
Warwick Mine	46
Gabel Mine	50
The Gabel Mine at Boyertown	52
John Rhoads Mining Company	54
Seminary Shaft	55
Later Mining Activities	55
Brecknock Township	59
Caernarvon Township	62
Iron Mines	62
Jones Mine	62
Kinney Mine	78
Byler Mine	78
Limestone Quarries	79
Styer Quarry	79
Daniel Mast Quarry	80
Hertzler Quarry	81
John Plank Quarry	82
John Stoltzfus Quarry	82
Christian Mast Quarry	83
P.W. Plank and Company	83
Brunner Farm Quarry	83
Byler Farm Quarry	84
Morgantown Roadcut Scapolite Occurrence	84

Centre Township 84
    Westley's Quarry 84
    Price's Quarry 85
    Shonour's Quarry 85
Colebrookdale Township 85
    Bechtelsville (Martin) Quarry 85
    Stauffer's Quarries 86
    Brower Mine 87
    Rhoades and Grim Mine 87
    New Berlinville Clay Mine 88
    Railroad Cuts South of Boyertown 89
    Graphite Mines 89
        Boyertown Graphite Co Mine (Betchel Farm) 89
        Dr. Funk's Fish Dam Graphite Mine 89
Cumru Township 92
    Fritz Island Mine 92
    Raudenbush Mine 100
    Opposite Fritz Island Locality 102
    Fehr and O'Rourke Stone Company Quarry 102
    U.S. Route 222 Roadcut Locality 105
    Schlegel's Farm Locality 105
    Schuylkill Copper Mining Company Mine 105
    Almshouse Quarry 106
    Mount Penn Furnace 107
District Township 107
    Early Iron Mines 107
    Mica Prospect Number 2 107
    Hoffman Prospect 107
    Peter Smith Farm Graphite Prospect 107
    District (German) Furnace) 107
Douglass Township 108
    Greshville Quarry 108
    J. Davidheiser Quarry 109
    Keely Quarry 109
    Wren Mine 109
    Pottstown Trap Rock Quarry 109
    Little Oley Reibeckite Locality 111
Earl Township 111
    Iron Mines 111
        Kauffman and Spang Iron Mine 111
        A Visit to Ore Mines in Earl Township 113
        Dotterer Mine 114
        Berks Development Company 116
    Limestone Quarries 116
        David Davidheiser Quarry 116
        Fryermuth Quarry 116
        Rapp Quarry 117
        Mengle's Quarry 117
        Boyer Quarry 117
    Other Mines and Localities 117
        Pennsylvania Uranium Mining Co Prospect 117
        Mica Prospect Number 1 117
    Riebeckite Localities 118
        Gabelsville Riebeckite Localities 119
        Shanesville Riebeckite Locality 119
        Oley Valley Electric Railway Cut Riebeckite
            Locality 119
Exeter Township 119
    Jacksonwald Occurrence 119
    Kinsey Hill Locality 120
    Bishop's Mill Locality 120
    Snydersville Malachite Occurrence 121
    Stonersville Magnetite Occurrence 121
    Bishop Mine 122
    Monocacy Hill 122
    Esterly Mine 122
    Guldin Hill Sandstone Quarry 122
    Limestone Quarries 123
        D. Snyder Quarry 123
        S. Kauffman Quarry 123
        Albert Knabb Quarry 124
        Benjamin Ritter Quarry 124
        Cornelius Tyson Quarry 124
        Abandoned Quarry at Jacksonwald 124
        Reifton Quarry 125
        Jonas DeTurk Quarry 125
    Klapperthal Park Oil Well 125
Greenwich Township 125
    Greenwich Manufacturing Company Quarry 125
    Maiden Creek Furnace 126
Heidelberg Township 127
    William Moore Quarry 127
    John Marshall Quarry 127
    Samuel Deppen Quarry 128
    W. Wenrich Quarry 128
    Reed Quarry 128
    Big Spring Quarry 128
    Ryland Road Quarry 129
    Sheetz Sand Quarry 129
Hereford Township 129
    Siesholtzville District Iron Mines 129
        Bittenbender Mines 133
        Christian Bittenbender Mine 138
        J. B. Gehman Mine 139
    Other Mines and Prospects 139
        Olafson Mine 139
        Huff Church Locality 140
        Rauch Mine 140
        Dale Mine 140
        Rush's Ore Pit 140
        Reitnauer Mine 141
    Limestone Quarries 141

  Hampton Furnace Quarries 141
   Trollinger Quarry 143
   Clemmer Quarry 143
 Seisholtzville Granite Quarry 143
 Gregory Graphite Prospect 145
 Hereford (Mayburry's) Furnace 145
Kutztown Borough 145
 Kutztown Furnace 145
Leesport Borough 145
 Leesport Iron Company (Leesport) Furnace 145
 Mining and Washing Iron Ore 146
Longswamp Township 148
 Rittenhouse Gap District 149
  The Catasauqua and Fogelsville Railroad 152
  Rock Mine 153
  Tunnel Mine 154
  Conrad's Slope 156
  Gap Mine 156
  Ginkinker Mine 157
  Thomas Iron Company Mines 157
  Other Mines 157
 Thomas Iron Company 158
 Red Lion Station Mines 160
  Trexler Mine 160
  Weiler Mine 160
  Wetzel Mine 160
  Dunkle Mines 160
  Miller Farm Mines 161
 Garner Station Mines 161
  Bethlehem Iron Co. Gardner Station Mines 161
  Thomas Iron Company Garner Station Mines 162
  Finley Mine 162
  Old Mickley Mine 162
  Smoyer Mine 162
  Peter Kline Mine 162
 Klines Corner Mines 162
  D.K. Kline Mine 166
  D.L. Trexler Mine 166
  Thomas Iron Company Klines Corner Mines 166
  Allentown Iron Company Mine 168
  Moatz and Schrader Mine 168
  Fritch and Brother's Mine 168
  Fenstermacher Mine 168
  Mining in Klines Corner in 1898 170
  Amos Fisher Mine 170
  Fleetwood Iron Company Mine 170
  Klines Corner Mine 170
  Other Iron Mines 170
 Farmington Area Mines 172
  Zeigler Mine 172
  Merkel Mine 172

  Fegley and Walbert Mine 172
  Other Iron Mines 173
 Shamrock Area Mines 173
  John Fegley Mine 174
  Solomon Boyer & Company Mine 174
  Fegley Mine 174
  Jesse Laros's Mine No. 3 174
  Jesse Laros's Mine No. 4 174
  Aaron Hertzog Mine 174
  Wescoe Mine 174
  Wagenhorst Mine 174
  Henry Stein Mine 175
  Dresher Mine 175
  Haas Mine 176
 Longsdale Area Mines 176
  Litzenberger Mine 176
  Lichenwallner Mine 176
  Long Mines 176
  East Penn Graphite Company Mine 177
  Star Clay Company 178
  Kaolin Mining In Longswamp Township 179
  Hancock Mud-Dam Deposit 181
  Fritch and Brother Iron and Clay Mines 181
  Frederick Mines 182
  Tatham Mine 182
 Topton Area Mines 183
  Henry Erwin and Sons Ocher Mine 183
  Atlas Ocher Mine 185
  Other Localities 185
  Walker Granite Company Quarry 185
  E.H. Trexler Quarry 186
 Graphite Prospects 186
  Franklin DeLong Graphite Prospect 186
  Charles Brensinger Graphite Prospect 186
  Schmeck's Farm Graphite Prospect 186
Lower Alsace Township 187
 Big Dam Quarry 187
 Antietam Reservoir Locality 187
 Railroad Cut Opposite Poplar Neck 189
 Neversink Mountain Sienna Mine 189
 Stony Creek (Ohlinger) Mills Locality 190
 David Knabb Mine 190
 Fischer Prospect 190
Lower Heidelberg Township 191
 Berkshire Furnace 191
 Berkshire Furnace Mine 191
 Benjamin Hull Quarries 191
 Hospital Creek Quarry 191
 Glen-Gery Quarry 192
 Old Quarry No. 1 192
 Old Quarry No. 2 192

Lyons Borough 193
    East Penn Furnace 193
Maidencreek Township 193
    Limestone Quarries 193
        Reading Cement Company Quarry 193
        Vindex Portland Cement Company 194
        Maidencreek Portland Cement Company 195
        Evansville Quarry and Cement Plant 196
        J.M. Meredith Quarry 201
        Pleasant Hill Road Quarry 201
        Maidencreek Quarry 201
    Iron Mines 202
        Crane Iron Company Mine 202
        The Crane Iron Company 204
        Shaeffer's Old Mine 206
        Shaeffer's New Mine 206
        Wade Mine 206
Marion Township 207
    Dishong (Stouchsburg) Quarry 207
    Smaltz Road Quarry 207
    Filbert Ocher Mine 207
Maxatawny Township 207
    Limestone Quarries 207
        Hinterleiter Quarry 207
        Hottenstein Quarry 210
        Keystone Quarry 210
        Berks Products Quarry 211
        Bowers Quarry 211
        Angstadt Quarry 211
        A.G. Smith Quarry 212
        Kohler Quarry 212
        Kohler Road Quarry 214
        Bailey Quarry 214
    Iron Mines 214
        Klines Corner—Farmington Area Iron Mines 215
        Samuel Lewis Mine 215
        Charles Miller Mine 215
        Schweyer and Liess Mine 215
        Other Iron Mines 215
    Lyons Area 216
Muhlenberg Township 216
    W. Hartman Farm Locality 216
    Barnhart's Dam Locality 216
    Brook's Quarry 216
    Laureldale Quarry 216
    Nolan's (Tuckerton) Cave 221
    South Temple Quarry 221
    G.W. Focht Stone Company Quarry 222
    Temple Sand Company Quarry 223
    Gossler Quarry 224
    Becker Quarry 224

    Blue Bell Lime Company 224
    Temple Furnace 225
New Morgan Borough 227
    Grace Mine 227
    A Very Brief History of the Bethlehem Steel Corp. 248
North Heidelberg Township 250
    Tulpehocken Stone Company Quarry 250
Oley Township 250
    Limestone Quarries 250
        Eastern Industries Quarry 250
        Lehigh Cement Company Quarries 254
        L. DeTurk's Quarries 255
        Houck Quarry 256
        Peter Guldin Quarry 256
        Levi Hartman Quarry 256
        S.P. Guldin Quarry 256
        E. Schaeffer Quarry 257
        Colonel J. Weaver's Quarry 257
        Seth Grim Quarry 257
        Kemmerer Quarries 258
        P. Sneider Quarry 258
        Hine Quarry 258
        Levi Knabb Quarry 258
        Raudenbusch Quarry 259
        Ezra Griesermer Quarry 259
        J.G. Fischer Quarry 259
        Levi Herbein Quarry 259
        L.J. Bertolet Quarry 259
        John Snyder Quarry 259
        Isaac Brumbach Quarry 259
        David Yoder Quarry 260
        Reuben Shearer Quarry 260
        Wilman Quarry 260
        Schollenberger Quarry 260
        William Weidner Quarry 260
        Deisher Quarry 260
        J.G. Bertolet Quarry 260
        D.F. Bertolet Quarry 261
        F.V. Kauffman Quarries 261
        Thomas P. Lee Quarry and Farm Locality 261
        Reiffe Quarry 261
        Yale Quarry 261
    Slate Quarries 261
        Israel Bertolet Quarry 261
        Wellington B. Griesemer Quarry 261
    Iron Mines 261
        Hunter Mine 262
        Weaver Mine 262
        Manwiller Mine 263
        Hertzel and Swoyer Mine 264
        Talley Mine 264

Other Localities 264	Hay Creek Prehnite Locality 304
Oley (Friedensburg) Area 264	Humphrey Sandstone Quarries 304
Oley Uranium Occurrence 264	Joanna Furnace 305
Oley (Shearwell) Furnace 265	Robesonia Borough 305
Ontelaunee Township 265	Albert Wenrich Quarry 305
Ontelaunee Quarry 265	Robesonia Furnace 306
Maidencreek Station Quarry 267	Rockland Township 307
Leesport Iron Company Quarries 267	Beitler Mine 307
Perry Township 268	Percival Brumbach's Mine 308
Glen-Gery Quarries 268	Green Hill Area 309
Jacob Leiby Flagstone Quarry 270	Old Millert Mine 309
Collier's Flagstone Quarry 270	New Jerusalem Mines 309
Weidman Farm Shale Prospect 270	Bieber Mine 309
Pike Township 270	Daniel Fisher Mine 310
Rohrback Mine 270	Schrading Mines 310
S. Yoder Mine 270	Flint Hill Locality 310
Mines South of Lobachsville 270	Sally Ann (Hunter's, Rockland) Furnace 310
Hill Church Riebeckite Localities 271	Ruscombmanor Township 312
Mica Prospect Number 3 271	Udree Mine 312
John Keim Quarries 271	Clymer Mines 313
Rolling Rock Building Stone Quarry 271	Old Tunnel Mine 314
Reading 272	Schittler (Hain) Mine 314
The Eckert Family 272	Schweiter and Kutz Mine 315
Mount Penn Iron Mines 273	Pricetown Area 316
Mount Penn White Spot Quarry 275	Other Localities 316
Whitman Quarry 275	Sand Hill Sand Quarries 317
The Pagoda 276	Blue Quartzite Quarry 317
Mount Penn Quarries 277	Bomegratz's Farm Quartz Locality 317
Long's Quarry 277	Blandon Area Wavellite Locality 317
Gring's Quarry 277	Jacob Fox Farm Locality 317
Gold in Reading 278	Sinking Spring Borough 318
Quarries in the City of Reading 278	Evans Quarry 318
Henry Clay Furnace 279	George Peipher Quarry 318
Keystone Furnace 280	Deckert Quarry 318
Reading Iron Company Furnace 280	Shillington Road Quarry 319
Richmond Township 281	Thierwechier Mine 319
Dragon Cave 281	Sinking Spring Area 319
C.K. Williams & Company Mines 282	South Heidelberg Township 320
Fleetwood Area 286	Levi Goul Quarries 320
Kirbyville Quarries 286	Daniel Seltzer Quarry 321
Moselem Mines 288	Huyett Quarry 321
Virginville Area 290	Limestone Quarry Number 2 321
Kirbyville-Moselem Springs Quartz Crystal Locality . . . 290	Fritztown Area 322
Noll's Mine 290	Cushion Mountain Locality 322
Charles Heffner's mine 290	Hain Mine 322
Merkel Quarries 291	Cushion Mountain Iron Mines 322
Richmond No. 5 Quarry 291	Spodumene from Cushion Mountain 323
Robeson Township 291	Spring Township 322
Dyer Gibraltar Quarry 291	Limestone Quarries 322
John T. Dyer 292	Ludwig Quarry 322
Dyer Trap Rock (Birdsboro Materials) Quarry 297	

Ruth Quarries 322	Landis Mine 345
Breneman Quarry 324	Barto Mine 346
Gring's Sand Quarry 324	Berthou Mine 347
Iron Mines 324	Dysher Mine 347
Wheatfield Mine 324	Other Iron Mines 347
Ruth Mine 329	L. Gilbert (Edison) Mine 347
Raub Mine 330	Gilbert Mine 349
Grill MIne 331	Stauffer Mine 350
Eureka Mine 331	Gilberg Mine 350
Montello Brick Works 331	Sparr Mine 350
Fritztown Area 333	Eline Mine 350
St. Lawrence Borough 333	Fegley mine 351
W.M. Stauffer Quarry 333	Quarries 351
Esterly Area 333	Schall Quarry 351
Tilden Township 333	Oberholtzer Quarry 351
Topton Borough 334	Dielh Quarry 351
Topton Furnace 334	Barr Quarry 352
Tulpehocken Township 334	Rush Quarries 352
Frystown Fetid Barite Nodule Localities 334	Gabel Quarry 353
McLain Farm Fetid Barite Occurrence 335	Eshbach Area 353
Burkholder Farm Fetid Barite Occurrence 335	Wernersville Borough 353
Moore Farm Fetid Barite Occurrence 335	Deppen Quarries 352
Gibble Farm Fetid Barite Occurrence 336	Witmoyer and Brother Quarry 353
Elvin Kurtz Farm Fetid Barite Occurrence 336	Henry Miller Quarry 353
Kurtz-Landis Farm Fetid Barite Occurrence 336	Wernersville Lime and Stone Company Quarry 354
Bohn Farm Fetid Barite Occurrence 336	West Reading Borough 355
Union Township 336	McQuade Brothers (Frill) Quarry 355
Dyer Monocacy Quarry 336	Drexel Quarry 357
Hopewell Furnace 342	Wyomissing Borough 357
The Iron-Making Process at Hopewell Furnace 343	Seitzinger Mine 357
Birdsboro Sandstone Quarries 344	Muhlenberg (Beidler) Mine 358
Hampton Furnace 345	William Gudlin Quarry 358
Washington Township 345	Bibliography 359
Mount Pleasant (Barto) Mines 345	

APPENDICIES

Appendix 1 Maps showing the locations of mines and mineral localities of Berks County 379
Appendix 2 Index to Berks County mines and mineral localities by name 417
Appendix 3 Index to Berks County mines and mineral localities shown in Appendix 1. 435

FIGURES

		Page
1.	Municipalities of Berks County.	2
2.	Comparison of aerial photography, topographic mapping, and LIDAR imagery for the same area	2
3.	John P. Heister's 1854 geologic map of Berks County	3
4.	Index to geological mapping in Berks County	4
5.	Officers of the Mt. Penn Association of Science in 1898	7
6.	Aerial photograph of the slate quarries at Quaker City, Albany Township, September 12, 1937	9
7.	George Deysher posing next to a Quaker City slate quarry boiler	10
8.	Slate quarry near Quaker City, Albany Township	10
9.	Quaker State Slate Company quarry, Albany Township, April 2015	10
10.	Advertisement for the Philadelphia Slate Mantle Company	11
11.	Letterhead of the Pittsburg Slate Company	12
12.	The J. Wilbur quarry, Greenawald, Albany Township, ca. 1925	13
13.	Concentric circles in shale with pyrite in the center, Greenawald, Albany Township	14
14.	Maximillian Wirtz at his desk	15
15.	Atlas Mineral Products plant and rail cars, Albany Township	15
16.	Atlas Mineral Products plant, Albany Township	16
17.	Atlas Mineral Products plant and quarry, Albany Township	16
18.	Atlas Mineral Products quarry, Albany Township	16
19.	Slate quarry, Albany Township	17
20.	Smoky quartz from Albany Township	18
21.	Aerial photograph of the Berks Silica Sand Company, Alsace Township, 2015	18
22.	Sketch map of the Berks Silica Sand Company quarry, Alsace Township	19
23.	Muscovite from the Trexler mica mine, Alsace Township	19
24.	Allanite from the Trexler mica mine, Alsace Township	20
25.	Enstatite, var. hypersthene, from Gottschall's mine, Alsace Township	20
26.	Hedenbergite from Gottschall's mine, Alsace Township	20
27.	Molybdenite from Valentine Hartman's mine, Alsace Township	21
28.	Goethite from the Salem Church locality, Alsace Township	21
29.	Gravel containing gold from McKnight's Gap, Alsace Township	22
30.	Allanite from McKnight's Gap, Alsace Township	22
31.	Bechtelsville Iron Company (Norway) Furnace, Bechtelsville	23
32.	Bond issues by the Bechtelsville Iron Company in 1876	24
33.	Bechtelsville (Norway) Furnace weekly report for the week ending March 11, 1889	24
34.	Plan of the Bechtelsville (Norway) Furnace	25
35.	Ore roaster at the Bechtelsville (Norway) Furnace	25
36.	Sketch of Ohnmacht cave, Bern Township	26
37.	Aragonite from the Ohnmacht cave, Bern Township	27
38.	Aerial photograph showing the Rickenbach area quarries, Bern Township, September 14, 1937	27
39.	Quartz crystal from the Red Bridge locality, Bern Township	28
40.	Barite nodule, Bethel Township	29
41.	Advertisement for the E & G Brooke Iron Company from the 1860 Boyd's Directory	29
42.	Letterhead of the E & G Brooke Iron Company, 1874	30
43.	Plan of the E & G Brooke Iron Company Furnace No. 3, 1912	30
44.	Edward Brooke	31
45.	George Brooke	31
46.	Birdsboro Steel Foundry and Machine Company, 1906	31
47.	E & G Brooke Iron Company Furnace No. 3	32
48.	Advertisement for the Pennsylvania Diamond Drill and Machine Company, 1888	32
49.	Unissued stock certificate of the Pennsylvania Diamond Drill and Machine Company, ca. 1890s	33

50.	Second Pennsylvania Geological Survey map of the Boyertown mines, 1883	34
51.	Surface plan of ore bodies mined in Boyertown	36
52.	Location of the Boyertown mines on the 1876 atlas	37
53.	Surface plan of the Boyertown mines	38
54.	Photographs taken underground in the Boyertown mines, ca. 1880s	39
55.	Photograph showing the flooded Eckert open cut, Boyertown	40
56.	Diagram showing the Eckert open cut on the Hagy ore vein, Boyertown	41
57.	Plan of the California mine (Phoenix mine lower slope), Boyertown	41
58.	Plan of part of the California (Phoenix mine lower slope) and Warwick mines, Boyertown	42
59.	Analyses of the Boyertown blue and black vein ore, September 5, 1882	42
60.	Plan of the Phoenix mine upper and middle slopes, Boyertown	43
61.	Plan of the Phoenix Iron Company mines, Boyertown, 1886	44
62.	Sketch of the Boyertown mines made by Griffith Jones, February 5, 1872	45
63.	Letterhead of the Warwick Iron Company, 1897	47
64.	Plan of the upper (500-foot) level of the Warwick mine, Boyertown	48
65.	Plan of the bottom level of the Warwick mine, Boyertown	48
66.	Warwick Iron Company Furnace, Pottstown, 1907.	48
67.	Analysis of Boyertown ore, April 14, 1881	49
68.	Plan of the Warwick, Gable, and Rhoades mines, Boyertown, 1886	50
69.	The Gabel mine, Boyertown, ca. 1880s	51
70.	Plan of the workings of the Gabel and Warwick mines, Boyertown	51
71.	Ore bodies intersected by the Gabel mine, Boyertown	53
72.	Letterhead of the John Rhoads Mining Company, 1886	54
73.	The Boyertown iron crane, Boyertown.	55
74.	Boyertown mine buildings still standing in 1948	57
75.	Analysis of magnetite ore and letter from Hiram W. Hollenbush to Griffith Jones offering to trade his chemical analysis services for magnetite specimens from the Boyertown mines, October 15, 1881	58
76.	Pyrite and calcite from the Boyertown mines	59
77.	Calcite from the Boyertown mines	59
78.	Calcite and pyrite from the Boyertown mines	60
79.	Native copper from the Boyertown mines	60
80.	Dolomite and calcite from the Boyertown mines	60
81.	Magnetite from the Warwick mine, Boyertown	60
82.	Chalcopyrite from the Boyertown mines	60
83.	Pyrite from the Boyertown mines	61
84.	Marcasite and quartz from the Boyertown mines	61
85.	Pyrite from the Boyertown mines	61
86.	Stilbite from the Boyertown mines	61
87.	Pyrite from the Boyertown mines	61
88.	Jones mine, Caernarvon Township, April 9, 2011	62
89.	Smokestack at the Jones mine, Caernarvon Township, 2011	63
90.	Second Pennsylvania Geological Survey map of the Jones and Kinney mines, Caernarvon Township. 1882	64
91.	Jones mine, Caernarvon Township, ca. 1880s	66
92.	Incline at the Jones mine, Caernarvon Township, ca. 1880s	67
93.	Jones mine, Caernarvon Township, ca. 1880s	68
94.	Jones mine, Caernarvon Township, ca. 1880s	69
95.	Jones mine, Caernarvon Township, 1910	70
96.	Jones Mine, Caernarvon Township	70
97.	Fishing in the flooded Jones mine, Caernarvon Township	70
98.	Geologic cross section of the Jones and Kinney mines, Caernarvon Township	71

99.	Native copper from the Jones Mine, Caernarvon Township	71
100.	Aragonite from the Jones Mine, Caernarvon Township	72
101.	Apatite and chlorite from the Jones Mine, Caernarvon Township	72
102.	Aragonite from the Jones Mine, Caernarvon Township	72
103.	Aragonite from the Jones Mine, Caernarvon Township	72
104.	Calcite from the Jones Mine, Caernarvon Township	73
105.	Azurite from the Jones Mine, Caernarvon Township	73
106.	Hoppered chalcopyrite crystals from the Jones Mine. Caernarvon Township	73
107.	Chalcopyrite crystals from the Jones Mine, Caernarvon Township	73
108.	Native copper from the Jones Mine, Caernarvon Township	74
109.	Dolomite and pyrite from the Jones Mine, Caernarvon Township	74
110.	Malachite from the Jones Mine, Caernarvon Township	74
111.	Malachite on quartz crystals from the Jones Mine, Caernarvon Township	75
112.	Magnetite from the Jones Mine, Caernarvon Township	75
113.	Malachite from the Jones Mine, Caernarvon Township	75
114.	Magnetite from the Jones Mine, Caernarvon Township	75
115.	Banded malachite from the Jones Mine, Caernarvon Township	75
116.	Octahedral magnetite crystals from the Jones Mine, Caernarvon Township	75
117.	Malachite from the Jones Mine, Caernarvon Township	76
118.	Malachite from the Jones Mine, Caernarvon Township	76
119.	Pyrite from the Jones Mine, Caernarvon Township	76
120.	Limonite (bombshell ore) from the Jones Mine, Caernarvon Township	76
121.	Malachite from the Jones Mine, Caernarvon Township	76
122.	Geologic cross section of the Jones and Kinney mines, Caernarvon Township	77
123.	Kinney mine, Caernarvon Township, January 2016	78
124.	Location of the Byler mine, Caernarvon Township, 1876	78
125.	Styer quarry, Caernarvon Township, April 2015	79
126.	Styer lime kilns, Caernarvon Township, April 2015	79
127.	Mast quarry, Caernarvon Township, April 2015	80
128.	Daniel Mast draw kiln, Mill Road, Caernarvon Town-ship	80
129.	Mast lime kilns, Caernarvon Township, April 2015	80
130.	Hertzler lime kiln, Caernarvon Township	81
131.	Team of oxen at the Hertzler lime kiln. Caernarvon Township	81
132.	John Plank lime kiln, Caernarvon Township, April 2015	81
133.	John Stoltzfus quarry, Caernarvon Township, December 2010	82
134.	John Stoltzfus lime kilns, Quarry Road, Caernarvon Township, December 2010	82
135.	C. Mast quarry, State Route 23, Caernarvon Township, April 2015	83
136.	Clouse & Plank Kiln, Twin Valley Road, Caernarvon Township, April 2015	83
137.	Schorl from the Morgantown Bypass, Caernarvon Township	84
138.	Marcasite from Centre Township	84
139.	Wesley's quarry, Centre Township, ca. 1890s	85
140.	Bechtelsville quarry, Colebrookdale Township, ca. 1890s	85
141.	Bechtelsville quarry, Colebrookdale Township	86
142.	Bechtelsville quarry, Colebrookdale Township, March 2016	86
143.	Advertisement for Martin Infield Mix	87
144.	Limonite from Colebrookdale Township	87
145.	Mines in the Boyertown area, 1882	88
146.	Second Pennsylvania Geological Survey map of the Fritz Island mine, Cumru Township, 1882	90
147.	Fritz Island mine shaft uncovered during construction for a sewer plant expansion, Cumru Township	92
148.	Fritz Island mine air shaft uncovered during construction for a sewer plant expansion, Cumru Township	93

149.	Geologic cross section in the vicinity of the Fritz Island mine, Cumru Township	94
150.	Andradite crystals from the Fritz Island mine, Cumru Township	95
151.	Apophyllite crystals from the Fritz Island mine, Cumru Township	95
152.	Tabular apophyllite crystals from the Fritz Island mine, Cumru Township	95
153.	Aragonite crystals from the Fritz Island mine, Cumru Township	95
154.	Apophyllite crystals from the Fritz Island mine, Cumru Township	95
155.	Azurite crystals from the Fritz Island mine, Cumru Township	96
156.	Brucite crystals from the Fritz Island mine, Cumru Township	96
157.	Scalenohedral calcite crystals from the Fritz Island mine, Cumru Township	96
158.	Calcite crystals from the Fritz Island mine, Cumru Township	97
159.	Chabazite crystals from the Fritz Island mine, Cumru Township	97
160.	Chabazite crystals from the Fritz Island mine, Cumru Township	97
161.	Chabazite crystals from the Fritz Island mine, Cumru Township	97
162.	Chabazite crystals in diabase from the Fritz Island mine, Cumru Township	97
163.	Native copper from the Fritz Island mine, Cumru Township	98
164.	Datolite crystals from the Fritz Island mine, Cumru Township	98
165.	Gismondine crystals from the Fritz Island mine, Cumru Township	98
166.	Mesolite crystals from the Fritz Island mine, Cumru Township	98
167.	Gismondine crystals from the Fritz Island mine, Cumru Township	98
168.	Octahedral magnetite crystals from the Fritz Island mine, Cumru Township	99
169.	Mesolite from the Fritz Island mine, Cumru Township	99
170.	Spherical cluster of stilbite crystals from the Fritz Island mine, Cumru Township	99
171.	Magnetite crystals from the Fritz Island mine, Cumru Township	99
172.	Natrolite crystals from the Fritz Island mine, Cumru Township	99
173.	Thompsonite from the Fritz Island mine, Cumru Township	101
174.	Thompsonite and chabazite from the Fritz Island mine, Cumru Township	101
175.	Vesuvianite crystal from the Fritz Island mine, Cumru Township	101
176.	Vesuvianite crystals from the Fritz Island mine, Cumru Township	101
177.	Stilbite from the Radenbush mine, Cumru Township	102
178.	The Fehr and O'Rourke Stone Company quarry, Cumru Township, 2015	103
179.	Advertisement for Fehr and O'Rourke, Inc., 1930	103
180.	Andradite crystal from the U.S. Route 222 roadcut locality, Cumru Township	104
181.	Andradite crystals from the U.S. Route 222 roadcut locality, Cumru Township	104
182.	Flattened quartz crystals from the U.S. Route 222 roadcut locality, Cumru Township	104
183.	Hematite from the U.S. Route 222 roadcut locality, Cumru Township	104
184.	Andradite from Shiloh Hills, Cumru Township	104
185.	Limonite from Schlegel's Farm, Cumru Township	105
186.	Limonite from Schlegel's Farm, Cumru Township	105
187.	The Berks County Almshouse farm, Cumru Township	106
188.	Goethite from the Berks County Almshouse quarry, Cumru Township	106
189.	The Greshville quarry, Douglass Township, 2015	108
190.	Quarries in the Greshville area, Douglass Township, 1882	108
191.	Map of the Greshville Cave, Douglass Township	109
192.	The H&K Group Pottstown Trap Rock Douglassville Quarry, Douglass Township, 2015	110
193.	Wavellite from Earl Township	111
194.	Stock certificate of the Manatawny Bessemer Ore Company, Earl Township, 1916	112
195.	Hematite from the Kauffman and Spang mine, Earl Township	113
196.	Bond issued by the Manatawny Railroad Company, Earl Township, 1912	114
197.	Plan of the Dotterer mine, Earl Township	115
198.	Analysis of ore from the Dotterer mine, Earl Township, by H.W. Hollenbush	115
199.	Geologic cross section of the Dotterer mine, Earl Township	116

200.	Thorite from the Pennsylvania Uranium Company mine, Shanesville, Earl Township	117
201.	Geologic map of the Jacksonwald area, Exeter Township	118
202.	Amydular and vesicular basalt from the Jacksonwald basalt, Exeter Township	119
203.	Prehnite from the Jacksonwald occurrence, Exeter Township	120
204.	Apophyllite from the Kinsey Hill locality, Exeter Township	121
205.	Prehnite from the Kinsey Hill locality, Exeter Township	121
206.	Stellerite from the Kinsey Hill locality, Exeter Township	121
207.	Garnets from Bishop's Mill, Exeter Township	122
208.	Quarries in the Oley Line area, Exeter Township, 1882	123
209.	Lime kiln on Limekiln Road near the D. Snyder quarry, Exeter Township	124
210.	Quarries in the Jacksonwald area, Exeter Township, 1882	125
211.	Pyrite nodule from Windsor Township	126
212.	Limestone quarries in Heidelberg Township, 1891	127
213.	Moore quarry, Heidelberg Township, March 2015	128
214.	Lime kilns at the Moore quarry, Heidelberg Township, March 2015	128
215.	Aragonite from the Seisholtzville mines, Hereford Township	129
216.	Second Pennsylvania Geological Survey map of the Seisholtzville iron mines, Hereford Township	130
217.	Incline at the Bittenbender mine, Hereford Township, 1901	132
218.	LIDAR image of the Seisholtzville mines, Hereford Township	132
219.	Stilbite from Seisholtzville, Hereford Township	133
220.	Limonite from the Gehman mine, Hereford Township	133
221.	The Seisholtzville District mines, Hereford Township, 1882	134
222.	Location of the Philadelphia and Reading Coal and Iron Company mine at Seisholtzville, Hereford Township, 1876	134
223.	The Bittenbender mine, Hereford Township, ca. 1890	135
224.	Stock certificate of the Philadelphia and Reading Coal and Iron Company, 1939	136
225.	Steel Ore Company advertisement for ore from the Seisholtzville mines, Hereford Township	137
226.	Section showing the method of mining soft ore at the Seisholtzville mines, Hereford Township	138
227.	Analyses of ore from the Bittenbender mines, Hereford Township	138
228.	Loellingite from the Olafson mine, Huffs Church, Hereford Township	140
229.	Dale mine, Hereford Township, March 2015	141
230.	Limonite from Dale Forge, Hereford Township	141
231.	Hampton Furnace quarries, Hereford Township, 1882	142
232.	Lime kiln at one of the Hampton Furnace quarries, Hereford Township, March 2015	142
233.	One of the Hampton Furnace quarries, Hereford Township, March 2015	143
234.	Trollinger quarry, Hereford Township, March 2015	143
235.	Lime kiln at the Trollinger quarry, Hereford Township, March 2015	143
236.	Thomas Edison with a group of men at the Seisholtzville granite quarry, Hereford Township	144
237.	Thomas Edison at the Seisholtzville granite quarry, Hereford Township	144
238.	Letterhead of the Leesport Furnace Company, Leesport	145
239.	The Bradford ore separator	146
240.	The Thomas ore washer	147
241.	Mining areas in Longswamp Township	148
242.	Mines at Rittenhouse Gap and Red Lion Station, Longswamp Township, 1882	149
243.	Stock certificate of the Southern Coal and Iron Company, 1924	150
244.	Map of the Rittenhouse Gap mines, Longswamp Township, 1882	151
245.	Catasauqua and Fogelsville Railroad bed at the Rittenhouse Gap mines, Longswamp Township, March 2015	152
246.	Rock mine open cut, Rittenhouse Gap, Longswamp Township, March 2012	153
247.	Thomas Iron Company open cut at Rittenhouse Gap, Longswamp Township, March 2012	153
248.	Tunnel mine, Rittenhouse Gap, Longswamp Township, March 2012	154

249.	Magnetite from the Gap (Moll and Gery) mine, Rittenhouse Gap, Longswamp Township	154
250.	Mine building ruins, Rittenhouse Gap, Longswamp Township, March 2012	155
251.	Ore wharf on the Catasauqua and Fogelsville Railroad at Rittenhouse Gap, Longswamp Township, March 2015	155
252.	Rittenhouse open cut, Rittenhouse Gap, Longswamp Township, March 2012	155
253.	Open shaft at Rittenhouse Gap, Longswamp Township, March 2012	155
254.	Magnetite and quartz from the Ginkinker mine, Rittenhouse Gap, Longswamp Township	156
255.	Thomas Iron Company open cut, Rittenhouse Gap, Longswamp Township, March 2012	156
256.	Open cut of the Thomas Iron Company, Rittenhouse Gap, Longswamp Township, March 2012	157
257.	Magnetite from the Thomas Iron Company open cut, Rittenhouse Gap, Longswamp Township	157
258.	Molybdenite from the Rittenhouse Gap mines, Longswamp Township	157
259.	Thomas Iron Company Furnace, Hokendauqua, Pa., 1903	158
260.	Billhead of the Thomas Iron Company	159
261.	Garner Station mines, Longswamp Township, 1882	161
262.	Quartz and Hematite from Maple Grove, Longswamp Township	162
263.	Klines Corner mines, Longswamp Township, 1882	163
264.	LIDAR image of the Klines Corner area, Longswamp Township	163
265.	Mines and ore washeries at Klines Corner, Longswamp Township, 1897	164
266.	Mines and ore washeries at Klines Corner, Longswamp Township, 1897	164
267.	Ore wharf at Klines Corner, Longswamp Township, 1897	165
268.	Mining and ore washing at Klines Corner, Longswamp Township, 1897	165
269.	Stock certificate of the Allentown Iron Company, 1863	166
270.	Thomas Iron Company ore washer at Klines Corner, Longswamp Township, 1897	167
271.	Incline at iron mine number 41 between Klines Corner and Oreville, Longswamp Township	167
272.	Magnetite from Fritch and Brother's mine, Longswamp Township	168
273.	Mining lease from John Fritch to Levi Fritch, November 19, 1879	169
274.	Mines in the Farmington area, Longswamp Township, 1882	171
275.	Ziegler's ore washer, Farmington, Longswamp Township, 1897	172
276.	Mines in the Shamrock area, Longswamp Township, 1882	173
277.	Location of the Nathan Haas mine, Shamrock, Longswamp Township	175
278.	Mines in the Longsdale area, Longswamp Township, 1882	175
279.	Mine number 53, Longsdale, Longswamp Township, March 2015	176
280.	Letterhead of the Penn Graphite Company, 1901	177
281.	Advertisement for the Star Clay Company, Longsdale, Longswamp Township	178
282.	Star Clay Company kaolin pits, Longswamp Township, 1915	179
283.	Digging kaolin from the pits, Star Clay Company, Longswamp Township, 1915	180
284.	Kaolin drying sheds, Star Clay Company, Longswamp Township, 1915	180
285.	Kaolin pits and drying sheds, Star Clay Company, Longswamp Township, 1915	180
286.	Fritch and Brother's mine, Longsdale, Longswamp Township, March 2015	181
287.	Goethite from Hancock, Longswamp Township	181
288.	Stalactitic pipe ore (limonite) from the Topton area, Longswamp Township	182
289.	Bombshell limonite ore from the Topton area, Longswamp Township	182
290.	Plan of the Henry Erwin and Sons ocher mine, Topton, Longswamp Township	183
291.	Henry Erwin and Sons open-pit ocher mine, Topton, Longswamp Township	184
292.	Henry Erwin and Sons boiler house and ocher drying sheds, Topton, Longswamp Township	184
293.	Henry Erwin and Sons ocher settling troughs, Topton, Longswamp Township	184
294.	Walker quarry, Longswamp Township, ca. 1932	185
295.	Walker quarry, Longswamp Township, April 2016	185
296.	Palm Schwenkfelder Church, Palm, Pa., constructed with Seisholtzville granite from the Walker quarry	186
297.	Geological setting of the Big Dam quarry, Lower Alsace Township	187
298.	Calcite from the Big Dam quarry, Lower Alsace Township	187

299.	Calcite from the big Dam quarry, Lower Alsace Township	188
300.	Antietam Reservoir locality, Lower Alsace Township, April 2015	188
301.	Geologic map of the Antietam Reservoir area, Lower Alsace Township	189
302.	Photograph of the Ohlinger Dam quarry, Lower Alsace Township, 1882	189
303.	Molybdenite from the Antietam Reservoir locality, Lower Alsace Township	190
304.	Hornblende from the Ohlinger Mills locality, Lower Alsace Township	191
305.	Limestone quarries in Lower Heidelberg Township, 1886	192
306.	Glen-Gery quarry, Lower Heidelberg Township	193
307.	Calcite from Maidencreek Township	194
308.	Zircon from Maidencreek Township	194
309.	Calcite from Maidencreek Township	194
310.	Limonite from Maidencreek Township	194
311.	Stock certificate of the Vindex Portland Cement Company, 1907	195
312.	Advertisement offering the Vindex Portland Cement Company plant for sale, 1916	196
313.	Aeriel view of the Evansville Quarry, Maidencreek Township, September, 2015	196
314.	Portrait of Charles A. Matcham	197
315.	Advertisement for the Allentown Portland Cement Company, 1914	197
316.	Power plant at the Allentown Cement Company Evansville plant, Maidencreek Township, ca. 1910	198
317.	Horse-drawn carts in the Evansville quarry, Maidencreek Township, early 1900s	198
318.	Evansville quarry, Maidencreek Township, early 1900s	198
319.	Evansville quarry and cement plant, Maidencreek Township, 1912	199
320.	Steam shovel loading stone at the Evansville quarry, Maidencreek Township, early 1900s	199
321.	Evansville quarry, Maidencreek Township, early 1900s	199
322.	Stone cars at the Evansville quarry, Maidencreek Township.	200
323.	Stone car dumping stone inside the Evansville cement plant, Maidencreek Township	200
324.	Evansville cement plant, Maidencreek Township, 1915	200
325.	Restored stone car at the Evansville cement plant, Maidencreek Township	201
326.	Evansville cement plant, Maidencreek Township, March 2016	201
327.	Unissued stock certificate of the National Gypsum Company	202
328.	Sketch of the Maidencreek quarry, Maidencreek Township, 1951	202
329.	Sketch of the Ontelaunee Orchards quarry, 1951	203
330.	Northern Crane Iron Company mine, Maidencreek Township, April 2015	203
331.	Crane Iron Company furnaces, Catasauqua, Pa.	204
332.	Stock Certificate of the Lehigh Crane Iron Company, 1872	205
333.	Southern Crane Iron Company mine, Maidencreek Township, April 2015	206
334.	Goethite from Schaeffer's mine, Maidencreek Township	206
335.	Dishong quarry and kiln, Marion Township, April 2015	206
336.	Limestone quarries in Maxatawny Township south and southeast of Kutztown, 1882	207
337.	Eastern Industries Hinterleiter quarry, Maxatawny Township, March 2016	208
338.	Aerial view of the Hinterleiter quarry, Maxatawny Township, September 2015	208
339.	Portrait of Dr. U.S.G. Bieber	209
340.	Portrait of Edward Hottenstein	209
341.	Sketch of the Hottenstein quarry, Maxatawny Township, 1951	210
342.	Crusher at the Bieber (Keystone) quarry, Maxatawny Township	211
343.	Aerial view of the Berks Products and Koller quarries, Maxatawny Township, September 2015	210
344.	Aragonite from the Baldy Street (Keystone) quarry, Maxatawny Township	211
345.	Lime kiln near the Angstadt quarry, Maxatawny Township, April 2015	212
346.	Sketch of the A.G. Smith quarry, Maxatawny Township, 1951	212
347.	Kohler Road quarry, Maxatawny Township, March 2015	213
348.	Lime kiln at the Kohler Road quarry, Maxatawny Township, March 2015	213
349.	Platform for loading rail cars at the Kohler Road quarry, Maxatawny Township, March 2015	213

350.	Samuel Lewis mine near Klines Corner, Maxatawny Township, March 2015	214
351.	Limonite from near Lyons, Maxatawny Township	215
352.	Limonite from Muhlenberg Township	216
353.	Calcite from Brook's quarry, Muhlenberg Township	216
354.	Aerial photograph of the limestone quarries in Muhlenberg Township, September 14, 1937	217
355.	Advertisement for the Reading Quarry Company Laureldale quarry, 1914	217
356.	Reading Quarry Company's Laureldale quarry lime kilns, Muhlenberg Township, 1913	218
357.	Reading Quarry Company's Laureldale quarry rail car loading area, Muhlenberg Township, ca. 1915	218
358.	Reading Quarry Company's Laureldale quarry and lime kilns, Muhlenberg Township, 1915	218
359.	Stock certificate of the Reading Quarry Company, 1907	219
360.	Aragonite from Nolan's (Tuckerton) Cave, Muhlenberg Township	220
361.	Calcite from Nolan's (Tuckerton) Cave, Muhlenberg Township	220
362.	Calcite from Nolan's (Tuckerton) Cave, Muhlenberg Township	220
363.	Calcite (cave pearls) from Nolan's (Tuckerton) Cave, Muhlenberg Township	220
364.	Aragonite from Nolan's (Tuckerton) Cave, Muhlenberg Township	220
365.	Calcite from Nolan's (Tuckerton) Cave, Muhlenberg Township	220
366.	Aragonite from the South Temple quarry, Muhlenberg Township	221
367.	Advertisement for the Berks Products Corporation, 1933	221
368.	Help wanted advertisement for laborers at the South Temple quarry, 1923	221
369.	Map of the South Temple Cave, Muhlenberg Township	222
370.	South Temple quarry, Muhlenberg Township, April 2015	222
371.	G.W. Focht quarry, Muhlenberg Township, 1882.	223
372.	Stock certificate of the Reading Rail Road Company, 1899	223
373.	Advertisement for the Temple Silica Sand Company, 1925	224
374.	Temple Iron Company Furnace, ca. 1860s	224
375.	Receipt for the purchase of Clymer Iron Company stock	225
376.	Stock certificate of the Temple Iron Company, 1902	225
377.	Limonite from Muhlenberg Township	226
378.	Quartz from Muhlenberg Township	226
379.	Portrait of Eugene C. Grace	227
380.	Sign at the entrance to the Grace mine on Pennsylvania State Route 10 north of Morgantown	227
381.	Part of an aeromagnetic map of the Morgantown area showing the anomaly at the Grace mine iron ore body	228
382.	Core drilling rig exploring for the Grace mine ore body, January 12, 1954	228
383.	Location of core holes used to delineate the Grace mine ore body	229
384.	Grace mine shaft A headframe during construction, June 14, 1956	229
385.	Aerial view of the Grace mine, April 13, 1970	230
386.	Aerial view of the Grace mine, ca. 1961	230
387.	Cross sections of the Grace mine shafts	231
388.	Grace mine shaft A hoists and compressors	231
389.	Grace mine shaft A hoists	231
390.	Grace mine 513B ramp drift, January 5, 1967	232
391.	Grace mine entry drifts to panel 51, December 6, 1966	232
392.	Grace mine sixth level, August 12, 1964	232
393.	Grace mine sixth level, January 6, 1966	232
394.	Grace mine 31 north production drift on the seventh level of the ore body, January 5, 1967	233
395.	Scoop tram in the Grace mine, August 3, 1968	233
396.	Grace mine sixth level conveyor drift, June 6, 1969	233
397.	Electric locomotive hauling ore cars in the Grace mine	233
398.	Grace mine ore cars, July 20, 1962	233
399.	Mining by the panel-caving method at the Grace mine	234

400.	Aerial view of subsidence at the Grace mine	234
401.	Grace mine ore-processing flow diagram	235
402.	Grace mine rod and ball mills	235
403.	Grace mine magnetic separators	235
404.	Grace mine tailings pond and dam	236
405.	Grace mine revolving balling cones	236
406.	Grace mine pellet furnaces	236
407.	Grace mine magnetite pellets	237
408.	Toppling of the Grace mine smoke stack, November 16, 1984	237
409.	Magnetite pellets being loaded into rail cars at the Grace mine	237
410.	Scenes around the Grace mine property, December 2012	238
411.	Mineralogical Society of Pennsylvania mineral-collecting field trip to the Grace mine, April 13, 1958	239
412.	Aerial view of the Grace mine dumps, April 13, 1970	239
413.	Block diagram of the Grace mine ore body	240
414.	Cross section of the Grace mine ore body	240
415.	Apophyllite on prehnite from the Grace mine, New Morgan Borough	241
416.	Apophyllite from the Grace mine, New Morgan Borough	241
417.	Augite crystals from the Grace mine, New Morgan Borough	242
418.	Augite crystals from the Grace mine, New Morgan Borough	242
419.	Calcite from the Grace mine, New Morgan Borough	242
420.	Zoned clinochlore crystals from the Grace mine, New Morgan Borough	242
421.	Chrysotile from the Grace mine, New Morgan Borough	242
422.	Datolite from the Grace mine, New Morgan Borough	243
423.	Galena crystal from the Grace mine, New Morgan Borough	243
424.	Heulandite from the Grace mine, New Morgan Borough	243
425.	Gypsum from the Grace mine, New Morgan Borough	243
426.	Laumontite from the Grace mine, New Morgan Borough	243
427.	Magnetite from the Grace mine, New Morgan Borough	244
428.	Magnetite crystals with striated faces from the Grace mine, New Morgan Borough	244
429.	Magnetite crystals coated with tremolite from the Grace mine, New Morgan Borough	245
430.	Magnetite crystals from the Grace mine, New Morgan Borough	245
431.	Magnetite crystal group from the Grace mine, New Morgan Borough	245
432.	Magnetite crystals from the Grace mine, New Morgan Borough	245
433.	Magnetite from the Grace mine, New Morgan Borough	245
434.	Natrolite on prehnite from the Grace mine, New Morgan Borough	246
435.	Cubic pyrite from the Grace mine, New Morgan Borough	246
436.	Prehnite from the Grace mine, New Morgan Borough	246
437.	Octahedral pyrite from the Grace mine, New Morgan Borough	246
438.	Pyrrhotite from the Grace mine, New Morgan Borough	246
439.	Pyrrhotite from the Grace mine, New Morgan Borough	247
440.	Talc pseudomorphs after magnetite from the Grace mine, New Morgan Borough	247
441.	Talc from the Grace mine, New Morgan Borough	247
442.	Talc pseudomorph after magnetite from the Grace mine, New Morgan Borough	247
443.	Tochilinite in calcite from the Grace mine, New Morgan Borough	247
444.	Tochilinite from the Grace mine, New Morgan Borough	248
445.	Stock certificate of the Bethlehem Steel Corporation, 1953	249
446.	Skip Colflesh and Scott Snavely collecting minerals in the Eastern Industries Oley quarry, Oley Township	250
447.	Aerial view of the Eastern Industries Oley quarry, Oley Township, April 2016	250
448.	Sphalerite from the Eastern Industries quarry, Oley Township	251
449.	Calcite and pyrite from the Eastern Industries quarry, Oley Township	251

#	Description	Page
450.	Barite from the Eastern Industries quarry, Oley Township	252
451.	Aurichalcite from the Eastern Industries quarry, Oley Township	252
452.	Calcite from the Eastern Industries quarry, Oley Township	252
453.	Palygorskite from the Eastern Industries quarry, Oley Township	252
454.	Calcite from the Eastern Industries quarry, Oley Township	252
455.	Fluorite from the Eastern Industries quarry, Oley Township	253
456.	Fluorite with uranium-bearing round spots from the Eastern Industries quarry,	253
457.	Hemimorphite from the Eastern Industries quarry, Oley Township	253
458.	Quartz, var. amethyst, on calcite from the Eastern Industries quarry	253
459.	Sphalerite from the Eastern Industries quarry, Oley Township	253
460.	Lehigh Cement Company Oley No. 2 quarry, Oley Township, April 2015	254
461.	Aerial view of the three Lehigh quarries, Oley Township, April 2016	254
462.	Stock certificate of the Lehigh Portland Cement Company, 1914	255
463.	Letterhead of the Lehigh Portland Cement Company, 1901	255
464.	Lime kiln on Bertolet Mill Road, Oley Township, April 2015	256
465.	Lime kiln near the E. Schaeffer quarry, Oley Township, April 2015	256
466.	Quarries and iron mines in central Oley Township, 1882	257
467.	Colonel J. Weaver's quarry, Oley Township, April 2015	257
468.	Lime kiln near the Levi Knabb Quarry, Oley Township, April 2015	258
469.	Limonite from the Hunter mine, Oley Township	262
470.	Weaver mine, Oley Township, April 2015	263
471.	Limonite from the Weaver mine, Oley Township	263
472.	Limonite from the Manweiller mine, Oley Township	264
473.	Magnetite from the Hertzel and Swoyer mine, Oley Township	264
474.	Polished section of petrified wood from Oley Valley, Oley Township	265
475.	Aerial view of the Ontelaunee quarry, Ontelaunee Township, April 2016	267
476.	Southernmost quarry of the Leesport Iron Company, Ontelaunee Township, April 2015	267
477.	Stock certificate of the Glen-Gery Shale Brick Company, 1929	268
478.	Ariel photograph of the Glen-Gery quarries and brick plant, Shoemakersville, Perry Township, 1958	269
479.	Ilmenite from Lobachsville, Pike Township	271
480.	Portraits of Issac Eckert, Henry Eckert, and George Eckert	272
481.	The Mount Penn iron mines, Reading, 1857	273
482.	Actinolite from Mount Penn, Reading	274
483.	Goethite from Mount Penn, Reading	274
484.	Limonite from Mount Penn, Reading	274
485.	Fibrous goethite from Mount Penn, Reading	274
486.	Limonite from near Reading	274
487.	Scepter quartz from Reading	274
488.	The Pagoda on Mount Penn, Reading, June 2016	276
489.	Gring's quarry, Reading, April 2015	277
490.	Advertisement for Kirschmann's Mount Penn Sand and Stone Works, 1914	277
491.	Eckert & Brother Henry Clay Furnace letterhead	279
492.	Henry Clay Furnace, Reading	279
493.	Reading Iron Company Furnace, Reading, 1907	280
494.	Reading Iron Company letterhead, 1897	280
495.	Calcite from Dragon Cave, Richmond Township	281
496.	Map of Dragon Cave, Richmond Township	281
497.	Aragonite from Dragon Cave, Richmond Township	282
498.	Letterhead of C.K. Williams & Company, 1901	282
499.	Plan of the Keystone Ocher Company underground workings, Richmond Township	283
500.	Occurrence of ocher and clay in the Keystone Ocher Company mine, Richmond Township	283

501.	Settling troughs and drying sheds, C.K. Williams ocher mine, Richmond Township, 1911	284
502.	Settling pond, drying shed, and mill, C.K. Williams ocher mine, Richmond Township, 1911	284
503.	Settling boxes, C.K. Williams ocher mine, Richmond Township, 1911	284
504.	Drying sheds filled with ocher, C.K. Williams ocher mine, Richmond Township, 1911	284
505.	Shaft house and settling troughs, C.K. Williams ocher mine, Richmond Township, 1911	284
506.	Engine house and tracks, C.K. Williams ocher mine, Richmond Township, 1911	284
507.	Removing partially dried ocher from the settling ponds, C.K. Williams ocher mine, Richmond Township, 1911	285
508.	Drying sheds and settling pond, C.K. Williams ocher mine, Richmond Township	285
509.	Drying shed, C.K. Williams ocher mine, Richmond Township	285
510.	Drying sheds, C.K. Williams ocher mine, Richmond Township, 1940s	285
511.	Ocher pits, C.K. Williams ocher mine, Richmond Township, April 2015	285
512.	Tom Pracher and the rock house, C.K. Williams ocher mine, Richmond Township, April 2015	285
513.	Kirbyville quarry, Richmond Township, April 2015	286
514.	Limonite from the Fleetwood area, Richmond Township	287
515.	Druzy quartz from the Fleetwood area, Richmond Township	287
516.	Quartz crystal from the Fleetwood area, Richmond Township	287
517.	Limonite "bog ore" from the Fleetwood area, Richmond Township	287
518.	Epidote from the Fleetwood area, Richmond Township	287
519.	Map of the Moselem mines, Richmond Township, 1882	288
520.	The Moselem mine, Richmond Township, April 2015	289
521.	Moselem Furnace, Richmond Township, ca. 1930s	289
522.	Goethite from the Moselem mine, Richmond Township	290
523.	Quartz crystals from Kirbyville, Richmond Township	290
524.	Quartz crystals from Noll's mine, Richmond Township	291
525.	Richmond No. 5 quarry, Richmond Township, March 2016	291
526.	Portrait of John T. Dyer	292
527.	Dyer Gibraltar quarry, Robeson Township, September 2010	293
528.	Apophyllite from the Dyer Gibraltar quarry, Robeson Township	293
529.	Actinolite from the Dyer Gibraltar quarry, Robeson Township	294
530.	Actinolite, var. byssolite, from the Dyer Gibraltar quarry, Robeson Township	294
531.	Babingtonite from the Dyer Gibraltar quarry, Robeson Township	294
532.	Chabazite from the Dyer Gibraltar quarry, Robeson Township	294
533.	Natrolite from the Dyer Gibraltar quarry, Robeson Township	294
534.	Calcite from the Dyer Gibraltar quarry, Robeson Township	295
535.	Opal from the Dyer Gibraltar quarry, Robeson Township	295
536.	Quartz, var. agate, from the Dyer Gibraltar quarry, Robeson Township	295
537.	Prehnite from the Dyer Gibraltar quarry, Robeson Township	295
538.	Stellerite from the Dyer Gibraltar quarry, Robeson Township	295
539.	Stilbite from the Dyer Gibraltar quarry, Robeson Township	296
540.	Thaumasite from the Dyer Gibraltar quarry, Robeson Township	296
541.	Birdsboro Materials (Trap Rock) quarry, Robeson Township, April 2015	297
542.	Stock certificate of the Wilmington & Northern Railroad Company, 1877	297
543.	Number 3 and 4 crushers at the Dyer Trap Rock quarry, Robeson Township, early 1900s	298
544.	Crusher Number 2 at the Dyer Trap Rock quarry, Robeson Township, early 1900s	298
545.	Dyer Trap Rock quarry and crusher Number 2, Robeson Township, early 1900s	298
546.	Diabase exposed in the Birdsboro Materials (Trap Rock) quarry, Robeson Township, April 2015	299
547.	Workers at the Dyer Trap Rock quarry, Robeson Township	299
548.	Advertisement for the Autocar Company featuring the Dyer Trap Rock quarry, 1920	300
549.	Mineralogical Society of Pennsylvania mineral collecting field trip at the Dyer Trap Rock quarry, August 17, 1952	301

550.	Apophyllite from the Dyer Trap Rock (Birdsboro Materials) quarry, Robeson Township	302
551.	Apophyllite and prehnite from the Dyer Trap Rock (Birdsboro Materials) quarry, Robeson Township	302
552.	Calcite from the Dyer Trap Rock (Birdsboro Materials) quarry, Robeson Township	303
553.	Prehnite from the Dyer Trap Rock (Birdsboro Materials) quarry, Robeson Township	303
554.	Natrolite from the Dyer Trap Rock (Birdsboro Materials) quarry, Robeson Township	303
555.	Sphalerite from the Dyer Trap Rock (Birdsboro Materials) quarry, Robeson Township	303
556.	Quartz from the Dyer Trap Rock (Birdsboro Materials) quarry, Robeson Township	303
557.	Humphrey sandstone quarry adjacent to Hay Creek, Robeson Township, April 2015	304
558.	Locations of the two Brusstar quarries, Robeson Township, 1876	304
559.	Sandstone quarries in Robeson Township advertising in the 1876 atlas	305
560.	Joanna Furnace, Robeson Township, April 2015	305
561.	Robesonia Furnace, ca. 1910	306
562.	Large slag piles at the Robesonia Furnace, ca. 1920s	306
563.	Letter from White Ferguson of the Robesonia Furnace to Griffith Jones of the Steel Ore Company concerning the use of ore from the Bittenbender mine, 1874	307
564.	Location of the Beitler mine, Rockland Township, 1882	308
565.	Locations of Brumbach's mine, J. Fisher's ore holes, and Green Hill, Rockland Township, 1882	308
566.	Goethite from Braumbach's mine, Rockland Township	309
567.	Limonite from the old Millert mine, Rockland Township	309
568.	"Psilomelane" from near New Jerusalem, Rockland Township	310
569.	Flint Hill, Rockland Township, April 2015	311
570.	Druzy quartz from Flint Hill, Rockland Township	311
571.	Quartz, var. chalcedony, from the Flint Hill area, Rockland Township	311
572.	Quartz, var. yellow jasper, from the Flint Hill area, Rockland Township	311
573.	Native American arrowheads made from yellow jasper	311
574.	Locations of the Udree, Clymer, and old Tunnel mines, Ruscombmanor Township, 1882	312
575.	Goethite and limonite pipe ore from the Udree mine, Ruscombmanor Township	313
576.	Clymer mine, Ruscombmanor Township, April 2015	313
577.	Limonite pipe ore from the Clymer mine, Ruscombmanor Township	314
578.	Limonite from the Clymer mine, Ruscombmanor Township	314
579.	Magnetite from the Old Tunnel mine, Ruscombmanor Township	314
580.	Magnetite and zircon from the Hain (Schittler) mine, Ruscombmanor Township	315
581.	Schweitzer and Kutz mine open cut and flooded open pit, Ruscombmanor Township, April 2015	315
582.	Limonite from the Schweitzer and Kutz mine, Ruscombmanor Township	316
583.	Location of the Pricetown area, Ruscombmanor Township, 1882	316
584.	Zircon from the Pricetown area, Ruscombmanor Township	317
585.	Zircon from the Pricetown area, Ruscombmanor Township	317
586.	Blue quartzite quarry, Ruscombmanor Township, April 2015	318
587.	Limestone quarries in the vicinity of Sinking Spring Borough, 1891	318
588.	Locations of the Thierwechler mine and Dechert quarry, Sinking Spring, 1876	319
589.	Actinolite from Sinking Spring	320
590.	Aragonite from Fritztown	320
591.	Brucite from Fritztown	320
592.	Limestone quarries in the vicinity of Wernersville, South Heidelberg Township, 1891	321
593.	Spodumene from Cushion (South) Mountain, Spring Township	323
594.	Grand View Sanatorium on South Mountain, Spring Township, 1908	323
595.	Gring's sand quarry, Spring Township	324
596.	Sketch of Gring's quarry, Spring Township	325
597.	Plan of the Wheatfield mine, Spring Township, 1954	325
598.	Map of the Wheatfield mine, Spring Township, 1879	326
599.	Cross section of part of the Wheatfield mine, Spring Township	326

600.	Locations of the Wheatfield and Ruth mines and the Reber shaft, Spring Township, 1882	327
601.	Garnet from the Wheatfield mine, Spring Township	327
602.	Brucite from the Wheatfield mine, Spring Township	328
603.	Hematite from the Wheatfield mine, Spring Township	328
604.	Limonite from the Wheatfield mine, Spring Township	328
605.	Quartz from the Wheatfield mine, Spring Township	328
606.	Wavellite from the Wheatfield mine, Spring Township	328
607.	North-south structure section 100 feet east of the Ruth mine, Spring Township	329
608.	Magnetite from the Ruth mine, Spring Township	329
609.	Limonite from the Ruth mine, Spring Township	329
610.	"Deweylite" and aragonite from the Ruth mine, Spring Township	330
611.	Limonite from Lincoln Park, Spring Township	330
612.	Albert A. Gery, 1890	331
613.	Montello Brick Works and quarry, Spring Township, 1890	331
614.	Montello Clay and Brick Company quarry, Spring Township, early 1900s	332
615.	Wyomissing plant of the Montello Brick Company, 1898	332
616.	Letterhead of the Glen-Gery Shale Brick Corporation, 1947	333
617.	Pyrite nodule coated with limonite from Tilden Township	333
618.	Topton Furnace, 1906	334
619.	Receipt for iron ore delivered to the Topton Furnace, 1889	334
620.	Fetid barite from Mt. Aetna, Tulpehocken Township	335
621.	Radially bladed barite nodule showing zoned tabular crystals from the Bohn farm, Tulpehocken Township	335
622.	Advertisement for Dyer quarry Birdsboro Trappe Rock, 1912	336
623.	Birdsboro Stone Company Monocacy quarry, Union Township, 1911	337
624.	Crusher and stone storage sheds at the Birdsboro Stone Company Monocacy quarry, Union Township, 1911	337
625.	Steam shovel, locomotive, and stone cars at the Birdsboro Stone Company Monocacy quarry, Union Township, 1911	337
626.	Steam shovel loading stone cars at the Birdsboro Stone Company Monocacy quarry, Union Township, 1911	338
627.	Stone car used at the Birdsboro Stone Company Monocacy quarry, Union Township, 1921	338
628.	Receipt for stone shipped from the Birdsboro Stone Company Monocacy quarry, Union Township, to Repampo, New Jersey, 1911	338
629.	Diagram showing blasting tunnels at the Dyer Monocacy quarry, Union Township, 1930	339
630.	Dyer Monocacy quarry, Union Township, prior to blasting, ca. 1950	340
631.	Blast at the Dyer Monocacy quarry, Union Township, ca. 1950	340
632.	Dyer Monocacy quarry, Union Township, August 14, 1955	341
633.	Apophyllite and pyrite on prehnite from the Dyer Monocacy quarry, Union Township	341
634.	Stoves cast at Hopewell Furnace, Union Township	342
635.	Cast house at Hopewell Furnace National Historic Site, Union Township	343
636.	Cross section of a charcoal iron furnace	343
637.	Stilbite from the Dyer Monocacy quarry, Union Township	344
638.	Stilbite from the Dyer Monocacy quarry, Union Township	344
639.	Iron mines in the vicinity of Barto, Washington Township, 1876	345
640.	Diagram showing the pinching and swelling of iron ore beds	346
641.	The Dysher mine in Barto, Washington Township, March 2015	347
642.	The L. Gilbert Mine, which was operated by Thomas Edison, Washington Township, June 2016	348
643.	Iron ore from the L. Gilbert (Edison) Mine, Washington Township	348
644.	Stock Certificate of Thomas Edison's New Jersey and Pennsylvania Concentrating Works, 1891	349
645.	Limonite from the Gilbert Mine, Washington Township	349

646.	Barr quarry, Barto, Washington Township	352
647.	Crusher at the Barr quarry, Barto, Washington Township	352
648.	Promotional ink blotter from the Wernersville Lime and Stone Company, Wernersville, ca. 1940s	354
649.	Limestone quarries in the vicinity of West Reading, 1891	355
650.	Locations of the Frill quarry, West Reading, 1876	356
651.	Fluorite from Leinbach's Hill, West Reading	357
652.	Aerial view of the Seitzinger mine (Weiser Lake), Wyomissing, 1925	358
653.	Pipe ore limonite from the Sitzinger mine, Wyomissing	358

APPENDIX 1 FIGURES

	Index map	379
1.	Mines on the southeastern quarter of the USGS New Ringgold 7.5-minute topographic quadrangle map	380
2.	Mines on part of the eastern part of the USGS Hamburg and western part of the Kutztown 7.5-minute topographic quadrangle maps	381
3.	Mines on the southeastern quarter of the USGS Kutztown 7.5-minute topographic quadrangle map	382
4.	Mines on the southwestern quarter of the USGS Topton 7.5-minute topographic quadrangle map	383
5.	Mines on the southeastern quarter of the USCS Topton and part of the southwestern quarter of the USGS Allentown West 7.5-minute topographic quadrangle maps	384
6.	Mineral localities on the central part of the USGS Bethel 7.5-minute topographic quadrangle map	385
7.	Mines on the northwestern quarter of the USGS Temple and southwest part of the Hamburg 7.5-minute topographic quadrangle maps	386
8.	Mines on the northeastern quarter of the USGS Temple 7.5-minute topographic quadrangle map	387
9.	Mines on the southwestern quarter of the USGS Temple 7.5-minute topographic quadrangle map	388
10.	Mines and mineral localities on the southeastern quarter of the USGS Temple 7.5-minute topographic quadrangle map	389
11.	Mines and mineral localities on the northwestern quarter of the USGS Fleetwood 7.5-minute topographic quadrangle map	390
12.	Mines and mineral localities on the northeastern quarter of the USGS Fleetwood 7.5-minute topographic quadrangle map	391
13.	Mines and mineral localities on the southwestern quarter of the USGS Fleetwood 7.5-minute topographic quadrangle map	392
14.	Mines and mineral localities on the southeastern quarter of the USGS Fleetwood 7.5-minute topographic quadrangle map	393
15.	Mines and mineral localities on the northwestern quarter of the USGS Manatawny 7.5-minute topographic quadrangle map	394
16.	Mines on the northeastern quarter of the USGS Manatawny 7.5-minute topographic quadrangle map	395
17.	Mines on the southwestern quarter of the USGS Manatawny 7.5-minute topographic quadrangle map	396
18.	Mines and mineral localities on the southeastern quarter of the USGS Manatawny 7.5-minute topographic quadrangle map	397
19.	Mines and mineral localities on the northwestern part of the USGS East Greenville 7.5-minute topographic quadrangle map	398
20.	Mines on the southwestern part of the USGS East Greenville 7.5-minute topographic quadrangle map	399
21.	Mines on the northwestern part of the USGS Wolmelsdorf and southwestern part of the Strausstown 7.5-minute topographic quadrangle maps	400
22.	Mines on the northeastern part of the USGS Wolmelsdorf and southeastern part of the Strausstown 7.5-minute topographic quadrangle maps	401
23.	Mines and mineral localities on the northwestern quarter of the USGS Sinking Spring 7.5-minute topographic quadrangle map	402

24.	Mines and mineral localities on the northeastern quarter of the USGS Sinking Spring and southeastern part of the Bernville 7.5-minute topographic quadrangle maps	403
25.	Mines and mineral localities on the southern part of the USGS Sinking Spring 7.5-minute topographic quadrangle map	404
26.	Mines and mineral localities on the northwestern quarter of the USGS Reading 7.5-minute topographic quadrangle map	405
27.	Mines and mineral localities on the northeastern quarter of the USGS Reading 7.5-minute topographic quadrangle map	406
28.	Mines on the southern part of the USGS Reading 7.5-minute topographic quadrangle map	407
29.	Mines and mineral localities on the northwestern quarter of the USGS Birdsboro 7.5-minute topographic quadrangle map	408
30.	Mines and mineral localities on the northeastern quarter of the USGS Birdsboro 7.5-minute topographic quadrangle map	409
31.	Mines and mineral localities on the southwestern quarter of the USGS Birdsboro 7.5-minute topographic quadrangle map	410
32.	Mines and mineral localities on the southeastern part of the Birdsboro and southwestern part of the Boyertown USGS 7.5-minute topographic quadrangle maps	411
33.	Mines and mineral localities on the northwestern quarter of the USGS Boyertown 7.5-minute topographic quadrangle map	412
34.	Mines and mineral localities on the northeastern quarter of the Boyertown and part of the northwestern Sassamansville USGS 7.5-minute topographic quadrangle maps	413
35.	Mines and mineral localities on the southeastern quarter of the USGS Morgantown 7.5-minute topographic quadrangle map	414
36.	Mines and mineral localities on the northern part of the USGS Elverson 7.5-minute topographic quadrangle map	415
37.	Mines and localities on the southwestern quarter of the USGS Elverson 7.5-minute topographic quadrangle map	416

TABLES

1.	Minerals reported from Berks County	6
2.	Generalized geologic log for the core hole drilled at the Warwick mine, Boyertown	47
3.	Vertical section through the Gabel mine No. 1 shaft, Boyertown	53
4.	Iron ore production from the Jones mine, Caernarvon Township, 1836 to 1852	67

THE MINES AND MINERALS OF BERKS COUNTY, PENNSYLVANIA

INTRODUCTION

Berks County has a rich mining and mineral history that spans more than two centuries. For much of the 18th and 19th centuries, the mining and smelting of iron was a dominant industry. Berks County possessed all the ingredients for a successful early iron industry—iron ore, limestone for flux, abundant forests for charcoal, and streams for water power. Pennsylvania's iron industry began near Boyertown when the Colebrook Dale Furnace was built in 1720. The iron industry in Berks County reached its peak in the 1880s. Montgomery (1884, p. 30) reported: "*In 1882 there were over one hundred mines in successful operation, whose annual production exceeded three hundred thousand tons. They then furnished constant employment to over a thousand men and brought into our county over a million of dollars.*"

The quarrying of limestone to produce agricultural lime lasted into the 20th century. Early German settlers established farms in the fertile Oley valley, which was underlain by limestone. Many farms had quarries and lime kilns. In the Morgantown area, lime was produced on a commercial scale and was shipped by wagon and rail to nearby towns and cities.

There are many mineral-specimen localities in Berks County; however, only three produced outstanding crystallized mineral specimens—the Jones, Fritz Island, and Grace mines. All three were iron mines. The Jones mine is best known for fine specimens of malachite, some of which were cut and polished into gemstones. The Fritz Island mine produced a variety of minerals, including native copper, azurite crystals, and zeolite minerals, especially chabazite and thompsonite. The Grace mine, which closed in 1977, is known for spectacular crystallized specimens of magnetite, pyrrhotite, and apophyllite.

Format of this Book

This book is organized alphabetically by municipality (fig. 1). Where the locations of mines or mineral localities are known with certainty or determined with some confidence, they are shown on maps in Appendix 1. Appendix 2 provides a cross-reference for mines and mineral localities. Appendix 3 provides a list of mines shown in Appendix 1. An extensive bibliography lists published information on the minerals, mines, mining history, and geology of Berks County.

The locations of most inactive mines were not field checked; the time required to determine property ownership and contact property owners was prohibitive. The locations of most mines were determined from geologic maps, property maps, published reports, aerial imagery and photographs, and LIDAR imagery. Active quarries were visited in 2016. Remnants of some mines, such as water-filled holes or tailings piles (mine dumps), remain visible; however, many mines have been lost to urbanization or have disappeared.

LIDAR imagery was especially useful in locating mines. LIDAR is a remote sensing technology that measures distance by illuminating a target with a laser and analyzing the reflected light. Acquisition of high-resolution data was competed for Berks County in 2008 for the Pennsylvania Geological Survey PAMAP project. LIDAR imagery shows features that are not apparent on aerial photographs or topographic maps (fig. 2).

For mines that were reliably located, a latitude and longitude (North American Datum of 1983) was determined using computer software. For some mineral localities, such as a farm, a general or approximate location is described. Distance and direction in the descriptions were measured using topographic maps. Distances generally were measured from the intersections of major roads or town centers. Names of roads were taken from street maps published by the Alexander Drafting Company (2006).

Deciding which name to use for a particular mine was at times problematic. While some mines, such as the Grace mine, had only one name, other mines changed owners and names many times over the course of their history. When a mine had multiple names or owners, either the original name of the mine or the most common name of the mine found in the literature was used. If the name of the mine did not appear in the literature, the name of the property owner or the name of the road on which the mine was located was used. Maps showing the location of mines and property owners include Fagan's maps of Berks County (Fagan, 1860 and 1862), the 1876 Atlas of Berks County (Davis and Kochersperger, 1876), and the topographic maps of part of Berks County published by the Second Geological Survey of Pennsylvania as the Atlas to Report D3 (Pennsylvania Geological Survey, 1883). These maps are referred to in this book as the 1860 map, the 1862 map, the 1876 atlas, and the 1882 topographic map, respectively.

INTRODUCTION

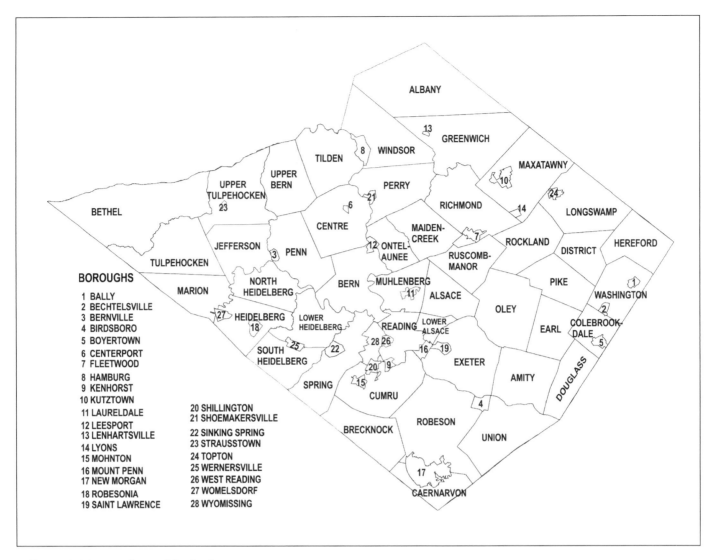

Figure 1. Municipalities of Berks County.

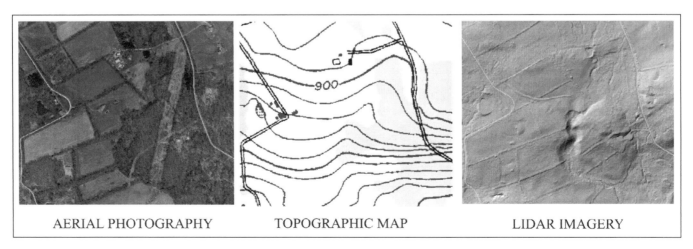

Figure 2. Comparison of aerial photography, topographic mapping, and LIDAR imagery for the same area. The location of the mine is clearly visible on the LIDAR imagery.

INTRODUCTION

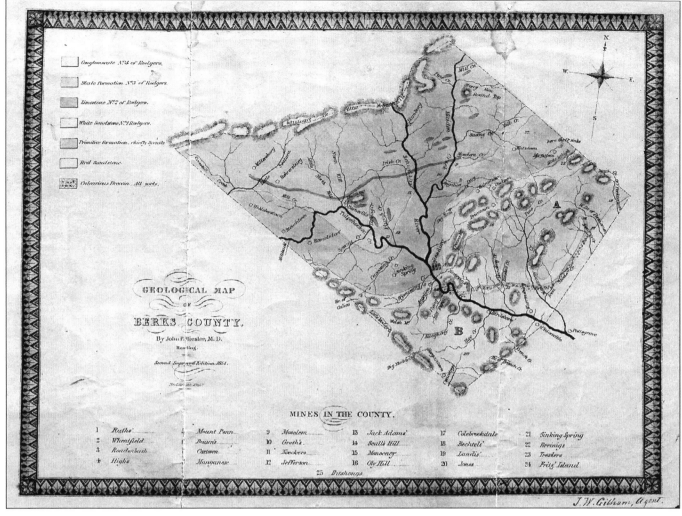

Figure 3. John P. Heister's 1854 geologic map of Berks County.

The geology of Berks County is complex. The first geologic map of Berks County was published by Dr. John Hiester in 1854 (fig. 3). The printed and hand colored map sold for 37-1/2 cents per copy. The map was not original, but was copied from the work of the First Pennsylvania Geological Survey. The geologic nomenclature used in this book was taken from the sources shown on figure 4.

Mineral Products of Berks County

Berks County's varied geology provides an abundance of mineral deposits. Production from the county's mineral deposits began before the Revolutionary War and continues today. During a history spanning more than 200 years, some of the deposits were successfully exploited, and some were never economically viable. Mineral products from Berks County include iron ore, lime, cement, copper ore, graphite, building stone, aggregate (crushed stone), sand, slate, clay, and brick.

Iron ore was mined at well over a hundred locations in Berks County. Iron mining began about 1720 and ended in 1977. Two types of iron ore deposits were mined—magnetite and limonite. The "magnetic" or magnetite ores predominantly were located in the southern part of the county and were associated with the intrusion of diabase during the early Jurassic period. Magnetite also was mined from gneiss in the northwestern part of the county. Magnetite was mined by open-pit and underground methods. The "brown" or limonite iron ores were found in many parts of the county, most commonly in areas underlain by carbonate rocks in the Oley Valley and in Longswamp Township. Limonite was mined mainly by the open-pit method.

One of Berks County's oldest mineral industries, which still continues today, is the quarrying of limestone and dolomite. Initially, small quarries were opened on farms, and the rock was burned in on-site kilns to produce agricultural lime for local use. The quarries also produced aggregate, building stone, furnace flux, and stone for lime and cement

INTRODUCTION

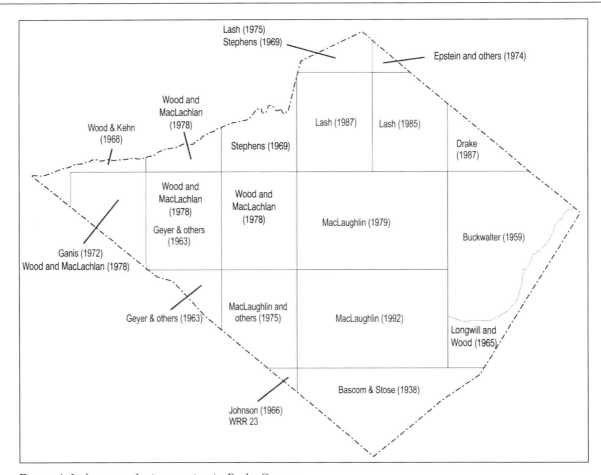

Figure 4. Index to geologic mapping in Berks County.

manufacture. Cement has been manufactured since 1900 from the Jacksonburg Formation cement rock.

Only one mine in Berks County, the Jones mine near Morgantown, was operated for copper ore. Copper ore was found at other mines, notably the Fritz Island and Boyertown mines, but the quantity was not economically significant. The Jones mine was worked for copper ore by several copper companies. Charles Wheatley mined the copper-rich strata in the 1870s to supply ore for his Chemical Copper Company in Phoenixville, Chester County (see Sloto, 2009, p. 125-126).

Graphite was mined in small quantities at several locations in Berks County. The graphite often was found in decomposed rock, and mining was done by pick and shovel. Generally, the mines were not economically viable. Despite its abundance, the cost to mine and mill graphite exceeded the sale value, and bankruptcy was usually the result.

The quarries of Berks County have produced aggregate and building stone from a variety of geological units. Building stone includes limestone, diabase, quartzite, gneiss, "granite," and sandstone. The stone was used to construct many houses, factories, churches, and public buildings in the county.

The Hardyston Formation quartzite weathers to sand, which was quarried in several places in the county. In some quarries, the weathered quartzite was crushed to produce sand. The sand usually was used for construction. Sand also was quarried from residual deposits.

In the mid to late 1800s, slate was mined from a number of quarries in Albany Township. The slate was used for roofing, fireplace mantles, filler, and for the manufacture of linoleum and roofing paper.

Clay deposits often were found associated with iron ore deposits. The clay was used for pottery and paper making. Residual clay derived from weathered shale was mined to manufacture bricks at several places in Berks County. The Martin Bechtelsville quarry produces a clay mix used for baseball infields.

Minerals of Berks County

Berks County's varied geology produces an abundance of minerals. Table 1 lists 125 mineral species found in Berks County plus an additional 8 mineral species that were reported, but not verified.

Mineral names from the literature were updated to conform to valid species listed by Back (2014) with the exceptions listed below.

(1) Biotite — Back (2014, p. 8) considered biotite to be the series name of dark-colored trioctahedral micas. Because of the long-time, popular usage of biotite to describe a black or dark-colored mica and because chemical analyses have never been performed to determine the exact species, the name biotite is used here.

(2) Hornblende — Back (2014, p. 88) recognized ferrohornblende and magnesiohornblende as species in a series, but did not recognize hornblende as a species. Because chemical analyses have not been performed to determine the exact species, the name hornblende is used here.

(3) Limonite is a mixture of iron oxides and is not a valid mineral species. Because of the long-time, popular usage of limonite, the name is used here.

(4) Plagioclase — Back (2014) did not list plagioclase as a distinct mineral species. Because of the long-time, popular usage of plagioclase as a member of the albite-anorthite series, the name is used here.

(5) Zeolite group minerals - Many minerals in the zeolite group form a series; for example, the chabazite series includes chabazite-Ca, chabazite-K, chabazite-Mg, and chabazite-Sr. The species can be determined only with a chemical analysis. Where chemical analyses are not available, the series name (chabazite) is used.

(6) Apophyllite — The apophyllite series includes fluorapophyllite, fluorapophyllite-Na, and hydroxylapophyllite. The species can be determined only with a chemical analysis. Because chemical analyses generally are not available, the series name (apophyllite) is used.

Minerals that are not considered valid species by Back (2014), other than those listed above, are in quotation marks, for example, "deweylite." Minerals reported from a locality that are not verified and are considered doubtful are followed by a question mark in parentheses, for example, wavellite (?).

In the early literature, microcline was sometimes called orthoclase (Montgomery, 1969, p. 75). The original descriptions from the literature were retained, and no attempt was made to determine if orthoclase was actually microcline.

The first state mineralogy for Pennsylvania was published by Genth (1875). Supplements to Genth's mineralogy were published by Eyerman (1889 and 1911). Gordon (1922) published the last complete mineralogy of Pennsylvania. Gordon's mineralogy was supplemented by Montgomery (1969) and Smith (1978).

Natural Science Organizations

Several organizations were devoted to the natural sciences in Berks County. Among the earliest was the Reading Youth and Apprentices Institute, which was founded in 1836. One of the objectives of the Institute was to promote public education and science by means of lectures and discussions. The lectures were open to the public.

The Reading Cabinet

The Reading Cabinet was organized in the fall of 1836 and was composed of the members of the short-lived Reading Youth and Apprentices Institute. The officers were Dr. John P. Hiester, president; H.H. Muhlenberg and James Whitaker, vice presidents; John S. Richards, secretary; Samuel Johnson, treasurer; and George W. Oakley and Powell Griscom, curators. Dr. Hiester delivered the first lecture, which was on chemistry, on January 3, 1840 (Heizmann, 1961).

One of the Cabinet's greatest achievements was to offer a series of five lectures by Professor Benjamin Silliman of Yale University. The cost to attend was one dollar. Silliman was the first person to distill petroleum and was a founder of the American Journal of Science, the oldest scientific journal in the United States. The mineral sillimanite was named after him in 1850. The topic of Silliman's first lecture, held on September 4, 1846, was "Objects, Means, and Ends of the Science of Geology." His concluding lecture on September 18 was "Diluvial Action, Glaciers and Moraines, Deluge and the Bible, with a Discourse on Metals."

The Reading Cabinet assembled a collection of minerals. Dr. Hiester donated a box of choice minerals acquired during his travels in Europe, and B.M. Keim donated a collection. The Cabinet disbanded in 1849 and bequeathed its books, minerals, and curiosities to the Reading Library (Heizmann, 1961).

Reading Society of Natural Sciences

The Reading Society of Natural Sciences existed from January 1869 to May 1885. The first meeting was held on the second floor of John B. Raser's drug store at Sixth and Walnut Streets. The first officers were Fred H. Strecker, president; Dr. John Heyl Raser, vice president; Dr. Walter J. Hoffman, recording secretary; and R.S. Turner, corresponding secretary.

The stated mission of the Reading Society of Natural Sciences was to disseminate science information through lectures. Some of Reading's most prominent citizens, such as Daniel M. Bertolette, David B. Brunner, and Fred H. Strecker, were founding members of the Society. The Society established a rigorous agenda that included lectures on ornithology, ichthyology, mineralogy, microscopy, paleontology, geology, astronomy, botany, herpetology, and other topics. The original membership included 80

INTRODUCTION

Table 1. Minerals reported from Berks County.

Confirmed Mineral Species				
actinolite	cerussite	gold	natrolite	sphalerite
albite	chabazite series	graphite	okinite	spodumene
allanite	chalcocite	grossular	olivine	staurolite
allophane	chalcopyrite	gypsum	opal	stellerite
almandine	chlorite group	hematite	orthoclase	stibnite
anatase	chrysocolla	heulandite series	palygorskite	stilbite series
andradite	chrysotile	hornblende	phillipsite series	stilpnomelane
anglesite	clinochlore	hydromagnesite	plagioclase	talc
anhydride	clinopyroxene	hypersthene	phlogophite	tenorite
anothite	cobaltite	ilmenite	pigeonite	thaumasite
antigorite	copper	jarosite	powellite	thomsonite series
apatite group	cuprite	kaolinite	prehnite	thorite
apophyllite group	datolite	laumontite	pyrite	thorogummite
aragonite	diginite	lepidocrocite	pyrolucite	titanite
arsenopyrite	diopside	limonite	pyroxene group	tochilinite
augite	dolomite	loellingite	pyrrhotite	tourmaline supergroup
aurichalcite	epidote	magnesite	quartz	tremolite
azurite	erythrite	magnetite	riebeckite	vermiculite
babingtonite	fayalite	malachite	rutile	vesuvianite
barite	feldspar group	magnesio-riebeckite	safflorite	wavellite
biotite	fluorite	marcasite	scapolite group	zircon
bornite	forsterite	melanterite	schorl	
brochantite	galena	mesolite	scolecite	
brucite	garnet group	microcline	serpentine group[1]	
calcite	gismondine	molybdenite	siderite	
celestine	goethite	muscovite	skutterudite	
Mineral species reported, but not validated				
brookite	chalcanthite	labradorite	scheelite	
bismuth	dravite	millerite	silver	

[1] kaolinite and serpentine group

individuals, but only a few people regularly attended the programs or the Society's meetings.

In 1876, the Society moved from Raser's drugstore to the second floor of the Reading Library. The Society arranged to use a room, formerly used as a reading room, on Franklin Street adjoining the library. Display cases housing the collections were arranged in the center of the room, and the room was open to the public one afternoon a week (*Reading Eagle*, January 6, 1876).

For 18 years, the viability of the Society fluctuated. Often, its officers and members displayed little interest in the Society. At times, when it appeared the Society would disband, renewed interest prolonged its life. A committee appointed in 1881 to study the viability of the Society recommended that it be disbanded because of lack of interest on the part of its members. One of the problems of dissolution was the disposition of the Society's collections, which were valued at $1,500, a considerable sum at that time (Heizmann, 1961). The Society disbanded in the spring of 1885. The collections were sold to the Reading School District, and some of these specimens eventually became part of the collection of the Reading Public Museum.

Reading Chapter of the Agassiz Association

The Agassiz Association, named after Alexander Agassiz, was a union of local societies organized for the study of nature by personal observation. Chapter number 258 of the Agassiz Association was organized in Reading in 1885 and lasted until 1890. Members included Edward Cox, Fitz-Daniel Ermentrout, Herbert Leaf, Dr. Levi Mengle, Charles H. Muhlenberg, Frederick W. Nichols, Herbert N. Sternberg, and Frederic W. Wilson. The Association met bi-monthly at member's homes to study natural history. However, emphasis on social, rather than scientific, activities created friction among the members. After its demise, the chapter's library was donated to the Reading Public Museum (Heizmann, 1961).

Spencer F. Baird Association

Spencer Fullerton Baird was born in Reading in 1823. He was a naturalist, ornithologist, ichthyologist, and herpetologist. He was the first curator at the Smithsonian Institution and served as the second Secretary of the Smithsonian from 1878 to 1887. He was dedicated to expanding the natural history collections of the Smithsonian and published over 1,000 works during his lifetime. The short-lived Spencer F. Baird Association was organized in Reading in 1887, the year of Baird's death. The association was composed largely of mineralogists; however, it disbanded because of internal problems (Heizmann, 1961).

Mt. Penn Association of Sciences

The Mt. Penn Association of Sciences was founded on November 22, 1895. It was headquartered at 618 Mulberry Street in Reading, the home of George Gehret, it's president. The Reading Eagle (October 2, 1898) described the meeting room: *"The large parlor has the appearance of a museum, 2 large cases, filled with minerals, occupying 2 sides of the room."* The membership was composed mostly of young men with an interest in minerals. In 1898, the officers were George Gehret, president (fig. 5); Ezra Gehret, vice president; James S. Keiser, treasurer; Charles F. Schweigert, financial secretary; John W. Wanner, recording secretary; and Daniel Emerich, James S. Keiser, and W.L. Beacher, trustees. Other members included Carl Arnold, J.L. Berger, Foster Biehl, Howard Billig, Amos Breneiser, D.S. Emerich, Edwin Fisher, Frank Fisher, August Fleckenstine, Lester Ganser, Mrs. George Gehret, Irvin Giles, Luther Heilman, A. Hepler, James Keene, Sylvester Keiser, Charles Lutz, Paul McKinney, Fred Mosser, Morris Mull, A. Mulllgan, George Runyeon, Samuel Schaeffer, John Schlotterbeck, Dr. John Schoenfeld, Henry Stein, Robert Toole, and Isaac Wolff.

Meetings of the Association were held every other week. A member read an essay at every meeting; the essays were collected into a reference volume in the Association's library. A brief summary of some of the meetings between 1898 and 1903 was printed in the Mineral Collector (1898, p. 78, 82, 114, 137, 138, 117; 1899, p. 116, 139, 154; 1900, p. 61, 145; 1901, 149, 191; 1902, p. 122; 1903, p. 112).

In 1898, the Association possessed a collection of 500 mineral specimens. About 50 specimens were provided by D.B. Brunner, including chabazite from the Fritz island mine, polished serpentines from the Fritz Island and Wheatfield mines, and fluorite from Leinbach's Hill. C.F. Gauker donated a specimen of drusy quartz on limonite from Fleetwood (*Reading Eagle*, October 2, 1898). Other specimens included molybdenite from near Gottshall's mine (*Mineral Collector*, 1899, vol. 6, no. 8, p. 139). George Gehret made his collection available for study by Association members. Gehret's collection of about 1,000 minerals was housed in a cabinet 6 feet long and 8 feet high containing 24 drawers. In addition, there was an open case containing museum-quality specimens.

The preamble of the Association's by-laws included a charge to the president to form a committee "*to appropriate a tract of land, to put up a suitable building for use of the association as a public museum... kept open to the public at least 2 evenings each week.*" The museum was to be called the Mt. Penn Museum of Sciences (*Mineral Collector*, 1898, vol. 5, no. 6, p. 82). On March 5, 1897, the Reading Eagle announced a sauerkraut supper to be held for the benefit of the Mt. Penn Association of Sciences at the residence of Charles Schweigert. A fund-raising drawing proposed at the November 24, 1898, meeting included a gold watch, a dinner set, and an umbrella as prizes (*Mineral Collector*, 1899, vol. 5, no. 11, p. 177). The museum never materialized. It is not known how long the Association was in existence. A brief

Figure 5. Officers of the Mt. Penn Association of Science in 1898 (*Reading Eagle*, November 2, 1898).

INTRODUCTION

note in the January 4, 1906, Reading Eagle listed a meeting of the Mt. Penn Association of Sciences that evening at 244 Pearl Street in Reading.

Berks County Mineralogical Society

The Berks County Mineralogical Society was founded on October 8, 1957. The founding was the culmination of months of meetings of a small group of people interested in mineralogy. While the society was still in the planning stage, the founders, George Daniels, Robert Eisenhauer, Leonard Gerhart, and Ralph Heller, met at the homes of Daniels and Gerhart. This group formed with the purpose of making a serious study of minerals and creating a complete list of minerals and mineral locations in Berks County. The founders thought that by limiting the membership to approximately 40 sincerely interested people, they would make a worthwhile contribution to the field of mineralogy while furthering their own knowledge of the subject. After two or three meetings, Ronald Cocroft, Harry Schaeffer, Howard Hays, and Samuel Gundy, director of the Reading Museum, also joined the group. The Reading Museum offered a classroom as a meeting place, and the first meeting of the Berks County Mineralogical Society was held on October 8, 1957. The meeting was open to the public.

Officers of the Society for the first year were Robert Eisenhauer, president; Leonard Gerhart, vice president; Howard Hays, secretary; and Harry Schaeffer, treasurer. The president also acted as the field trip chairman. After the first year, the field trip chairman became a separate position that was filled by Robert Eisenhauer. The first field trip was held in May 1958, when members were guests of the Bethlehem Steel Company at the Bridgeport quarry in Montgomery County.

The Berks Mineralogical Society newsletter, The Geode, made its debut in December 1959, with George Petro as editor. Prior to publication of the newsletter, meeting and field trip notices were sent first by postcard and later by a letter copied on a duplicating machine. Dues in 1959 were $1 per year.

From April 30 to May 28, 1961, the Society sponsored its first mineral show at the Reading Museum. Members brought their best specimens for display in four large cases provided by the museum. Samuel Gundy and the members of several committees carefully labeled and arranged the specimens. The success of the show and exhibit was evidenced by the large turnout and the many complimentary remarks overheard at the display cases (Woodeshick, 2007).

The Society drew up a new constitution and became incorporated in October 1962. The Society continued to grow and celebrated it's 50th anniversary in 2007. The Society sponsors a gem and mineral show in Leesport every May and offers field trips throughout the year.

Reading Public Museum

Levi Walter Mengel was born in Reading on September 27, 1868. From his early childhood, he developed a strong interest in the natural sciences of Berks County. At the age of six, he started collecting rocks, minerals, butterflies, and other natural objects. By the time he graduated from the Reading High School for Boys in 1886, his collection numbered over 5,000 objects and specimens. Following his graduation, Mengel went to work for Jones' Drug Store at Ninth and Penn Streets in Reading. Later, he worked as an apprentice at Steinmetz's Drug Store at Sixth and Penn Streets. Mengel then entered the Philadelphia College of Pharmacy, from which he graduated in 1891.

While studying at the College of Pharmacy, Mengel was introduced to Dr. Angelo Heilprin, curator at the Academy of Natural Sciences of Philadelphia. Dr. Heilprin was so impressed with Mengel's knowledge of insects that he was invited to join the staff of the Academy as its entomologist for an expedition to West Greenland. On June 6, 1891, Mengel departed for Greenland with the Peary expedition. This trip marked the beginning of Commander Robert Peary's exploration of the northern Polar Regions. On board were two teams of scientists whose goal was to explore the interior of Greenland.

For a short period after his return from Greenland, Mengel worked as a chemist for the Reading Company. His love of the natural sciences, however, soon led him to a career as a teacher. From 1894 to 1902, Mengel taught natural history, chemistry, and physics at the Reading High School for Boys. During class, he encouraged his students to interact with objects and specimens from his personal collection.

At the St. Louis World's Fair in 1904. Mengel obtained nearly 2,000 museum-quality items from China, Japan, India, Ceylon, the Philippines, and Central and South American countries. They formed the nucleus for the Reading Museum's collection. In 1907, the former High School for Boys became the Reading School District's administration building, and the third floor was made available for exhibition of Mengle's collection.

By 1911, the collection was open to school children and the public. As the collection grew, the museum expanded to include all floors of the building. In 1913, the museum received its first painting, "Desolate Winter" by Victor Shearer. This donation set the stage for additional gifts of fine art, and, by the end of 1913, the museum became the Reading Public Museum and Art Gallery.

In 1924, the Reading School District asked the citizens of Reading to approve a loan to construct a new school building. Included were provisions for a modern museum building. After a campaign in which the school children participated, the loan was approved, and plans for a new building were drawn. The building was designed by Alexander Forbes Smith and built by Irvin F. Impink. The site of the museum was selected and donated to the school district by Ferdinand Thun, Henry Janssen, and Gustave Oberlaender. On April 1, 1927, Mengel donated his personal collections to the Reading Public Museum and Art Gallery, and in 1929, the new building opened to the public. The mineral collection included more than 25,000 specimens.

In 1992, administration of the museum was transferred from the Reading School District to a private, non-profit foundation. With the foundation's leadership and partnership with the County of Berks, the Reading School District, and the City of Reading, there has been a rededication to the museum's mission. As a result of this restructuring, the Reading Public Museum is enjoying a renewed vitality (Reading Public Museum, 2014; Natural History Museum, 2014).

ACKNOWLEDGEMENTS

The author gratefully acknowledges and thanks the many people and organizations that provided assistance in producing this book. I especially appreciate the generosity of the following (in alphabetical order) for allowing me to study and photograph minerals in their collections: Academy of Natural Sciences of Drexel University (with assistance from Douglas Klieger, Ned Gilmore, and Ted Daeschler), Bryn Mawr College (with assistance from Maria Luisa Crawford), Carnegie Museum of Natural History (with assistance from Marc and Debra Wilson), Steve Carter, Skip Colflesh, Ron Kendig, Reading Public Museum (with assistance from Jacquelynn Accetta), and Scott Snavely.

The author also gratefully acknowledges and thanks the following people and organizations: Lisa Adams and the Historical Society of Berks County for providing photographs and the use of their library; Charlie Miller and the Amity Heritage Society for providing photographs and the use of their library; Lindsay Dierolf and the Boyertown Historical Society for providing photographs and the use of their library; Larry Eisenberg for providing information and mineral specimens from the Grace mine; Mark Eschbacher and the H&K Group for providing photographs and a visit to the H&K Group quarries; John Grubb for providing photographs and a guided tour of the Rittenhouse Gap and Edison mines; Hagley Museum and Library for providing photographs and the use of their library; Anne Kelhart and Martin Stone Quarries, Inc for providing photographs and information; Peggy Light and the Albany Historical Society for providing photographs; Edward Mosheim and the Hereford Township Heritage Society for providing photographs; Dr. Fred Munson of West Chester University for providing assistance with the X-ray diffractometer and scanning electron microscope energy dispersive spectrometer (SEM-EDS); Tom Pracher for providing photographs and a guided tour of the C.K. Williams ocher mines; Rebecca Ross and Hopewell Furnace National Historic Park for the use of their library; the Smithsonian Institute for providing photographs; Hunt Schenkel and the Swenkfelder Library and Heritage Center for providing photographs; Suzanne Smith for giving permission to publish her grandfather's photographs of Klines Corners mines; the Tr-County Heritage Society in Morgantown for providing photographs; and the U.S. Geological Survey in Reston, Virginia, for the use of their library.

A special thank you goes to Lana Dickinson and Allison Sloto for spending countless hours editing this manuscript.

Figure 6. Aerial photograph of the slate quarries at Quaker City, Albany Township, September 12, 1937. Courtesy of the Bureau of Topographic and Geological Survey, PennPilot Historical Aerial Photo Library. 1, Pittsburg quarry; 2, unnamed quarry 1; 3, Centennial quarry; 4, unnamed quarry 2; 5, Quaker City quarry.

INTRODUCTION

Figure 7. George Deysher posing next to a Quaker City slate quarry boiler. Undated photograph. Photograph courtesy of the Albany Historical Society.

Figure 8. Slate quarry near Quaker City, Albany Township. Undated photograph. Photograph courtesy of the Albany Historical Society.

Figure 9. Quaker State Slate Company quarry, Albany Township, April 2015.

ALBANY TOWNSHIP

Albany Township is at the western end of the southeastern Pennsylvania slate region, which extends from northern Berks County eastward across Lehigh County. The first slate quarry in the area, the Excelsior Slate Works, was opened in Lynnport, just across the Lehigh County line, by Daniel Jones, James Porter, and Robert McDowell in 1844. Behre (1933) divided the slate region quarries in Berks County into the Lynnport group and the Greenawald group. The Lynnport group is centered around Quaker City; the Greenawald group is centered around Greenawald.

Lynnport Group Slate Quarries

The Lynnport group slate quarries produced roofing slate and flagstone. Some slate formed smooth, flat sheets, which could be used for roofing when expertly cut by striking parallel to the foliation with a specialized tool. Slate was a popular roofing material up to the 1920s because it is waterproof, fireproof, and durable; it can last for several hundred years with minimal maintenance. All slate quarried at the Lynnport group quarries was from the Martinsburg Formation.

The Philadelphia Slate Mantel Company was chartered by the Commonwealth of Pennsylvania on May 25, 1870. The company office was located at 139 North 7th Street in Philadelphia. The company obtained slate from a quarry in Berks County (fig. 10); however, the name of the quarry is not known. Berks County deed records indicate that the quarry was leased from Isaac D. Guyer.

Quaker City Slate Company Quarries

The Quaker City Slate Company quarries are approximately 0.25 mile northwest of Quaker City and west of Riegel Lane at 40° 39′ 37″ latitude and 75° 52′ 52″ longitude on the USGS New Ringgold 7.5-minute topographic quadrangle map (Appendix 1, fig. 1, location A-1). They are shown as mine number 4 on the geologic map of Behre (1933, plate 25). They are abandoned and flooded (fig. 9).

The Quaker City Slate Company operated three quarries west of a secondary road. According to Behre (1933, p. 340), the northern quarry was small, measuring about 30 feet square and 10 feet deep. The middle quarry was the original Quaker City Slate Company quarry. It was 80 feet south of the northern quarry and was about 300 feet by 100 feet. The southern quarry was 70 feet west of the middle quarry and was 40 feet square and about 10 feet deep.

The Quaker City Slate Company, which gave its name to the village of Quaker City, was originally organized by Philadelphia businessmen. The company office was located at 906 Filbert Street in Philadelphia. It was chartered with $250,000 in capital by the Commonwealth of Pennsylvania for *"buying, selling and leasing slate and slate properties, and quarrying, manufacturing and selling the same."*

Figure 10. Advertisement for the Philadelphia Slate Mantle Company (*The Manufacturer and Builder*, July 1870, p. 221).

The Quaker City Slate Company's first quarry was opened on the properly of Charles Faust (*Reading Times*, August 24, 1883). James Lutz was the superintendent of the quarry in 1883 (*Reading Eagle*, December 10, 1883). During 1883-85, the company acquired the properties of James S. Jones, William R. Granger, and William B. Shaffer. During the same period, the company obtained mortgages of $7,000 from Henry Albertson and $12,000 from the Guarantee Trust and Safe Deposit Company of Philadelphia.

The Quaker City Slate Company proposed to lay out a new town to be named Hoyt in honor of ex-governor Henry M. Hoyt, who also was the president of the company (*Reading Eagle*, May 2, 1885). Hoyt was governor of Pennsylvania from 1879 to 1883. He planned to construct a summer residence near the quarries. The town of Hoyt was never built.

In 1885, the Quaker City Slate Company was purchased by a group of Reading and Allentown businessmen (*Reading Eagle*, August 5, 1925). General W.P. Snyder of Allentown was the secretary of the company, and J.M. Collingswood was the assistant secretary. Jacob L. Farr of Allentown was the treasurer. The company directors also included Dr. H.B. Hartzell, Robert E. Wright, Jr., William B. Schaeffer, Col. B.K. Jamison (a Philadelphia banker), and Col. William M. Stewart (also of Philadelphia). The company office was located at 108 South 4th Street in Philadelphia. The new owners added $50,000 to the working capital. Two new quarries were opened on land owned by the company where the existing quarry was located (*Reading Eagle*, May 2, 1885). Four years later in 1889, the company was

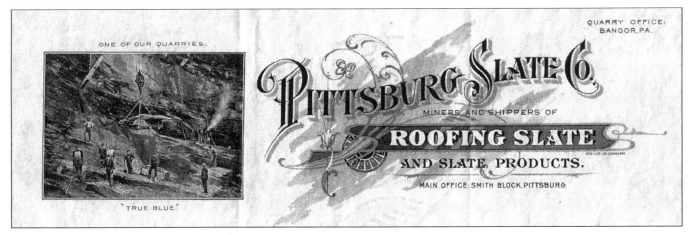

Figure 11. Letterhead of the Pittsburg Slate Company.

bankrupt, and the property of the Quaker City Slate Company was sold at sheriff's sale (*Stone*, August 1889, vol. 2, no. 4, p. 66).

The quarry continued to operate after 1889. The Reading Eagle (March 27, 1895) reported that a bolt of lightning "*struck the triangle at the Quaker City Slate quarry, passing along the cable, and thence to a derrick rope at the Big Bed slate quarry. From there it leaped to the hoisting rope and shot into the engine room. The engineer, C.L. Correll, who was sitting at the lever, was partly stunned, and his right hand singed. His arm and leg were rendered useless for two hours.*" All three quarries were abandoned before Behre's visit in 1927.

Hemerly Quarry

The Hemerly quarry was east of Schucker Lane about 0.7 mile west of Quaker City at about 40° 39' 25" latitude and 75° 53' 21" longitude on the USGS New Ringgold 7.5-minute topographic quadrangle map (Appendix 1, fig. 1, location A-2). It is mine number 1 on the geologic map of Behre (1933, plate 25). The Hemley slate quarry was located on the north slope of a small, hogback-like hill. Behre (1933, p. 340) described the quarry in 1927 as an old, largely filled, elliptical open cut in the south bank of the stream. The strike of the bedding was N. 75° E., and the dip was 53° S.

Mammoth Quarry

The Mammoth quarry was east of Schucker Lane about 0.6 mile west of Quaker City at about 40° 39' 26" latitude and 75° 53' 25" longitude on the USGS New Ringgold 7.5-minute topographic quadrangle map (Appendix 1, fig. 1, location A-3). It is mine number 2 on the geologic map of Behre (1933, plate 25). The Mammoth slate quarry was located 600 feet east of the Hemerly quarry. Behre (1933, p. 340) described the quarry in 1927 as 40 feet square, probably 40 feet deep, and filled with water. Slate was quarried in 1866 and was used for flagstone.

Oswold Quarry

The Oswold quarry was east of Schucker Lane about 0.45 mile west of Quaker City at about 40° 39' 27" latitude and 75° 53' 19" longitude on the USGS New Ringgold 7.5-minute topographic quadrangle map (Appendix 1, fig. 1, location A-4). It is mine number 3 on the geologic map of Behre (1933, plate 25). The Oswold slate quarry was about 200 feet east of the Mammoth quarry. Behre (1933, p. 340) described the quarry in 1927 as very small, shallow, and flooded.

Centennial Quarry

The Centennial quarry was east of Riegel Lane about 0.15 mile north-northeast of Quaker City at 40° 39' 41" latitude and 75° 52' 46" longitude on the USGS New Ringgold 7.5-minute topographic quadrangle map (Appendix 1, fig. 1, location A-5). It is mine number 6 on the geologic map of Behre (1933, plate 25). The Centennial quarry was about 300 feet south of unnamed quarry 1.

The Centennial quarry may have been opened as early as 1850. In 1876, the quarry was operated by David Heinly and Charles Faust under the name Faust, Heinly & Brothers. It also was operated by or known as the Centennial Slate Company. The quarry produced roofing slate, which was marketed as Guarantee Blue Roofing Slate and sold through an agent named F.P. Semmel in Lehighton (*Carbon Advocate*, July 28, 1877). In 1877, the quarry employed 18 men.

In 1881, the quarry was sold to the Standard Slate Company of Slatington, Lehigh County. The quarry was later operated by the Big Bed Slate Company (ca. 1890). At the height of its operation, the Centennial quarry employed 40 men (Kempton Centennial Committee, undated). The quarry last operated in 1905.

In 1883, Sanders (1883, p. 125) noted that the quarry was 150 feet by 50 feet by 80 feet in size. Behre (1933, p. 340-341) stated that in 1927 the quarry was 80 feet by 300 feet in size and was separated into two parts by a waste pile. The waste pile dividing the quarry can be seen on figure 6. The strike of the bedding was N. 70° E. and the dip was 80° N.

Pittsburg Quarry

The Pittsburg quarry (spelled Pittsburgh by Behre) was east of Riegel Lane about 0.3 mile northeast of Quaker City at about 40° 39′ 45″ latitude and 75° 52′ 35″ longitude on the USGS New Ringgold 7.5-minute topographic quadrangle map (Appendix 1, fig. 1, location A-6). It is mine number 8 on the geologic map of Behre (1933, plate 25).

The Pittsburg quarry was worked by the Pittsburg Slate Company of Bangor, Lehigh County (fig. 11). The quarry was about 800 feet east of unnamed quarry 1 at the head of a small tributary to Ontelaunee Creek. Behre (1933, p. 341) noted that in 1927 the quarry was 200 feet by 50 feet and exposed 18 feet of slate above the water level. The strike of the beds was N. 83° E., and the dip was 60° N. This is likely the abandoned quarry east of the Centennial quarry noted by Sanders (1883, p. 125), who described a quarry with a bed 20 feet thick dipping 70° N. with vertical cleavage.

Unnamed Slate Quarry 1

Unnamed slate quarry 1 was east of Riegel Lane about 0.2 mile north-northeast of Quaker City at 40° 39′ 38″ latitude and 75° 52′ 45″ longitude on the USGS New Ringgold 7.5-minute topographic quadrangle map (Appendix 1, fig. 1, location A-7). It is mine number 5 on the geologic map of Behre (1933, plate 25).

Behre (1933, p. 340-341) described an unnamed quarry to the east and immediately across the road from the Quaker City Slate Company quarries in 1927 as a flooded pit 225 feet by 100 feet exposing 10 feet of slate above the water level. In general, the strike of the bedding was N. 77° E., and the dip varied from 5° N. to 20° S. Behre (1933) noted: *"The slate on the dumps appears to be of fair quality, but much of it bears closely-spaced calcareous joints."*

Unnamed Slate Quarry 2

Unnamed slate quarry 2 was east of Riegel Lane about 0.1 mile north-northeast of Quaker City at about 40° 39′ 37″ latitude and 75° 52′ 44″ longitude on the USGS New Ringgold 7.5-minute topographic quadrangle map (Appendix 1, fig. 1, location A-8). It was about 100 feet south of unnamed quarry 1. It is mine number 7 on the geologic map of Behre (1933, plate 25). Behre (1933, p. 340-341) described the quarry in 1927 as a small prospect pit 75 feet square filled with waste material and water.

Greenawald Group Slate Quarries

Slate produced from the Greenawald group slate quarries was crushed and granulated for use as chips in roofing tar and as a filler with pigmenting qualities. All quarries in the Greenawald group are in shale and graywacke of the Hamburg sequence. Red and green slate beds occur interstratified with coarse sandy layers. The workable beds were thin, slabby beds 1 to 3 inches thick. The beds were slatey to shaley and fine grained with a texture approximating a fine-grained sandstone. They exhibited various shades of purple, green, and ocher yellow. The red slates were colored by hematite. The quarries of the Greenawald group were worked intermittently from about 1907 to 1927 (Behre, 1933, p. 353-355).

J. Wibur Company Quarry

The J. Wilbur Company quarry (fig. 12) was east of Kirk Road in Greenawald at about 40° 36′ 12″ latitude and 75° 52′ 11″ longitude on the USGS Kutztown 7.5-minute topographic quadrangle map (Appendix 1, fig. 2, location A-9). It is mine number 3 on the geologic map of Behre (1933, plate 25).

The J. Wilbur Company quarry, which was opened in 1895, was a large quarry 500 feet southeast of the Greenawald railroad station. The quarry measured 400 feet by 80 feet and was 60 feet deep in 1927. The strike of the beds was N. 60° E., and the dip was 65° S. The dominant color was bright brick red. The red, light green, and dark green beds were all distinctly banded. The brown beds were sandy and calcareous.

Figure 12. The J. Wilbur quarry, Greenawald, Albany Township, ca. 1925. From Behre (1933, p. 354, plate 62A). Courtesy of the Pennsylvania Geologic Survey.

Figure 13. Concentric circles in shale with pyrite in the center, Greenawald, Albany Township. Reading Public Museum collection 2005C-3-510.

Slate was transported from the quarry on a spur line to the Philadelphia and Reading Railroad and was shipped to the Wilbur Job Company in Providence, Rhode Island, where it was crushed for use in roofing paper. When finely ground, it was used as filler for coarse oil cloth and roofing paper (Behre, 1933, p. 356). Some of the shale also was shipped to Philadelphia. Miller (1911, p. 81) stated: *"The shale is quarried and shipped to Philadelphia on the Philadelphia and Reading Railroad, a branch of which passes through Greenawald, where it is ground to a very fine powder. Its principal use is in the manufacture of oil cloth and linoleum. The amount of material obtainable is practically unlimited and the cost of production low so that the output could easily be greatly increased if the market demanded it."*

Miller (1911, p. 79) noted: *"In the vicinity of Greenawald the J. Wilbur Company of Providence, Rhode Island, has been working some red shales of this formation since 1895. The quarry where the greatest amount of material has been obtained lies about three-eighth mile southeast of the station at Greenawald. At that locality the shale bed suitable for use as a pigment includes a thickness of about 75 feet of fine grained dark brick red shales. The beds are tilted to the south at an angle of about 55°. The quarry was extended to the boundary lines of the property and excavated to a depth of 55 to 60 feet on the up-hill side of the opening."* The quarry was abandoned about 1910.

A second quarry was opened around 1910. The quarry was near the top of a high hill about one-eighth mile east of the Greenawald railroad station. Only about 600 tons had been shipped from this quarry at the time of Miller's visit in September, 1910. The quarry exposed about 35 feet of good material. James S. Focht, the general manager of the quarry, reported that exploration showed about 250 feet of workable material farther back in the hill (Miller, 1911, p. 80). The quarry operated as late as 1927 (*Reading Eagle*, August 30, 1927).

The shale, especially in the quarry abandoned about 1910, has concentric discoloration spots that are irregularly distributed throughout the material (fig. 13). They range up to about 2 inches in length and 1.5 inches in width but most are less than half that size. Their thickness is variable, but the discolored parts are seldom more than one-quarter inch thick. In the center of most spots there is a small rounded cavity and sometimes a small pyrite concretion (Miller, 1911, p. 81).

The shale contains a great amount of fine-grained muscovite (sericite) sufficient to produce a soapy or talcose feeling when rubbed. Because of this, the material produced by the company was sold under the trade name of Talckene. Talckene came in shades of red and green; it was finely ground and sprinkled on heavy roof coverings (*Reading Eagle*, May 25, 1957). Miller (1911, p. 492) provided an analysis of Talckene. The J. Wilbur Company also quarried a light yellow sericite shale that was ground and used for linoleum; this deposit was worked less than the red strata. Quartz crystals have been found in the abandoned shale pits (*Reading Eagle*, May 25, 1957).

Atlas Mineral Products Albany Quarry

The Albany quarry was 0.4 mile southwest of Albany, on the west bank of Maiden Creek at 40° 36' 46" latitude and 75° 52' 17" longitude on the USGS Kutztown 7.5-minute topographic quadrangle map (Appendix 1, fig. 2, location A-10). It is mine number 1 on the geologic map of Behre (1933, plate 25).

The Atlas Mineral and Machine Company of Lincoln, New Jersey, operated the Albany quarry on the farmstead of Andrew Kunkle. In 1892, the company was founded in Lincoln by Maximillian F. Wirtz (fig. 14). The company initially manufactured and exported bicycle parts. As the bicycle craze of the 1890s subsided, Wirtz began grinding and selling red shale for coloring pigment. Prior to the fall of 1908, red shale was shipped from the quarry to the company's mill in Lincoln for grinding. The quarry suspended operations in the summer of 1908.

Around 1914, a large processing plant was built on-site (figs. 15-18), and quarrying resumed after about 6 years of inactivity. The plant consisted of a 60 foot by 146 foot one-story factory building, a 36 foot by 66 foot two-story annex, three motor houses, and an office building. The plant, locally known as the paint mill, was operated by Maximillian Wirtz and his son, George L. Wirtz. The operation was complete with a company store and frame "bed, stove, and shovel" shanties for

Figure 14. Maximillian Wirtz at his desk. Photograph courtesy of the Albany Historical Society.

bachelor lodging. At times, as many as 24 men were employed (Behre, 1933, p. 356; Miller, 1911, p. 81; *Reading Eagle*, June 4, 1925, and December 16, 1981).

On June 3, 1925, a fire destroyed the plant; only a few sheds were saved. The fire destroyed a large tube mill, a stone crusher, two driers, two screeners, three disintegrators, a brand new fuller mill (just installed at a cost of $1,000), six motors, 200 tons of finished granules, 200 tons of powered paint, 150 tons of raw material, several carloads of coal, and thousands of empty bags. The firemen were able to save a shed containing 95 gallons of gasoline. The loss was estimated at $25,000. The plant was not rebuilt, and the quarry was abandoned. After the fire, Wirtz and his son moved to their farm in Monterey near Mertztown (*Reading Eagle*, June 4, 1925). The Atlas Mineral and Machine Company, now known as Atlas Minerals and Chemicals, Inc., still (2016) maintains a facility in Mertztown.

The Albany quarry was located on a bluff about 50 feet above Maiden Creek. A double inclined track ran from the railroad up the hill to the quarry, and the cars loaded with slate moved downhill by gravity. A descending loaded car pulled an empty car up to the workings.

Behre (1933, p. 356) described the quarry in 1927 as 300 feet wide along strike and 100 feet long downdip with a maximum depth of 80 feet. The rock was thin bedded, olive-green clay slates and shales and brick red to purple shales. The strike of the beds was N. 75° E., and the dip was 60-70°. Sericite was abundant in the shale, and concentric discolorations were absent. Much of the red iron oxide, however, had been removed, and the shales were blotched irregularly. Percolating water carrying organic acids was likely responsible for the removal of the coloring matter (Miller, 1911, p. 81).

B. Frank Ruth and Company Quarry

The B. Frank Ruth and Company quarry was east of Kirk Road in Greenawald at 40° 36′ 13″ latitude and 75° 52′ 14″ longitude on the USGS Kutztown 7.5-minute topographic quadrangle map (Appendix 1, fig. 2, location A-11). The quarry was near the Greenawald railroad station on the east side of Maiden Creek. It is mine number 2 on the geologic map of Behre (1933, plate 25).

Figure 15. Atlas Mineral Products plant and rail cars, Albany Township. Photograph courtesy of the Albany Historical Society.

Figure 16. Atlas Mineral Products plant, Albany Township. Photograph courtesy of the Albany Historical Society.

Figure 17. Atlas Mineral Products plant and quarry, Albany Township. Photograph courtesy of the Albany Historical Society.

Figure 18. Atlas Mineral Products quarry, Albany Township. Photograph courtesy of the Albany Historical Society.

After working for the Wilhem Paint Works in Reading from 1873 to 1884, Ruth organized a company of his own, B. Frank Ruth and Company of Reading, to manufacture paint (Meiser and Meiser, 1982, p. 40). The B. Frank Ruth quarry was opened before 1910. The quarry, which was worked intermittently, was a small pit 60 feet square and about 50 feet deep. The slate was red and green. The strike of the beds was N. 70° E., and the dip was 60° S. The slate was hauled by horse team to the railroad station (Behre, 1933, p. 356).

Focht Quarry

The Focht quarry was east of Kirk Road in Greenawald at 40° 36′ 11″ latitude and 75° 52′ 11″ longitude on the USGS Kutztown 7.5-minute topographic quadrangle map (Appendix 1, fig. 2, location A-12). It is mine number 4 on the geologic map of Behre (1933, plate 25).

Figure 19. Slate quarry, Albany Township. Photograph taken in November 1924 by Ralph W. Stone. Courtesy of the Pennsylvania Geological Survey

The Focht quarry was a large pit about 450 feet northeast of the J. Wilbur Company quarry. The quarry was a series of partly connected openings totaling about 500 feet along strike, 100 feet wide, and 60 feet deep. The southern edge exposed green slate, but the rock was mostly slightly gritty red beds. The property was owned and operated by James S. Focht of Greenawald. Focht worked the quarry as late as 1926 (*Reading Eagle*, August 16, 1926).

Operations were begun in 1890 but were later shifted to the J. Wilbur Company quarry (Behre, 1933, p. 356), where Focht was the general manager. Production in 1895 was 1 to 2 car loads of red shale daily (*Reading Eagle*, April 19, 1895). In 1899, the Reading Eagle (February 2, 1899) reported that Focht attempted to thaw frozen dynamite in a small fire. The dynamite exploded, and fortunately no one was injured. In 1905, the Reading Eagle (May 11, 1905) reported that Focht shipped 9 car loads of red shale to Philadelphia during the previous week.

Unnamed Slate Mine

An unnamed slate mine was northeast of the intersection of Stoney Run Valley and Donat Roads at about 40° 36′ 36″ latitude and 75° 50′ 38″ longitude on the USGS Kutztown 7.5-minute topographic quadrangle map (Appendix 1, fig. 2, location A-13). The mine was in shale and graywacke of the Hamburg sequence. Sanders (1883, p. 128) reported a mine on Stone Run, 1.5 miles above its mouth. The shaft, about 20 feet deep, was sunk in red slate. Most of the slate was red, but some also was green. Some of the red slate had spots of green. According to Sanders: *"The slate does not have good cleavage and was not suitable for roofing slate."* The mine is shown as an unnamed and unlabeled mine on the geologic map of Behre (1933, plate 25). The mine was east of the Greenawald group and is not associated with either the Lynnport or Greenawald group.

Other Quarries

Greenawalt Quarry

The Greenawalt quarry was in the vicinity of Albany. The 1876 atlas shows a lime kiln on present day Pennsylvania State Route 143 opposite the railroad depot. In 1870, the Berks County Orphans Court ordered the sale of the property of the estate of Jacon Greenawalt. The 139-acre farm contained a large limestone quarry (*Reading Eagle*, February 5, 1870).

Kunkle Quarry

The Kunkle quarry was north of Mountain Road at about 40° 39′ 22″ latitude and 75° 54′ 34″ longitude on the USGS New Ringgold 7.5-minute topographic quadrangle map (Appendix 1, fig. 1, location A-14). This small quarry was operated for building stone by Thomas J. Kunkle, a stone mason. Production was small (Barnes, 2011).

Gilt Flagstone Quarry

Sanders (1883, p. 126) described the location of John Gilt's flagstone quarry as 2 miles from the Kempton railroad station. Sanders (1883) stated: *"The sandstone comes out of the quarry with rough faces, but after being dressed it looks good."* The sandstone in the quarry dipped 65° S.

Legler Flagstone Quarry

Around 1892, Louis H. Legler leased a quarry at the foot of the Pinnacle in Albany Township. The Pinnacle, located on Blue Mountain, is the second highest elevation in Berks County. Legler operated the quarry for 10 years. He produced flagstones and curbstones, which were sawed and dressed by hand. He shipped them to Reading, as well as other locations (*Reading Eagle*, January 24, 1910).

Berg's Farm Locality

Eyerman (1889, p. 14) reported smoky quartz crystals "*on Berg's farm in Albany.*" There are two Berg's farms in the western part of Albany Township shown in the 1876 atlas, one owned by Daniel Berg and the other owned by John Berg.

Brobst (Union) Furnace

The Brobst Furnace was located on Pine Creek, north of Hamburg, in Albany Township. The exact date of construction and the builder is unknown. Montgomery (1884) stated that Michael Brobst purchased two tracts of land in Albany Township in 1780 for £11,000. No iron works were mentioned in the deeds. The furnace was not mentioned in Samuel Potts' 1789 list of active furnaces. Michael Brobst died in 1814, and the furnace is mentioned in his will, so the furnace likely was built sometime between 1790 and 1814. At the time of Michael Brobst's death, his two sons, George Michael and Johannes Brobst were actively running the family iron business.

By 1818, John Brobst (disinherited son of Michael Brobst) was full owner of the furnace and also was bankrupt. The furnace property (two forges, a furnace, and 1,330 acres of land) was seized and sold by the sheriff in October of that year. About 1820, George Reagan obtained one of the properties and changed the name to Union Forge. By 1826, Reagan built a small furnace at the site for use in conjunction with the forge, and it became known as the Union Ironworks. In 1828-30, Union Furnace, owned by George Reagan, employed 18 workmen, had 12 horses, used 4,500 cords of wood, and produced 250 tons of pig iron and 100 tons of castings, the smallest production of any furnace in Berks County for those years. Reagan operated the Union Ironworks for several years. Union Furnace was not listed by Lesley (1859); however, D'Invilliers (1883, p. 236) stated that the Union Furnace was a cold-blast charcoal furnace owned by H.B. Fisher & Company. The furnace had one stack, a 1,000 ton per year capacity, and made car wheel iron.

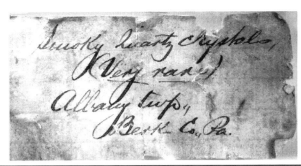

Figure 20. Smoky quartz from Albany Township, 6cm. Carnegie Museum of Natural History collection CM 31221 (Brookmeyer collection 2508).

Figure 21. Aerial photograph of the Berks Silica Sand Company, Alsace Township, 2015.

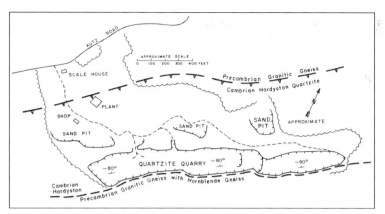

Figure 22. Sketch map of the Berks Silica Sand Company quarry, Alsace Township. From Berkheiser (1984, p. 2). Courtesy of the Pennsylvania Geological Survey.

ALSACE TOWNSHIP

Temple Quarry

The large, active (2016) Temple quarry (fig. 21) is northwest of Pennsylvania State Route 12, about 0.8 mile northwest of Alsace Manor at 40° 23′ 50″ latitude and 75° 52′ 30″ longitude on the USGS Temple and Fleetwood 7.5-minute topographic quadrangle maps (Appendix 1, fig. 13, location AL-1). The quarry is in the Hardyston Formation.

The quarry was most likely opened prior to 1920 (Pennsylvania Department of Labor and Industry, 1920, p. 359). The quarry was operated by a number of companies, including the Temple Crushed Stone Company (ca. 1920), the Temple Slag Company (ca. 1932-1952), and the Berks Silica Sand Corporation of Kutztown, which ceased operations and dissolved on November 10, 1980. At the time of dissolution, R.E. Jordan was the president of the corporation (*Reading Eagle*, November 20, 1980). The quarry later was operated by A&J Aggregates (ca. 1985) and Temple Crushed Stone, Inc. (ca. 1991-1998). When Temple Crushed Stone, Inc. purchased the quarry in 1981, Alsace Township ordered them to remove 10,000 tons of slag stored on the property. The slag was a steel manufacturing by-product produced by the Carpenter Technology Corporation in Reading (*Reading Eagle*, August 8, 1991).

In November 1998, the 134-acre quarry property was sold to Highway Materials, Inc. (*Reading Eagle*, February 16, 1999) and is currently (2016) known as the Temple quarry. Highway Materials, Inc. is owned by the DePaul Group of Blue Bell in Montgomery County. Quarry products in 2016 ranged in size from sand to rip-rap. Highway Materials, Inc. supplies bunker sand for golf courses and aggregate for ready-mix and pre-cast concrete products.

Berkheiser and Smith (1984) noted a light-colored kaolinitic clay in the quarry suitable for medium- to high-duty refractories. This was discovered during a car-borne scintillometer survey for uranium and thorium in the Reading Prong. During the survey, some areas of extremely low gamma activity were noted over white soils. Clay zones, with an apparent thickness greater than 25 feet, occur about 160 feet stratigraphically above the steeply dipping basal conglomerate and an unconformable contact with the underlying Precambrian gneiss. The clay may be continuous over the length of the quarry. The stratabound clay marks a boundary between massive, fractured, indurated quartzite to the south and a more friable sandstone to the north. Berkheiser (1985, p. 55) reported zircon grains in the quartzite.

Trexler Mica Mine

The Trexler mica mine was east of Skyline Drive about 0.3 mile east of McKnight's Gap. Gordon (1922, p. 152) placed the location around 40° 22′ 18″ latitude and 75° 53′ 14″ longitude on the USGS Reading 7.5-minute topographic quadrangle map (Appendix 1, fig. 27, location AL-2). The mine is underlain by felsic to mafic gneiss. In 1897, David Bechtel leased land from Joel Trexler for prospecting and discovered mica. Bechtel was offered $13,000 for the lease, but refused to sell it. He spent a considerable sum of money developing the property. However, the mica proved to be speckled, and the sheets were not large enough for commercial use. He was unable to sell any mica (*Reading Eagle*, October 16, 1897).

Minerals
Autunite (Eyerman, 1911, p. 20)
Allanite - large masses (Gordon, 1922, p. 152) (fig. 24)
Muscovite - silvery books (fig. 23)
Torbernite (Eyerman, 1911, p. 20)
Zircon (Gordon, 1922, p. 152)

Figure 23. Muscovite from the Trexler mica mine, Alsace Township, 10 cm. Bryn Mawr College collection Heyl 2017. Collected by Allen Heyl in the 1930s.

Figure 24. Allanite from the Trexler mica mine, Alsace Township, 1 cm. Bryn Mawr College collection Heyl 1983. Collected by Allen Heyl in the 1930s.

Figure 25. Enstatite, var. hypersthene, from Gottschall's mine, Alsace Township, 13 cm. Carnegie Museum of Natural History collection ANSP 25864. Collected by Sam Gordon in 1922.

Figure 26. Hedenbergite from Gottschall's mine, Alsace Township, 7.7 cm. Bryn Mawr College collection Rand 6532.

Gottschall's Mine

Gottschall's mine was northeast of the intersection of Antietam and Mexico Roads about 0.3 mile south of Alsace Manor. Gordon (1922, p. 152) placed the location at about 40° 23′ 33″ latitude and 75° 51′ 20″ longitude on the USGS Fleetwood 7.5-minute topographic quadrangle map (Appendix 1, fig. 13, location AL-3). Gottschall's mine was on the west side of Capella Hill (Gordon, 1918) on H.S. Gottschall's farm. The mine was in hornblende gneiss.

Minerals

Augite - abundant (Gordon, 1918)

Enstatite, var. hypersthene (fig. 25)

Garnet group (D'Invilliers, 1883, p. 397)

Hedenbergite - massive (fig. 26)

Pyroxene, var. diallage - dark, greenish-black, sometimes showing a sub-metallic bronze luster; cleavage masses often greater than 1 inch across and several inches long (Genth, 1876, p. 219, analysis)

Pyrrhotite - slightly nickeliferous; in disseminated masses (Genth, 1876, p. 219); abundant (Gordon, 1918)

Gottschall's Farm Locality

Gottschall's farm was south of Alsace Manor, and northeast of the intersection of Antietam and Mexico Roads on the USGS Fleetwood 7.5-minute topographic quadrangle map (Appendix 1, fig. 13, location AL-4). The locality is underlain by the Hardyston Formation.

Minerals

Pyroxene (Eyerman, 1889, p. 15)

Quartz, vars. agate and jasper (D'Invilliers, 1883, p. 401)

Valentine Hartman's Mine

Valentine Hartman's mine was on the west bank of Antietam Creek about 0.7 mile west of Spies Church. It was northwest of the intersection of Antietam, Simmons, and Gauby Roads at about 40° 22′ 14″ latitude and 75° 51′ 36″ longitude on the USGS Birdsboro 7.5-minute topographic quadrangle map (Appendix 1, fig. 29, location AL-5). The mine is underlain by felsic to mafic gneiss. This locality also is known as Zion Church (Dana, 1892, p. 1067; Weatherill, 1853, p. 345) and Reading (Dana, 1854, p. 488).

Weatherill (1853, p. 345-346) described molybdenite from the mine: "*This mineral, specimens of which were given to me by Dr. Bischoff and Geo. M. Keirn, Esq., of Reading, is found in abundance at the Zion Church, Alsace, in the neighbourhood of that city. It occurs of considerable purity in plates and scales in a quartz matrix. The colour of the latter is like plumbago, but more brilliant. The streak on paper, that of plumbago; on porcelain, olive green. It is impressible to the nail, giving a hardness of 1 by Mohs' scale. A calculation of the per centage relations of the molybdenum and sulphur, without reference to the other ingredients gives*, S_2 *40.668% and Mo 69.332%."*

Minerals

Epidote - associated with garnet (Genth, 1875, p. 220)

Garnet group - massive (Genth, 1875, p. 75)

Molybdenite - crystallized and foliated masses (D'Invilliers, 883, p. 399); plates and scales; analysis by Weatherill (1853, p. 345) (fig. 27)

Siderite - Genth (1875, p. 158) stated: *"on V. Hartman's farm, near Reading, in Berks County, where the siderite is found in the lower part of the beds."*

Salem Church Ocher Pit Locality

The Salem Church ocher pit was 0.7 mile north of Alsace Manor, north of the intersection of Pennsylvania State Route 12 and Laurel Road at about 40° 24' 18" latitude and 75° 51' 37" longitude on the USGS Fleetwood 7.5-minute topographic quadrangle map (Appendix 1, fig. 13, location AL-6). This likely is the ocher pit described by Gordon (1922, p. 152) 0.5 mile east of Salem Church, now known as Shalters Church, on the south side of a hill. The pit was sunk in a residual deposit of goethite and limonite in Cambrian quartzite. Goethite occurs as geodes up to 2 feet across. Limonite occurs as ocher filling the geodes and in stalactitic and mammillary forms (Gordon, 1918, p. 164).

Figure 27. Molybdenite from Valentine Hartman's mine, Alsace Township. Steve Carter collection.

Weist School Locality

The Weist School locality was 0.75 mile east of Five Points, north of Friedensburg Road at about 40° 21' 47" latitude and 75° 49' 18" longitude on the USGS Birdsboro 7.5-minute topographic quadrangle map (Appendix 1, fig. 29, location AL-7). The locality is in felsic to mafic gneiss. Gordon (1922, p. 152) described a small quarry 0.75 mile west of Weist School where reibeckite, var. crocidolite, occurred in Precambrian gneiss.

McKnights Gap Locality

McKnight's Gap is located at the intersection of Skyline Drive, McKnights Gap Road, Oak Lane, and Bingaman Road at about 40° 22' 11" latitude and 75° 53' 27" longitude on the USGS Reading 7.5-minute topographic quadrangle map (Appendix 1, fig. 27, location AL-8). There were reports in the mid-1800s that traces of gold were found in the McKnight's Gap area (*Reading Eagle*, August 5, 1962). An exact location for the gold occurrence has never been verified.

The Reading Public Museum collection contains a sample of gold from McKnight's Gap. The label (fig. 29A) indicates that the sample was provided to Dr. J. Schoenfeld by Dr. Charles M. Weatherill. The sample consists of gravel (fig. 29B). Al least one piece of gravel contains visible gold (fig. 29C). Most gold specimens labeled "McKnight''s Gap" seen by the author do not appear to be from Berks County.

Mount Laurel (Alsace) Furnace

The Mount Laurel cold blast charcoal furnace, originally called Alsace Furnace, was built in 1836 on Laurel Creek. In 1845, the furnace was owned by J. & S. Kauffman. It was

Figure 28. Goethite from the Salem Church locality, Alsace Township, 6 cm. Ron Kendig collection.

Figure 29. Gravel containing gold from McKnight's Gap, Alsace Township. Figure C is magnified. Reading Public Museum collection 2005C-3-123.

Figure 30. Allanite from McKnight's Gap, Alsace Township, 8.8 cm. Collected by Edgar Wherry in 1920. Carnegie Museum of Natural History collection ANSP 9283.

capacity of 4,013 tons and produced 2,336 tons of foundry pig iron. In 1883, the furnace was owned by the Clymer Iron Company. It was dismantled in 1892.

AMITY TOWNSHIP

J. Rhodes Sandstone Quarry

The J. Rhodes sandstone quarry was on the north side of Pennsylvania State Route 52, north of its intersection with Blacksmith Road at about 40° 18′ 54″ latitude and 75° 43′ 17″ longitude on the USGS Boyertown 7.5-minute topographic quadrangle map (Appendix 1, fig. 33, location AM-1). The quarry is underlain by the Leithsville Formation. A sandstone quarry was opened by J. Rhoades close to the north side of the road. D'Invilliers (1883, p. 116) described the quarry: "*The quality of the stone is poor here and greatly broken up with cleavage joints and stained brown. The dip was apparently S. 60° E. 52°, but was rendered doubtful by the steeply S. E. dipping cleavage joints.*"

BECHTELSVILLE BOROUGH

Miller Quarry

The Miller quarry was at the west end of Green Street, north of Franklin Street in Bechtelsville at about 40° 22′ 05″ latitude and 75° 37′ 57″ longitude on the USGS Boyertown 7.5-minute topographic quadrangle map (Appendix 1, fig. 34, loca-

rebuilt in 1847. Lesley (1859) stated that in 1856, it was owned by W.H. Clymer & Company of Reading. During 45 weeks in 1855, the furnace produced 954 tons of car wheel iron, and during 44 weeks in 1856, the furnace produced 952 tons of car wheel iron. The furnace used a mixture of limonite from the Moselem, Dumm, and Hefner mines mixed with magnetite from the Wheatfield mine. The furnace was converted to a hot-blast anthracite furnace in 1873, but was not blown in until February 1, 1880. D'Invilliers (1883) stated it had an annual

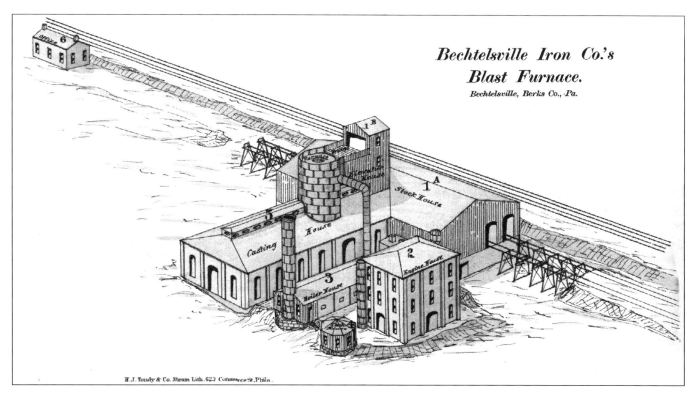

Figure 31. Bechtelsville Iron Company (Norway) Furnace, Bechtelsville. From Hexamer and Son (1872-1911).

tion BT-1). The quarry also has been known as the H. Geist quarry. The quarry is shown on the 1882 topographic map as the Miller quarry.

D'Invilliers (1883, p. 151) reported: *"This quarry has not been worked for some time, and exposes about 30 feet of blue and dove-colored limestone, interstratified with inches of limestone shale. It is very silicious. The blue limestone occurs in massive beds from 2 to 6 feet thick, but is filled with nodules of white feldspar, which considerably deteriorates its commercial value and increases the percentage of insoluble matter. The dip is about S. 85° E. 10° to 15°."* An analysis of the rock was published by D'Invilliers (1883, p. 151): 19.82 percent calcium carbonate, 14.53 percent magnesium carbonate, and 64.37 percent silica.

Bechtelsville Furnace

The Bechtelsville Iron Company was chartered in 1875 with $60,000 in capital stock for *"Mining, preparing for market and selling iron ore, and also for manufacturing and selling iron."* Eli S. Bechtel was president, and William S. Berlin was secretary and superintendent. The company purchased property for construction of an iron furnace (fig. 31) from David H. Bechtel on November 9, 1874, and obtained iron ore leases from John Rush and Nathan Landis on January 7, 1875 (Berks County deed book). The furnace also was known as the Norway Furnace.

Construction of the one-stack furnace began in 1875. The annual capacity was 11,400 tons. It was a hot-blast furnace using anthracite coal and coke. In 1876, the Bechtelsville Iron Company issued bonds (fig. 32) to raise additional capital. Although construction of the furnace (fig. 34) was completed by the Bechtelsville Iron Company, the company went bankrupt and never put the furnace into blast.

The furnace was acquired by the Philadelphia and Reading Coal and Iron Company. It was leased to Gabel, Jones & Gabel of the Pottstown Iron Company, which renamed the furnace the Norway Furnace, and put it into blast in 1880. In 1881, it produced 6,104 tons of foundry iron. The furnace was operated under the superintendence of Levi Yocum. Substantial improvements were made to the furnace, and the annual capacity was boosted to 19,000 tons per year. The furnace produced foundry and mill pig iron under the brand name "Norway" (American Iron and Steel Association, 1884 and 1892). The ore used by the furnace came from the Gabel mine in Boyertown, which was owned by Gabel, Jones & Gabel. Daily production in 1885 was forty tons of foundry iron, and 45 men were employed (Montgomery, 1886, p. 1002). The furnace was out of blast by 1894 (American Iron and Steel Association, 1894).

Figure 32. Bond issues by the Bechtelsville Iron Company in 1876. The bond was signed by Eli S. Bechtel.

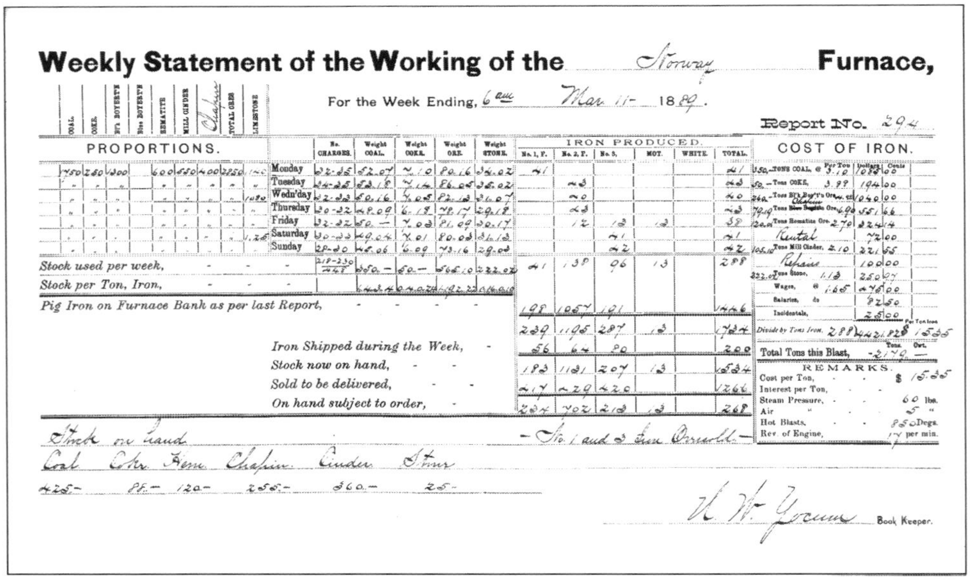

Figure 33. Bechtelsville (Norway) Furnace weekly report for the week ending March 11, 1889. From the Griffith Jones Papers Concerning Iron Ore Furnaces (3615), Historical Collections and Labor Archives, Special Collections Library, Pennsylvania State University.

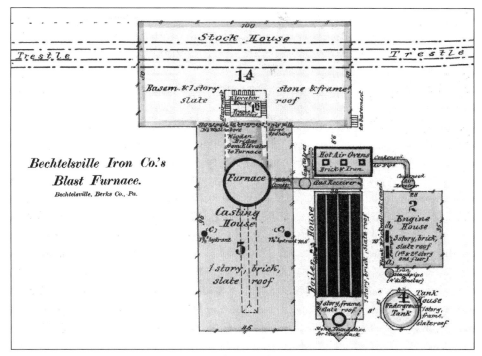

Figure 34. Plan of the Bechtelsville (Norway) Furnace. From Hexamer and Son (1872-1911).

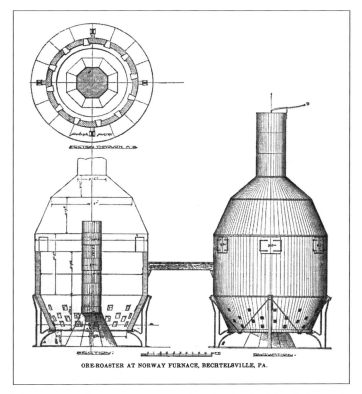

Figure 35. Ore roaster at the Bechtelsville (Norway) Furnace. From Birkinbine (1884, p. 371).

BERN TOWNSHIP

Essig Quarry (Ohnmacht Cave)

The Essig quarry was west of the intersection of Palisades Drive and County Welfare Road at about 40° 23′ 00″ latitude and 76° 01′ 56″ longitude on the USGS Bernville 7.5-minute topographic quadrangle map (Appendix 1, fig. 24, location B-1). The Essig quarry was 200 feet north of Tulpehocken Creek on the property of Adam and Katie Ohnmacht. The quarry was in the Epler Formation.

The Essig quarry was opened sometime in the 1800s. Joseph G. Barr purchased the property in 1887 for $74 an acre. In the late 1800s, the quarry was operated by John and Robert Essig for limestone, which was burned in two nearby kilns to produce lime for local farmers. The Essig brothers abandoned the quarry around 1900. In the early 1900s, Charles Balthaser, roadmaster of Bern Township, operated a stone crusher at the quarry. In 1906, Barr sold his 141-acre farm to Adam Ohnmacht for $44.25 an acre. The property included the limestone quarry and kilns (*Reading Eagle*, September 10 and November 11, 1906).

The Essig quarry was reopened by the Essig brothers in 1932 and operated until the late 1960s. In 1947, the quarry workers discovered a cave when they drilled into a void. It took several months of quarrying until the cave opening was large enough to enter. The quarry workers considered the cave a nuisance and avoided the area. The cave was named the Ohnmacht cave because it was on the Ohnmacht farm (Snyder, 2000, p. 52).

By the late 1950s, the entrance to Ohnmacht Cave had been enlarged to a two-foot wide hole at the base of the quarry wall. The cave was formed along a five-foot vertically displaced normal fault that was visible across the quarry face. The main passage was 6 feet wide, 4 to 25 feet high, and 120 feet long and ran east-northeast along the fault plane (fig. 36). Twenty feet from the entrance there was a 38-foot long side passage that contained draperies and other speleotherms. At the far end of the cave, a 15-foot mud chute led to an upper level room. The room was 9 feet by 25 feet and had a 12-foot high ceiling. The northwest corner of the room had a 4-foot diameter passage that extended 26 feet northwest. The passage was filled with speleotherms (Fisher, 1957). Ohnmacht cave was richly decorated with various forms of speleotherms, including bacon, rimstone, draperies, stalactites, and stalagmites. Some of these were pure white aragonite (Snyder, 2000, p. 52).

The Ohnmacht farm, including the Essig quarry and Ohnmacht cave, was purchased from Katie Ohnmacht by the U.S. Government on June 28, 1972. Construction of the Blue Marsh Dam by the U.S. Army Corps of Engineers began in 1974 and was completed in 1979. The quarry was used as a source of rock for rip rap for the upstream face of the dam. During this time, the front part of the cave was destroyed. After quarrying ended, the remaining part of the cave was sealed with sand, and the entrance was covered by rock. Today Ohnmacht cave lies under 25 feet of water in Blue Marsh Lake (Snyder, 2000, p. 55-56).

Prior to the construction of the Blue Marsh dam on Tulpehocken Creek, Samuel Gundy, a biology professor at Kutztown University and former director of the Reading Public Museum, collected aragonite specimens (fig. 37) and sawed off stalactites and flowstone that decorated the cave. The stalac-

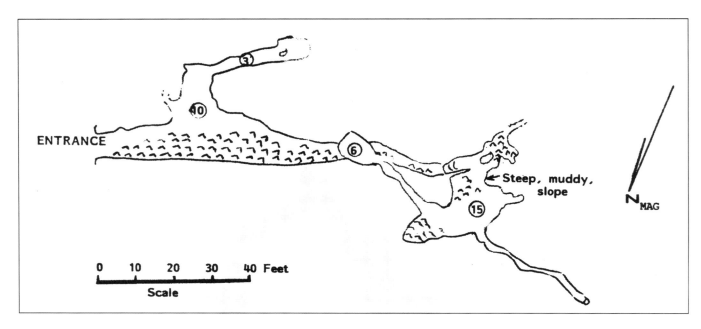

Figure 36. Map of Ohnmacht cave, Bern Township. From Mostardi and Durant (1991, p. 54). Courtesy of the Mid-Appalachan Region of the National Speleological Society.

Figure 37. Aragonite from the Ohnmacht cave, Bern Township, 4 cm. Sloto collection 3458.

tites were given to his friends, and the flowstone was sold to a rock shop. Loring Emery also removed a few broken speleothems from the cave; some were given to the geology department at Albright College and Penn State University (Snyder, 2000, p. 55). Some aragonite specimens from Ohnmacht cave are labeled as State Hill, which is just south of the cave location.

Rickenbach Area Quarries

The Rickenbach area quarries (fig. 38) are a group of quarries in and around the village of Rickenbach on the USGS Temple 7.5-minute topographic quadrangle map (Appendix 1, fig. 9, location B-2). The quarries are in the Ontelaunee Formation.

In 1842, Alexander and William Peacock opened a quarry and lime manufacturing plant near what is now the intersection of Cross Keys and Rickenbach Roads. They shipped lime by canal boat down the Schuylkill River to dealers in Spring City, Phoenixville, and Pottstown (*Reading Eagle*, June 3, 1973).

From about 1876 to 1916, one of the quarries supplied stone for a lime kiln on a 50-acre farm adjoining the railroad station at Rickenbach. The 1,000-bushel capacity lime kiln was located adjacent to the quarry. One of the quarries was opened in 1890 by J.B.K. Rickenbach of Leesport. In 1904, the crusher at the quarry was operated by Aaron Schaeffer, a Bern Township supervisor (*Reading Eagle*, September 7, 1904, and October 17, 1916).

Miller (1934, p. 216) described a quarry 0.5 mile south of West Leesport that was connected by a spur to the Reading Railroad. Six lime kilns were located at the quarry. The quarry was worked on two levels. The lower level had a 10-foot face, and the upper level had a 20-foot face. The length of the quarry was about 300 feet. Miller (1934, p. 216) also reported that a quarry west of the Rickenbach railroad station along the Reading Railroad was worked for 400 feet on the line of strike. The strike of the beds was N. 34° E. and the dip was 66° NW. The quarry was 50 feet wide at the west end and narrowed to 20 feet at the east end. The quarry exposed about 40 feet of gray to grayish blue, somewhat siliceous limestone.

The Reading Crushed Stone Company operated a quarry at Rickenbach in the 1930s. In 1937, it produced 100,000 tons of crushed limestone (Pit and Quarry, 1937). The company operated a Lima model 101 power shovel and used trucks to transport stone to the crusher. Stone was crushed by a 28-inch by 36-inch Traylor crusher and a Kennedy-Van Saun reduction crusher. In 1936, the Reading Eagle (August 11, 1936) reported that a workman operating a conveyor at the quarry was struck by lightning and thrown 8 feet to the ground. In 1939, the Reading Eagle (June 16, 1939) reported that a workman fractured his arm at the Rickenbach station stone quarry.

Epler Quarries

The Epler quarries were located on the east side of the railroad track, just south of West Leesport, on the west side of the Schuylkill River at 40° 23′ 12″ latitude and 75° 57′ 50″ lon-

Figure 38. Aerial photograph showing the Rickenbach area quarries, Bern Township, September 14, 1937. Arrows point to the quarries. Courtesy of Bureau of Topographic and Geologic Survey, PennPilot Historical Aerial Photo Library.

Figure 39. Quartz crystal from the Red Bridge locality, Bern Township, 4.5 cm. Ron Kendig collection.

gitude on the USGS Temple 7.5-minute topographic quadrangle map (Appendix 1, fig. 9, location B-3). The quarries are in the Ontelaunee Formation. The quarries were in the Ontelaunee Formation.

In the 1860s, Epler and Leinback quarried limestone and burned it on kilns on the property (1860 and 1862 maps). In the late 1800s, H.H. Epler operated the quarry and lime business. The 1876 atlas listed H.H. Epler as a lime dealer that shipped lime on the Pennsylvania and Reading Railroad below Leesport. The 1876 atlas also listed J.V. Epler, M.D. of West Leesport as a manufacturer and dealer in lime. The 1877 Berks County business directory listed John V. Epler as a lime producer (Phillips, 1877). Eyerman (1889, p. 6) reported marcasite from Epler's quarry at Leesport.

Reading Railroad Quarry

The Reading Railroad quarry was on the west bank of the Schuylkill River near Felix's dam. Felix's dam was located south of where present day U.S. Route 222 crosses the Schuylkill River between Bern and Muhlenberg Townships. The quarry was north and south of the U.S. Route 222 bridge around 40° 23' 35" latitude and 75° 58' 22" longitude on the USGS Temple 7.5-minute topographic quadrangle map (Appendix 1, fig. 9, location B-6). Although not visible on aerial photographs, the quarry is easily seen on LIDAR imagery. The quarry is known to have been worked by the Reading Railroad Company in the 1860s and 1870s. The quarry was in the Ontelaunee Formation.

Red Bridge Quartz Crystal Locality

The Red Bridge spans Tulpehocken Creek between Bern and Spring Townships. The bridge is on Red Bridge Road, at its intersection with Tulpehocken Road at 40° 21' 46" latitude and 75° 58' 06" longitude on the USGS Reading 7.5-minute topographic quadrangle map (Appendix 1, fig. 26, location B-4). The locality is in the Allentown Formation. The area on the southwest side of Tulpehocken Creek is the Berks County Red Bridge Recreation Area. The bridge, originally known as Wertz's Bridge, is a 220-foot covered bridge built in 1867. It is the longest single span covered bridge in Pennsylvania. Quartz crystals (fig. 39) have been found in the vicinity of the bridge.

BERNVILLE BOROUGH

Bernville Area

Bernville is located on the USGS Bernville SW 7.5-minute topographic quadrangle map. It is surrounded by Jefferson Township to the northwest, Penn Township to the northeast, and North Heidelberg Township to the south. Minerals reported from the Bernville area are listed below.

Minerals
Barite - fetid, columnar, radiating masses (Gordon, 1922, p. 157)
Malachite (Eyerman, 1889, p. 45)
Pyrite - radiating nodules, altered on the surface to limonite (Gordon, 1922, p. 157)

BETHEL TOWNSHIP

Frystown Fetid Barite Nodule Localities

The Frystown barite nodule localities are in Bethel and Tulpehocken Townships. Only the localities in Bethel Township are discussed in this section. The localities in Bethel Township are east of Frystown, and the localities in Tulpehocken Township are southeast of Frystown. The localities in the 6-square mile area were described in detail by Berkheiser (1984).

The barite generally consists of nearly pure, fist-size nodules or lensoidsal float, which is medium gray to medium dark gray. The fragments are made of dense, compact, 2-mm size, subrounded to subangular barite crystals, which generally are surrounded by an envelope of radially-bladed barite crystals. Some core fragments appear to have a faint lamination with a pale yellowish orange to gray to grayish black argillaceous material occurring at crystal boundaries and as occasional interstitial fillings (Berkheiser, 1984, p. 24).

The host rock is mapped as the Martinsburg Formation, but it may include older, transported carbonate rocks. The nodules were formed in deep, stagnant, sulfate-bearing marine sediments. They were apparently permeated by barium chloride-bearing hydrothermal solutions of unknown origin. The barite is more resistant to weathering than the shale, so it is found loose in the soil. Some barite occurs in calcite veins cutting limestone. The barite emits a foul odor similar to that of rotten eggs (Gyer and others, 1976, p. 42).

Figure 40. Barite nodule, Bethel Township, 9 cm. Reading Public Museum collection 2004c-16-619.

Huber Farm Fetid Barite Occurrence

The Huber farm fetid barite occurrence is southwest of the intersection of Pennsylvania State Route 501 and Frystown Road at about 40° 27′ 05″ latitude and 76° 18′ 31″ longitude on the USGS Bethel 7.5-minute topographic quadrangle map (Appendix 1, fig. 6, location BE-1).

Miller Farm Fetid Barite Occurrence

The Miller farm fetid barite occurrence is southeast of the intersection of Pennsylvania State Route 501 and Frystown Road at about 40° 27′ 12″ latitude and 76° 17′ 49″ longitude on the USGS Bethel 7.5-minute topographic quadrangle map (Appendix 1, fig. 6, location BE-2).

Sadler Farm Fetid Barite Occurrence

The Sadler farm fetid barite occurrence is located in the vicinity of Bordner Road, on the west side of Little Swatara Creek at about 40° 26′ 53″ latitude and 76° 17′ 24″ longitude on the USGS Bethel 7.5-minute topographic quadrangle map (Appendix 1, fig. 6, location BE-3). Barite occurs in an area about 200 feet by 75 feet elongated along strike in black shale (Berkheiser, 1984, p. 22-24).

Kline Umber Prospect

An umber deposit was found on the farm of Dr. W.C. Kline about 1 mile west of Bethel. The umber was a surface deposit on brecciated chert and was formed by the decomposition of sandy layers of the Martinsburg Formation. Several test pits were sunk, and umber was discovered at a depth of 3 to 8 feet. The umber was dark brown and generally free from impurities. It was reported that the umber was examined by paint manufacturers who stated that it was of good quality. Because the deposit was far from any railroad and the nearest shipping point was 8 miles away in Myerstown, the deposit was never developed (Miller, 1911, p. 45).

BIRDSBORO BOROUGH

The borough of Birdsboro was incorporated in 1872 on land that was formerly part of Union and Robeson Townships. Birdsboro was named for iron maker William Bird, who established a forge on Hay Creek about 1740. The E & G Brooke Iron Company operated furnaces in Birdsboro. The company became the Birdsboro Iron Foundry Company and later the Birdsboro Steel Company. The iron industry was the principal employer in Birdsboro for 120 years until the steel plant closed in 1988 following a lengthy strike.

A Very Brief History of the E & G Brooke Iron Company

The iron industry in Birdsboro began in 1740 when William Bird purchased land at the confluence of Hay Creek with the Schuylkill River. He built several forges in the area. William Bird died in 1761 and was succeeded by his son, Mark, who expanded his fathers' iron works. In 1770, Mark Bird built Hopewell Furnace in Union Township, which became one of the largest American producers of iron during the American Revolution. In 1788, creditors took ownership of the Bird holdings and leased them to John Louis Barde. In 1796, Barde purchased the iron works. Barde died in 1799, and the following year, Daniel Buckley, Thomas Brooke, and Matthew Brooke acquired the property.

Matthew Brooke assumed control of the iron works at the request of Barde's widow and brother-in-law and later married Barde's daughter. Matthew Brooke became sole owner of Hopewell furnace and the Hopewell mines in 1809. He died of malaria in 1821 leaving behind five minor children—Ann,

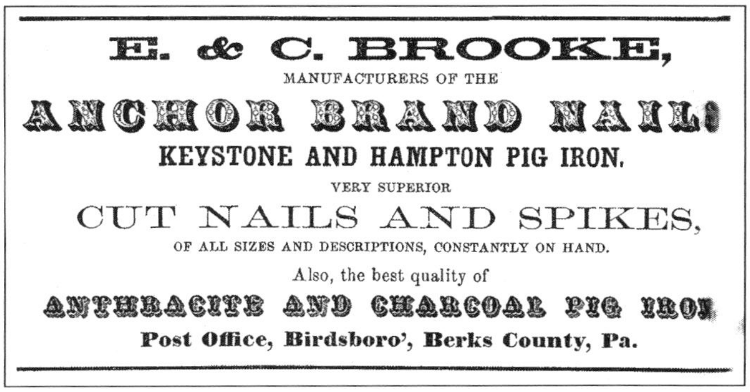

Figure 41. Advertisement for the E & G Brooke Iron Company from the 1860 Boyd's Directory.

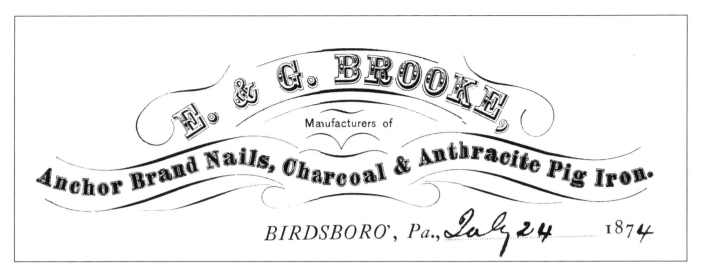

Figure 42. Letterhead of the E & G Brooke Iron Company, 1874.

Sarah, Edward, George, and Elizabeth. Edward (fig. 44) and George (fig. 45) Brooke were given the best education possible, attending schools in Reading, Litiz, Philadelphia, and the West Chester Academy. They assumed control of the family business interests on April 1, 1837. Those business interests included two forges in Birdsboro, a tilt hammer, a grist mill, a saw mill, and many acres of farm and woodland. Under the name of E & G Brooke, they expanded the business with the completion of a large grist mill in 1844, Hampton Furnace in 1846, and a nail factory in 1848. They were the first in the country to galvanize nails. The arrival of the railroad allowed the Brookes' iron products to be transported nationwide.

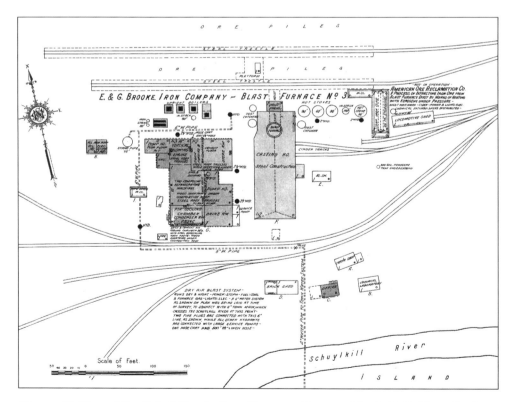

Figure 43. Plan of the E & G Brooke Iron Company Furnace No. 3, 1912. From Sanborn Map Company, Birdsboro, sheet 2, December 1912.

Figure 44. Edward Brooke. From Montgomery (1909, p. 338).

Figure 45. George Brooke. From Montgomery (1909, p. 336)

Figure 46. Birdsboro Steel Foundry and Machine Company, 1906.

The E & G Brooke Iron Company anthracite Keystone Furnace, located on the Schuylkill River, was built in 1834. In 1856, the furnace used a mixture of 2/3 magnetite from the Jones and Warwick (Chester County) mines and 1/3 limonite from the mines in Chester Springs (Chester County) to produce 3,885 tons of iron (Lesley, 1859).

As part of the expanding iron industrial complex, E & G Brooke purchased the French Creek mine in Chester County. They also acquired interests in the Warwick (Chester County) and Jones mines. E & G Brooke acquired the William Penn Colliery in Schuylkill County and marketed all sizes of commercial coal. In 1867, the holdings of the two Brooke brothers were incorporated as the Birdsboro Iron Foundry Company. The company manufactured stoves and machinery castings from steel.

Anthracite Furnace No. 2, near the first furnace, was erected in 1872 and razed in 1916. In 1873, anthracite Furnace No. 3 (figs. 43 and) was built along the railroad tracks north of Birdsboro in Exeter Township. It ceased to operate in 1963. The furnaces produced forge pig iron under the brand name "Keystone" (fig. 41). The annual capacity was 50,000 tons. In 1896, the annual capacity reached 63,000 tons, and the furnace produced foundry and forge pig iron under the brand name "Brooke" (American Iron and Steel Association, 1884, 1896, and 1902).

In order to continue the business after the death of Edward Brooke in 1878, the holdings were organized into two companies, under the names Edward and George Brooke Iron Company (E & G Brooke Iron Company) and Edward and George Brooke Land Company (E & G Brooke Land Company).

Figure 47. E & G Brooke Iron Company Furnace No. 3. Courtesy of the Amity Heritage Society.

George Brooke became president of the two companies; George W. Harrison was treasurer, and Richard T. Leaf was secretary.

The Pennsylvania Diamond Drill Company (fig. 48) moved from Pottsville to the property of the Birdsboro Iron Foundry in 1885 and continued to make mining tools and machinery. In 1894, the Birdsboro Iron Foundry Company merged with the Pennsylvania Diamond Drill Company to form the Pennsylvania Diamond Drill and Machine Company (fig. 49). A year later, the new company sold its assets to the E & G Brooke Iron Company. In 1905, the company was renamed the Birdsboro Steel Foundry and Machine Company (fig. 46).

George Brooke died in 1912. George's sons Edward and George II continued the family iron-making business as did his

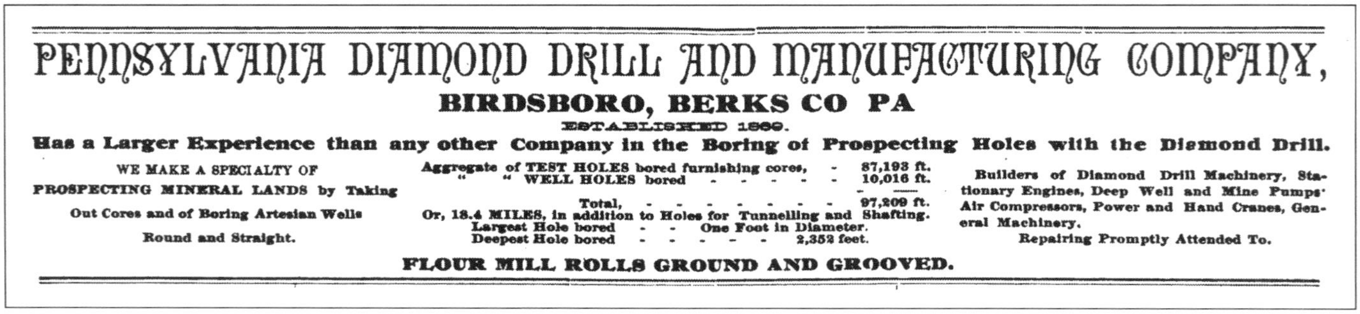

Figure 48. Advertisement for the Pennsylvania Diamond Drill and Machine Company, 1888. From the Colliery Engineer (September, 1888).

grandsons George III and Edward, Jr., and Edward's grandson, G. Clymer Brooke.

In 1938, the Birdsboro Steel Foundry and Machine Company purchased the Reading Iron Company's former Scott Foundry on North Eighth Street in Reading, operating it as its Reading plant. Around the same time, it formed a new engineering division dedicated to the design and production of rolling mills, industrial machinery used in the production of steel and the shaping of it into various sizes and shapes, and hydraulic machinery, such as industrial presses.

During World War II, the government expressed an interest in acquiring steel for the U.S. Navy from The Birdsboro Steel Foundry and Machine Company. Birdsboro's reputation for quality work played a key role in the U.S. Navy's decision to build the Armorcast plant adjacent to Birdsboro's sprawling works during World War II. The large steel mill constructed in 1944 consisted of nine massive bays up to 1,480 feet long and 50 feet high. The role of the plant was later shifted to manufacturing tanks for the Army. Sherman and Patton tanks were the main product of the plant from World War II through the Korean War. The company ran the huge plant, whose products included tank parts, for the government.

In 1952, the company became a subsidiary of the Colorado Fuel and Iron Company. In 1960, the Birdsboro Steel Foundry and Machine Company was reorganized as the Birdsboro Corporation. It continued to produce castings, including side frames and other parts for railroad cars, and others of enormous size and weight, such as anchors and pump housings. In 1968 a controlling interest in the Birdsboro Corporation was sold to the Pennsylvania Engineering Company in New Castle, Lawrence County. It was one of Berks County's largest employers with 1,400 workers in 1968 (Hoffman, 1976).

In 1972, Armorcast was sold to the Birdsboro Corporation in a lease-purchase agreement with Greater Berks Development Fund. In 1975, Armorcast failed to win a government contract to continue production of tanks, and the plant was closed in 1988 after a lengthy strike. A few businesses still operated within the space until 2002, when a five-alarm fire damaged the building. Plans were made to demolish the structure, which then became known as the Armorcast Redevelopment Area.

Figure 49. Unissued stock certificate of the Pennsylvania Diamond Drill and Machine Company, ca. 1890s.

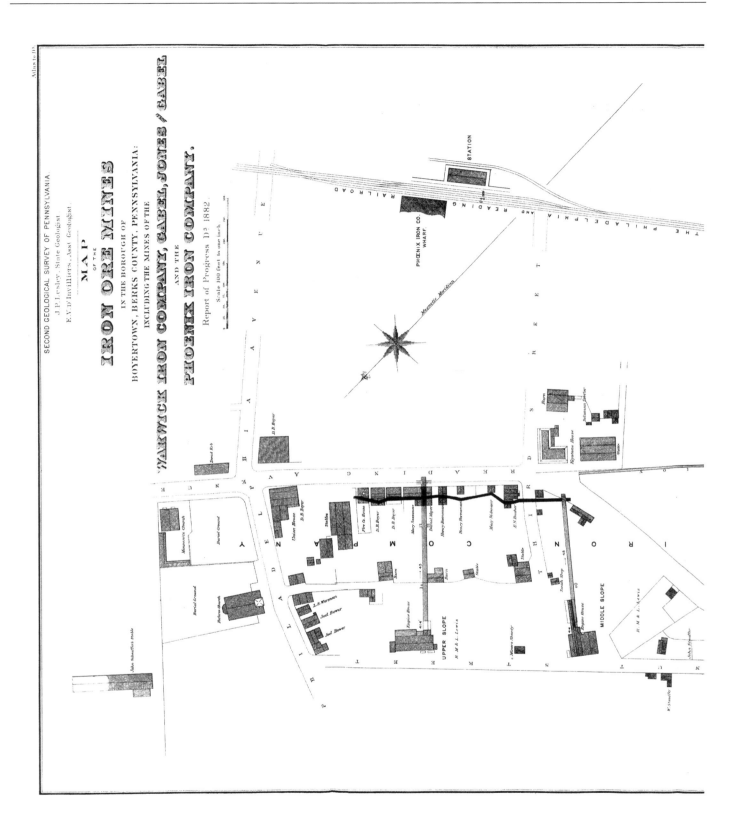

Figure 50. Second Pennsylvania Geological Survey map of the Boyertown mines, 1882. From Pennsylvania Geological Survey (1883).

BOYERTOWN BOROUGH

BOYERTOWN BOROUGH

Five separate magnetite ore bodies were mined at Boyertown (fig. 51). Spencer (1908, p. 43-61) provided the most detailed description of these deposits. Willis (1886) described and published maps of the mines. D'Invilliers (1883, p. 304-333) described the individual mines and published analyses of the ores. Hawkes and others (1953) provided the results of geological and geophysical studies in the Boyertown vicinity that utilized core drilling, as well as ground and airborne magnetic surveys. Most of the information on the Boyertown mines presented below was taken from those sources.

Four mines in Boyertown were major ore producers between 1850 and 1900—the Phoenix, California, Warwick, and Gable mines (fig. 52). The first two mines were worked by the Phoenix Iron Company of Phoenixville, the third by the Warwick Iron Company of Pottstown, and the fourth by Gabel, Jones & Gabel, also known as the Steel Ore Company, of Pottstown. No production figures are available; however, Rose (1970, p. 8) estimated, on the basis of the extent of the workings, that production was roughly 1 million tons of ore worth several million dollars.

Early History

On July 29, 1718, David Powell secured a grant of 200 acres from the Penn family. On June 4, 1719, Powell sold the 200-acre tract to pioneer ironmaster Thomas Rutter for £24 (Graham, 1996), and the tract became known as the Furnace Tract. In 1719, Rutter, with financial backing from Philadelphia Quaker merchants, formed Rutter, Coates & Company and began building Colebrook Dale Furnace on Ironstone Creek at the present day site of Morysville. The furnace, named after Abraham Darby's famous iron furnace in Shropshire, England, began producing iron in 1720. It was the first iron furnace in Pennsylvania. At that time, the furnace was located in Philadelphia County because Berks County was not created until 1752.

An iron ore deposit was known to exist on the Furnace Tract prior to construction of the furnace. In a letter written in 1717, Jonathan Dickinson wrote that Thomas Rutter was mining iron ore along Manatawny Creek in 1716. Ore initially was mined from pits sunk on surface outcrops. The ore was mined by men using picks and shovels and was carried out of the pits in wicker baskets, often by women because of a shortage of labor. The ore was loaded onto wagons and hauled to the Colebrook Dale Furnace. The mine pits were located about 2,500 feet northeast of the furnace, in present day Boyertown. The

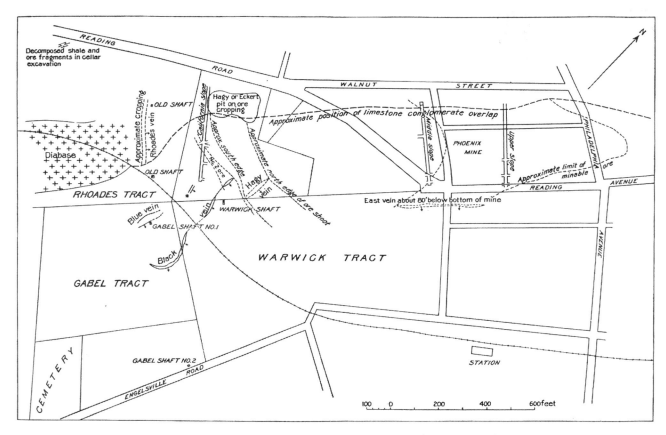

Figure 51. Surface plan of ore bodies mined in Boyertown. From Spencer (1908, plate 8).

1 Gabel mine
2 Warwick Iron Company
3 Rhoade's mine
4 Phoenix Iron Company California mine (lower slope)
5 Eckert Iron Company Eckert slope
6 Phoenix Iron Company ?
7 Phoenix Iron Company middle slope
8 Phoenix Iron Company upper slope

Figure 52. Location of the Boyertown mines on the 1876 atlas (Davis and Kochersperger, 1876).

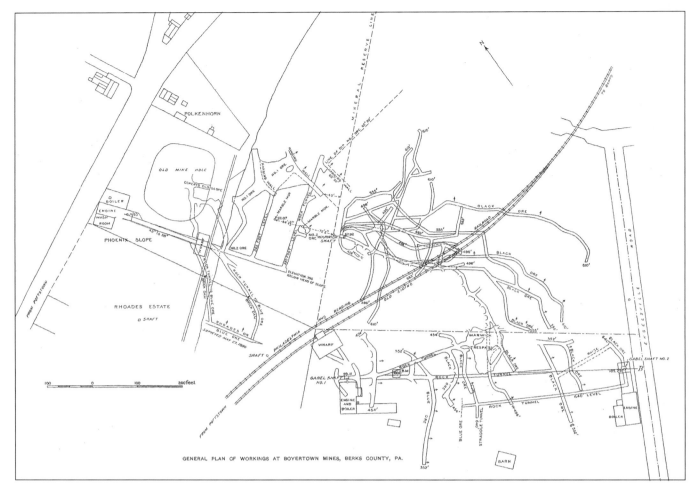

Figure 53. Surface plan of the Boyertown mines. From Spencer (1908, plate 10).

pits were known as Rutter's iron mine, and it is considered to be the first iron mine in Pennsylvania.

Thomas Rutter and his sons, Joseph and John, managed the furnace until 1725, when Rutter rented the property to a company composed of Evan Owen, Maurice Morris, James Lewis, Robert Griffith, Thomas Marke, and later Anthony Morris. They in turn rented it to Thomas Potts, who remained there as the ironmaster until his death in 1752. By 1737, Potts became the majority owner (Graham, 1979; Graham, 2010).

By 1750, ore production began to decline when the pits and open cuts became too deep to mine safely. Ore was brought in from other mines; however, that arrangement proved too expensive. The furnace was dormant from 1765 to 1770 and went out of blast in 1770. The mine lay idle for many years afterwards.

In the early 1770s, the Furnace Tract was divided and sold to local farmers Mathias Roth (also spelled as Rhoades) and Henry Stauffer (numerous spellings exist). Thomas Rutter III and Samuel Potts retained the mineral rights in addition to one acre of land where the iron ore vein, known as the red bank, cropped out. Their descendants sold the mineral rights in the 1830s (Graham, 1996, p. 39-41). This parcel, including the mineral rights, later became the property of Robert and Morris Lewis of Philadelphia, who also acquired an interest in the Eckert slope. They leased the tract to the Phoenix Iron Company, and the mine became known as the Phoenix mine.

In 1847, William Rowe, Sr. sank the first mine shaft in Boyertown, possibly on Walnut Street, which was originally called Ore Street. It is likely that the earliest mining took place at the southeastern corner of Third and Walnut Streets, and this site may be the red bank mine of Montgomery (1886). Iron mining increased substantially when the Colebrookdale Branch of the Reading Railroad, also known as the Colebrookdale Railroad, was completed from Pottstown to Boyertown in 1870. The railroad greatly facilitated shipping ore.

Figure 54. Photographs taken underground in the Boyertown mines, ca. 1880s. These lantern slides (photographs) were taken underground in one of the Boyertown mines in the 1880s using "flashlight." Magnesium power was lit on fire, producing a brilliant flash of white light followed by copious smoke. The flash powder was ignited by hand. This series of lantern slides are the only known underground mine photographs taken in southeastern Pennsylvania in the 19th century. They were taken by A.B. Parvin, who was active in the Photographic Society of Philadelphia.

Geology

The Boyertown mines were situated near the northwest edge of the Mesozoic Basin. The magnetite ore deposit is a Cornwall-type replacement ore body that occurs at the contact of Cambrian and Ordovician limestones and Jurassic diabase of the Quakertown diabase sheet. The ore generally is on top of the diabase. Limestone lying within 200 feet of the lower contact of Triassic sedimentary rocks is the host for the ore bodies. The contact with Triassic rocks serves as the upper limit for ore (Hawkes and others, 1953, p. 140).

The average dip of the northwest contact of the Triassic sedimentary rocks is about 35° SE. The rocks directly beneath this contact are a complex of diabase and Cambrian-Ordovician limestones that have been subjected to varying degrees of metamorphism. The limestones are white to steel gray, fine grained, and massive. Where it has been locally metamorphosed by diabase intrusions, the limestone largely has been altered to a complex of secondary silicate minerals—chlorite, garnet, green amphibole, and pyroxene. Fracture fillings of clear gypsum and large spangles of black specular hematite replacing limestone pebbles are locally conspicuous in drill cores (Hawkes and others, 1953, p. 138-139).

The three best developed ore bodies lie immediately beneath the conglomerate, which is composed of limestone fragments up to an inch in diameter in red clay. This conglomerate bed, with a general southwest-northeast strike, dips toward the southeast. The two other ore bodies occur in the limy beds of Paleozoic rocks at the contact with diabase (fig. 51).

Mining at Boyertown developed five separate bodies of magnetite iron ore, all of which occur in somewhat irregular layers of varying thickness. The two principal ore bodies were the Gabel-Warwick and the Phoenix. These bodies may join at depth, but exploration did not provide a conclusive answer. The other smaller ore bodies were either isolated deposits or faulted segments of one of the principle deposits.

The relative positions of the veins were plotted by Spencer (1908, p. 44) (fig. 51). The ore bodies just beneath the conglomerate were called the east vein, the Hagy vein (sometimes called as the Eckert vein), and the Warwick or black vein (Spencer, 1908, p. 43). The deposits in contact with the diabase were known as the Rhoades and blue veins. The east vein was mined from two inclined shafts southeast of Walnut Street known as the Phoenix upper and middle slopes. The Hagy vein, outcropping on the outskirts of town, was first mined by an open cut known as the Hagy pit or Eckert open pit, and afterwards by the Eckert slope and the Phoenix mine lower slope or California mine. The Warwick vein was mined from the Warwick shaft and the two shafts of the Gabel mine. In the Gabel mine, the Warwick vein was called the black vein. A body of ore encountered in the lower workings of the California mine, known as ore No. 2, was believed by Spencer (1908, p. 43-44) to represent the northward extension of the Warwick vein. The blue vein was encountered only in the Gabel mine. The Rhoades vein was mined at the surface in several places and in the California mine.

The diabase dips with the bedding of the intruded rock, and its edge dips steeply beneath the Warwick and Gabel mines. The blue vein in the Gabel mine appears to follow the upper surface of the diabase, but the Rhoades vein seems to lie against its blunt edge. In the Gabel mine, more than 100 feet of limy strata lie between the diabase and the conglomerate that caps the black vein (Spencer, 1908, p. 59-60).

Brown- and blue-veined rock, which occurs in considerable quantities in the mines, is a mixture of feldspars, hornblende, and epidote. It was called greenstone by the miners. The greenstone is, for the most part, a crystalline mass, which is yellow to dark green when freshly broken and brown when weathered. Occasionally, it contains radiating crystals of stilbite (D'Invilliers, 1883, p. 319).

Figure 55. Photograph showing the flooded Eckert open cut, Boyertown. Courtesy of John Grubb.

Eckert Open Cut and Slope

Eckert's open cut (figs. 55 and 56) (mine number 5 on figure 52) was east of Reading Avenue and south of Second Street at 40° 19' 53" latitude and 75° 38' 31" longitude on the USGS Boyertown 7.5-minute topographic quadrangle map (Appendix 1, fig. 34, location BB-5). The open-pit mine also is known as the Hagy pit. The open cut is shown on figures 50 and 53.

Eckert's open cut was mined by Issac Eckert of the Henry Clay Furnace in Reading beginning in the 1850s and ending in the early 1870s. The Phoenix Iron Company had a joint interest in the mine. The open cut was mined to a depth of 100 feet before underground operations commenced. The pit flooded after mining ceased (*Reading Eagle*, April 30, 1898). A slope, sometimes referred to as the Eckert slope, was sunk on the ore bed from the open pit to the southeast (fig. 53).

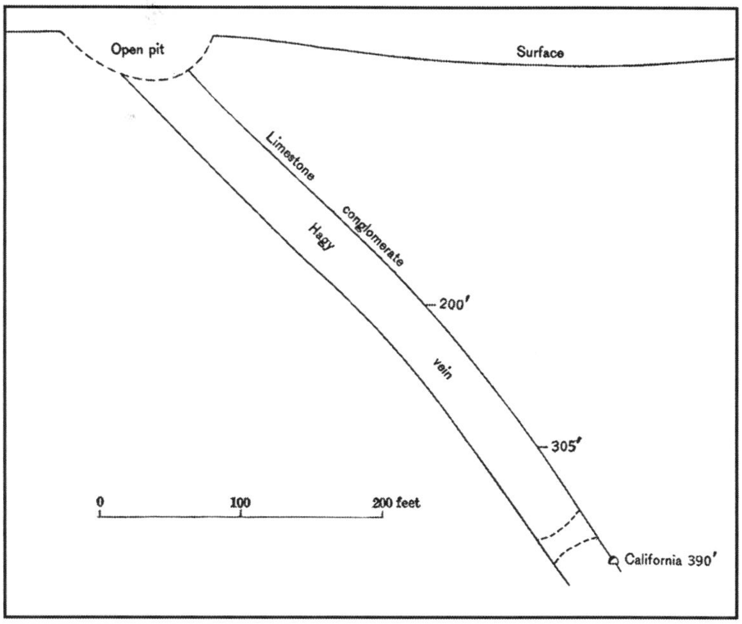

California Mine
(Phoenix Mine Lower Slope)

The California mine (mine number 4 on figure 52) was east of Reading Avenue (Pennsylvania State Route 562) at 40° 19' 52" latitude and 75° 38' 33" longitude on the USGS Boyertown 7.5-minute topographic quadrangle map (Appendix 1, fig. 34, location BB-4). The Phoenix lower slope was nicknamed the California mine by William Rowe, Sr. around 1854 when a

Figure 56. Diagram showing the Eckert open cut on the Hagy ore vein, Boyertown. From Spencer (1908, plate 15).

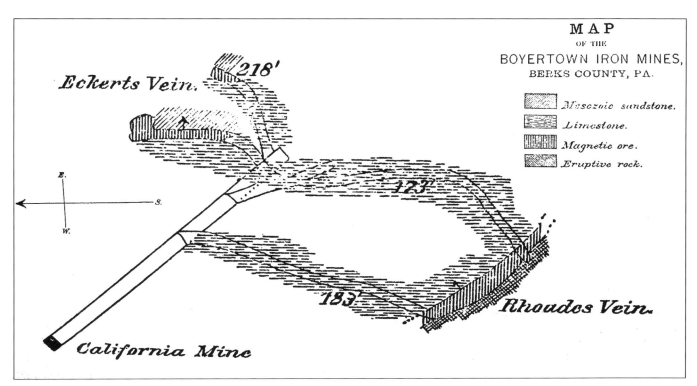

Figure 57. Plan of the California mine (Phoenix mine lower slope), Boyertown. From Willis (1888, p. 230).

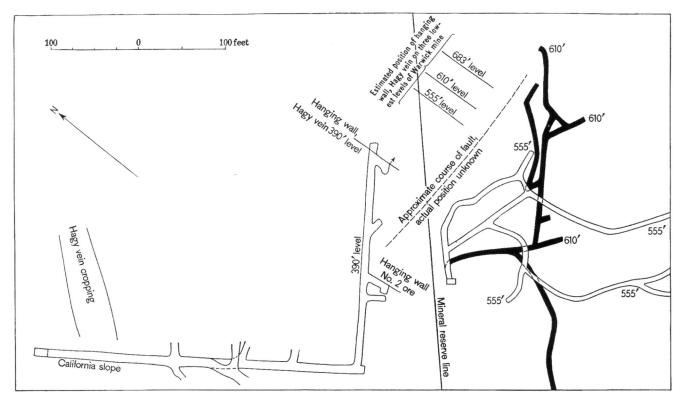

Figure 58. Plan of part of the California (Phoenix mine lower slope) and Warwick mines, Boyertown. From Spencer (1908, plate 16).

scuffle that broke out among the miners reminded him of some of the free-for-alls he witnessed in California during the 1849 gold rush (Meiser and Meiser, 1982, p. 120). The name stuck, and the mine became known as the California mine.

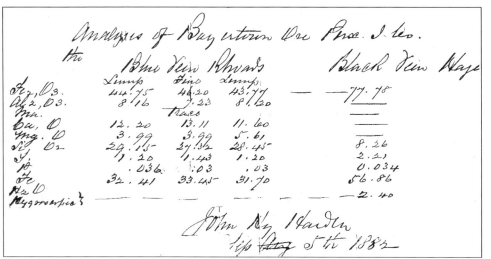

Figure 59. Analyses of the Boyertown blue and black vein ore, September 5, 1882. Signed by John H. Hardin. From the Griffith Jones Papers Concerning Iron Ore Furnaces (3615), Historical Collections and Labor Archives, Special Collections Library, Pennsylvania State University.

The Phoenix mine was located on two adjoining tracts of land; the northern tract was owned by the Lewis Brothers of Philadelphia, and the southern tract was owned by Hagy and Rhoades. The Phoenix Iron Company leased the mineral rights on both tracts. The lease on the Lewis Brothers tract began in 1852 and expired on November 26, 1882. The lease on the Hagy and Rhoades tract expired in 1883 (D'Invilliers, 1883, p. 307). The 1880 Census (Pumpelly, 1886, p. 960) listed the Phoenix Iron Company as the operator of the mine. Production between July 1, 1879 and June 30, 1880, was 8,200 tons of ore.

Development to 1883 consisted of a 36° slope about 300 feet long, from which three levels were driven toward the northeast to the Eckert (Hagy) vein. Two of the levels also were driven toward the southwest to the Rhoades vein (fig. 57).

The upper level was about 176 feet below land surface. By 1883, the upper level had been about 80 feet along the south line of the Eckert open cut; however, the ore body pinched out toward the northeast. Exploration work was done south of the slope, where the workings consisted of a tunnel driven southwest for about 200 feet through micaceous and quartzose rock to the Rhoades vein, which had a strike of about S. 25° E. and a dip of 50° NE. The ore on the upper level was a hard, compact, massive magnetite with a sub-metallic luster, very slightly crystallized, practically free from phosphorus, and containing less than 1 percent sulfur (D'Invilliers, 1883, p. 308-311).

The middle level was about 40 feet above the lower level. It was not parallel to the lower level, but turned toward the northeast. For the first 30 feet, it passed through rock and then into the ore body (D'Invilliers, 1883, p. 308). The ore body was 30 feet thick on the middle level, but, because it was lower in iron content than the Hagy vein ore, mining was suspended in 1883.

In 1883, the lowest working level, the 200-foot level, was 218 feet below land surface. This level was driven for 110 feet to intersect the ore bed and then driven through ore for more than 50 feet.

The California mine produced about 50 tons of ore per day when operating. About 20 men were employed under superintendent Richard Richards. The engine house was equipped with one pair of 8 by 12 hoisting engines and one 12 by 24 single pumping engine with two sets of 10-inch pumps. The ore was hauled by contract teams to the company's wharf at the Boyertown station on the Colebrookdale Railroad 0.25 mile away at a cost of 22 cents per ton of 2,240 pounds (D'Invilliers, 1883, p. 313).

Between 1883 and 1893, the workings of the California mine were considerably extended. The workings reached a vertical depth of 390 feet with two additional levels below the 200-foot level. On the 305-foot level, ore body No. 2 was encountered about 20 feet from the slope; it extended for about 30 feet along the drift and was reported to be 18 to 20 feet thick. The Hagy vein was encountered about 45 feet beyond ore body No. 2, or 95 feet from the slope. It was mined along the hanging wall for 125 feet.

On the 390-foot level, ore body No. 2 was encountered about 75 feet from the slope. It was mined for about 30 feet; the maximum thickness was about 20 feet. The Hagy vein was encountered 50 feet beyond the No. 2 ore body and was mined for about 90 feet. The Hagy vein was stoped upward from the 390-foot level for about 500 feet along dip (Spencer, 1908, p. 56-57).

The southernmost workings in ore body No. 2 on the 390-foot level were only about 60 feet from the Warwick shaft and about 30 feet higher than the 416-foot level of the Warwick mine (fig. 58). Spencer (1908, p. 57) believed that because of the proximity of the workings, ore body No. 2 and the Warwick vein were both parts of the same ore body. Spencer (1908, p. 57) also believed that ore body No. 2 and the Hagy vein were probably parts of a single vein that had been separated by faulting.

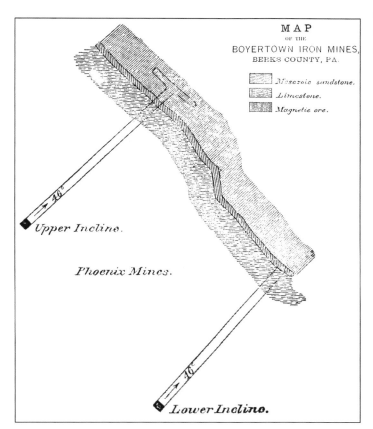

Figure 60. Plan of the Phoenix mine upper and middle slopes, Boyertown. The middle slope is labeled as the lower incline on the diagram. From Willis (1888, p. 230).

The California mine was shut down in early August 1893. The Phoenix iron Company continued pumping the mine until it could determine if the mine would remain shut down or re-open (*Reading Eagle*, August 9, 1893). The mine did not re-open. About 1930, a miniature golf course was built on the site of the mine on South Walnut Street on property owned by Daniel and Horace Boyer. The golf course was designed for boys 9 to 12 years old. It cost one cent per game of 9 holes (*Reading Eagle*, August 15, 1930).

D'Invilliers (1883, p. 310-312) published analyses of the ore from the California mine: 41.80-54.85 percent metallic iron, 0.47-1.67 percent sulfur, and a trace to 0.053 percent phosphorus. An analysis of blue vein ore was 34.55 percent metallic iron, 0.167 percent metallic copper, 1.64 percent sulfur, 0.034 percent phosphorus, and 0.03 percent cobalt oxide. Analyses also were furnished by the Phoenix and the Pottstown Iron Companies: 29.56-33.45 percent metallic iron, 1.02-1.43 percent sulfur, and 0.03-0.04 percent phosphorus.

Phoenix Mine Upper and Middle Slopes

The Phoenix mine upper slope (mine number 8 on figure 52) was east of Walnut Street between Philadelphia Avenue and Third Street at 40° 20′ 01″ latitude and 75° 38′ 21″ longitude on the USGS Boyertown 7.5-minute topographic quadrangle map (Appendix 1, fig. 34, location BB-8). The Phoenix mine middle slope (mine number 7 on figure 52) was at the intersection of Walnut and Third Streets at about 40° 19′ 59″ latitude and 75° 38′ 23″ longitude (Appendix 1, fig. 34, location BB-7).

The Phoenix Iron Company leased the Lewis mine in 1852 and converted the shallow shaft to a slope (Meiser and Meiser, 1982, p. 118). John Ellis was the first manager of the mine. Ore was hauled from the mine in wagons by six-mule teams. In 1855, the Phoenix Iron Company introduced steam power, which was used to pump water and raise ore (Fegley, 1935).

The Phoenix mine slopes were inclines (fig. 60) with an average slope of 46° sunk on the ore bed. The hanging wall was Triassic red sandstone, and the foot wall was dark gray limestone (Willis, 1886, p. 229-231). The slopes were 350 feet apart, and both were in the same ore body, which averaged from 12 to 15 feet thick. The dip of the ore was about 45° S. 45° E. (D'Invilliers, 1883, p. 314).

The upper slope ended 353 feet below land surface. The lowest level was driven from the bottom of the slope for about 150 feet northeast and southwest. The middle level was 43 feet vertically above the lower level. From the middle level, gangways were driven on each side of the slope 300 feet to the northeast and 300 feet southwest to intersect the middle slope. The upper level was 267 feet below land surface. It was driven each way about 300 feet, connecting on the southwest with the middle slope (D'Invilliers, 1883, p. 314). This level was later extended about 500 feet northeast of the slope and reached beyond Philadelphia Avenue (Spencer, 1908, p. 49). The ore body in the upper slope was much softer than in the middle slope, and a great deal of expensive work was necessary to keep the slope from collapsing.

The middle slope was about 310 feet long, with two levels corresponding to and connecting with the upper and middle levels of the upper slope. The middle slope was driven in rock. A gangway was driven northward from the slope in an attempt to intersect the underlying blue ore vein; however, ore was not encountered (D'Invilliers, 1883, p. 315-316; Spencer, 1908, p. 49-55). The middle slope was equipped with a pair of 8 by 12 engines, one 16 by 36 pumping engine, and four 36 by 30 boilers.

During the economic depression of 1880-81, the Phoenix mine upper and middle slopes closed, and the slopes flooded. In 1889, the Phoenix mine engine house caught fire, and some of the machinery was destroyed (*Reading Eagle*, February 13, 1889).

The foot wall in both slopes was an altered rock containing thin seams of earthy magnetite with a dull luster showing little crystallization. The foot wall rock was filled with pink feldspar nodules, hornblende, and epidote, all distinctly stratified. In the upper slope, a dark, greenish-black, unctuous limestone layer was encountered in many places between the main ore body and the foot wall; much of it contained large masses of chert. The

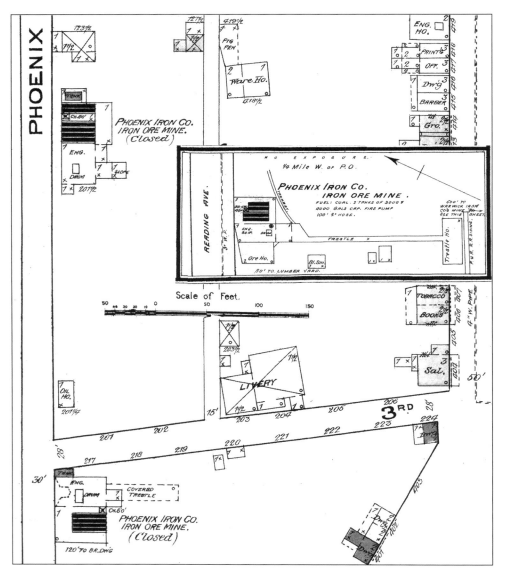

Figure 61. Plan of the Phoenix Iron Company mines, Boyertown, 1886. From Sanborn Map Company, Boyertown, 1886, sheet 1.

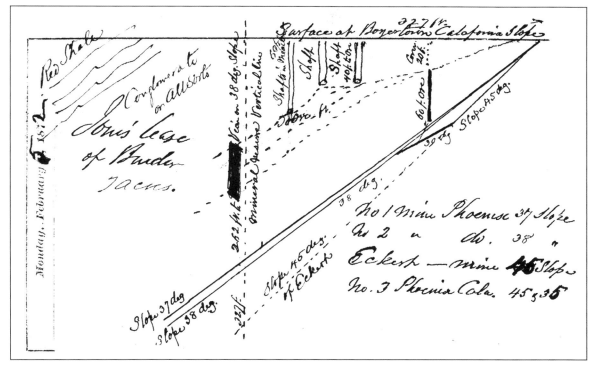

Figure 62. Sketch of the Boyertown mines made by Griffith Jones, February 5, 1872. From the Griffith Jones Papers Concerning Iron Ore Furnaces (3615), Historical Collections and Labor Archives, Special Collections Library, Pennsylvania State University.

hanging wall was a decomposed, light greenish gray, serpentinized limestone, which was slaty and contained pyrite crystals. The host rock generally was an impure conglomerate limestone containing masses of dull-colored crystalline limestone, serpentine, and magnetite; however, the bulk of it was a green to black dolomite with calcite coatings. The typical ore was magnetite with limestone and minute pyrite crystals diffused through the mass. Locally, the pyrite occurred in large and well-defined crystals (D'Invilliers, 1883, p. 315-316). Ore from the No. 1 ore body was a fine, earth-black, partially crystallized magnetite. It was nearly free from sulfur. Analyses of the ore were published by D'Invilliers (1883, p. 316): 37.37-49.79 percent metallic iron, a trace to 1.29 percent sulfur, and 0.03-0.14 percent phosphorus.

The Great Trespass

The "Mineral Reservation Line" (shown on fig. 53) divided the Phoenix Iron Company's mineral rights from those of the Warwick and Gabel mines. The Warwick Iron Company's miners encountered the north edge of an immense ore body 40 feet thick quite close to the property line. This bonanza was tempting, and the absence of mine surveys led to a continuation of the drifts into the Gabel property. D'Invilliers (1883, p. 322) noted: "A magnificent pillar of excellent black ore, considerably crystallized, was left as a support to the roof of the immense chamber."

While mining on the upper level, the owners of the Gabel mine, became convinced that a trespass had occurred onto its property through the 500-foot level of the Warwick mine. From the 474-foot level of the Gabel mine, a gangway was started toward the southeast. The blue vein was encountered 30 feet from the shaft; it extended for 33 feet. The gangway was in limestone past the blue vein. The black vein was encountered 175 feet from the shaft. The gangway then entered the Warwick Iron Company's stopes, which had been driven up from its 500-foot level (D'Invilliers, 1883, p. 329). The location of the trespass is shown on figure 53.

At the insistence of the Warwick Iron Company, the Kendell Brothers of Reading—under the direction of civil engineer Zacharias, who was assisted by L.M. Koons in the presence of mine superintendent Richard Richards—surveyed the Warwick mine to determine whether or not the company had mined past its limits. After remaining underground for 36 hours, they ascertained that the Warwick Iron Company had mined 111 feet past the Mineral Reservation Line (*Reading Eagle*, March 29, 1879). The Warwick Iron Company then had to pay restitution to Gabel, Jones & Gabel.

A TRIP INTO THE BOYERTOWN MINES

Sunday Morning Star (Wilmington, Delaware), August 16, 1885

"Mr. Jacob Shupp is foreman of the Warwick Iron Company's mines at Boyertown. After presenting our letter from Mr. Jacob Fegley, the president of the company, and stating our wish and objective, Mr. Shupp kindly consented to take us down the mines, and we at once prepared ourselves according to his directions. First we were shown to the miners' dressing shanty and ordered to select a pair of boots for our temporary use. We chose a pair of the solid number tens, exchanged our light tramping shoes for them, and then hobbled along on a pretty wide base to the next place to get oil suits and gum hats. We now looked like the Esquimaux one sees in pictures. Once in the car, the descent began. Down, down, with a sickening rapidity that made me cling with a child-like fear to the experienced foreman, through impenetrable darkness, with falling streams of water pattering upon the tin roof of the car--down, down, to a depth of 613 feet and 6 inches. It was the nearest approach to sheol we had ever made. We landed at the terminus of the underground railway, in a narrow tunnel that led away to the east. There is a general sameness here below. Gangways, or tunnels, cutting through masses of ore and rock, and branching off in every direction; darkness, relieved only by the dim light of tallow candles or miner's lamps; lowness and narrowness of way, often compelling one to stoop and crawl or stand close to the side to leave an ore car pass; water, under foot and overhead; growths of a soft fungus that hang from the protecting boards above like stalactites in a cave; shoots opening into the gangways; and chambers, large and small, known to the miners as "stoops," from which the ore is taken, make up the structure of these subterranean works.

Under the guidance of our leader, we were conducted through many tunnels to all points of interest to visitors. Occasionally we met a gang of men at work, while at other times the sound of the pick and drill were scarcely ever lost to our ears. We were led from the lowest level upon which we landed, to the one above, which we called the second story, and from this we crawled and climbed over heaps of ore into the third. We were still 560 feet below the vegetation point. On this level we were brought to the spot where the hammer and shovel in the next door mine could be distinctly heard. Several years ago by accident the two mines came together. By measurement it was found that one company had unknowingly but extensively trespassed upon the property of the other and a heavy payment was required to give satisfaction for the encroachments. We were also shown the stoop in which Mr. Brown while at work came to a sad death in September 1883. Though the mines have been worked for years and to a stranger they may seem to be dangerous, yet this is the only case of accidental death that has ever occurred here. This speaks much for the care and precautions the company employs in protecting its laborers.

By this time we had been beneath the ground about two and one half hours, and though we assured all inquiring miners that we would like it "immensely" to work with them, yet we were glad when we again reached the shaft and ordered the "cage" to stop at our level. The ascent varied somewhat from the descent. In going down the cage had a tin covering, and we only heard the pattering of the waters. In ascending, we were not sheltered in this way and the heavy streams of water fell upon us thick and fast. But our uniforms were a sufficient protection, for when we landed again where the azure sky could be seen, and we had laid aside our miner's regulation garments, we as dry as a linen shirt fresh from Hop Lee's laundry."

Warwick Mine

The Warwick mine (mine number 2 on figure 52) was south of Second Street and adjacent to the Colebrookdale Railroad tracks at about 40° 19′ 49″ latitude and 75° 38′ 27″ longitude on the USGS Boyertown 7.5-minute topographic quadrangle map (Appendix 1, fig. 34, location BB-2). The Warwick mine was located on the south side of the Mineral Reserve Line about 500 feet from the Phoenix lower slope and immediately northeast of the Gabel mine property (fig. 53).

The Henry Binder family sank a shaft on Warwick Street; the shaft later became the Warwick mine. The street received its name from the mine (*Reading Eagle*, November 6, 1947). The land was jointly owned by Henry M. Binder, Franklin G. Binder, and Clara G. Hartman, heirs of William Binder, who purchased the 14-acre property from James Ellis in 1859. On March 6, 1872, the mineral rights were leased to Jacob Gabel, Griffith Jones, Jacob Fegley, and Isaac Fegley of the Warwick Iron Company for a period of 20 years (Fegley, 1935). Jacob Fegley was treasurer, and Griffith Jones was secretary of the company. The company constructed an engine house equipped with a 30-horsepower steam engine and a 10-inch pump (*Reading Eagle*, December 5, 1872).

In 1873, the Warwick Iron Company sank a shaft 62 feet without striking ore. Before continuing further, they hired the Pennsylvania Diamond Drill Company to drill a core hole. A detailed geologic log was not recorded; however, a generalized log is available (table 2). D'Invilliers (1883, p. 317) indicated that the total depth of the core hole was 588 feet. The bottom 21 feet of the hole was left as a water sump to catch mine and surface drainage. The drilling log (table 2) indicated that the core hole was drilled to 376 feet, 7 inches before ore was

Table 2. Generalized geologic log for the core hole drilled at the Warwick mine, Boyertown, in 1873. From D'Invilliers (1883, p. 318).

Depth (feet below land surface)	Thickness (feet)	Geologic description
62	0	Depth of shaft where core began
187.3	125.3	Red and white conglomerate
324.4	137.1	Brown rock, veined
376.7	52.3	Blue rock, veined
404.4	27.7	Magnetic iron ore, blue
429.8	25.4	Magnetic iron ore, black
454.1	24.3	Magnetic iron ore, blue
480.6	26.5	Magnetic iron ore, hard black
578.1	97.5	Magnetic iron ore, blue, mostly soft

encountered. The drilling log also indicated that an ore bed 212 feet thick was penetrated. It was not possible for the ore bed to be 212 feet thick, and the driller likely mistook diabase and dark, impure limestone gangue for ore.

The results of the core hole drilling project were favorable to the Warwick Iron Company, and sinking of the shaft continued. In 1875, the shaft struck ore at 391 feet. By 1876, there were 24 men working 3 shifts, mining 30 to 50 tons of ore per day (Meiser and Meiser, 1982, p. 120).

The 1880 Census (Pumpelly, 1886, p. 960) listed the Warwick Iron Company as the operator of the Binder mine. Production between July 1, 1879, and June 30, 1880, was 14,400 tons of ore. In 1881, The Warwick Iron Company signed an agreement with Henry M. Binder to continue mining (Berks County deed book, recorded on January 19, 1882).

In 1885, when Jacob Schupp was the foreman, an accident occurred at the mine. A cage loaded with 8 to 9 tons of water and ore broke loose and fell to the bottom of the sump. Repairs to the mine and equipment were estimated at $25,000, which the company did not wish to spend, and, thus, mining and pumping of water ceased. Ironically, the framework for the large engine house was completed by contractor John Schealer that same year (*Reading Eagle*, April 1, 1885, and February 11, 1907). Nine years later, Schealer received the contract to tear down the buildings and stacks at the Warwick mine. In 1895, the shaft house was torn down, and the heavy timbers were shipped to the Warwick Iron Company in Pottstown. The remaining wood and several of the smaller buildings were offered for sale. The boards, square timber, tin roofing, post and rails, and bricks by the thousand were sold at the mine by former superintendent Jacob Schupp (*Reading Eagle*, October 18 and 31, 1894, and May 15, 1895).

Figure 63. Letterhead of the Warwick Iron Company, 1897.

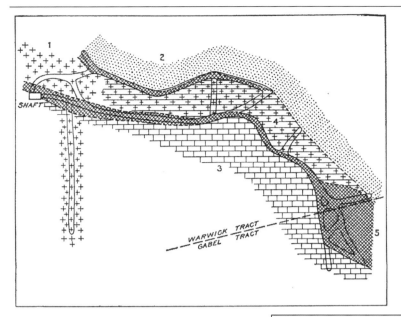

Figure 64. Plan of the upper (500-foot) level of the Warwick mine, Boyertown. From Willis (1888, p. 87), modified by Spencer (1908, p. 48). 1, diabase; 2 Mesozoic sandstone; 3, limestone; 4, limestone mixed with ore; 5 magnetite ore.

Figure 65. Plan of the bottom level of the Warwick mine, Boyertown. From Willis (1888, p. 87).

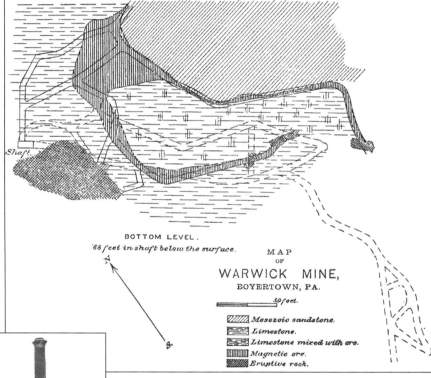

Figure 66. Warwick Iron Company Furnace, Pottstown, 1907. Ore from the Warwick mine was sent to this furnace for smelting.

The Warwick mine had three levels with numerous gangways. The first level was at 420 feet below land surface; the second level was at 500 feet; and the third level was at 567 feet; this corresponds to the 416-, 496-, and 555-foot levels shown on figure 53. Little development was done on the first level; because of the dip of the ore, stoping would have gone past the Mineral Reservation Line (D'Invilliers, 1883, p. 324).

The 469-foot level proved to be one of the most lucrative and extensive; it was developed for 500 feet and continued into the Gabel mine property across the Mineral Reservation Line a considerable distance (see "The Great Trespass" on page 45). The Warwick Iron Company mined about 8,000 tons of ore from this part of the mine. Masses of calcite with pyrite clusters were often found with the ore. The ore body attained a great thickness here, necessitating gangways on both the foot and hanging walls. The gangway on the hanging wall, which generally was in Triassic conglomerate, was the northernmost gangway; it ended to the southeast under the Colebrookdale Railroad tracks. A second gangway extended southeast and then south into the Gabel mine property. The average horizontal distance between gangways was about 50 feet, which would give a thickness of about 25 feet of ore measured at right angles to dip. The space between the two gangways frequently contained horses of serpentine, limestone, and greenstone, as well as several occurrences of pinching where the foot and hanging walls came together. This was especially true in the north gangway, where the ore was only 3 inches to 1 foot thick about 150 feet from the shaft; and again at end of the gangway, where the ore pinched out (D'Invilliers, 1883, p. 320-322).

Most of the stoping was confined to the middle level. Because the ore body swelled, timbering could not be used, and immense pillars of ore were left standing to support the roof. One pillar connecting the lower and middle levels consisted of nearly two-thirds ore and was 30 feet long by 40 feet wide by 80 feet high (D'Invilliers, 1883, p. 324).

The 567-foot level was the lowest level in 1883. Two parallel gangways, which extended southeast close to the property line, were connected in two places by cross cuts. The two gangways were driven east from the shaft to the ore body through 125 feet of limestone. The last 60 feet of the southernmost of the two gangways was in ore, but the ore was truncated by limestone and Triassic rocks (D'Invilliers, 1883, p. 322).

The ore body gradually thinned to less than 1 foot to the south-southeast along the hanging-wall gangway, about 60 feet below the south cross cut. The pinch in the ore body corresponded closely in position and character to that on the 496-foot level. The pinch extended for 50 feet along the gangway to the point where the gangway turned S. 40° E. and encountered a 10-foot thick ore body. The ore swelled and then pinched again in about 35 feet.

The S. 40° E. course of the gangway extended 120 feet to a point under the Colebrookdale Railroad, where the gangway turned south. Sixty feet beyond the turn, a greenstone dike was encountered, similar to that occurring in several places on the 567-foot and 497-foot levels (D'Invilliers, 1883, p. 323).

The foot-wall gangway on the 567-foot level was almost entirely in ore, which dipped steeply. The foot-wall gangway was nearly parallel with the foot-wall gangway on the 497-foot level above. On both levels, the gangways had a general curve toward the east. Toward the south, the dips decreased until they were

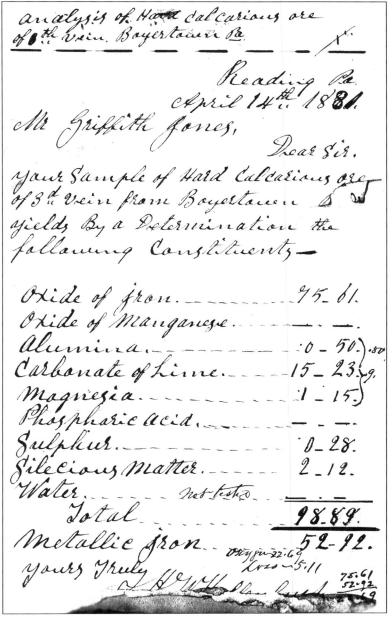

Figure 67. Analysis of Boyertown ore, April 14, 1881. From the Griffith Jones Papers Concerning Iron Ore Furnaces (3615), Historical Collections and Labor Archives, Special Collections Library, Pennsylvania State University.

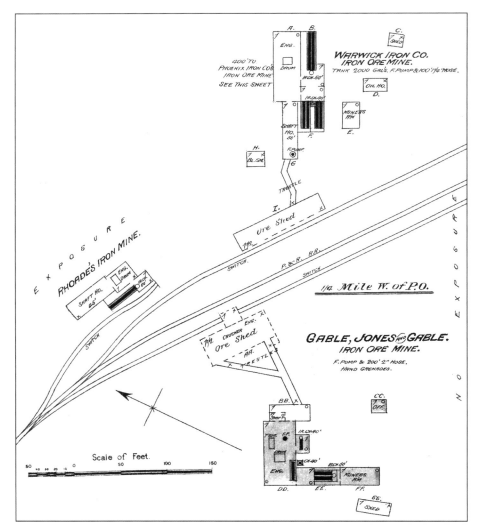

Figure 68. Plan of the Warwick, Gable, and Rhoades mines, Boyertown, 1886. From Sanborn Map Company, Boyertown, 1886, sheet 2.

38 miners, 1 blacksmith, 6 trammers, 1 laborer, 2 engineers, 2 firemen, 8 wharfers, 1 pump tender, and 2 landers. Jacob Schupp was the mine superintendent, and William Clark was the night superintendent. The mining equipment consisted of a 30-horse power steam engine with an 8-inch cylinder and an 18-inch stroke, which D'Invilliers (1883, p. 319) deemed "*hardly adequate for the work it is called upon to perform.*" A pump located in the shaft on the lowest level ran continuously to keep the mine dewatered. The ore was loaded directly on to the company's cars from the stock house on the Colebrookdale Railroad siding. All ore went to the Warwick Furnace (fig. 66) in Pottstown (D'Invilliers, 1883, p. 319).

D'Invilliers (1883, p. 324-325) published analyses of the ore: 43.4-51.2 percent metallic iron, 0.008 percent metallic copper, 0.008 percent sulfur, 0.434-3.78 percent phosphorus, 1.44-7.08 percent iron pyrite (FeS_2), and 0.01 percent cobalt oxide. The black ore was 35-51.73 percent metallic iron, 0.29-2.3 percent sulfur, and 0.113 percent phosphorus.

Gabel Mine

The Gabel mine (mine number 1 on figure 52) was west of the intersection of present day Engelsville Road and Front Street at about 40° 19' 46" latitude and 75° 38' 30" longitude on the USGS Boyertown 7.5-minute topographic quadrangle map (Appendix 1, fig. 34, location BB-1).

The land on which the Gabel mine was located adjoined the Warwick Iron Company property to the east (fig. 68). The land was owned by H. and J. Gabel and Griffith Jones. The original tract of land was enlarged by the addition of 60 acres from the Samuel Shaner estate (D'Invilliers, 1883, p. 327).

Henry H. Gabel was born on March 5, 1808, in Colebrookdale Township. He started work at the Colebrookdale grist mill, later known as the Boyertown roller mills. In 1856, he purchased the Pottstown roller mills in partnership with his brother Jacob under the name of H. & J. Gabel. Around 1873, Gabel formed the firm of Gabel, Jones & Gabel, also known as the Steel Ore Company, to mine iron ore in Boyertown. The firm leased the Bechtelsville Furnace, which they operated in connection with their mines until 1890 (*Berks County Democrat*, January 29, 1901).

20° to 35° near the Gabel mine property line. Because of the low dip in the southern part of the mine, most of the stoping was confined to the 497-foot level (D'Invilliers, 1883, p. 323-324).

Two additional lower levels were developed after 1883. On the 610-foot level, the ore vein was found to be a double vein (as on the 567-foot level), and both the foot-wall and hanging-wall leads were developed by drifts extending southward nearly to the property line. On the 610-foot level, a 200-foot cross cut that extended nearly to the property line failed to intersect the blue vein. Two gangways were run about 100 feet east of the shaft in an unsuccessful search for the Hagy vein. The workings on the 683-foot level consisted of a 200-foot long gangway on the foot wall of the ore vein and two cross cuts (Spencer, 1908, p. 56).

An average of about 85 tons of magnetite ore per day were mined. The work force in October 1882 consisted of 61 men—

Figure 69. The Gabel mine, Boyertown, ca. 1880s. From the Griffith Jones Papers Concerning Iron Ore Furnaces (3615), Historical Collections and Labor Archives, Special Collections Library, Pennsylvania State University.

By October 1878, the Gable mine shaft No. 1 reached a depth of 300 feet. In June 1880, the hoisting of ore was suspended so that the shaft could be deepened and additional gangways could be driven. The 1880 Census (Pumpelly, 1886, p. 960) listed the Steel Ore Company as the operator of the Gabel mine. Production between July 1, 1879, and June 30, 1880, was 14,792 tons of ore. The No. 1 shaft reached a final depth of 525 feet. In 1883, Gabel, Jones & Gabel sunk a second shaft along Englesville Road near the Unionville cemetery on what was originally the Ritter tract (Fegley, 1935). The Gable mine No. 2 shaft, which reached a depth of 640 feet, was still open in 1947 (*Reading Eagle*, November 6, 1947).

In 1883, the Gabel mine had two levels; the upper level was 180 feet below land surface, and the lower level was 474 feet below land surface. Some development took place on the upper level close to the shaft, but the ore was mixed with a light green limestone gangue, and the vein was discontinuous. The underlying limestone was similar to that found in the Warwick mine, mostly a light-colored, serpentine-green limestone; however, the gangue-rock limestone was much darker and, in places, was black from included magnetite (D'Invilliers, 1883, p. 328).

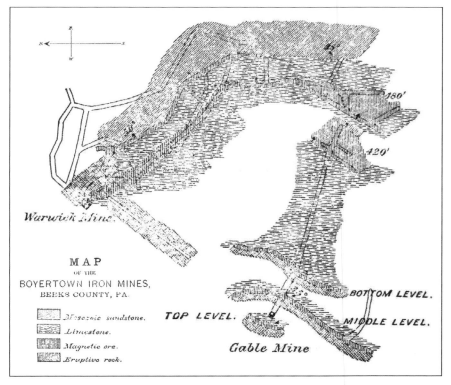

Figure 70. Plan of the workings of the Gabel and Warwick mines, Boyertown. From Willis (1888, p. 87).

The Gabel Mine at Boyertown

The Reading Eagle, November 26, 1881

"An Eagle reporter visited Boyertown yesterday for the purpose of seeing the extensive iron ore mines of the Steel Ore Company of Pottstown and noting the great improvements and facilities for mining ore, that have been introduced there. The mine, which is known as the Gabel mine, is located on a tract of land owned by H. and J. Gabel, of Pottstown, and situated about a quarter of a mile from the depot, along the Colebrookdale Railroad.

[The shaft] is a double shaft 8 X 10 feet, outside of the timbers, 463 feet in depth, and has two working levels leading off from it, one at the depth of 400 feet, which is 214 feet in length, the other at a depth of 464 feet, which is 375 feet in length. It is well and securely timbered from top to bottom, and is covered by a large frame shaft house.

Immediately adjoining the shaft house is a brick engine house 40 X 70 feet, with a slate roof in which are located the larger hoisting engines, which are monsters of power, as well as models of beauty. These engines which were designed and built at the Reading Iron Works, are horizontal engines, with steam cylinders 18 X 48 inches, and a 4 foot stroke and are connected with a 6 foot drum which they drive on the first motion, and which winds two cast steel ropes 1-1/4 inches in diameter and 650 feet in length. The cylinders and steam chests are encased in paneled walnut, with brass bands. The steam pipes leading to it are also encased, so as to prevent as much as possible, the escape of steam and the rapid condensing of it. A large tubular boiler, 5 X 32 feet, with 14 six-inch flues, furnishes the steam for the engines, and is so arranged with a heating arch over the top, that the heat passes under it back through the flues, and the through the arch over the top, thus utilizing the heat for 60 feet in length.

In connection with this boiler, there is a locomotive boiler standing outside the engine house, that can be used at a moment's notice in case of emergency, also in the engine room, a 30 horse power West engine, that can be brought into requisition should necessity demand it. For bringing the ore and water from the bottom of the shaft to the surface, two iron cages, with water tank attached, made after the manner of those used in the Nevada silver mines, are used. A Nevada detaching hook is fastened to the rope and holds the cage, and in case the cages should be overwrought, the moment the hook strikes a certain point, the jaws are thrown open and it becomes detached from the cage, which is held in position and kept from going to the bottom by four grippers attached to the top of the cage. These grippers, the moment the hook becomes detached, or the wire rope breaks, sink their teeth into the wooden guides by means of four spiral springs with which they are adjusted. Nothing but the simultaneous breaking of the four springs in case of the rope becoming detached or breaking, would drop the cage to the bottom. The moment the cage reaches the top, four hydraulic jacks connected with a lever and placed the proper distance from the mouth of the shaft, are thrown under it and the cage is let down upon them as upon springs thus preventing any jar by the too sudden letting down of the cage.

Everything is now in readiness for taking out ore in large quantities, yesterday being the first time that ore was hoisted since the introduction of the new cages. Each cage as it comes up, has an iron car on it that will hold about two tons of ore. Under the old system of hoisting ore with buckets, the company mined about 100 tons daily, but with the present facilities, they expect to take out not less than two hundred tons daily, and on the first of the month propose putting on men sufficient to mine that quantity.

The ore taken out of this mine is similar to that taken out from the other mines in and about Boyertown, being of two kinds, black and blue; the former which lies uppermost, containing about 50 per cent of iron, and the latter about 40 per cent. The company have numerous orders on hand and at present are shipping about 200 tons daily. Among the heaviest purchasers of their ores are the Ringgold furnace, Pottsville Iron and Steel Company, (Atkins), Leesport Iron Company, Sheridan Iron Company, Kutztown Iron Company, Topton Iron Company, Reading Iron Works, (Trexler), Monocacy Iron Company, Pottstown Iron Company, and The Montgomery Iron Company at Port Kennedy. The miners are in charge of Griffith Jones, of Pottstown, who is the superintendent, and whose vast knowledge in that direction especially adapts him for the position."

In the Gabel mine, the two ore beds and the strata between them dipped about 45° S. 55° E. The ore occurred in masses and bunches up to 40 feet thick, which pinched out and was not continuous. The ore was harder and more compact than in the other mines, probably because of their proximity to the Gabel Hill diabase dike. From the 474-foot level, two gangways were driven in the foot wall of the black vein. Because of its great thickness and large limestone partings, stoping was carried up almost to the shaft, where the black vein was encountered 180 feet below land surface. In June 1883, a cross cut was driven from the 474-foot level toward the southwest into the hill, but the ore was found to be cut off by a dike that cropped out on the north side of Gabel Hill (D'Invilliers, 1883, p. 330).

Between 1883 and 1893, the Gabel mine shaft No. 1 was deepened to about 570 feet (fig. 71). On the 552-foot level, a crosscut was run to the blue and black veins. The Gabel mine

Table 3. Vertical section through the Gabel mine No. 1 shaft, Boyertown. From D'Invilliers (1893, p. 328) modified by Spencer (1908, p. 53).

Depth (feet below land surface)	Geologic description	Thickness (feet)
0 - 180	Banded shales, sandstones, and conglomerate, with a layer of altered mud rock about 5 feet thick at the bottom	180
180 - 206	Black bedded ore, folded and broken, and consisting mostly of calcareous breccia	26
206 - 356	Limestone	150
376 - 411	Blue ore bed (measured across the bedding)	55
411 - 485	Chloritic rock or greenstone	74

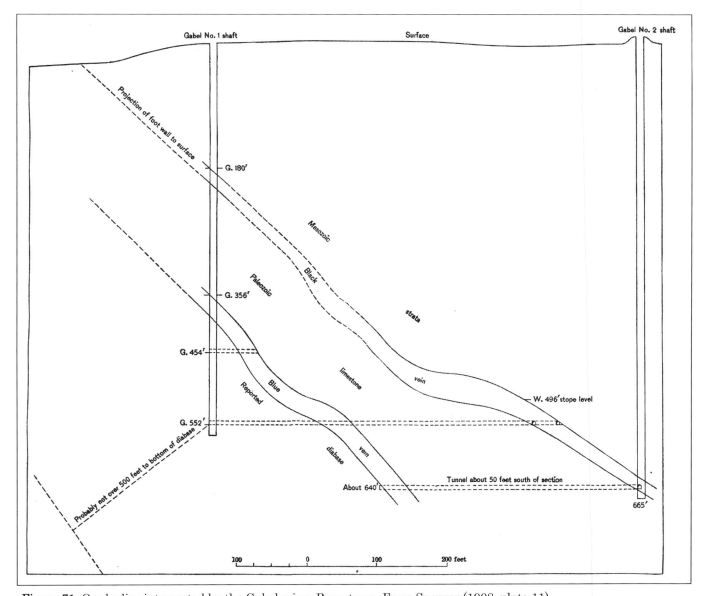

Figure 71. Ore bodies intersected by the Gabel mine, Boyertown. From Spencer (1908, plate 11).

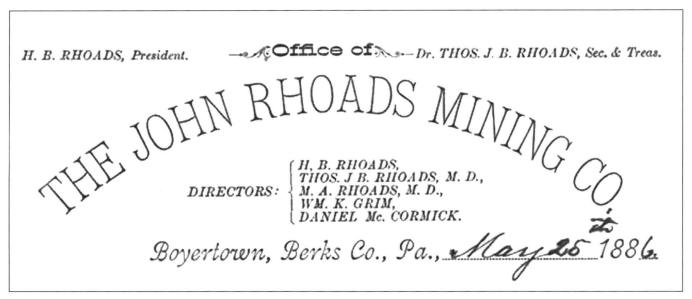

Figure 72. Letterhead of the John Rhoads Mining Company, 1886.

shaft No. 2 was sunk to a depth of 665 feet. The shaft encountered the black vein at about 640 feet, but the thickness of the ore body is unknown. Two short drifts ran from the shaft to the ore body. The northern drift, about 55 feet long, was connected by a raise with the 552-foot level of the No. 1 shaft. The southern drift was 25 feet long with a 365-foot long cross cut that penetrated the foot wall. About 35 feet from the end of the cross cut, short drifts were reported to follow the hanging wall of the blue vein. Between shaft No. 1 (where the ore was first encountered at a depth of about 180 feet) and shaft No. 2, the black vein was a continuous layer for about 800 feet in the direction of dip. On the 474-foot level, the drifts in the ore body were about 250 feet long. The blue vein was more than 100 feet below the black vein, measured normal to the bedding. It was mined along strike for about 250 feet. On the 640-foot level, neither vein had been fully developed when mining was suspended (Spencer, 1908, p. 55).

The mining machinery consisted of a 140-horsepower steam engine with two 50-inch cylinders and a 50-inch stroke used for both hoisting ore and pumping water. The engine was made by John West of the Scott Foundry in Reading. The flywheel was 12 feet in diameter and made 46 revolutions per minute. The 6-foot diameter winding drum was furnished with 1/2-inch steel rope and hoisted directly from the shaft. The engine house used a No. 1 Cameron pump to supply water to the boilers. The boilers were 22 feet long and 5 feet in diameter. Water was provided by the Boyertown Water Works (D'Invilliers, 1883, p. 327).

The ore was magnetite of two varieties—black ore and blue ore. The black ore was the same ore mined in the Warwick mine, and the blue ore was similar to the Rhoades vein ore in the Phoenix mine lower slope. In 1883, an average of 125 lifts of ore and water were made per day, one lift comprising 1 ton of ore and nearly 2 tons of water. The shaft was equipped with self-acting safety cages onto which the mine cars were directly loaded. Ore was shipped from a siding on the Colebrookdale Railroad; the cost for loading the cars was 8 cents per ton. The Pottstown Iron Company and the Reading Iron Works were the principal consumers, although a considerable amount of ore was shipped to the Bechtelsville Furnace (D'Invilliers, 1883, p. 327-328).

Analyses of ore from the blue vein were published by D'Invilliers (1883, p. 331-333): 29.76-38.72 percent metallic iron, 0.151 percent metallic copper, 1.55 percent sulfur, 0.037 percent phosphorus, 22.85-28.95 percent silica, and 0.02 percent oxide of cobalt. Analyses of ore from the back vein were published by D'Invilliers (1883, p. 331): 42.54-47.10 percent metallic iron and 8.20-11.55 percent silica.

John Rhoads Mining Company

The John Rhoads mine (fig. 68) (mine number 3 on figure 52) was southwest of the intersection of Second Street and Englesville Road at about 40° 19′ 49″ latitude and 75° 38′ 30″ longitude on the USGS Boyertown 7.5-minute topographic quadrangle map (Appendix 1, fig. 34, location BB-3).

Little is known about the John Rhoads Mining Company. John Rhoads (1788-1880) was a farmer in Colebrookdale Township. The John Rhoads Mining Company was formed when iron ore was discovered on his property, and a shaft was sunk. Most of the company officers were related (fig. 72). H.B. Rhoads was the president, and Dr. Thomas J.B. Rhoads was the secretary-treasurer. The board of directors included Dr. M.A. Rhoads, W.M.K. Grim, and Daniel McCormick. A article in the Reading Eagle (April 18, 1888) stated that Gabel, Jones & Gabel purchased 15,000 tons of ore lying on the warf at the Rhoads mine. The ore was shipped to the Bechtelsville Furnace. The Sanborn insurance map of 1891 indicated that the mining operation was closed.

BOYERTOWN BOROUGH

Figure 73. The Boyertown iron crane, Boyertown. The crane, located in the former Reading Railroad yard at Third and Washington Streets in Boyertown, is 20 feet tall. It was built by the Phoenix Iron Company in the late 1800s or very early 1900s. The crane, which is based on a Phoenix column, is hand cranked. It was used to hoist containers of iron ore onto flatbed railroad cars.

Seminary Shaft

The Seminary shaft (mine number 9 on figure 52) was north of West Philadelphia Avenue and College Street at about 40° 20′ 07″ latitude and 75° 38′ 27″ longitude on the USGS Boyertown 7.5-minute topographic quadrangle map (Appendix 1, fig. 34, location BB-9). The seminary shaft is shown on the 1882 topographic map. It was called the seminary shaft because it was on the property of the Mt. Pleasant Seminary.

Spencer (1908, p. 45-46) reported that a prospecting shaft was sunk northeast of Philadelphia Avenue along the strike of the east vein. The shaft showed the presence of carbonaceous shale, which was reported to have been similar to some of the material associated with the ore.

Later Mining Activities

The Boyertown Mining Company was formed around 1886. George F. Baer was a director of the company. Baer, a lawyer, also was president of the Reading and Philadelphia Railroad Company, Reading Iron Company, Temple Iron Company, Reading Paper Mills, and the Allentown Terminal Company and a director of the Reading Railroad, Reading Fire Insurance Company, Clymer Iron Company, Penn National Bank, Reading Hospital, and the Keystone Furnace Company (*Reading Eagle*, November 3, 1889) (see "Temple Furnace"). The Boyertown Mining Company was in litigation for over a decade over disputed mining claims.

The Reading Times (August 23, 1916) reported that during World War I, the Boyertown Mining Company received several large boilers, a smoke stack 100 feet long, two large pumps, several rolls of cable, a hoisting engine, and other equipment to be used in the Boyertown mines. The Pennsylvania and Reading Railroad wreck crew, with the aid of a large crane, unloaded the material from cars on the railroad company property near the Warwick mine shaft, where it was supposed to be erected. It is not known if the Boyertown Mining Company did any actual mining.

In 1901, William G. Rowe, Jr. of Reading purchased the Phoenix, Gabel, and Warwick mines and the Lewis tract along with the mineral rights and formed the Boyertown Ore Company with Rowe as the general manager (*Reading Eagle*, February 11, 1907). In 1902, Rowe and George W. Lex of Philadelphia reorganized the company, which was then capitalized at $300,000. Richard Richards of Boyertown was the mine superintendent. William Galley and William H. Harvey, New York financiers associated with the U.S. Steel Company, both served as presidents of the company. David Powell Wilson of Philadelphia also served as president. The Boyertown Ore Company operated the Boyertown mines from 1902 to 1907 (*Reading Eagle*, November 6, 1947).

In 1902, the mines were dewatered. Water was pumped using two cages measuring about 4 feet by 4 feet by 6 feet. While one cage was dumping water into Ironstone Creek, the other was in the mine being filled. The reported pumping rate was 1,000 gallons per minute. In 1902, all of the Boyertown mines were interconnected by tunnels (Meiser and Meiser, 1982, p. 120).

In 1907, Rowe traveled to New York City to confer with Charles N. Schwab, president of the Bethlehem Steel Company, on the possible use of Boyertown ore at the Bethlehem Steel plant (*Reading Eagle*, February 8, 1907). Various newspaper articles reported that Schwab purchased the Boyertown mines for amounts ranging from $400,000 to $1,000,000. The New York Times (February 11, 1907) reported that Schwab "*purchased the mines of the Boyertown Ore Company for a consideration said to be between $400,000 and $500,000.*" In reality, negotiations broke down, and the Bethlehem Steel Company never obtained an interest in the mines.

From 1902 to 1907, an estimated 17,000 tons of ore were mined. The ore was shipped to the Warwick Furnace at Pottstown. The Boyertown Ore Company suspended mining in March 1907. Richard Richards was appointed caretaker of the properties (*Reading Eagle*, February 23, 1912). Spencer (1908, p. 47) reported that the mines were abandoned and flooded at the time of his visit in 1907.

In 1916 and 1917 during World War I, the Eastern Steel Company of Pottstown, successor to the Warwick Iron Company, operated the Boyertown mines for 13 months. In 1916, they acquired the mines of the Boyertown Ore Company (*Iron Trade Review*, August 3, 1916). J. Ross Corbin was the mine superintendent. The Eastern Steel Company averted a strike in July 1916 by granting a 10 percent increase in the wages of its 900 employees (*Standard Corporate Service*, July, 1916). They spent a considerable sum of money dewatering the mines, making repairs, sampling, and core drilling. Water was pumped from the mines at the rate of 900,000 gallons per day. Work continued around the clock in three 8-hour shifts. The miners found the slopes caved in a number of places. The Gabel mine No. 2 shaft was connected to the Warwick shaft 2,000 feet away at a depth of 600 feet in the Gabel mine and 500 feet in the Warwick mine. The Warwick shaft was deepened to 735 feet. No production was reported for this period. Operations were abandoned in July 1917 (*Reading Eagle*, April 1, 1917, and November 6, 1947).

The Eastern Steel Company drilled four exploratory core holes. The geologic logs of the holes were provided by J. Ross Corbin and published by Hawkes and others (1953, p. 141, table 2). The locations of the core holes are shown on plate 19 of Hawkes and others (1953). The holes were drilled vertically. Core hole 1, drilled to a depth of 1,415 feet, penetrated a trace of magnetite at 721 feet. Core hole 2, drilled to 1,343 feet, penetrated magnetite ore from 954.5 to 967 feet and magnetite and limestone from 967 to 1,002 feet. Core hole 3, drilled to 1,006 feet, and core hole 4, drilled to 1,679 feet, did not penetrate ore (Hawkes and others, 1953, p. 141).

During a major fire at the Boyertown Casket Company on May 28, 1926, water for fire fighting started to deplete the Boyertown reservoir. Water was pumped from the Warwick mine shaft for 2 hours to help contain the blaze (*Reading Eagle*, May 29, 1926). The Eckert open cut was used as a dump for junk cars and caskets damaged by the Boyertown Casket Company fire.

In 1941, Henry K. Grim and a local financier partnered with J. Ross Corbin in an unsuccessful attempt to reopen the mines (*Reading Eagle*, November 6, 1947). On February 6, 1943, the Boyertown Iron Mines, Inc., received a state charter to "*acquire, own, hold, lease and let, encumber, sell or otherwise dispose of, work and operate the mines and lodes, properties and deposits, and to do all things incident and necessary to the general business of mining iron and related deposits, and to treat and market the product of the ore mines.*" L.S. Wentz, president of the Whitehall Cement Company, was chairman of the board of directors of the company. The War Production Board approved reopening the mines. However, the company needed the approval of the Defense Plant Corporation before mining could begin (*Reading Eagle*, April 16, 1943). Because mining never took place, it is assumed that approval was never granted.

An exploration project by the U.S. Government was conducted in 1943-44 based on the premise that undiscovered deposits of magnetite may exist in promising calcareous host rocks along the northwest margin of the Mesozoic Basin near Boyertown. The exploration program was described by Hawkes and others (1953). Two deep core holes were drilled by the Defense Plant Corporation and logged by the U.S. Geological Survey. A magnetic survey was made in the immediate vicinity of the Boyertown deposits by the U.S. Bureau of Mines. A geologic and magnetic survey of a belt 6 miles west and 3.5 miles northeast of Boyertown was made by the U.S. Geological Survey.

In April 1944, the U.S. Geological Survey made five aerial aeromagnetic transverses over the Boyertown area. This was the first use of the airborne magnetometer in the Western Hemisphere as a geophysical prospecting instrument over

Figure 74. Boyertown mine buildings still standing in 1948. The buildings were demolished for construction of an industrial plant. Photographs taken by the Boyertown Autobody Works, Inc. Courtesy of the Boyertown Historical Society.

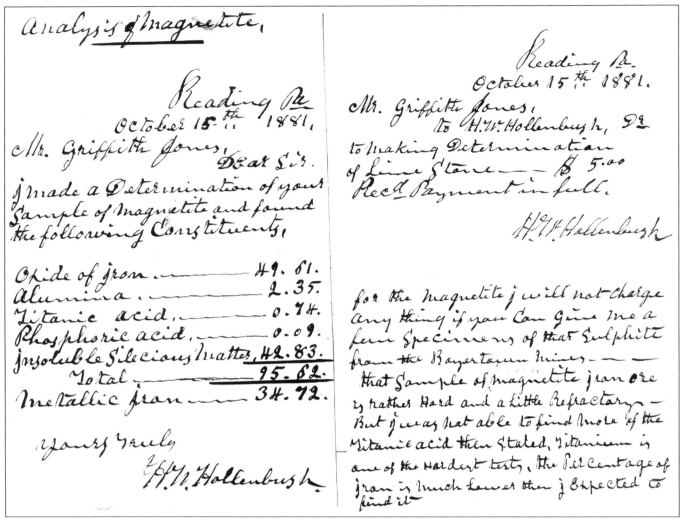

Figure 75. Analysis of magnetite ore and letter from Hiram W. Hollenbush to Griffith Jones offering to trade his chemical analysis services for magnetite specimens from the Boyertown mines, October 15, 1881. From the Griffith Jones Papers Concerning Iron Ore Furnaces (3615), Historical Collections and Labor Archives, Special Collections Library, Pennsylvania State University. Hiram W. Hollenbush (1805-1858) was born in Ruscombmanor Township, Berks County. He learned carpentry and cabinetmaking from his father and worked in that trade for a number of years. As a youth, he began to study mineralogy and spent his spare time in Dr. T.G. Bertolet's library. Despite having only 15 days of formal schooling, he adopted mineralogy and chemistry as his full-time profession some time after 1871 and became one of the best known mineral collectors in Pennsylvania. He provided ore analyses for mining companies and others. He was a founding member of the Reading Society of Natural Sciences in 1869. D'Invilliers (1883, p. 393), in his "*List of the Minerals of Berks county with their localities,*" stated: "*In the compilation of this table I am especially indebted to Mr. H. W. Hollenbush and Dr. D. B. Brunner of Reading, not only for permission to inspect their beautiful cabinets, but for a very full statement of the specimens and localities known to them.*" After his death, Hollenbush's mineral collection was purchased by Nicholas H. Muhlenberg of Reading (Mengel, 1923).

known geological and magnetic conditions. The research crew was lead by J.R. Balsley of the U.S. Geological Survey. They used an AN/ASQ-3A Magnetic Airborne Detector, which was developed at the Bell Telephone Laboratory under a Naval Ordnance Laboratory contract and subsequently modified under supervision of the U.S. Geological Survey to be a geophysical mapping instrument. Two airborne traverses were run northeast of Boyertown to coincide roughly with ground traverses.

Three other short traverses without magnetic ground control were run across the old mines at Boyertown. The 600-foot and 900-foot aeromagnetic traverses over Boyertown apparently reflected the magnetite ore bodies of the Gabel, Warwick, and the Phoenix mines. On the 300-foot flight, these same magnetic features were partially masked by the industrial anomalies that rendered much of the ground survey essentially useless (Hawkes and others, 1953, p. 144-146).

The locations of the two vertical core holes (A and B) drilled by the Defense Plant Corporation in 1943 and 1944, are shown on plate 19 of Hawkes and others (1953). Core hole A was drilled to 713 feet and encountered ore with 7.8 to 23.62 percent metallic iron from 613 to 636 feet. Core hole B was drilled to 1,204 feet and encountered ore with 13.05 to 43.53 percent metallic iron from 1,167.2 to 1,200.2 feet. The core hole encountered diabase below 1,200 feet. Hawkes and others (1953, p. 142, table 3) summarized the geologic record of the cores, which was prepared by the U.S. Geological Survey. Hawkes and others (1953, p. 142, table 4) also provided the results of analyses of the cores. A weighted composite sample of the 32-foot mineralized section between 1,167.9 and 1,200.2 feet in core hole B contained 31.47 percent magnetic iron, 0.02 percent cobalt, and 0.04 percent titanium oxide.

In 1948, the 16-acre property of the Boyertown Ore Company was sold to the Boyertown Auto Body Works, Inc. for construction of an industrial plant. When the property was developed, several dilapidated mine buildings were torn down and one open shaft was filled. Photographs of the existing mine buildings taken in 1948 prior to demolition are shown in figure 74. Boyertown Auto Body built truck bodies. It declared bankruptcy in 1990 (*Reading Eagle*, December 2, 1997).

Minerals

Amphibole, vars. byssolite and mountain leather (D'Invilliers, 1883, p. 393)

Azurite - associated with malachite

Calcite - light yellow to light tan scalenohedral crystals, often associated with pyrite (figs. 76-78); green byssolitic calcite

Chalcopyrite - massive, associated with calcite and magnetite (fig. 82)

Chlorite group (D'Invilliers, 1883, p. 395)

Chrysocolla - coatings

Copper - native, rare; up to 2.8 cm (fig. 79)

Cuprite - rare (D'Invilliers, 1883, p. 396)

Dolomite - white rhombs to 1.5 cm associated with calcite (fig. 80)

Epidote - massive (D'Invilliers, 1883, p, 396)

Gypsum (Eyerman, 1889, p. 42)

Hematite - red, also var. specular

Magnetite - octahedral and dodecahedral crystals (D'Invilliers, 1883, p. 398).

Malachite (Smith and others, 1988, p. 328-329)

Marcasite - tin white metallic (fig. 84)

Pyrite - cubic and octahedral crystals reembling those from the French Creek mine in Chester County (figs. 83, 85, and 87)

Pyroxene (Eyerman, 1889, p. 15)

Pyrrhotite (D'Invilliers, 1883, p. 400)

Quartz, vars. agate, jasper, basanite, flint, and hornstone

Serpentine group

Stilbite - light yellow "wheat sheaf" crystals, sometimes in radiating groups (fig. 86)

"Wad" - globular (Eyerman, 1889, p. 13)

BRECKNOCK TOWNSHIP

Sandstone for the 1838 Berks County courthouse was quarried by a group of men (John Trostel, Abraham Trates, William Griffith, Henry Trostel, Samuel Rathman, Simon Kohl, Abraham Scales, Issac Adams, and Peter Traitt) who lived in the vicinity of Knauers. The courthouse was razed in 1931; however, the sandstone columns were stored at the Almshouse quarry in Shillington. In the 1960s, the Reading City Council voted to sell the columns. They were sold to Nathan Reimert, owner of Jacob Knabb's grist mill, located west of Oley, in a sealed bid for $50. Around 1847, white sandstone quarries on the Agustus W. Schweitzer and Henry J. Ziemer farms provided some of the stone for the Berks County prison (Davis and Kochersperger, 1876).

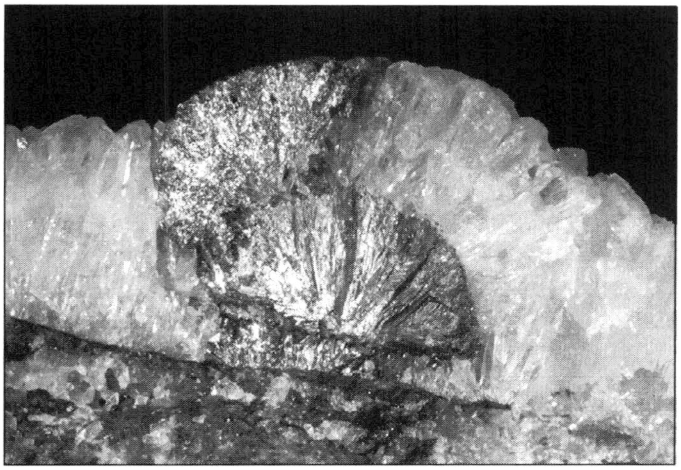

Figure 76. Pyrite and calcite from the Boyertown mines. Carnegie Museum of Natural History collection ANSP-25667.

Figure 77. Calcite from the Boyertown mines. Reading Public Museum collection 2005c-3-693.

MINERALS FROM THE BOYERTOWN MINES

 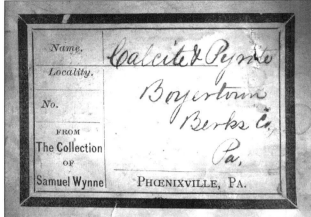

Figure 78. Calcite and pyrite from the Boyertown mines, 10 cm (left). Carnegie Museum of Natural History collection ANSP-25667. Originally in the collection of Samuel B. Wynne (1846-1905) of Phoenixville. Although Wynne's label was of a design similar to that of other local collectors, the pink paper made it distinctive.

Figure 79. Native copper from the Boyertown mines, 1.5 cm. Reading Public Museum collection 2005c-3-225.

Figure 80. Dolomite and calcite from the Boyertown mines. Reading Public Museum collection 2004c-7-515.

Figure 81. Magnetite from the Warwick mine, Boyertown, 14 cm. Reading Public Museum collection 2004C-2-181.

Figure 82. Chalcopyrite from the Boyertown mines, 7 cm. Reading Public Museum collection 2005C-2-928.

Figure 83. Pyrite from the Boyertown mines, 1.7 cm. Ron Kendig collection.

Figure 86. Stilbite from the Boyertown mines. Bryn Mawr College collection Rand 7518.

Figure 84. Marcasite and quartz from the Boyertown mines, 5.5 cm. Ron Kendig collection.

Figure 85. Pyrite from the Boyertown mines, 2.4 cm. Sloto collection 3234.

Figure 87. Pyrite from the Boyertown mines, 3 cm. Sloto collection 3234.

CAERNARVON TOWNSHIP

Iron Mines

Jones Mine

The flooded Jones open-pit mine (fig. 83) is between Pennsylvania State Route 82, Hopewell Road, and Reed Hill Road at 40° 10′ 19″ latitude and 75° 50′ 54″ longitude (Appendix 1, fig. 37, location C-1). It is labeled as "Jones Millpond" on the USGS Elverson 7.5-minute topographic quadrangle map. The mine is near the contact of the Vintage Formation and diabase. The Jones mine was both an iron and a copper mine; it was the only mine worked for copper ore in Berks County. The Jones mine consisted of two adjoining mine properties—the Jones mine and the Jones Good Luck.

David Jones (1709-1782), the son of Rev. William Jones, an Episcopalian minister, emigrated by himself at the age of 12 from Merionetshire in Wales in 1721. After arriving in America, he resided with relatives in an area known as the "Welsh tract," now Radnor, in Delaware County. He married Elizabeth Davies when he was 26. He bought 1,000 acres in the upper Connestoga Valley using money inherited from his mother's estate. In 1735, he moved there and lived at the foot of Welsh Mountain. The historical record is not clear, but it is assumed that David Jones discovered the iron deposit on his property.

David Jones' son, Jonathan Jones (1738-1782), acquired a patent for 39-3/4 acres on part of the Jones mine tract on December 22, 1773. The patent likely was obtained for the sole purpose of selling the property because two months later on February 14, 1774, Jones sold the property to Mark Bird, owner of the nearby Hopewell Furnace (Harden, 1886, p. 33). This part of the mine property became known as the Jones Good Luck. Contract miners for Hopewell Furnace working at the Jones mine received 5 shillings per ton of ore mined during the Revolutionary War (Pierce, 1950, p. 13).

In 1783, the Jones mine was visited by Samuel Hermelin, a Swede traveling through America, who published the following description: *"The Jones ore field is situated on a low mountain, on top, a level sandy field. The first stratum below the sandy earth consists of a red clay, partly yellow alternately mixed with sand with lumps of iron ore. The first stratum is 7 ft thick. The second stratum, 2 to 3 ft thick, is a reddish brown and somewhat dark-grey iron ore mixed with clay and potstone. The third stratum, 1 to 2 ft thick, consists of a white to*

Figure 88. Jones mine, Caernarvon Township, April 9, 2011.

Figure 89. Smokestack at the Jones mine, Caernarvon Township, 2011. Courtesy of Jim Woodeshick.

grey potstone mixed with a small quantity of iron ore. The fourth stratum, 6 ft thick, is a brown iron ore mixed with grey iron ore and some fine white clay; this is the stratum that is worked for iron ore. The fifth stratum, which has not been gone through is a reddish-yellow clay intermixed with iron ore. The stratum are mostly inclined, dipping from 15 to 20 degrees to the south. There are several ore quarries worked by stooping and a timbered adit has been driven 200 fathoms to carry away the water, which hardly ever has a greater depth than 20 ft. However, a hand pump is being used" (Hermelin, 1783, p. 30-31).

Johaan Schoepf, a German traveling through America, visited the Jones mine in 1783. He wrote of his visit: *"...made another little detour to see Jones's Mine-holes, iron-mines very little different from those just mentioned. Brown, sandy, and soft iron-stone lies shallow beneath the surface, but is very productive; 3000 pounds of ore are said to yield 2000 pounds of iron. Beneath and above the ironstone there is a bed of grey, soft, clayey earth which is called by the workmen soapstone. The work is carried on as mentioned above. That is to say, they dig here or there deep and wide, open pits, and when these grow inconvenient on account of depth, water, or other circumstances, they begin new ones"* (Schoepf, 1788).

The boundary between the two mine properties was in dispute as early as 1801. According to Berks County court records, an amicable agreement was reached by the owners of the Jones mine tract—Joanna Potts (wife of ironmaster Samuel Potts; Joanna Furnace was named for her), Martha Rutter, and Sarah May (daughter of Samuel Potts)—and the owners of the adjoining Jones Good Luck tract—Daniel Buckley, Thomas Brooke, and Matthew Brooke—who were associated with Hopewell Furnace. The owners of the Jones mine tract were the plaintiffs. Historically, the dispute is noteworthy because it was very unusual for women to own property and be plaintiffs in a court case at that time; however, these women were members of very wealthy and powerful iron families. On December 17, 1801, the referees Robert Coleman, Benjamin Marclay, John Ralston, Patterson Bell, and James Gibbons, Esquires, submitted their report establishing what became known as the compromise line (Harden, 1886, p. 33-34). The compromise line became the boundary between the Jones mine tract and the Jones Good Luck.

When Samuel Potts died in 1793, his will provided a mining right to Rebecca Furnace in Chester County. His will also reserved the right for his children to mine copper ore from the Jones mine and the right to erect a furnace to smelt it, a right that none of his descendents exerted (*Reading Eagle*, April 1, 1917).

A letter from W.L. Trewick published by Harden (1886, p. 30-31) stated that the Pennsylvania Copper Company began mining at the Jones mine in 1805. Captain Henry Thomas from Falmouth County, Cornwall, England, was the mine manager and treasurer of the company. Captain Thomas died in May of 1808, and Richard Trewick, also from Falmouth County in Cornwall, was elected manager and treasurer of the company. Trewick was assisted by two Englishmen named Ryfert and Oldfield. A building was erected at the mine. The slate for the roof and some of the building materials, as well as the mining machinery, were shipped from England to New York City, and then were transported to the mine by wagon.

The Pennsylvania Copper Company sank a shaft 100 feet deep, ran a drift from the shaft to the northeast, and sank a second shaft 80 feet deep. The company was in existence from 1805 to 1811, but never made a profit. They found it difficult to separate the copper from the iron ore. In addition, their steam engine was expensive to operate because it was fired with wood and consumed many acres of timber. The company went bankrupt, and the assets were sold at sheriff's sale. The steam engine, boilers, and other equipment were shipped to Pittsburgh in 1814.

The Jones mine was a major supplier of iron ore to Hopewell Furnace. In 1824, 418 loads of Jones mine ore were sent to Hopewell Furnace where it was smelted to produce 325 tons of iron.

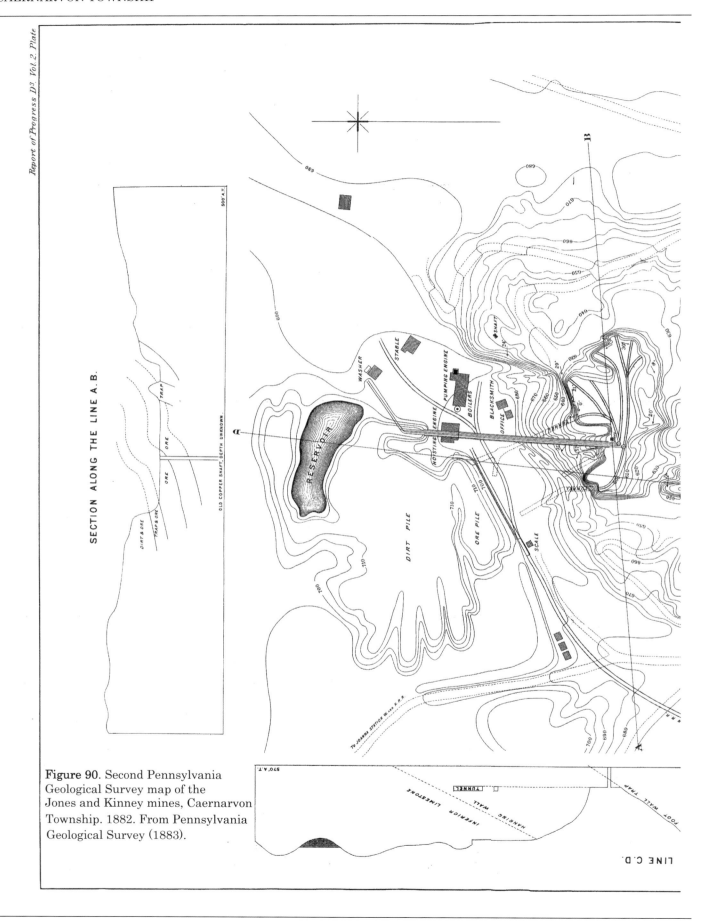

Figure 90. Second Pennsylvania Geological Survey map of the Jones and Kinney mines, Caernarvon Township. 1882. From Pennsylvania Geological Survey (1883).

CAERNARVON TOWNSHIP

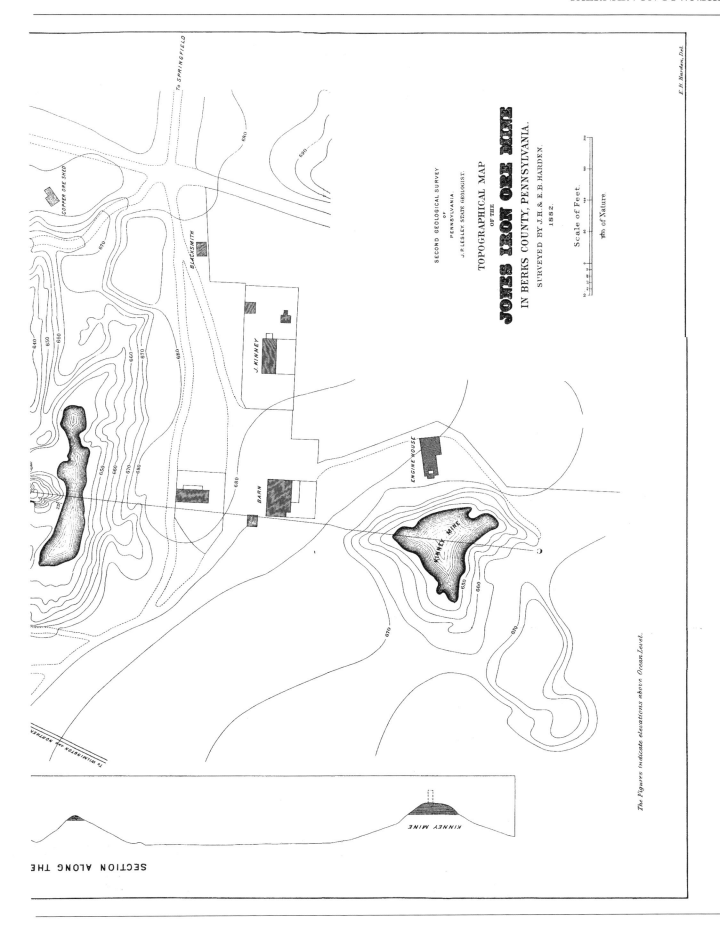

Figure 91. Jones mine, Caernarvon Township, ca. 1880s. The Jones mine map (fig. 90) published by the Second Pennsylvania Geological Survey was used to identify the buildings at the mine. Identifiable features include the scale house, ore pile, ore loading area, hoisting engine house, office, pumping engine house, blacksmith shop, and stables.

Prior to 1836, each furnace owner mined their own ore with no established system of mining. About January 1836, the furnace owners formed a loosely organized company called the Berks and Chester Mining Company. The right to mine ore belonged to the following ironmasters and furnaces: Clement Brooke and Company of Hopewell Furnace (1,000 ton annual allotment); David Potts of Warwick Furnace (1,500 ton annual allotment); H. and D. Potts of Isabella Furnace (1,600 ton annual allotment); Smith and Darling of Joanna Furnace (1,200 ton annual allotment); John Schwartz of Mount Penn Furnace (2,500 ton annual allotment); Keims, Jones & Company of Monroe Furnace (1,500 ton annual allotment); and later the owners of the Phoenixville and Hampton Furnaces. Each ironmaster had the right to mine a specified tonnage of ore annually and was required to pay 50 cents per ton to the others for any additional ore mined. In addition to the Jones mine, the mining company operated the Warwick mine in Chester County (Harden, 1886, p. 24).

When the Berks and Chester Mining Company was formed, ore was mined both by miners working directly for the furnaces and by contract miners who sold the ore they mined to the furnaces. Ore mined on contract was sold for $1.25 to $1.50 per ton. In 1836, a miner's wages ranged from $18 to $20 a month. A strong boy with a horse and cart earned $1.25 a day. By 1839, miner's wages rose to $20 to $22 per month (Harden, 1886, p. 25).

In 1836, William McIlvain (1807-1890) was hired as the manager of the Jones mine. McIlvain was born in Delaware County and moved to the Jones mine in 1836, where he served as the manager for 10 years. On December 3, 1844, McIlvain resigned and moved to Gibraltar, where he formed a partnership with Henry Seyfert to manufacture boiler plate iron under the name of Seyfert, McIlvain & Company (Biographical Publishing Company, 1898, p. 39-40). McIlvain later became president of the Second National Bank of Reading.

Prior to McIlvain's tenure, there was no regular mine opening or system of mining. Miners from each furnace would dig a hole with shovels, remove the ore, and cast the dirt to the side or fill in a previously dug hole. Water was pumped by hand. McIlvain established an orderly system of mining where dirt was moved away from the mining area. He installed a horse-powered iron pump. On October 30, 1836, the company directed that the "coppery" ore be kept separate, and mining be conducted to avoid it. The copper and sulfide-rich ore was removed from the "good" iron ore and deposited in large heaps. At a later time, the iron ore was separated from the copper ore by magnets (Harden, 1886, p. 25).

McIlvain invented the log washer when he was the manager of the Jones mine. Miners were docked for dirt in the ore delivered to the furnace, so an efficient method of washing ore was necessary. The log washer, built from timber, was 10 to 15 feet long and 9 to 12 inches in diameter. Inside were wrought iron blades that formed a crude propeller driven by horse power. The "wash ore" was fed in at the lower end and the water at the upper end. The propeller moved the ore in the direction opposite that of the flowing water. The heavier ore

Figure 92. Incline at the Jones mine, Caernarvon Township, ca. 1880s. The ore cars (bottom right) were hauled up the incline and out of the mine by a stationary steam engine.

was moved up and out the top, and the water washed the mud and dirt out the lower end (Harden, 1886, p. 32).

On January 14, 1841, Reeves and Whitaker, the predecessor to the Phoenix Iron Company, were admitted to the mining company on payment of one-seventh of the value of the stock ($1,350), which was later reduced to a one-sixth share. By 1844, Reeves and Whitaker were mining 1,600 tons of ore annually from the Jones and Warwick mines for their iron furnace in Phoenixville. On January 1, 1843, Jonathan Sidle was admitted to the company on payment of one-seventh of the cost of tools and machinery and one-seventh of the cost for finding "red ore" (hematite) at the Jones mine ($575.04) (Harden, 1886, p. 26).

Table 4. Iron ore production from the Jones mine, Caernarvon Township, 1836 to 1852. From Harden (1886).

Year	Tons of ore mined	Year	Tons of ore mined
1836	9,300	1844	4,400
1837	6,900	1845	4,200
1838	6,600	1846	9,900
1839	2,400	1847	11,400
1840	6,800	1848	12,900
1841	6,800	1849	8,050
1843	5,000	1850	8,100
1844	4,400	1851	7,800

Figure 93. Jones mine, Caernarvon Township, ca. 1880s. The photograph shows a tunnel, miners, and a mine car. From the collection of the Historical Society of Berks County Museum and Library, Reading, Pa.

It 1842, it was feared that the mine would have to be abandoned because the ore was too "coppery." During the summer of 1842, miners encountered a new iron ore vein south of the old workings that laid those fears to rest (Pierce, 1950, p. 15). This southern vein may be the Kinney mine.

On April 1, 1845, Hartley Potts was hired as the manager of the Jones mine at an annual salary of $550 (Harden, 1886, p. 26). On February 13, 1849, John Kenny took charge of shafts and drifts. Kenny was paid a salary of $8 per week. The Kinney mine south of the Jones mine likely bears his name. In 1856, the steam engine at the Warwick mine was moved to the Jones mine, presumably to aid in dewatering (Harden, 1886, p. 28-30).

By 1857, the location of the compromise line was lost, and Clement Brooke and Edward S. Buckley, the owners of the Jones Good Luck tract, brought suit for trespass against David Potts, Jr., the owner of the Jones mine tract. Brooke and Buckley contended that Potts had mined across the compromise line and had taken 15,000 tons of ore valued at $7,500. The trial resulted in a verdict for the plaintiffs for $1,024 and costs. The defendant took the case to the Pennsylvania State Supreme Court, where the decision of the lower court was reversed. In 1882, both parties agreed to hire surveyors Henry T. and Joseph V. Kendall of Reading to locate the compromise line. They set the compromise line on November 3, 1882 (Harden, 1886, p. 33-34).

The ownership of the mining rights became very complicated when shares of the rights were subdivided with each succeeding generation. By 1845, the mining rights were divided into 704 parts. The August 4, 1878, deed from the E & G Brooke Iron Company to Henry S. and George B. Eckert was for "*one undivided half part of one fourth part of one undivided one hundred and seventy-sixth part of the ores in the Jones mine.*"

The Jones mine was described by Rogers (1858, vol. 1, p. 181): "*The chief mine is an open excavation covering rather more than 5 acres, and there is another* [the Kinney mine] *to the south of it covering about 1 acre. Magnesian limestone bounds the ore on the northern edge of the principal excavation. Here there is a mine shaft 180 feet deep. The shaft enters the limestone at a depth of 50 feet, and a boring 20 feet from the bottom of the shaft is still in this rock.*"

"*A dike of trap rock* [diabase] *cuts the ore-bearing strata near the southern side of the pit and produces phenomena precisely identical with those caused by the trap dikes in the Cornwall-Lebanon mines, converting the ore to a more highly crystalline form and endowing it partially with magnetism. As in every such Instance, the ore is richest and purest adjacent to the trap rock. This is equally the case in the southern* [Kinney] *or smaller mine. The strata dip N. 30° W. at about 20°; and in the northern bank of the large* [Jones] *mine we may perceive the Auroral limestone regularly overlying the upper beds of the Primal slate, containing or consisting of the ore. In this mine, as in*

Figure 94. Jones mine, Caernarvon Township, ca. 1880s. From the collection of the Historical Society of Berks County Museum and Library, Reading, Pa.

that of Cornwall and that of Lebanon, some of the ore contains a small amount of copper."

The mine was idle and flooded when J.P. Lesley examined the workings on July 23, 1864. The workings were inaccessible, and he considered the mine to be in poor condition. However, he concluded that the ore body was not exhausted, and he recommended that the inefficient mining methods of the past be abandoned and capital be invested in proper mining and pumping machinery to work the ore body profitably (Lesley, 1864).

A letter from William McIlvain published by Harden (1886, p. 32) indicated that Nicholas Simons (also spelled Symons) mined copper ore sometime between 1836 and 1845. Simons reportedly hauled the ore in wagons to Jersey City, New Jersey, and was able to make a profit. The ore was first taken to the Conestoga Creek where it was placed in sieves immersed in barrels. The dirt was washed out by jigging, and then the ore was raised to the surface where the remaining refuse was scraped off. The clean ore was packed in barrels and hauled to Jersey City where it was smelted. From 1850 to 1854, the mine was worked by the American Mining Company, which maintained an office in New York City. The American Mining Company, under the superintendence of A. Parker, installed expensive machinery for crushing and grinding copper ore. Part of the process used magnets to separate the iron ore from the copper ore. The ore was shipped to Baltimore and Jersey City. However, the venture was unprofitable (Pierce, 1950, p. 29).

Lesley (1864) noted: *"A great deal of money has been fruitlessly spent at this mine in an endeavor to obtain copper. The northwest side of the mine yields ore containing a small amount of copper, and a shaft was sunk 192 feet deep to work ore lying against limestone on that side. A large four story building was erected; the upper rooms filled with separating machines, the ore being reduced to powder and subjected to revolving magnets to take out the magnetic iron, after which the copper was collected and smelted. The effort was a failure and the pumping and twisting machinery was sold for $2000."*

On one side of the open pit, irregular layers of a greenish, earthy-looking material contained considerable oxidized copper. It was called clay-carbonate of copper and "copper dirt." The copper ore occurred in layers 6 to 8 feet thick averaging 3 to 5 percent copper, with some layers 1/2 inch thick yielding as much as 10 to 12 percent copper. These layers were mined by Charles Wheatley, who obtained several thousand tons of ore containing 6 to 7 percent copper from shallow pits and drifts. In 1875, Wheatley mined 306 tons of "copper dirt" for a royalty of $1 a ton. Copper was extracted from this ore at Wheatley's Chemical Copper Company in Phoenixville (see Sloto, 2009, p. 125-126). Silliman (1886) published a description of a double muffle furnace used at Wheatley's copper works to smelt the "clay ore" from the Jones mine. T. Sterry Hunt isolated a copper silicate mineral from this ore. The mineral was analyzed by George W. Hawes of Yale University. Hunt determined that it was a new mineral and proposed the name venerite (Hunt,

JONES MINE AFTER 1892

Figure 95. Jones mine, Caernarvon Township, 1910.

Figure 96. Jones Mine, Caernarvon Township. This undated photograph was taken after the Jones mine was abandoned. The mine buildings appear to be in good condition. Courtesy of the Tri-County Heritage Society, Morgantown.

Figure 97. Fishing in the flooded Jones mine, Caernarvon Township. Undated photograph courtesy of the Tri-County Heritage Society, Morgantown.

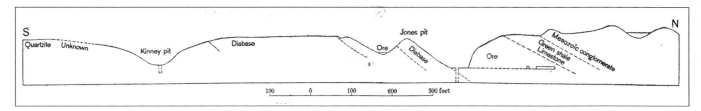

Figure 98. Geologic cross section of the Jones and Kinney mines, Caernarvon Township. From Spencer (1908, p. 67).

1876, p. 325-328). Venerite was later discredited as a valid mineral species.

In 1867, the E & G Brooke Iron Company entered into an agreement to mine ore at the Jones mine for a $2.50 per ton royalty paid to Hopewell Furnace. L.H. Smith of Joanna Furnace also mined ore at the Jones mine. According to the 1880 census (Pumpelly, 1886, p. 981), Smith mined 1,680 tons of ore between July 1, 1879, and June 30, 1880. During the same period, the E & G Brooke and Phoenix Iron Companies mined 5,080 tons of ore. Hopewell Furnace went out of blast in 1883.

By 1886, the Jones mine was operated in partnership by the E & G Brooke and Phoenix Iron Companies, the former taking 40 percent and the latter taking 60 percent of the ore mined and paying the expenses of mining in the same proportion. From 1880 to 1882, a total of 37,000 tons of ore was mined (Harden, 1886, p. 34). In 1889, the mine property was sold by Maria Clingan, Edward S. Buckley, and Mary Vaux Buckley, who were associated with Hopewell Furnace, for $250 to the E & G Brooke and Phoenix Iron Companies. The mine was abandoned and allowed to flood on January 9, 1892. Mining ceased because of (1) the high sulfur content of the ore, (2) the high cost of pumping water from the open pit, and (3) the abundance of cheap iron ore from the Mesabi Range in Minnesota. The mine property passed out of mining company hands and into private ownership in 1949.

There were two shafts and three tunnels at the bottom of the open pit. One of the tunnels can be seen in figures 92 and 93. Tracks were laid at the bottom of the pit for four-wheeled cars called hoppers (figs. 92 and 93). The cars were hauled out of the mine up an incline (fig. 91) by a steam engine. A second steam engine was used for dewatering the mine.

Ore was transported by rail car and mule teams. Mule teams were used to get the rail cars rolling; they rolled the rest of the way to the Kenny railroad station by gravity. The empty rail cars were hauled back to the mine by mule teams. Ore not shipped by rail was hauled by six- and eight-mule teams (Pierce, 1950, p. 27).

The Jones mine is a Cornwall-type magnetite ore deposit. The host rock of the deposit is the Cambrian Vintage Formation dolomite and shale (fig. 98). The dolomite and shale are a reentrant in the north edge of a thick diabase dike. Two diabase bodies occur on the south side of the Jones mine open pit and were exposed along the edge of the excavation (fig. 98). Triassic sediments bound the carbonate unit on the north, probably by a fault, although some writers indicate an unconformity. The diabase dips northwestward under the ore, and a small diabase dike occurs within the ore body (fig. 98). The carbonate bedding is considerably contorted, but has a general dip to the west-northwest.

The Jones mine ore is unusual because of the high copper content. Copper concentrations in samples of ore analyzed by Sloto and Reif (2011, p. 8) ranged from 9 to greater than 10 percent. A sample of ore analyzed by Smith and others (1988, p. 330) contained 1.8 percent copper.

The ore was a back magnetite containing pyrite, calcite, and chalcopyrite interstratified with light-green slate. The upper layers passed into crystalline marble with grains of magnetite (Willis, 1886, p. 225). Blair (1886, p. 566-567) published analyses of the ore. Magnetite from the north shaft contained 45.26 percent metallic iron. Decomposed magnetite from the south shaft contained 51.32 percent metallic iron. Rose (1970, p. 10) estimated a total production of 500,000 tons of ore.

Figure 99. Native copper from the Jones Mine, Caernarvon Township, 6 cm. Reading Public Museum collection 2005C-003-223.

MINERALS FROM THE JONES MINE

Figure 100. Aragonite from the Jones Mine, Caernarvon Township. Aragonite crystals to 7 mm. Carnegie Museum of Natural History collection CM 4635 (Jefferis 2260). Acquired by William Jefferis on June 2, 1857 (label on right).

Figure 101. Apatite and chlorite from the Jones Mine, Caernarvon Township. White, hexagonal apatite crystal is 1.5 cm. Steve Carter collection.

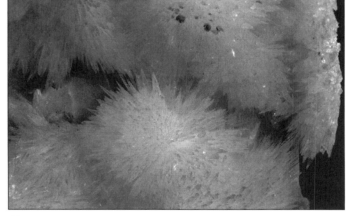

Figure 102. Aragonite from the Jones Mine, Caernarvon Township. Bryn Mawr College collection Vaux 4052.

Figure 103. Aragonite from the Jones Mine, Caernarvon Township. Public Museum collection 2005C-003-646.

Figure 104. Calcite from the Jones Mine, Caernarvon Township. Calcite crystals to 1.2 cm. Reading Public Museum collection 2004C-10-964.

Figure 105. Azurite from the Jones Mine, Caernarvon Township, magnified. Reading Public Museum collection 2004C-7-880.

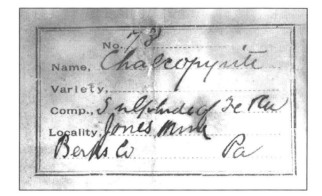

Figure 106. Hoppered chalcopyrite crystals from the Jones Mine. Caernarvon Township, 5.5 cm. Carnegie Museum of Natural History collection ANSP 18882. Originally in the collection of Edward D. Drown (1861-1919) of Philadelphia (label on right).

Figure 107. Chalcopyrite crystals from the Jones Mine, Caernarvon Township, 8.5 cm. Bryn Mawr College collection Vaux 1036.

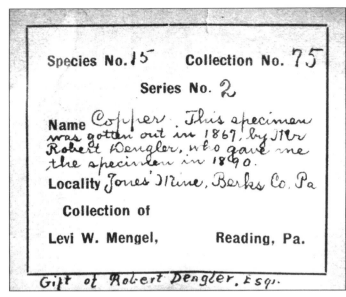

Figure 108. Native copper from the Jones Mine, Caernarvon Township, 3 cm. Collected in 1867 by Robert Dengler. Reading Public Museum collection 2005C-003-219.

Figure 109. Dolomite and pyrite from the Jones Mine, Caernarvon Township. Reading Public Museum collection 2004C-10-703.

Figure 110. Malachite from the Jones Mine, Caernarvon Township, 8 cm. Carnegie Museum of Natural History collection ANSP 6913.

Figure 111. Malachite on quartz crystals from the Jones Mine, Caernarvon Township, 8.4 cm. Carnegie Museum of Natural History collection ANSP 6471.

Figure 112. Magnetite from the Jones Mine, Caernarvon Township. Magnetite crystals to 8 mm. Carnegie Museum of Natural History collection CM 2940 (Jefferis 3337). Acquired by William Jefferis on August 26, 1861.

Figure 113. Malachite from the Jones Mine, Caernarvon Township, 12 cm. Union College collection 11.2.1-13.

Figure 114. Magnetite from the Jones Mine, Caernarvon Township. Magnetite crystals to 8 mm. Academy of Natural Sciences of Drexel University collection Vaux 12348.

Figure 115. Banded malachite from the Jones Mine, Caernarvon Township, 8.5 cm. Carnegie Museum of Natural History collection CM 4894.

Figure 116. Octahedral magnetite crystals from the Jones Mine, Caernarvon Township. Magnetite crystals to 8 mm. Steve Carter collection.

Figure 117. Malachite from the Jones Mine, Caernarvon Township, 8.5 cm. Union College collection 11.2.1-46. Originally in the Charles Wheatley collection (Wheatley 4402).

Figure 118. Malachite from the Jones Mine, Caernarvon Township, 5.7 cm. Carnegie Museum of Natural History collection CM 32022 (Brookmeyer 3422).

Figure 120. Limonite (bombshell ore) from the Jones Mine, Caernarvon Township, 4.9 cm. Reading Public Museum collection 2000C-027-150.

Figure 119. Pyrite from the Jones Mine, Caernarvon Township. Steve Carter collection.

Figure 121. Malachite from the Jones Mine, Caernarvon Township, 8.4 cm. Bryn Mawr College collection Rand 1249.

Minerals

The Jones mine was listed as "Morgantown" by Dana (1844, p. 545) in his *Catalog of American Localities of Minerals* in the second edition of A System of Mineralogy. Several specimens in the Carnegie Natural Museum of History collection and other collections labeled "Morgantown" are undoubtedly from the Jones mine. However, some specimens labeled "Morgantown" may actually be from the Morgantown area, possibly the Byler mine.

Actinolite (D'Invilliers, 1883, p. 394)

Actinolite, var. byssolite (Wheatley, 1882, p. 36)

Allophane - white and sky blue coatings and mammilary and stalactitic masses (Genth, 1875, p. 107)

Almandine - reported in crystals to 1.5 cm associated with staurolite in the host rock

Apatite-(F) - white crystals associated with magnetite (Genth, 1875, p. 139); slender hexagonal prisms (Gyer and others, 1976, p. 82): uncommon; in vuggy magnetite ore associated with chlorite, chalcopyrite, and magnetite (Smith, 1976, p. 43); crystals in the Steve Carter collection are up to 2 cm long (fig. 101)

Anglesite (Dana, 1868, p. 776-778); very rare

Aragonite - (Dana, 1868, p. 777); divergent sprays of acicular crystals; crystals to 1.5 cm (figs. 100, 102, and 103)

Aurichalcite (Wheatley, 1882, p. 36); blue-green crusts and tufts of tiny acicular crystals

Azurite - small blue crystals; very rare (fig. 105)

Bornite - scarce (D'Invilliers, 1883, p. 395)

Brochantite (?) - thin, emerald-green coatings on magnetite and chalocpyrite (Gyer and others, 1976, p. 82)

Calcite (Genth, 1875, p. 154); slender crystals and rounded groups (fig. 104)

Cerussite (Dana, 1850, p. 653); very rare

Chalcocite - occurs sparingly in granular and compact masses (D'Invilliers, 1883, p. 395); tetrahedral crystals (Gyer and others, 1976, p. 82)

Chalcopyrite - (Dana, 1844, p. 545); large tetrahedra, often tarnished and sometimes coated with malachite; sometimes hoppered; crystals to 1.5 cm across (figs. 106 and 107)

Chlorite group - green flakes (Smith, 1976, p. 43) (fig. 101)

Chrysocolla - Dana (1844, p. 545); greenish-blue, massive; botryoidal or stalactitic coatings and globullar aggregates (Genth, 1875, p. 105)

Clinochlore - (Genth, 1875, p. 132), small green translucent crystals

Copper - native copper in both crystallized and arborescent forms (Cleaveland, 1822, p. 555); some fine specimens taken out in 1859 were covered in some places with cuprite giving them a dull color (D'Invilliers, 1883, p. 395) (figs. 99 and 108)

Cuprite (D'Invilliers, 1883, p. 396)

Cyanotrichite - reported as small blue tufts

Dolomite, var. ferroan - white to tan crystals and cleavages (fig. 109)

Galena - reported; very rare

Garnet group (Spencer, 1908, p.68)

Graphite (D'Invilliers, 1883, p. 397)

Grossular - yellow, massive

Gypsum - acicular crystals, rare (D'Invilliers, 1883, p. 398); beautiful hexagonal prisms (Eyerman, 1889, p. 42)

Kaolinite - occurs in considerable quantities; contains up to 10 percent copper (Genth, 1875, p. 121)

Limonite - bombshell ore (fig. 120)

Magnetite - dodecahedral and octahedral crystals (Dana, 1844, p. 545); stepped dodecahedra (Smith, 1976, p. 43); tetrahedrons (Gyer and others, 1976, p. 82); crystals to 1.4 cm (figs. 112, 114, and 116)

Malachite - (Dana, 1844, p. 545); green, fibrous and botryoidal masses; some were cut into gems over 2 inches across (Kunz, 1885); sometimes banded (figs. 110, 111, 113, 115, 117, 118, and 121)

Natrolite - small gray crystals; rare

Pyrite (Dana, 1850, p. 653); octahedral, cubic, and complex crystals to 7 mm (figs. 109 and 119)

Quartz - crystals to 2.5 cm

Serpentine group (D'Invilliers, 1883, p. 401)

Scheelite (?) - Gordon (1922, p. 152) stated that the label on a specimen of scheelite in the Lehigh University collection indicated that it came from the Jones mine

Schorl - reported

Siderite (?) - white to tan cleavages (Gyer and others, 1976, p. 82)

Staurolite - reported in crystals to 1.5 cm associated with almandine in the host rock

Stilbite (D'Invilliers, 1883, p. 402)

Talc (Dana, 1868, p. 777)

Tenorite, var. melaconite (Wheatley, 1882, p. 36)

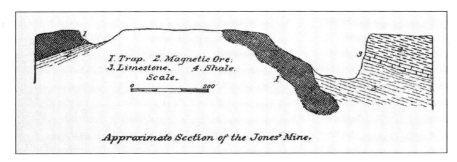

Figure 122. Geologic cross section of the Jones and Kinney mines, Caernarvon Township. The Jones mine is on the right; the Kinney mine is on the left. From Willis (1886, p. 225).

Figure 123. Kinney mine, Caernarvon Township, January 2016. Photograph courtesy of Dominic Richard.

Kinney Mine

The Kinney mine (fig. 123) is about 900 feet directly south of the Jones mine (fig. 90), south of Pennsylvania State Route 82 at 40° 10′ 12″ latitude and 75° 50′ 57″ longitude on the USGS Elverson 7.5-minute topographic quadrangle map (Appendix 1, fig. 37, location C-2). The Kinney mine was considered to be the southern opening of the Jones mine, rather than a separate mine, and it was referred to as such by Rogers (1858, vol. 1, p. 181), Lesley (1864), and Willis (1886, p. 225). The Kinney mine was in existence before 1858. It may have been the iron ore vein south of the Jones mine that was discovered in 1842.

The name of the mine is sometimes spelled Kenny. It was labeled as the Kinney mine by J.H. Harden on the 1882 Jones mine map (fig. 90), which also shows the residence of J. Kinney. However, the same residence is labeled J. Kenny in the 1876 atlas. J.H. Harden (1886) stated that John Kenny was in charge of shafts and drifts in 1849. In addition, the nearest station on Wilmington and Northern Railroad is the Kenny station, which is located south of the mine.

The Kinney mine was idle and flooded when J.P. Lesley (1864) examined the workings on July 23, 1864. He called the Kinney mine "the lower mine" and noted that the mine followed an 18-foot thick bed of "iron red rust," which was the local name for the ore. The bed continued downward under the northwestern face of the pit beneath a diabase dike dipping about 25° SE. (fig. 122). A shaft 105 feet deep was sunk to strike the ore bed at a greater depth. The ore was more oxidized and contained less pyrite than the Jones mine ore (Willis, 1886, p. 225).

The Kinney mine was later pumped out, and mining resumed. Harden (1886, p. 32) reported that a log washer was used at the Kenny mine, "*within the past five or six years*" (1880 to 1886), by Mr. H. Harvey, who was mining iron ore for Col. Smith of the Joanna Furnace.

The author has never seen a mineral specimen labeled "Kinney mine" in any collection. Because it was considered the lower pit of the Jones mine, specimens from the Kinney mine, if any were recovered from the "iron red rust" ore bed, may have been labeled as the Jones mine.

Byler Mine

The Byler mine, also known as the Lyken-Byler mine, was located between Swamp Road and the Pennsylvania Turnpike, about one mile northwest of Morgantown at about 40° 09′ 46″ latitude and 75° 54′ 12″ longitude on the USGS Morgantown 7.5-minute topographic quadrangle map (Appendix 1, fig. 35, location C-3). The mine is shown on the southwest side of the

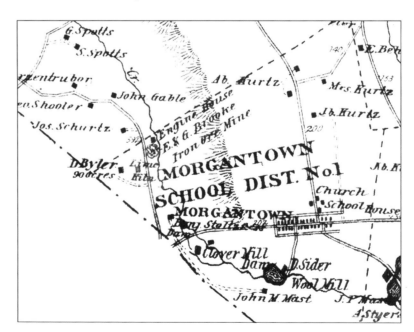

Figure 124. Location of the Byler mine, Caernarvon Township, 1876. From the 1876 atlas (Davis and Kochersperger, 1876).

creek as the E & G Brooke Iron Ore Mine in the 1876 atlas (fig. 124).

The Byler mine is a Cornwall-type iron ore deposit located at the contact between diabase and Cambrian limestone. The deposit was discovered by David Lykens, who sold the mining rights to the E & G Brooke Iron Company. The mine was opened about 1860 as an open-pit mine and was worked until the late 1880s by E & G Brooke. Some of the ore was sent to Joanna Furnace. In 1887, the Reading Eagle (May 15, 1887) reported that the E & G Brooke Iron Company was sinking a shaft, which was 49 feet deep at the time of the report.

Hezekiah Noble was the last mining engineer to work at the Byler mine. When the mine was determined to be too dangerous to work, he was transferred to Shirk's mine in Lancaster County where he lost his life in a mining accident. The mine was 100 feet deep when it was abandoned (Mast and Simpson, 1942, p. 472). The pit was about 500 feet wide and 70 feet deep in the 1930s. The mine dumps contained ore similar to that of the nearby Jones mine (Bascom and Stose, 1938, p. 123). Rose (1970, p. 10) estimated production at several hundred thousand tons of ore based on the size of the pit.

Figure 125. Styer quarry, Caernarvon Township, April 2015.

Limestone Quarries

Styer Quarry

The Styer quarry (fig. 125) was west of the intersection of Mill and Valley Roads, about 0.6 mile southeast of Morgantown at 40° 09′ 05″ latitude and 75° 52′ 53″ longitude on the USGS Morgantown 7.5-minute topographic quadrangle map (Appendix 1, fig. 35, location C-4). The quarry, in the hillside west of the creek, was in the Zooks Corner Formation. Adam Styer's lime kiln and quarry are shown in the 1876 atlas. The kilns were still standing in 2016 (fig. 126).

Adam Styer owned one of the largest quarries in the Conestoga Valley. The quarry was opened by Abraham Foreman, Thomas Ballentine, and George Byler. Adam Styer was listed as a lime dealer in the 1876 atlas. Styer operated three kilns for lime production. He delivered lime locally by horse team, and shipped lime by rail from the Joanna railroad station. The Conestoga wagons that hauled lime from the kilns to the railroad station were backed under the railroad trestle, and the carter dumped the entire contents of a coal car into the wagon as a back load. The coal was unloaded at the top of the

Figure 126. Styer lime kilns, Caernarvon Township, April 2015.

Figure 127. Mast quarry, Caernarvon Township, April 2015.

Daniel Mast Quarry

The Daniel Mast quarry (fig. 127) was north of the intersection of Mill and Valley Roads, about 0.6 mile southeast of Morgantown at about 40° 09′ 05″ latitude and 75° 52′ 47″ longitude on the USGS Morgantown 7.5-minute topographic quadrangle map (Appendix 1, fig. 35, location C-5). The quarry was in the Zooks Corner Formation. The quarry was east of the Styer quarry and north of the Hentzler quarry. A lime kiln is shown as the Levi Mast lime kiln in the 1876 atlas. The Daniel Mast farm is now a residential development.

There were three lime kilns on the Mast property on Mill road. Two of the kilns remain (2016) (fig. 129). Limestone was first quarried from an outcrop near the creek directly behind the kilns. A larger quarry is south-

kiln by removing loose boards laid cross-wise across the bottom of the wagon and allowing the coal to discharge through the bottom. This method of unloading also was used when delivering lime for agricultural or building purposes (Mast and Simpson, 1942, p. 369 and 373).

In 1909, two additional kilns were constructed after the farm was purchased by Mast Stoltzfus. Stoltzfus resided on the Styer property and operated the limestone quarry. He used a steam drill in the quarry. Arthur Sheetz furnished the steam for the drill by connecting it to the boiler of his 20-horsepower steam tractor. The supervisors of Caernarvon Township had a crusher in the quarry driven by a gasoline engine owned by John Hertzler (*Reading Eagle*, February 13 and July 30, 1909). Limestone and lime producers listed by Hice (1911) included Abram Foreman and Mast Stoltzfus in Morgantown.

Figure 128. Daniel Mast draw kiln, Mill Road, Caernarvon Township. Courtesy of the Tri-County Heritage Society, Morgantown.

east of the kilns (fig. 127); both quarries are visible from Mill Road.

In the early 1900s, Daniel Z. Mast believed that there would be a ready market for wood-burned lime. He constructed a draw kiln (fig. 125) at considerable expense to produce wood-burned lime. A draw kiln is taller than the typical pot kiln. A draw kiln is partially filled and then fired. Additional quantities of limestone and wood are added through the top each day, and the lime is removed from the bottom.

Shortly after construction of the draw kiln, a derrick used to transfer stone from the quarry to the kiln broke, killing one of his sons. To provide stone for the kiln, Mast drilled shot holes in the quarry with a well-drilling rig. He packed a large quantity of dynamite in the holes, and the resulting blast loosened enough stone to supply the kiln for a year. The blast also fractured the surrounding rock, allowing the

Figure 129. Mast lime kilns, Caernarvon Township, April 2015.

cesspool to contaminate the family's water supply, and the entire family contracted typhoid fever. The father, mother, and a daughter died. After the passing of the family in the winter of 1904-05, the kiln fell into disuse (Mast and Simpson, 1942, p. 369).

The farm passed into the hands of Stephen H. Mast, son of Daniel Z. Mast. In 1910, Stephen Mast purchased a New Holland stone crusher and elevator, which he used to crush rock for concrete work. In 1924, the 103-acre farm was sold for $171.15 an acre to Mast Stoltzfus, an adjoining property owner. The limestone quarry on the property had not been worked for many years prior to the sale (*Reading Eagle*, June 4, 1910, and September 24, 1924).

Hertzler Quarry

The Hertzler quarry was east of Mill Road, near the southern boundary of Caernarvon Township, about 1.4 miles southeast of Morgantown at 40° 08′ 31″ latitude and 75° 52′ 11″ longitude on the USGS Elverson 7.5-minute topographic quadrangle map (Appendix 1, fig. 37, location C-6). The

Figure 130. Hertzler lime kiln, Caernarvon Township. Courtesy of the Tri-County Heritage Society, Morgantown.

Figure 131. Team of oxen at the Hertzler lime kiln. Caernarvon Township. Courtesy of the Tri-County Heritage Society, Morgantown.

quarry was in the Zooks Corner Formation. The farm, originally the David K. Plank farm, later became the property of Henry M. Hertzler (Mast and Simpson, 1942, p. 367). The farm is now a residential development. The flooded quarry is located in a wooded area in the development.

A large quarry and several lime kilns and were located on the farm of David K. Plank, a former Berks County Treasurer. His sons, Daniel and William, were in charge of the quarry and kilns (*Reading Eagle*, January 11, 1872). In 1880, the 164-acre farm with the limestone quarry and three kilns in operation, was offered at public sale (*Reading Eagle*, October 15, 1880). The farm was acquired by David M. Hertzler.

In 1897, the Mine and Quarry News Bureau reported that Henry M. Hertzler operated his lime kilns for 230 days per year. Annual production from the kilns was 40,000 bushels of lime. He employed three people, including Frank D. Byler, the superintendent.

In 1907, Henry Hertzler employed six men quarrying limestone and burning lime. George Byler of Morgantown was the foreman and manager at the kilns. He was the son of David R. Byler, who also was a lime burner. George started work at the kilns about 1874 at the age of 14. He was in charge of blasting, filling the kilns, and burning lime (*Reading Eagle*, June 1, 1907, and June 18, 1915).

Hertzler operated four kilns (fig. 130) with a capacity of 700 bushels each. A team of oxen (fig. 131) hauled limestone 2,000 feet from the quarry to the top of the kilns. It took 4 days

Figure 132. John Plank lime kiln, Caernarvon Township, April 2015.

Figure 133. John Stoltzfus quarry, Caernarvon Township, December 2010.

The John Plank farm was originally bounded on the north by Pennsylvania State Route 23 and on the east by Twin Valley Road. Plank operated a quarry and two lime kilns. The kilns were located along Twin Valley Road south of Pennsylvania State Route 23. The quarry was 1,500 feet west of the kilns. The kilns were fired with coal and produced lime for the Plank farm and other customers. The lime was used for fertilizer, mortar, and whitewash (Kurtz, 1999). Plank's lime had a reputation for producing the finest whitewash of any lime produced in the Conestoga Valley. Stone for lime production was originally obtained from a quarry immediately across the road from the kilns. The quarry was abandoned when it became too deep to work. Over the years it filled with soil eroded from the surrounding farm fields. A new quarry was opened in the field east of the kilns (Mast and Simpson, 1942, p. 367).

to fill a kiln with alternating layers of limestone and coal and 9 days to burn the limestone into quicklime. In 1907, annual production was 55,000 bushels of lime, which was sold at the kiln for 8.5 cents per bushel. Hertzler also delivered building lime to Birdsboro and Reading, which were the largest local markets. Lime also was shipped to Wilmington, Downingtown, Frackville, and other places. Lime was shipped to Johnstown to help with rebuilding following the flood of 1889. Demand usually exceeded the supply (*Reading Eagle*, June 1, 1907). In 1915, Hertzler constructed an additional kiln to add to his battery of three lime kilns. The stone to build the kiln was brought from the farm of James Roberts, east of the Jones mine (*Reading Eagle*, June 18, 1915). In 1916, Hertzler erected a 20 foot by 40 foot building and installed machinery to manufacture hydrated lime. He also purchased a compressed air drill for use in the quarry (*Reading Eagle*, October 4, 1916).

According to Mast and Simpson (1942, p. 367), Hertzler eventually operated several quarries and six lime kilns on his farm. The most popular type of stone came from the large quarry near the battery of kilns in the meadow on the northwest corner of the farm. Hertzler's lime was especially popular for building and plastering. He used a six-mule team to deliver lime to his customers.

John Plank Quarry

The John Plank quarry wss southeast of the intersection of Pennsylvania State Route 23 and Valley Road, and southwest of the Pennsylvania Turnpike at about 40° 08′ 51″ latitude and 75° 52′ 33″ longitude on the USGS Morgantown 7.5-minute topographic quadrangle map (Appendix 1, fig. 35, location C-7). The 1876 atlas shows the John Plank lime kiln. The John Plank farm was purchased by C.F. Farms Investments for an industrial park. The ruins of a small lime kiln remain (2016) on the property (fig. 132).

John Stoltzfus Quarry

The John Stoltzfus quarry (fig. 133) was northeast of the intersection of the Pennsylvania Turnpike and Morgan Way, 0.7 miles northeast of Morgantown at 40° 09′ 33″ latitude and 75° 52′ 14″ longitude on the USGS Morgantown 7.5-minute topographic quadrangle map (Appendix 1, fig. 37, location C-9). Three kilns remain on Quarry Road next to the quarry (fig. 134). The quarry was in the Buffalo Springs Formation.

In the early 1900s, the Caernarvon Township supervisors operated a crusher belonging to the township at the quarry. They used the stone for road surfacing. Multiple heavy dynamite charges were used to break up the rock. A blast occurring in 1907 loosed an estimated 100,000 bushels of rock. Smaller dynamite blasts were used to break up large pieces of rock. Men with sledges then broke the rock into smaller pieces. The 117-acre John Stoltzfus farm and quarry was offered for sale in 1907 (*Reading Eagle*, October 12 and December 6, 1907, November 9, 1909, October 5, 1911, and November 9, 1912).

Figure 134. John Stoltzfus lime kilns, Quarry Road, Caernarvon Township, December 2010.

During the Great Depression, stone from the quarry was used by the Works Progress Administration (WPA) in a project that removed curves and straightened Pennsylvania State Route 23 (*Reading Eagle*, November 5, 1938). This is likely the quarry described by Bascom and Stose (1938, p. 108) where Walter Stoltzfus quarried limestone to pulverize in a small plant. Pulverized stone was delivered to local farmers and used as agricultural lime.

A cave was discovered in the quarry, but was never explored because it was flooded. In 1939, an effort was made to pump out the water so that the cave could be explored. After pumping for 24 hours, the water level was not perceptibly lowered, and the effort was abandoned (Mast and Simpson, 1942, p. 371).

The quarry eventually grew to be about 100 feet by 450 feet. During construction of the Pennsylvania Turnpike, part of the quarry was filled. It now (2016) measures about 100 feet by 200 feet.

Figure 135. C. Mast quarry, State Route 23, Caernarvon Township, April 2015.

Christian Mast Quarry

The Christian Mast quarry was on the north side of Pennsylvania State Route 23, about 0.95 mile west of Elverson at 40° 09' 21" latitude and 75° 50' 55" longitude on the USGS Elverson 7.5-minute topographic quadrangle map (Appendix 1, fig. 37, location C-10). The quarry was in the Vintage Formation. The quarry is visible on the 1937 aerial photograph, which shows Route 23 curving around the quarry to the south. When Route 23 was later straightened, it was routed through the quarry, which was mostly filled. The remains of the quarry are visible north of the intersection of Pennsylvania State Routes 23 and 401 (fig. 135).

In 1852, Hopewell Furnace purchased 169 loads of limestone from the quarry at 50 cents per load. Wagons hauling pig iron from the furnace to the forges at Churchtown and Spring Grove carried limestone as a return load (Weiler, 1976). The quarry is shown as the B. Wilson lime quarries on the 1860 map. The Christian Mast quarries are shown on the 1876 atlas. Christian Mast in Blue Rocks (the former name of Elverson) was listed as a lime dealer in the 1876 atlas.

P.W. Plank and Company

P.W. Plank and company operated a coal yard and two lime kilns on the property adjacent to the Kenny station on the Wilmington Branch of the Reading Railroad (Mast and Simpson, 1942, p. 366). The lime kilns are on North Twin Valley Road at 40° 09' 57" latitude and 75° 51' 05" longitude on the USGS Elverson 7.5-minute topographic quadrangle map (Appendix 1, fig. 37, location C-8). The kilns are shown as shown as the Clouse and Plank lime kilns on the 1876 map. The 1877 Berks County Business Directory listed Clous [SIC] and Plank as a lime producer (Phillips, 1877). The kilns remain standing (fig. 136).

Figure 136. Clouse & Plank Kiln, Twin Valley Road, Caernarvon Township, April 2015.

Brunner Farm Quarry

The Brunner farm quarry was south of the Pennsylvania Turnpike and west of Twin Valley Road, about 1.6 miles southeast of Morgantown. There were two lime kilns and a quarry on the property known as the Brunner farm (Mast and Simpson, 1942, p. 367). The Brunner farm is now an industrial park.

Figure 137. Schorl from the Morgantown Bypass, Caernarvon Township, 10 cm. Schorl crystals to 5 cm. Ron Kendig collection.

Byler Farm Quarry

The Byler farm was west of Swamp Road about one mile northwest of Morgantown. The 1876 atlas shows a lime kiln on the D. Byler farm (fig. 124).

Morgantown Roadcut Scapolite Occurrence

The Morgantown roadcut scapolite occurrence is on the northeast side of Interstates Route 176, about 0.2 mile from an overpass across the Pennsylvania Turnpike, and 0.8 mile northwest of Morgantown at about 40° 09′ 54″ latitude and 75° 54′ 04″ longitude on the USGS Morgantown 7.5-minute topographic quadrangle map (Appendix 1, fig. 35, location C-11). The road cut is in diabase. A scapolite-bearing vein is exposed about 1,000 feet north of the diabase contact in a 1.6-inch wide vein (Smith, 1978, p. 219-222).

Minerals

From Smith (1978, p. 219 and 222) and Smith and others (1988, p. 324-325)

Actinolite
Albite - chalky white
Apatite
Biotite
Chalcopyrite
Chlorite Group
Enstatite, var. bronzite
Ilmenite
Labradorite

Magnetite
Malachite
Marialite - occurs as interlocking, white to grayish-green laths that exhibit two distinct prismatic cleavages; X-ray diffraction data matches that of marialite
Pyrite
Titanite - yellow to golden grains

CENTRE TOWNSHIP

D'Invilliers (1883, p. 399) reported pyrite in radiating nodular concretions from Centre Township (fig. 138). No specific locality was given.

Building stone, known as Mohnsville stone, was quarried from the sides of a valley about 2 miles south of Mohnsville. Three quarries active in 1896 were operated by John Westley, Amos Price, and Daniel Shonour. The nearest railway station was 9 miles away, and the stone was hauled by wagon to the train station for shipment. Most of the stone was used in Reading, but stone also was shipped to Pottstown, Minersville, and Columbia. Buildings in Reading constructed from Mohnsville stone include the Catholic church on Perkiomen Avenue, Keystone National Bank, the Stevens building, and several schools and private residences (Hopkins, 1897, p. 70).

Westley's Quarry

John B. Westley's quarry (fig. 139) was 2 miles southwest of Mohnsville. Westley's quarry was the closest of the three

Figure 138. Marcasite from Centre Township, 4.5 cm. Bryn Mawr College collection Rand 310.

quarries to Mohnsville. The quarry, which was opened in 1880, was on the slope of a ridge about 150 to 200 feet above the valley. The quarry followed the strike of the rock up and down the slope. The quarried bed ranged from 10 to more than 12 feet thick. The stone was dark brown, porous, medium to coarse grained, and contained many pebbles. The overlying and underlying beds were conglomerate (Hopkins, 1897, p. 70). On January 3, 1909, Peter D. Wanner purchased the quarry from the estate of John B. Westley for $2,800 (*Reading Eagle*, January 3, 1939).

Price's Quarry

Amos Price's quarry was 2 miles southwest of Mohnsville on the opposite side of the valley from Westly's quarry. It was opened about 1890. From about 1882 to 1889, Price operated another quarry about 1/4 mile away. The face of Price's quarry was about 25 to 30 feet high. Most of the rock contained pebbles. The quarry was shallow because the rock dipped into the hill, and the overburden greatly increased with depth (Hopkins, 1897, p. 70-71).

Figure 139. Wesley's quarry, Centre Township, ca. 1890s. From Hopkins (1897, plate 18).

Shonour's Quarry

Daniel Shonour's quarry was 1 2.5 miles southwest of Mohnsville, and about 0.5 mile west of Price's quarry, on the south side of the valley. Hopkins (1897, p. 71) stated: *"As at the other quarries, there is much conglomerate and some nice sandstone."* The Calvary Reformed Church at the corner of Center Avenue and Oley Streets in Reading was built with brown sandstone from Shonour's quarry. The quarry was active in 1911 (Hice, 1911).

This quarry likely is the Eppler and Rischville sandstone quarry. The 1880 census stated that the Eppler and Rischville quarry was opened in 1780 (Shaler, 1883, p. 70-71). Massive, reddish brown sandstone of a uniform texture was quarried from surface rocks to produce sills, caps, fronts, base courses, and trimmings, used chiefly in Reading where several churches were constructed from the stone (Shaler, 1883, p. 157).

COLEBROOKDALE TOWNSHIP

Colebrookdale Township was settled around 1711 and became a township in 1741. It was named after Colebrook, England. The discovery of iron ore in 1717 along Ironstone Creek, within the present boundary of Boyertown, prompted the ear;y settlers to acquire large tracts of land.

Bechtelsville (Martin) Quarry

The Martin Stone Quarries, Inc. Bechtelsville quarry (figs 141 and 142) is in both Colebrookdale and Washington Townships, but lies mostly in Colebrookdale Township. The Bechtelsville quarry is east of North Avenue (old Pennsylvania State Route 100), just south of the Colbrookdale-Bechtelsville border at 40° 21′ 38″ latitude and 75° 37′ 56″ longitude on the USGS Boyertown 7.5-minute topographic quadrangle map (Appendix 1, fig. 34, location CB-4). The quarry is underlain by the Hardyston Formation and felsic to mafic gneiss.

Harvey Nester first produced screened sandstone from this site in 1897 (fig. 140). He sold a buckboard load of gravel for 25 cents. In 1953, Henry and Dorthy Martin leased the site and produced stone using hand tools and manual labor. First a steam powered shovel was acquired, and then a portable crusher, which was later replaced by a sta-

Figure 140. Bechtelsville quarry, Colebrookdale Township, ca. 1890s. Courtesy of Martin Stone Quarries, Inc.

Figure 141. Bechtelsville quarry, Colebrookdale Township. Courtesy of Martin Stone Quarries, Inc.

tionary crusher. In 1969, Henry Martin purchased the property. In 2016, the company operated five crushing plants and several mixing plants. Average annual production exceeds 1.5 million tons. Martin Stone Quarries, Inc. is operated by two of Henry Martin's sons and two grandsons—Glenn, Tom, Rod, and Trevor. The company employs about 65 people (Anne Kelhart, Martin Stone Quarries, Inc., written communication, 2016).

Products produced for construction range in size from granular (often used for septic systems) to stone more than 6 inches in size. Martin produces two specialty products—Martin Infield Mix (fig. 143) and Martin Track Mix. Track Mix is specially blended to create a running surface that allows traction while providing cushioned support for the runner. Infield Mix is a product designed for baseball diamonds and is widely used throughout the Mid-Atlantic States.

The overlying Hardyston Formation (feldspatic quartzite and sandstone) formerly was crushed into fine and coarse aggregate. Precambrian gneiss is now the major rock quarried as the overlying Hardyston has been mined out (Berkheiser, 2003). Minor lenses and discontinuous veins (up to 2.5 inches wide) of white and pink barite contain minor calcite, trace amounts of pyrite, chalcopyrite, galena, and quartz (Martin Stone Quarries, Inc., 1984).

Stauffer's Quarries

Stauffer's quarries were located in a small hill northwest of Boyertown, north of Pennsylvania State Roue 73 and west of Schaeffer Road. The smaller western quarry was located at 40°

Figure 142. Bechtelsville quarry, Colebrookdale Township, March 2016.

20′ 13″ latitude and 75° 38′ 25″ longitude, and the larger eastern quarry was located at 40° 20′ 15″ latitude and 75° 38′ 20″ longitude on the USGS Boyertown 7.5-minute topographic quadrangle map (Appendix 1, fig. 34, location CB-7). The quarries are shown on the 1882 topographic map (fig. 145). The quarries were underlain by the Hardyston Formation.

The quarries were owned and worked by P. Stauffer. The rock taken from Stauffer's quarries was filled with quartz and feldspar grains, which decomposed to clay. D'Invilliers (1883, p. 117) stated: *"The stone is not compact, and the ready decomposition of the feldspar causes it to quickly decrepitate into a sandy mass. It has however been considerably used for street ballast and the porous nature of its sandy decomposition allows of a thorough surface drainage of the streets, and so presents a neat appearance wherever used."* In the western quarry, the sandstone dipped S. 6° E. 40°, and in the eastern quarry, the sandstone dipped to the northwest.

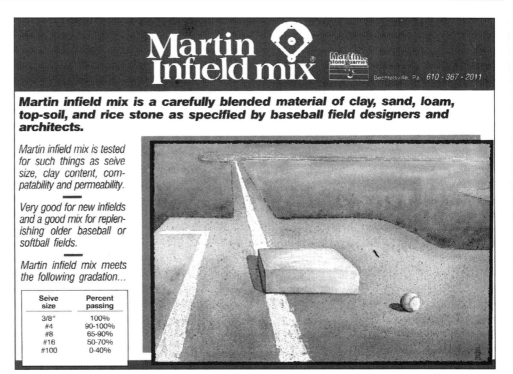

Figure 143. Advertisement for Martin Infield Mix. Courtesy of Martin Stone Quarries, Inc.

noted: *"The shaft is now cased over with concrete and used as a well. No rock dump is in evidence."*

Brower Mine

The Brower Mine was west of Mill Street between Mill Street and Ironstone Creek about 1.05 mile southwest of the center of Boyertown at about 40° 19′ 16″ latitude and 75° 38′ 55″ longitude on the USGS Boyertown 7.5-minute topographic quadrangle map (Appendix 1, fig. 34, location CB-1). The mine was in limestone fanglomerate. The location of the Brower mine, shown on plate 18 in Hawkes and others (1953), coincides with a magnetic anomaly on their airborne magnetometer survey. The deposit was a Cornwall-type iron ore body.

Iron ore was discovered while a post hole was being dug. The mine was worked by a tunnel and two shafts. Ore was stoped from two levels to a depth of about 70 feet and along strike for about 50 feet. The ore bed occurred very near, if not in contact with, the upper side of a diabase sill, under a hanging wall of baked shale or sandstone. Richard Richards of Boyertown, who was in charge of mining during 1857-58, reported that more than 2,000 tons of magnetite were mined from an irregular ore bed with a northeast-southwest strike and a dip of 35° or 40° SE. The maximum thickness of the ore bed was about 8 feet (Spencer, 1908, p. 62). When the property was visited by Bever and Liddicoat (1954, p. 86) in 1952, they

Rhoades and Grim Mine

The Rhoades and Grim mine was between Englesville Road and the railroad tracks just south of the Boyertown Borough boundary at about 40° 19′ 38″ latitude and 75° 38′ 39″ longitude on the USGS Boyertown 7.5-minute topographic quadrangle map (Appendix 1, fig. 34, location CB-5). The mine

Figure 144. Limonite from Colebrookdale Township, 8.7 cm. Reading Public Museum collection 2000C-027-156.

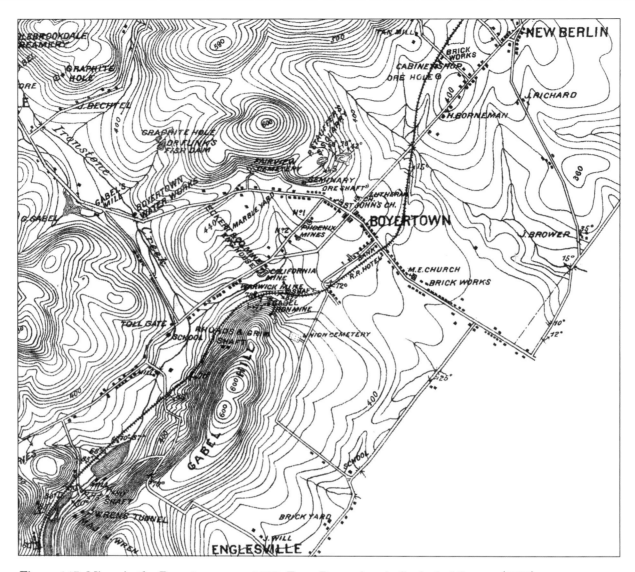

Figure 145. Mines in the Boyertown area, 1882. From Pennsylvania Geological Survey (1883).

is shown on the 1882 topographic map (fig. 145). The mine was in diabase.

The Rhoades and Grim mine was about 0.25 mile southwest of the Rhoades mine on a hillside about 50 feet above the railroad tracks. A shaft, known as the Rhoades and Grim shaft, was sunk to a depth of 40 or 50 feet. D'Invilliers (1883, p. 333) noted: "*About 300 yards southwest of the Gabel shaft, and about 30 feet up hill from railroad on south side of small ravine, there is an abandoned horse and windlass shaft, originally opened by Messrs. Rhoades and Grim, of Boyertown, who claim to have found ore there. The rock on dump is a greatly decomposed close-grained grey diorite, showing crystals of albite feldspar.*" Spencer (1908, p. 62) believed that the shaft was sunk in diabase and never encountered iron ore.

New Berlinville Clay Mine

The New Berlinville clay mine is the water-filled depression southeast of the intersection of Henry Avenue and Rothermel Roads in New Berlinville at about 40° 20′ 35″ latitude and 75° 37′ 55″ longitude on the USGS Boyertown 7.5-minute topographic quadrangle map (Appendix 1, fig. 34, location CB-6). The location was across from the brickworks on Landis Lane (also known as Brickyard Lane). The mine is shown on the 1882 topographic map (fig. 145). The mine was in limestone fanglomerate.

In the early 1880s, a clay deposit was exploited about 1,200 feet west-southwest of the New Berlin railroad station close to the east side of the Colebrookdale Railroad. The land was owned by Landis, Mutard, and Eshbach, and was leased by Schaeffer and Gresh. The flat-lying clay bed was a honey yellow color, which was caused by the iron content. About 10 feet of clay had been exposed when D'illvinners visited the mine on March 1, 1882. D'illvinners (1883, p. 152) noted that mining had only been recently begun, and the bottom rock beneath the clay bed had not been reached.

The clay was mixed with sand and made an excellent quality building brick. There was one brick oven on the property with a capacity for making 150,000 bricks. The daily output was 13,000 bricks, which sold for $6.50 per thousand at the oven. Eight men and eight boys were employed. It took 14 days, exclusive of rainy weather, to make finished bricks from the raw material. A shaft was sunk through the clay bed a little further south on the property. After sinking the shaft 70 feet through a mixture of clay and slaty limestone, and a little iron ore, the shaft was abandoned.

Railroad Cuts South of Boyertown

D'Invilliers (1883, p. 399) reported pyroxene and stilbite in veins in diabase in railroad cuts on the Colebrookdale Branch of the Philadelphia and Reading Railroad south of Boyertown. Eyerman (1889, p. 32) reported stilbite in radiating crystalline masses south of Boyertown.

Graphite Mines

D'Invilliers (1883, p. 397) reported graphite on Daniel Himmelreich's farm near Boyertown. Eyerman (1889, p. 3) reported graphite on the farms of Fegley and J. Bechtel and at Dr. Funk's fishpond.

Boyertown Graphite Company Mine

(Betchel Farm)

The Boyertown Graphite Company mine was on the Bechtel farm, east of the intersection of Pennsylvania State Route 73 and Ironstone Drive, 1.1 miles northwest of the center of Boyertown at about 40° 20′ 28″ latitude and 75° 39′ 24″ longitude on the USGS Boyertown 7.5-minute topographic quadrangle map (Appendix 1, fig. 34, location CB-2). The mine was in graphitic felsic gneiss. The mine is shown as the "black lead mine" on the 1860 map and as a "graphite hole" on the 1882 topographic map (fig. 145). The mine also was known as the Gablesville Graphite Works mine.

The graphite deposit on the John Betchel farm was discovered while prospecting for iron ore (*Reading Eagle*, July 1, 1905). The mine had many operators over the years. About 1820, John Betchel took a four-horse load of graphite ore to sell in Philadelphia. The property was idle until about 1850, when William Weaver and Sons sank a shaft and then gave up mining. In 1869, a Philadelphia party leased the mine, but did not operate it. In 1885, Wainright and Son of Spring City, Chester County, leased the mine and operated it for several years. Simon Romig of Longswamp Township next operated the mine and shipped the ore to Wainwright and Son. Later, John Schuler of Barto sunk another shaft. He abandoned the mine after he was unable to sell the ore. Next, Lewis Kutz of Longswamp Township leased the mine. He sold the lease to Major Wren of Boyertown who ran a drift from the shaft (*Reading Eagle*, August 12, 1899).

In the summer of 1899, the mine was leased by the Boyertown Graphite Company and operated by Reuben Kramm and Col. J.H. Bramwell (*Reading Eagle*, July 1, 1905). The company constructed a mill where the ore was crushed and the graphite was separated, washed, and dried. The refined graphite was sacked and hauled by wagon to Boyertown for shipment. Initially, graphite was mined from a 54-foot deep shaft in partially disintegrated mica schist. The ore was hoisted with a windllass and transferred by wheelbarrow to the crusher. The Boyertown Graphite Company operated from 1899 to about 1901.

In 1905, the mine was acquired by the Columbia Graphite Company of Philadelphia, which worked it for a short time. The Columbia Graphite Company, capitalized at $250,000, leased the mine and mill for 30 years. At the time of acquisition, the mine consisted of a 125-foot deep shaft and three slopes. The graphite-bearing bed was reported to be 15 feet thick. J.H. Bramwell represented the Columbia Graphite Company. He stated that the company was planning to install new and improved machinery and construct new buildings, one of which was to be 60 feet by 100 feet (*Reading Eagle*, July 1, 1905).

At the time of Benjamin Miller's visit in 1911, the shaft was open but partly filled with water. He noted that the mill was elaborately equipped, but much of the machinery had been removed prior to 1911. Miller (1912a, p. 112) noted: "*Much difficulty was encountered in cleaning the* [graphite] *flake on account of the mica present and by the appearance of the material on the waste heap it seems that much of the graphite was lost. The mine was abandoned for this reason and has now been idle for several years.*"

The rock is a fine-grained graphitic gneiss composed of kaolinized feldspar, quartz, graphite, and biotite. The graphite flakes are small, rarely larger than 1/4 inch in diameter. They are dull in appearance and friable. Some of the graphite flakes were compressed to form matted sheets of fibrous and foliated graphite. Miller estimated that the graphite content of the rock averaged about 7 to 10 percent. Mica is relatively abundant. Some vein quartz and pegmatites containing graphite and pyrite occur in the gneiss (Miller, 1912a, p. 111-112).

Dr. Funk's Fish Dam Graphite Mine

Dr. Funk's fish dam graphite mine was near where Funk Road crosses an unnamed tributary to Ironstone Creek, northeast of Pennsylvania State Route 73. Many years ago, there was a dam and pond at this location. The mine is shown as a "graphite hole" on the northwest side of Dr. Funk's fish dam on the 1882 topographic map (fig. 145). The mine was at about 40° 20′ 22″ latitude and 75° 39′ 02″ longitude on the USGS Boyertown 7.5-minute topographic quadrangle map (Appendix 1, fig. 34, location CB-3). The mine was in graphitic felsic gneiss.

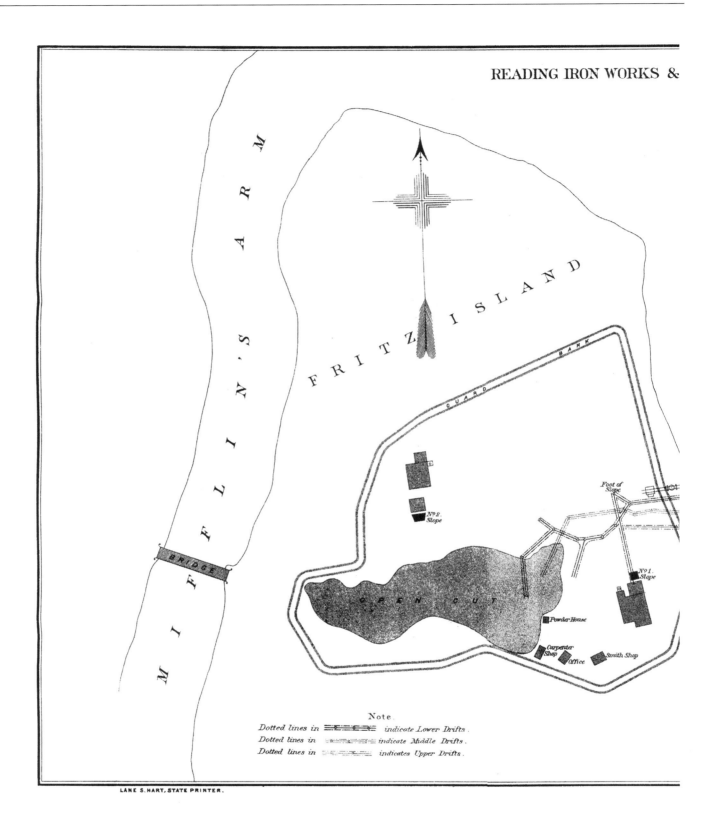

Figure 146. Second Pennsylvania Geological Survey map of the Fritz Island mine, Cumru Township, 1882. From Pennsylvania Geological Survey (1883).

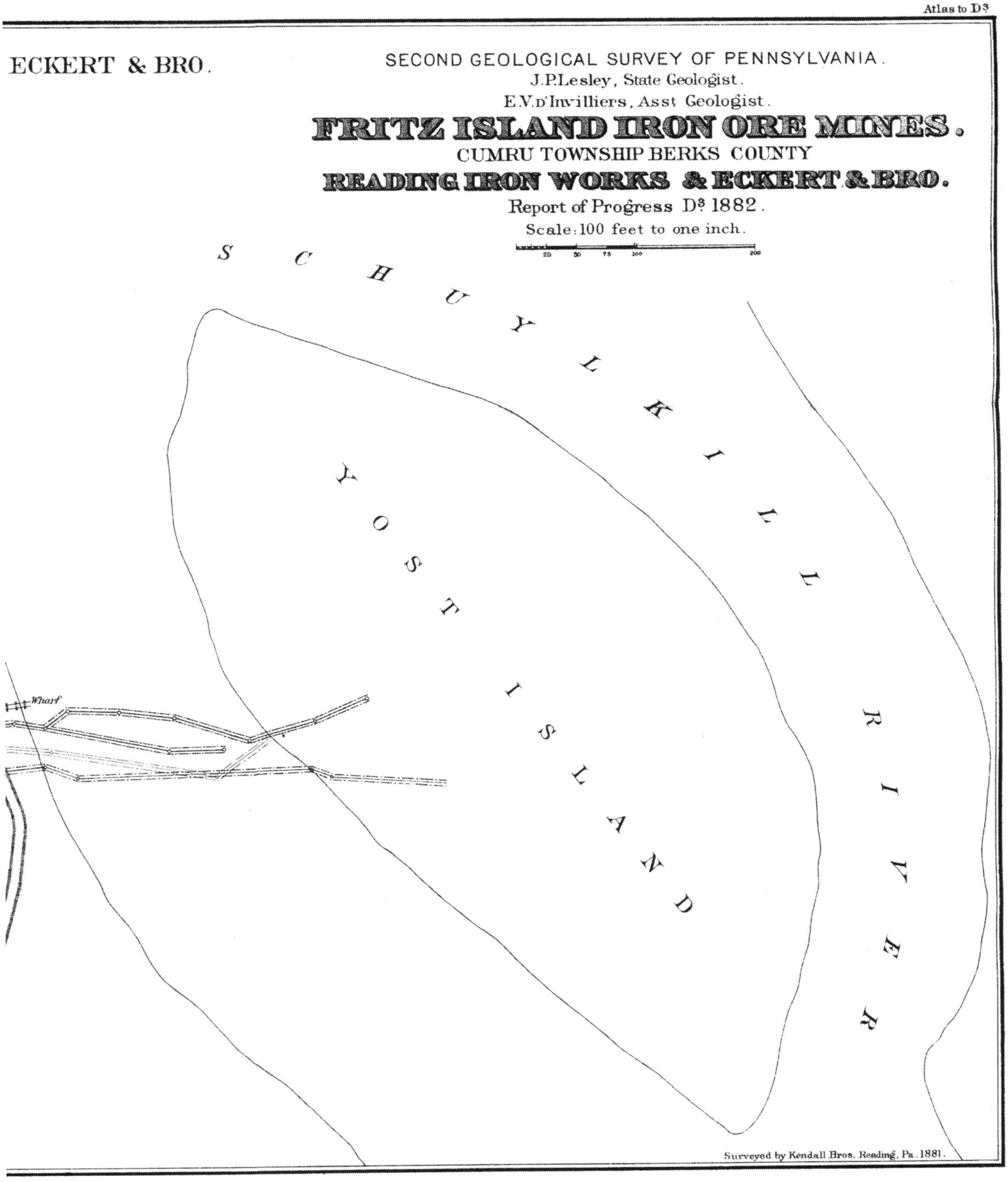

Atlas to D3
ECKERT & BRO.
SECOND GEOLOGICAL SURVEY OF PENNSYLVANIA.
J.P.Lesley, State Geologist.
E.V.D'Invilliers, Asst Geologist.
FRITZ ISLAND IRON ORE MINES.
CUMRU TOWNSHIP BERKS COUNTY
READING IRON WORKS & ECKERT & BRO.
Report of Progress D3 1882.
Scale: 100 feet to one inch.
25 50 75 100 200
SCHUYLKILL RIVER
YOST ISLAND
Wharf
Surveyed by Kendall Bros. Reading, Pa. 1881.

CUMRU TOWNSHIP

Fritz Island Mine

The Fritz Island mine was on an island in the Schuylkill River; this was perhaps the most unusual location for an iron mine in Berks County. The location ultimately led to the mine's demise. The Fritz Island mine was on the north end of Fritz Island, once known as Mifflin Island, in the Schuylkill River near the intersection of Pennsylvania State Routes 10 and 724 at 40° 18′ 02″ latitude and 75° 55′ 09″ longitude on the USGS Reading 7.5-minute topographic quadrangle map (Appendix 1, fig. 28, location C-1). While it once was an island, the channel on the western side has largely been filled. Today Fritz Island is the site of the City of Reading sewage treatment plant. The Fritz Island mine, also known as the Island mine, was worked for iron ore by an open pit and two inclined slopes.

Two accounts exist for the discovery of iron ore on Fritz Island. The most popular account is a Reading Eagle (January 3, 1882) article written by D.B. Brunner. This account states that erosion exposed a vein of iron ore on the island during a flood in the winter of 1850-51. The flood destroyed the Penn and Bingham Street bridges and also ended transportation on the 1827 Union Canal below the Penn Street lock. However, it is known that an attempt to smelt the ore from Fritz Island was made at the Mount Penn Furnace in 1843 (*Reading Eagle*, October 17, 1954).

The second account of the discovery of iron ore is more probable. Martin Fritz, the owner of the island, discovered iron ore while farming on the island and dug a small quantity with a pick and shovel. He soon leased the property to three furnace owners—Eckert & Brother of the Henry Clay Furnace; John Schwartz of the Mount Penn Furnace; and Seyfert, McManus & Company of the Reading Furnace (Reading Iron Works). Horse power was used to raise the ore, which was deposited in three equal heaps, one for each furnace. Schwartz was not able to use the ore in the Mount Penn Furnace and sold his interest to the other two furnace owners (*Reading Eagle*, January 3, 1882). In 1846, the mine was leased jointly by the Reading Iron Works and Eckert & Brother. The mine was managed by the Reading Iron Works. The ore was near the surface in a thick vein, and about 3,000 tons were mined by the fall of 1851.

The Fritz Island mine was called the Island mine by Rogers (1858, vol. 2, p. 716), who wrote: "*The Island Mine, situated on an island in the Schuylkill, one mile below Reading, has been wrought by Eckert, Syfert, and Company, but is now, perhaps temporarily, abandoned. The vein dips about 40° N.W. It is overlaid by dense brecciated limestone, locally known as "all sorts" limestone. This is, no doubt, the Mesozoic conglomerate, which appears in the opposite bank of the river in situ. The N.W. dip of this rock has no doubt regulated the*

Figure 147. Fritz Island mine shaft uncovered during construction for a sewer plant expansion, Cumru Township. From the collection of the Historical Society of Berks County Museum and Library, Reading, Pa.

dip of the injected material. The under-rock of the vein we do not certainly know, but from the specimens seen, it appears to be an impure silicious limestone, or that usually termed "bastard." The surface of the island is strewn with igneous rocks, but we are informed that none are found in contact with the vein. The iron-ore vein, which is in thickness from a few inches to 15 feet, is a heavy, fine-grained slate-blue rock, containing lime in its constitution, and decomposing rapidly on exposure to the atmosphere. When decomposing, it assumes a deep sea-green colour, and develops copperas on the surface, from the sulphuret of iron in the ore. In some specimens the pyrites have so much the aspect of sulphuret of copper that chemical evidence is required to correct the impression. A slope has been sunk upon the vein 90 feet below the surface, and, 28 feet above its foot, gangways are driven along the vein 20 feet towards the N.E., and 250 feet S.W. "

Iron ore initially was mined to a depth of 35 to 40 feet from an open cut south of the No. 2 slope (fig. 146). The open cut was oriented east-west and was nearly 300 feet long. The ore body later was mined from two slopes. The No. 1 slope started at a 62° dip; the dip averaged about 46°. The ore body dipped to the north. Three levels were driven east from the slope beneath the Schuylkill River and Yost's Island. West of the slope, the two lower levels were driven about 150 feet, which brought them beneath the east end of the open pit. The gangways were about 60 feet apart vertically; the upper one was about 50 feet below land surface. The gangways were likely more extensive than shown on figure 146. Mining was stopped at 450 feet from the top of the slope because of problems with ventilation. The same problems with ventilation occurred in the eastern gangways. Considerable stoping was done above the lowest level. At the bottom of the mine, 125 feet east of the slope, the ore branched and was followed by two drifts, which came together 175 feet beyond the branch. There, the ore was stoped up to the upper level (D'Invilliers, 1883, p. 337-338).

The No. 2 slope was about 300 feet west-northwest from the No. 1 slope. In 1883, the No. 2 slope, which was sunk in rock, was 122 feet deep on a average dip of about 45° (Spencer, 1908, p. 39). The No. 2 slope was idle in 1882 (D'Invilliers, 1883, p. 335).

In 1883, the average daily production of the Fritz Island mine was 25 to 30 tons, totaling about 10,000 tons per year. At that time, only the No. 1 slope was in operation. The total yield up to 1883 was estimated by D'Invilliers (1883, p. 341) at 250,000 tons. During its period of operation, the mine was idle twice, once during the panic of 1857 and again from 1873 to 1879. In 1882, there were 32 men employed under superintendent Henry Taylor—22 miners, 4 engineers, 1 carpenter, 1 blacksmith, and 4 laborers.

According to D'Invilliers (1883, p. 342): *"Like most of the mines in Berks county the slope is badly situated, and the arrangements for pumping water and raising ore very awkward and primitive."* One 20-horsepower steam engine was used for pumping, and one 15-horsepower steam engine was

Figure 148. Fritz Island mine air shaft uncovered during construction for a sewer plant expansion, Cumru Township. From the collection of the Historical Society of Berks County Museum and Library, Reading, Pa.

used for pumping and hoisting ore. Both engines were located on the surface, with two donkey engines for emergencies, as the mine was prone to flooding with groundwater beneath the surface and flooding from the Schuylkill River at the surface. The surface of the mine was only 4 to 6 feet above the river-water level. To protect against flooding by the Schuylkill River, the company built a guard bank around the surface plant (fig. 146).

The ore was shipped by canal boat from a wharf at the mine in equal parts to the Henry Clay Furnace and the Reading Iron Works. The ore cost 36 cents per ton to ship. The cost of mining a ton of ore was about $2 per ton (D'Invilliers, 1883, p. 342). Mr. Fritz hauled 9,000 tons of ore by horse team and boat in 1883, the mine's last year of operation.

The Fritz island mine was abandoned on January 10, 1884, after it collapsed, and the Schuylkill River flooded the mine. Prior to the collapse, there was a cave-in in the No. 1 slope. The mine foreman suspended all work until an evaluation could be made. Shortly thereafter, a drift only 20 feet below the bottom of the Schuylkill River collapsed, and the river flooded the mine. Two miners on watch in the mine, Thomas Miles and Samuel Johnson, escaped through the air shaft. There were no casualties (*Reading Eagle*, July 21, 1982).

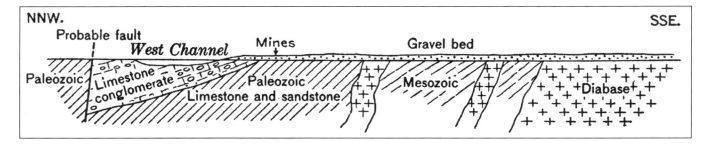

Figure 149. Geologic cross section in the vicinity of the Fritz Island mine, Cumru Township. From Spencer (1908, p. 39).

The City of Reading acquired part of Fritz Island for the location of its sewage treatment plant. In 1895, ground was broken for the first plant (*Reading Eagle,* May 15, 1895). In 1906, the city purchased an additional 36.3 acres of the northern part of the island from Lewis G. Fritz. Several generations of sewage treatment plants have been built on the island. No mine dumps remain.

Some of the old mine workings were uncovered during a $30 million upgrade and expansion of the sewage treatment plant. One of the slopes was discovered at the base of a 20- to 25-foot deep excavation for a new digestion tank. A water-filled, 8 foot by 10 foot near vertical slope reinforced by timbers in "*remarkably good shape*" was uncovered (fig. 147). The slope extended 10 feet below the excavation and ended in rubble. The 3 foot by 2 foot air shaft (fig. 148) was uncovered near the larger shaft.

The Fritz Island ore body is a Cornwall-type deposit. The host rock is a magnesian limestone or dolomite in contact with diabase (fig. 149). The iron ore occurs beneath a calcareous and brecciated Mesozoic conglomerate, which generally forms the hanging wall. Diabase sometimes forms the foot wall, but beneath the ore, there often was a decomposed sandstone or quartzite. Diabase was encountered beneath the ore in all the drifts. The diabase dike was nearly vertical and about 80 feet thick.

There were two ore beds with 90 feet of limestone between them (Willis, 1886, p. 231). The ore occurred in lenticular bodies from 18 inches to 22 feet thick. The ore often pinched out and then bulged into immense swells that yielded thousands of tons of excellent ore. Very little timbering was done in the mine, and large pillars were left to support the overlying rock. The ore south of the dike dipped to the south and was thought to be the same vein as that found in the No. 2 slope (D'Invilliers, 1883, p. 336-338). The strike of the ore bed was nearly east and west, and the general dip was toward the north, in places at an angle as great as 40°. Locally, the ore bed was vertical or dipped toward the south (Spencer, 1908, p. 38). The color of the gangue on the mine dump was a dull sea-green; the rock had a soapy, unctuous surface and contained much serpentine.

Analyses of the ore were published by D'Invilliers (1883, p. 339-341): 38.1-54 percent metallic iron, 0.25-0.86 percent metallic copper, 0.53-3.4 percent sulphur, 7.3-16.1 percent silica, and 0.019-0.038 percent phosphorus. The ore was 0.37 percent pyrite and 0.99 percent chalcopyrite.

Minerals

Andradite - tiny, golden dodecahedra zoned from dark to light; light brown crystals to 4 cm (fig. 150)

Apophyllite - colorless to white, tetragonal tabular or pyramidal crystals, often in twins or groups of rosettes; associated with calcite and zeolites in a granular garnet rock (Genth, 1875, p. 107); tetragonal tablets modified by octahedral planes, colorless and pearly white (Sadtler, 1883, p. 357; analysis); crystals to 1.2 cm (figs. 151, 152, and 154)

Aragonite - acicular crystals; botryoidal and fibrous coatings (Genth, 1875, p. 162) (fig. 153)

Aurichalcite (D'Invilliers, 1883, p. 394)

Azurite - small blue crystals, uncommon (Genth, 1875, p. 168) (fig. 155)

Bornite - scarce (D'Invilliers, 1883, p. 395)

Brucite - pearly white to yellow crystals, crystalline masses or seams in dolomitic limestone (D'Invilliers, 1883, p. 395; Genth, 1885, p. 40; analyses); coatings of indistinct crystals 3-4 mm in diameter and crystalline masses on a granular limestone (Genth, 1886, p. 40); large masses and distinct crystals on dolomite (Brunner and Smith, 1883, p. 281; analysis) (fig. 156)

Calcite - granular masses (Genth, 1875, p. 154); scalenohedral crystals to 2.9 cm (figs. 157 and 158)

Cerussite - reported

Chabazite - colorless crystals (Genth, 1875, p. 109); well-developed crystals associated with thompsonite, apophyllite, and calcite (Sadtler, 1883, p. 356-357, analysis); translucent, milky-white to colorless rhombohedral crystals to 9 mm (figs. 159-162)

Chalcocite - granular, compact (D'Invilliers, 1883, p. 395)

Chalcopyrite (Genth, 1875, p. 21); mainly massive; crystals to 8 mm

Chlorite group (Genth, 1875, p. 133)

Chrysocolla (D'Invilliers, 1883, p. 395)

Chrysolite (Eyerman, 1889, p. 17)

Clinochlore (Eyerman, 1889, p. 3817)

Copper - native, arborescent (fig. 163)

MINERALS FROM THE FRITZ ISLAND MINE

Figure 150. Andradite crystals from the Fritz Island mine, Cumru Township. Crystals to 4 cm. Ron Kendig collection.

Figure 151. Apophyllite crystals from the Fritz Island mine, Cumru Township. Reading Public Museum collection 2004C-10-635.

Figure 152. Tabular apophyllite crystals from the Fritz Island mine, Cumru Township. Reading Public Museum collection 2004c-10-634.

Figure 153. Aragonite crystals from the Fritz Island mine, Cumru Township, 7.5 cm. Ron Kendig collection.

Figure 154. Apophyllite crystals from the Fritz Island mine, Cumru Township. Reading Public Museum collection 2004C-10-634.

Figure 155. Azurite crystals from the Fritz Island mine, Cumru Township. Reading Public Museum collection 2004C-7-876. This is the finest known azurite specimen from the Fritz Island mine.

Figure 157. Scalenohedral calcite crystals from the Fritz Island mine, Cumru Township. Calcite crystals to 2.6 cm. Reading Public Museum collection 2004C-7-456.

Figure 156. Brucite crystals from the Fritz Island mine, Cumru Township. Largest crystal is 3.5 cm. Bryn Mawr College collection Vaux 3254. This is the finest known brucite specimen from the Fritz Island mine.

Figure 158. Calcite crystals from the Fritz Island mine, Cumru Township, 1.9 cm. Reading Public Museum collection 2004C-7-493.

Figure 159. Chabazite crystals from the Fritz Island mine, Cumru Township, with D.B. Brunner label. Carnegie Museum of Natural History collection ANSP 6858.

Figure 160. Chabazite crystals from the Fritz Island mine, Cumru Township. Crystals to 7 mm. Reading Public Museum collection 2004C-10-752.

Figure 161. Chabazite crystals from the Fritz Island mine, Cumru Township. Crystals to 7 mm. Reading Public Museum collection 2004C-10-754.

Figure 162. Chabazite crystals in diabase from the Fritz Island mine, Cumru Township. Bryn Mawr College collection Vaux 6573.

Figure 163. Native copper from the Fritz Island mine, Cumru Township, 2 cm. Ron Kendig collection.

Figure 164. Datolite crystals from the Fritz Island mine, Cumru Township, 7.5 cm with crystals to 1.1 cm. Reading Public Museum collection 2005c-2-136.

Figure 165. Gismondine crystals from the Fritz Island mine, Cumru Township. Carnegie Museum of Natural History collection ANSP 26253.

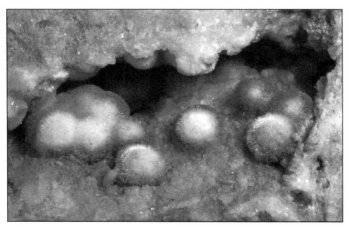

Figure 166. Mesolite crystals from the Fritz Island mine, Cumru Township. Carnegie Museum of Natural History collection ANSP 18889.

Figure 167. Gismondine crystals from the Fritz Island mine, Cumru Township. Bryn Mawr College collection Rand 7597.

Figure 168. Octahedral magnetite crystals from the Fritz Island mine, Cumru Township; crystals to 5 mm. Reading Public Museum collection 2004C-2-183.

Figure 170. Spherical cluster of stilbite crystals from the Fritz Island mine, Cumru Township, 1.5 cm. Ron Kendig collection.

Figure 172. Natrolite crystals from the Fritz Island mine, Cumru Township, 6 cm. Ron Kendig collection.

Figure 169. Mesolite from the Fritz Island mine, Cumru Township. Most mesolite from the Fritz Island mine is in this form. Bryn Mawr College collection Heyl 2021.

Figure 171. Magnetite crystals from the Fritz Island mine, Cumru Township, 5.5 cm. Ron Kendig collection.

Datolite - white crystals to 1.1 cm (fig. 164)

"Deweylite" - white, yellowish-white, or brownish amorphous masses, stalactitic or botryoidal coatings; also pseudomorphs after aragonite(?) (Genth, 1885, p. 41; analysis)

Erythrite - reported

Fluorite - pale yellow cubic crystals associated with calcite (D'Invilliers, 1883, p. 397; Eyerman, 1889, p. 7); colorless crystals to 4 mm

Galena (D'Invilliers, 1883, p. 397)

Gismondine - (Genth, 1875, p. 110; Genth, 1885, p. 42; Eyerman, 1911, p. 12; analyses); 1 to 3 mm twinned psuedo-octrahedra with deep, sharply cut, reentrant grooves running along the edges of the octahedra (Montgomery, 1972a). The reentrant twinning groves give the triangular faces of the octahedra a single-step appearance. Translucent, milky-white core and colorless, transparent rims with a glassy luster (Smith, 1976, p. 135); larger, crude, roughly pyramidal shapes consisting of aggregates of pyramidal individuals massed together in subparallel, partly radiating orientation. The outer surfaces of some of these shapes display tiny, glassy, parallel-oriented elongate to stubby-prismatic crystals (Montgomery, 1972a); occurs with other zeolites (figs. 165 and 167)

Goethite

Grossular - grayish-green, granular (Genth, 1875, p. 73); crystals to 1 mm

Hematite - micaceous and specular (D'Invilliers, 1883, p. 398)

Heulandite - white crystals to 1.2 cm

Limonite (D'Invilliers, 1883, p. 398)

Magnetite - octahedral and dodecahedral crystals (Genth, 1875, p. 38); often containing considerable pyrite (Dewey, 1891, p. 123); crystals to 5 mm (figs. 168 and 171)

Malachite - fibrous, radiating, and botryoidal coatings (Genth, 1875, p. 167; D'Invilliers, 1883, p. 398); associated with azurite

Marcasite (Eyerman, 1889, p. 6)

Mesolite - minute white tufts, radiating needles, very fine white fibers (Genth, 1875, p. 108); also as globular concretions (Sadtler, 1883, p. 357, analysis); radiating groups of acicular crystals to 5 mm (figs. 166 and 169)

Muscovite

Natrolite - radiating groups of white acicular crystals (fig. 172)

Phillipsite - clear, colorless, psuedohexagonal tabular crystals to 1 mm (Montgomery, 1972a); on altered diabase and as overgrowths on gismondine crystals (Montgomery, 1972d; Smith, 1976, p. 186); crystals to 7 mm

Pyrite - crystals and masses (D'Invilliers, 1883, p. 399); fine octahedral crystals (Eyerman, 1889, p. 6)

Quartz - small crystals (D'Invilliers, 1883, p. 400); clear to milky crystals to 4 mm

Quartz, var. chalcedony (Eyerman, 1889, p. 14)

Scapolite, var. wernerite - resinous, white crystals (D'Invilliers, 1883, p. 492)

Serpentine group - yellowish brown to dark olive green (Genth, 1875, p. 114; D'Invilliers, 1883, p. 401)

Stibnite - minute, dark, lead-gray prismatic crystals, longitudinally striated; observed with zeolites by Samuel Tyson (Genth, 1875, p. 9)

Stilbite - associated with mesolite (Brunner and Smith, 1883, p. 280); white, pale yellow, and yellow-orange; spherical crystal clusters; crystals to 4 mm (fig. 170)

Talc (D'Invilliers, 1883, p. 401)

Tenorite - massive

Thomsonite - very small, white spherical concretions of a very fine radiated structure with a waxy to pearly luster (Genth, 1875, p. 108; analysis); gray to bluish-gray mammillary crusts with a radiating, concentric structure; milky white with a clear outermost layer, associated with chabazite (Smith, 1976, p. 138); spherical concretions to 4 mm (figs. 173 and 174)

Vesuvianite - yellow and orange crystals (D'Invilliers, 1883, p. 402); yellow tetragonal prisms with pyramids in limestone, initially mistaken for garnet (Brunner and Smith, 1883, p. 280); well-developed prisms (fig. 175); rare (Eyerman, 1889, p. 20); crystals generally less than 7 mm; a specimen in the Ron Kendig collection is 3 cm (fig. 176)

Raudenbush Mine

The Raudenbush mine was on the north side of Mountain View Road, southwest of its intersection with Pennsylvania State Route 10 at about 40° 18′ 19″ latitude and 75° 56′ 03″ longitude on the USGS Reading 7.5-minute topographic quadrangle map (Appendix 1, fig. 28, location CU-2).

In 1904, the story of the discovery of the Raudenbush mine, as told by John Robarts, was published in the Transactions of the Historical Society of Berks County (Robarts, 1904). The Raudenbush mine was discovered in June 1851 by Thomas Robarts and his son, John Robarts. Thomas Robarts was born in St. Columb, Cornwall, England, on October 29, 1800. John Robarts was born on May 29, 1835, in Plymouth, Devonshire, England. In 1849, David Robarts, an older son of Thomas, emigrated to the United States. He wrote one letter aboard the ship and was never heard from again. In 1850, Thomas and John emigrated to the United States to learn what became of David. They found that David had died of "ship fever" and was buried on Staten Island. Thomas and John shortly thereafter moved to Reading.

One day in June 1851, Thomas and John set off on a walk. In the middle of a road, Thomas found a piece of magnetite. John traced the source to the road bank, where a vein of magnetite had been exposed by recent road construction. They walked to the nearest farmhouse and learned that the property was owned by Daniel Raudenbush, a butcher and farmer, who lived on Penn Street in Reading.

Thomas obtained permission from Daniel Raudenbush to begin exploration for iron ore. If sufficient ore were present, Raudenbush agreed to lease the mineral rights to Thomas for a royalty of 25 cents per ton. The Robarts secured picks, shovels,

Figure 173. Thompsonite from the Fritz Island mine, Cumru Township. Carnegie Museum of Natural History collection CM 8763.

Figure 174. Thompsonite and chabazite from the Fritz Island mine, Cumru Township. Bryn Mawr College collection Rand 2576.

Figure 175. Vesuvianite crystal from the Fritz Island mine, Cumru Township, 1 cm. Reading Public Museum collection 2004c-2-6338.

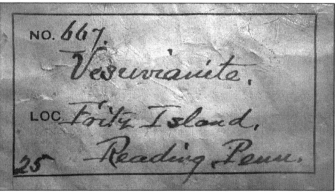

Figure 176. Vesuvianite crystals from the Fritz Island mine, Cumru Township, 7 cm with crystals to 3 cm. John Manley collection label (ca 1890s) on right. Ron Kendig collection.

Figure 177. Stilbite from the Radenbush mine, Cumru Township. The red crystals are distinctive. Reading Public Museum collection 2004c-10-699.

and wheelbarrows and began mining. The ore was near the surface, abundant, and easily mined. When Raudenbush realized the value of the deposit, he raised his royalty to 40 cents per ton. The mine was leased to Thomas Robart on July 4, 1851.

Shortly after the Robarts began work, Raudenbush leased an adjoining field to William Rowe, Sr., to prospect for iron ore. However, the ore body dipped east, and Rowe was exploring to the west. Rowe's shaft never encountered iron ore.

By June 1852, the Robarts had mined 1,000 tons of ore, which they sold for $2,000. The iron deposit was about a mile south of Eckert & Brother's Henry Clay Furnace and about a mile north of John Schwartz's Mount Penn Furnace. Isaac Eckert offered some unoccupied lots at Third and Pine Street in Reading in exchange for the mining lease, but his offer was not seriously considered. It is not known if Schwartz made an offer. On June 25, 1852, Thomas Robart transferred his lease to Reeves, Buck & Company of Phoenixville. Thomas became superintendent of the mine for Reeves, Buck & Company and its successor, the Phoenix Iron Company.

Rogers (1858, vol. 2, p. 717) described the mine: "*The vein ranges a little N. of E. Its foot-wall is a white metamorphic limestone or marble, and its hanging wall or roof a dull sea-green serpentine-like rock, which on exposure soon crumbles down like ordinary shale. The vein, dipping 36° S., is followed by a slope 280 feet beneath the surface. At the bottom, gangways are driven 200 feet W. and 400 feet E., to a fault cutting out the vein. A higher level, 160 feet from the surface, is driven 300 feet E. Like all others, this vein is exceedingly variable; while wholly or almost entirely absent in places, in others it has been found 30 feet thick. Its average bulk will not exceed 12 feet. The gangue-stone of the ore is a light-blue rotten limestone, from which the ore is scarcely distinguishable, except by its greater weight and deeper tint. Of the entire ground wrought, about one-half the material is sufficiently rich in iron for the furnace; the remaining rubbish is used as stopping in the old workings.*"

The mine was abandoned before 1883. D'Invilliers (1883, p. 342-343) found a slope and two shallow shafts on the property. He observed diabase along with some light grey to white limestone in a 50-foot deep shaft. When the property was visited by Bever and Liddicoat (1954, p. 73) in 1953, they noted: "*Virtually all traces of mining activity have now disappeared. That a mine existed here is evidenced now only by some shallow soil and rubble filled depressions. No ore fragments or outcrops are present.*"

Rogers (1858, vol. 2, p. 716-717) stated that the Phoenix Iron Company mined 5,000 tons of ore per year from the Radenbush mine. Rose (1970, p.10) estimated that total production was less than 100,000 tons of ore.

Minerals

From D'Invilliers (1883, p. 395-400) unless noted.

Chlorite group

Magnetite - octahedral crystals

Pyroxene

Pyrrhotite

Stilbite - radiating crystals in veins in diabase (Genth, 1875, p. 109; Smith, 1885, p. 414, analysis); red crystals to 4 mm (fig. 177)

Opposite Fritz Island Locality

The locality called "opposite Fritz Island" by Gordon (1922, p. 153) is on the east bank of the Schuylkill River at about 40° 17' 58" latitude and 75° 17' 58" longitude on the USGS Reading 7.5-minute topographic quadrangle map (Appendix 1, fig. 28, location CU-3). D'Invilliers (1883, p. 396) described the locality as a diabase outcrop on the east side of the Schuylkill River at the dam opposite Fritz Island. D'Invilliers reported the occurrence of feldspar.

D'Invilliers (1883, p. 337), in his description of the Fritz Island mine, stated: "*Ore has likewise been found on east bank of* [the Schuylkill] *river near tow-path, on High's property, in direct extension with these* [Fritz Island mine] *gangways.*" Spencer (1908, p. 39) noted that the iron ore was discovered near the towpath on the east side of the Schuylkill River in a prospect shaft.

Fehr and O'Rourke Stone Company Quarry

The Fehr and O'Rourke Stone Company quarry (fig. 178) was northwest of the intersection of Pennsylvania State Route 724 (Lancaster Pike) and Museum Road at about 40° 18' 26" latitude and 75° 58' 36" longitude on the USGS Reading 7.5-minute topographic quadrangle map (Appendix 1, fig. 28, location CU-5). The quarry was in the Richland Formation.

The quarry was opened on the 83-acre Jeremiah S. Hill farm to produce building stone and lime. The quarry and a lime kiln are shown on the J. Hill property in the 1876 atlas. In 1876, Hill operated four kilns with a capacity of 50,000 bushels per year.

James R. Trout is known to have operated the quarry as early as 1897 (Mine and Quarry News Bureau, 1897). Trout operated three lime kilns. One had a capacity of 1,800 bushels, and two had a capacity of 1,200 bushels. He burned 30 to 50 kiln loads per year and sold coal-burned lime for 7.5 cents per bushel. The quarry produced stone for the lime kilns, building stone for masonry work, and crushed stone for road work. During the busy season, as many as 15 men were employed in the quarry (*Reading Eagle*, December 3, 1904).

Figure 178. The Fehr and O'Rourke Stone Company quarry, Cumru Township, 2015.

In 1902, the Cumru Township supervisors purchased a new Climax stone crusher, rated at 100 tons per day, to crush stone for road work. Charles Bossler and Mr. Stafford, supervisors, had the crusher installed in the Trout quarry. John S. Gring used his traction engine to power the crusher (*Reading Eagle*, April 7, 1902, and August 6, 1903).

In 1905, the farm was sold to the Knights of the Maccabees, a fraternal organization, for a home for the aged and infirm, which was called the Maccabee Home. Trout continued to operate the quarry, and in 1910, he purchased a steam drill for use in the quarry (*Reading Eagle*, October 14, 1905, and March 5, 1910). Limestone and lime producers listed by Hice (1911) that were operating in 1911 included James R. Trout in Shillington.

Horace Fehr and James O'Rourke became partners in a contracting business in the late 1890s. They were engaged in general contracting, paving, and quarrying, as well as the feed and coal business (fig. 179). In 1919, a state charter was granted to the Fehr and O'Rourke Stone Company, Inc. of Reading to quarry and market stone and other minerals. The capital stock was $25,000. The equal partners in the venture were Horace Fehr, James O'Rourke, Allen E. Hildebrand of Reading, and Jesse S. Hildebrand of Shillington. Jesse S. Hildebrand served as the manager of the company. The Trout quarry was acquired by the company around 1923, and they operated the quarry from about 1923 to about 1936. In March 1930, all of the Fehr and O'Rourke business concerns were merged with the Davis Coal and Supply Company and incorporated as the Berks Products Company. James O'Rourke was president of the company from 1930 until his death in 1938 (*Reading Eagle*, April 29, 1919; February 6, 1923; and January 21, 1938).

In 1932, Andrew Larson was the superintendent of the quarry, and Walter H. Fehr was the general superintendent. Production was 100,000 tons per year. Stripping and loading was done by two P & H steam shovels. Stone was moved from the quarry to the crusher by trucks. The primary crusher was a Buchanon crusher, and the secondary crushers were Champion and Kennedy-Van Saun crushers (Pit and Quarry, 1932). The local residents continually complained about blasting in the quarry (*Reading Eagle*, November 15, 1927; September 30, 1936).

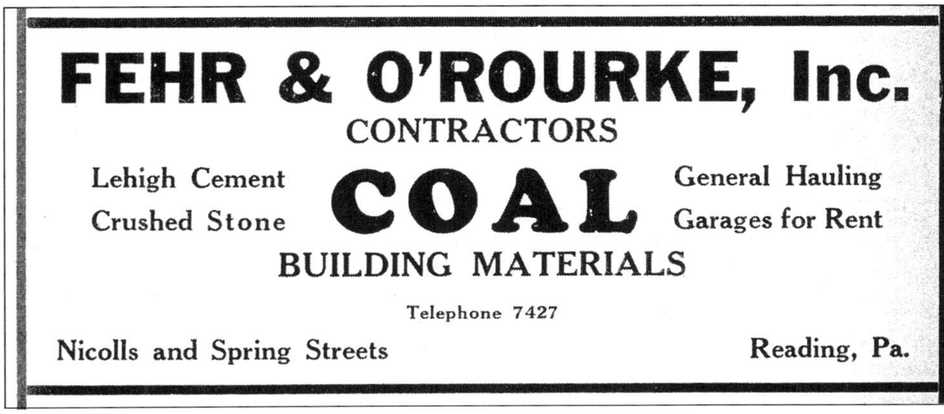

Miller (1934, p. 218) reported that the beds were massive and relatively flat. The stone varied in color from gray to dark blue and varied in composition from dolomite to low magnesian limestone. Calcite veins were numerous in parts of the quarry. In 1934, the quarry was 500 feet long, 350 feet wide, and had a face 65 feet high. Daily production was

Figure 179. Advertisement for Fehr and O'Rourke, Inc., 1930. Boyd's Reading City Directory.

Figure 180. Andradite crystal from the U.S. Route 222 roadcut locality, Cumru Township, 1.2 cm. Sloto collection 3229A.

Figure 181. Andradite crystals from the U.S. Route 222 roadcut locality, Cumru Township, 1.1 cm. Sloto collection 3229C.

Figure 183. Hematite from the U.S. Route 222 roadcut locality, Cumru Township, 5.5 cm. Sloto collection 3304.

Figure 182. Flattened quartz crystals from the U.S. Route 222 roadcut locality, Cumru Township. Carnegie Museum of Natural History collection CM 32349 (Brookmeyer 3821).

Figure 184. Andradite from Shiloh Hills, Cumru Township. Ron Kendig collection.

1,000 tons of crushed stone. The quarry eventually grew to a size of about 650 feet by 450 feet.

Around 1994, a 9-acre parcel containing the abandoned quarry was converted to soccer fields (fig, 178). The site, owned by the Wyomissing park system, was leased to the Wyomissing Soccer Club (*Reading Eagle*, August 7, 2001).

U.S. Route 222 Roadcut Locality

The U.S. Route 222 roadcut locality was open during construction of the Route 222 bypass near Gouglersville about 2003. The locality is on the USGS Sinking Spring 7.5-minute topographic quadrangle map, The locality is often called Shillington. Minerals recovered during road construction are listed below. Similar minerals on an identical matrix in older collections are labeled "Shiloh Hills" (fig. 184), which is a residential development northwest of Gouglersville. Those specimens may have been collected during construction of the development.

Minerals

Actinolite

Andradite - sharp, green-brown crystals on an actinolite matrix; crystals to 2.5 cm (figs. 180 and 181)

Epidote - small green crystals

Hematite - plates (fig. 183)

Quartz - crystals to 1.4 cm; small flattened crystals (fig. 182)

Tremolite - reported

Schlegel's Farm Locality

Schlegel's Farm was southwest of the intersection of U.S. Route 422 and U.S. Business Route 222 at about 40° 19′ 30″ latitude and 75° 56′ 13″ longitude on the USGS Reading 7.5-minute topographic quadrangle map (Appendix 1, fig. 27, location CU-8). The farm was underlain by the Buffalo Springs Formation. The farm is now Schlegel Park. The name was spelled Slegel by D'Invilliers (1883, p. 398), who reported botryoidal and dendritic limonite or goethite (figs. 185 and 186) and pseudomorphs after pyrite on S. Slegel's farm.

Schuylkill Copper Mining Company Mine

The Schuylkill Copper Mining Company mine was between Pennsylvania State Route 10 and the Schuylkill River, opposite Fritz island at about 40° 18′ 08″ latitude and 75° 55′ 31″ longitude on the USGS Reading 7.5-minute topographic quadrangle map (Appendix 1, fig. 28, location CU-4). The mine was in diabase.

The Schuylkill Copper Mining Company was incorporated in New York on May 13, 1864, with $500,000 in capital stock. Its office was at 54 William Street in New York City. The officers of the company were Gustavus A. Sacchi, president, and William Kenneys, secretary. In addition to these two men, the

Figure 185. Limonite from Schlegel's Farm, Cumru Township, 5.4 cm. Reading Public Museum collection 2000C-027-107.

Figure 186. Limonite from Schlegel's Farm, Cumru Township, 6.5 cm. Reading Public Museum collection 2000c-27-003.

board of trustees included Abel Easton, E.G. Burling, and George E. Ring (Schuylkill Copper Mining Company, 1864).

The Schuylkill Copper Mining Company secured a 100-year lease on a "copper mine" on the banks of the old Union Canal about 2 miles south of Reading in Cumru Township. The company's prospectus (Schuylkill Copper Mining Company, 1864) contained a report by a geologist named James T. Hodge. The mine was on a farm, which was partly on Fritz Island. The ore vein was directly across from the island at N. 85 E. west of the canal, the vein ran up a steep hill. The mine was located 250 feet south of the Fritz Island mine. According to the prospectus, the company was starting an adit on the vein from the canal side and was sinking a shaft to strike the ore at depth of 60 feet. The shaft was located 150 feet from the canal.

The company's prospectus stated that a vein of "pyritous copper ore" was exposed at the foot of a hill 4 feet below limestone at the edge of the canal. The vein was 1 foot wide at the outcrop and was estimated to be 200 feet long. An assay of the ore by John Torry stated that it was 23.3 percent metallic copper worth $60 per ton. The prospectus stated: *"The quantity appears to be inexhaustable."* The mine captain, Z. Williams, stated that there was a *"probability of the ore being arentiferous"* and that the ore contained a *"notable"* proportion of silver and a trace of gold. These statements and the assay were presented in order to sell the company's stock.

Figure 187. The Berks County Almshouse farm, Cumru Township.

Almshouse Quarry

The Almshouse quarry was located on the property of the Berks County Almshouse, which was northeast of the intersection of U.S. Business Route 222 and Pennsylvania State Route 724 at about 40° 18′ 05″ latitude and 75° 57′ 35″ longitude on the USGS Reading 7.5-minute topographic quadrangle map (Appendix 1, fig. 28, location CU-6). In 1957, the old almshouse complex was demolished and replaced with apartment housing, a shopping center, and the Governor Mifflin School. The quarry was in the Zooks Corner Formation.

The Berks County Almshouse, commonly called the poorhouse, was opened in 1826, after the Commonwealth of Pennsylvania enacted a law requiring all counties to have one. The almshouse stood along Pennsylvania State Route 724 on land that once belonged to Governor Thomas Mifflin. Mifflin's 417.5-acre Angelica Farm (fig. 187) was purchased at public sale for $16,690 ($40 per acre). The county later added an additional 116 acres. In 1826, the first resident moved in, and 130 people were admitted during the first year. It was a thriving community with a creamery, bakery, and slaughterhouse. Every physically capable adult that lived there was required to work. In 1878, the almshouse population peaked at 1,630. In 1932, the new Berks County prison and a tuberculoses sanatorium were opened on the tract. The almshouse closed in 1952 and was razed in 1957 (*Reading Eagle*, October 6, 1957).

A sandstone quarry was located near the tuberculosis sanatorium. It is not known when the quarry was opened, but it was in operation in the mid 1800s. An article in the September 28, 1871, Reading Eagle stated that a workman was injured while quarrying stone for the new hospital building. In 1934, the Berks County Prison furnished a motor from an old dismantled truck to run the crusher at the quarry. Under an agreement with the sanatorium, crushed stone was used to repair roads around the prison and sanatorium. The prison furnished the labor for crushing stone (*Reading Eagle*, February 8, 1939).

Figure 188. Goethite from the Berks County Almshouse quarry, Cumru Township, 5.5 cm. Reading Public Museum collection 2000C-027-098.

Mount Penn Furnace

The Mount Penn cold-blast charcoal furnace was built in 1825 by Simon Seyfert (also spelled Seifert) and John Schwartz and run under the firm name Seyfert & Schwartz. The furnace was built on the west side of the Schuylkill River and was operated in conjunction with Gibralter Forge. In 1828-30, the Mount Penn Furnace employed 220 workman, had 120 horses, and used 15,000 cords of wood to produce 1,700 tons of pig iron and 500 tons of castings. In 1835, the partnership dissolved with Schwartz taking ownership of the furnace, and Seyfert taking ownership of the forge.

In 1856, the Mount Penn Furnace was owned by Shatler & Kauffman of Reading and made 500 tons of iron (Lesley, 1859). About 1860, the furnace was acquired by William M. Kauffman & Company. Kauffman had served as a clerk at the Moselem and Leesport Furnaces. D'Invilliers (1883) indicated that the property was eventually bought back by William K. Shalter, who ran it for several years. The furnace was out of blast in 1875 (American Iron and Steel Association, 1876) and was abandoned in 1883 (Montgomery, 1884).

DISTRICT TOWNSHIP

Early Iron Mines

James Thompson purchased land containing an iron-ore deposit around 1782. He furnished ore to the District Furnace, which was located nearby. George and John Oyster (also spelled Eyster) also furnished ore from their mine to the District Furnace prior to 1820 (*Reading Eagle*, September 30, 1917).

On April 9, 1813, James Thompson's "iron lands" were sold to Reuben Trexler. John J. Thompson, brother of James, also owned tracts of iron-ore land in District Township prior to 1820. They were sold to ironmaster Thomas Bull Smith, who was part owner of Joanna Furnace. The properties were probably too far from Smith's furnace because he sold them to Reuben Trexler after only 5 months of ownership.

Mica Prospect Number 2

Mica prospect Number 2 of Buckwalter (1954) is east of the intersection and between Conrad and Forgedale Roads, in Landis Store at about 40° 25' 37" latitude and 75° 39' 26" longitude on the USGS Manatawny 7.5-minute topographic quadrangle map (Appendix 1, fig. 18, location DS-1). The prospect is 300 to 400 feet east of and behind the Landis Store School on an east-facing slope. It is a few hundred feet from both Conrad and Forgedale Roads. Mica is found in books up to 1 inch thick and 2 inches in diameter. The mica resembles muscovite in color, but has the optical properties of biotite. It is found in a coarse, buff-weathered pegmatite composed of quartz, orthoclase, and microcline (Buckwalter, 1954, p. 19-20).

Hoffman Prospect

The Hoffman prospect was west of Landis Store Road, and southwest of its intersection with Deer Lane, 0.8 mile south of Landis Store at about 40° 24' 58" latitude and 75° 39' 50" longitude on the USGS Manatawny 7.5-minute topographic quadrangle map (Appendix 1, fig. 18, location DS-2). The 1882 topographic map shows the locations of two shafts. The prospect was near the contact of the Hardyston Formation and hornblende gneiss. D. Hoffman sunk several shafts prospecting for iron ore on the hill west of Landis Store Road. The shafts were sunk to a depth of 60 feet through 4 to 5 feet of sandstone and conglomerate and into a decomposed micaceous gneiss. The ore was magnetite, but very little was encountered (D'Invilliers, 1883, p. 108).

Peter Smith Farm Graphite Prospect

The Peter Smith farm graphite prospect was located 1.5 miles southwest of Rittenhouse Gap. Several pits were dug, and several shafts were sunk while prospecting for graphite on the farm. Several different parties held leases on the property at various times and conducted investigations to determine the character and extent of the graphite ore; however, no attempts were made to open a mine. Some of the prospect shafts were 35 feet deep (Miller, 1912a, p. 114-115). In 1900, the East Penn Graphite Company leased the property (*Reading Eagle*, April 14, 1900).

The graphite forms part of a decomposed graphitic gneiss in which the feldspar has almost entirely weathered to kaolin. Both orthoclase and plagioclase occur in the rock with much quartz and smaller amounts of graphite, biotite, and pyrite. The graphite flakes are about 1/8 inch in diameter on average, and are roughly parallel to the banding of the gneiss. Pegmatites, composed of quartz, feldspar, and large flakes of graphite, some as large as one inch in diameter, are abundant. In some of the pegmatite, the graphite is enclosed within the quartz, and the flakes are sharply bent as though first formed about a quartz crystal that later grew by addition of more quartz and enclosed the graphite. Miller estimated that the average graphite content was more than 3 percent, although no analyses were available. There was enough biotite present to cause considerable difficulty in processing the graphite (Miller, 1912a, p. 114-115).

District (German) Furnace

District Furnace, also called German Furnace, was built on Pine Creek, probably by John or Jacob Lesher prior to 1784. It was listed as the German Furnace on Samuel Potts' 1789 list of active furnaces. In 1791, John Lesher sold a 1,582-acre property in District Township containing a furnace, gristmill, and sawmill to his son Jacob. The furnace was rented by Jacob Focht until December 1796. Montgomery (1884, p. 71) stated that Jacob Lesher abandoned the furnace about 1797 because of difficulty smelting the local ores. D'Invilliers (1883, p. 231), however, stated that the furnace was abandoned about 1814. Adjustment of the District-Pike Township boundary now places the furnace, which historically had been in District Township, in Pike Township.

Figure 189. The Greshville quarry, Douglass Township, 2015.

DOUGLASS TOWNSHIP

Greshville Quarry

The Greshville quarry (fig. 189) was south of Pennsylvania State Route 562, between Greshville Road and Farmington Avenue, about 0.3 mile northeast of Greshville at about 40° 19′ 11″ latitude and 75° 39′ 31″ longitude on the USGS Boyertown 7.5-minute topographic quadrangle map (Appendix 1, fig. 34, location D-2). The 1876 atlas shows the limestone quarry at Greshville (fig. 190). The quarry was in the Leithsville Formation.

Greshville was originally called Limestone. When a post office was established at the village inn and store, the name was changed to Greshville after innkeeper Adam Gresh. The limestone quarry and kilns east of Greshville are shown on the 1862 map. The quarry was known as the Jesse Bechtel quarry and Gresh and Bechtel's quarry (ca. 1870). The quarry was in operation as late as 1925 (*Reading Eagle*, September 5 and 21, 1870, and June 1, 1925).

In 1882, the quarry was owned by J. Livingood of Reading and leased by Levi and David Gresh. It was called Levi Gresh's quarry by D'Invilliers (1883, p. 153). According to D'Invilliers: "*It is one of the largest and best quarries in this part of the county, though its stone is very hard and rather expensive to quarry, having been somewhat metamorphosed and crystallized by the proximity of the trap dyke.*" There were 60 to 80 feet of blue and white limestone exposed in 1882. The strata dipped S. 50° E. 54° on the south side of the quarry and N. 40° W. 60° on the north side. Several adits were driven from the quarry into the nearby diabase in search of iron ore without success.

Levi and David Gresh produced about 30,000 bushels of lime per year in the kilns at the foot of the hill. In 1883, the selling price was 12 cents per bushel. Before the railroad was extended to Boyertown, 75,000 bushels of lime per year were sold for 16 cents per bushel (D'Invilliers, 1883, p. 153-154). In 1892, Henry Livingood of Reading began burning lime in his father's kilns (*Reading Eagle*, July 25, 1892). An analysis of the rock was was published by D'Invilliers (1883, p. 151): 87.3 percent calcium carbonate, 8.2 percent magnesium carbonate, and 8.5 percent silica.

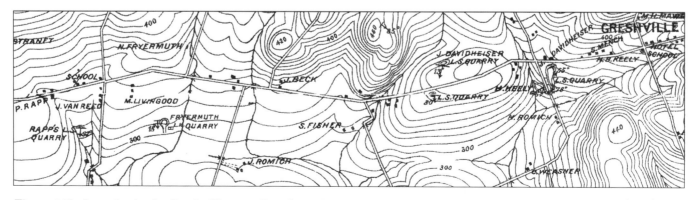

Figure 190. Quarries in the Greshville area, Douglass Township, 1882. From Pennsylvania Geological Survey (1883).

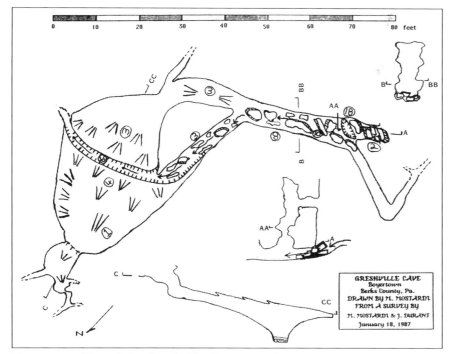

Figure 191. Map of the Greshville Cave, Douglass Township. From Mostardi and Durant (1991, p. 34). Courtesy of the Mid-Appalachan Region of the National Speleological Society.

The Greshville cave (fig. 191) was located beneath the face of the Greshville quarry. The cave was approximately 140 feet long. A passage 45 feet long led to a room about 25 feet by 50 feet with a ceiling 6 to 10 feet high. During heavy rains, the cave completely filled with water (Mostardi and Durant, 1991, p. 32-35).

J. Davidheiser Quarry

The Davidheiser quarry (fig. 190) was on the north side of Pennsylvania State Route 562, near its intersection with Wilcox Road at about 40° 18′ 59″ latitude and 75° 40′ 44″ longitude on the USGS Boyertown 7.5-minute topographic quadrangle map (Appendix 1, fig. 34, location D-3). The quarry was in the Leithsville Formation.

In 1882, the Davidheiser quarry was owned and worked by J. Davidheiser. His largest quarry was idle when visited by D'Invilliers that year. The quarry exposed about 25 to 30 feet of white and blue magnesian limestone, which dipped S. 40° E. 55° at the north end of the quarry. At the south end of the quarry, a slaty limestone dipped S. 42° W. 70° (D'Invilliers, 1883, p. 154-155). D'Invilliers presented an analysis of the limestone: 55.1 percent calcium carbonate, 38.1 percent magnesium carbonate, and 6.3 percent silica.

J. Davidheiser also opened two small quarries further west on the summit of a hill, one on the north side of the road in a field, and the other near the south side of road. In both of these quarries, a gray magnesian limestone was exposed. The limestone was slaty in the southernmost quarry (D'Invilliers, 1883, p. 154).

Keely Quarry

The flooded Keely quarry (fig. 190) is southwest of the intersection of Pennsylvania State Route 562 and Douglass Drive, 2.2 miles southwest of the center of Boyertown at 40° 18′ 59″ latitude and 75° 40′ 16″ longitude on the USGS Boyertown 7.5-minute topographic quadrangle map (Appendix 1, fig. 34, location D-4). The quarry was in the Leithsville Formation.

This quarry was owned by Henry Keely. The 1877 Berks County business directory listed Henry B. Keely as a lime producer (Phillips, 1877). The quarry was located close to Davidheiser's quarry on the west side of Douglass Drive. A branch of Ironstone Creek ran through its center. It was abandoned and flooded when visited by D'Invilliers in 1882. About 15 feet of blue, quartzose limestone with streaks of white calcite were exposed. The limestone was very hard and free from cleavage. The dips varied from S. 15° E. to S. 4° W. 40° to 64° (D'Invilliers, 1883, p. 155). D'Invilliers presented an analysis of the stone: 54.6 percent calcium carbonate, 41.9 percent magnesium carbonate, and 3.4 percent silica.

Wren Mine

The Wren mine was located between Mill Street and Pennsylvania State Route 562, just south of the boundary between Douglas Township and Boyertown at about 40° 19′ 12″ latitude and 75° 39′ 01″ longitude on the USGS Boyertown 7.5-minute topographic quadrangle map (Appendix 1, fig. 34, location D-1). The mine was in limestone fanglomerate near the contact with diabase.

The Wren shaft and tunnel are shown on the 1882 topographic map (fig. 145). The Wren mine was behind Major Wren's house about 900 feet southwest of the Brower mine, which was in Colebrookdale Township. Major Wren was an officer in the Civil War. In the early 1870s, he moved to Boyertown and began prospecting for iron ore along Ironstone Creek. He sunk a shaft and drove a short tunnel into hornblendic rock (D'Invilliers, 1883, p. 333). Major Wren's mining ventures were unsuccessful.

Pottstown Trap Rock Quarry

The Pottstown Trap Rock quarry (fig. 192) is on Rattlesnake Hill, northwest of the intersection of Squirrel Hollow and Trap Rock Roads, 1.75 miles east of Douglassville at 40° 15′ 31″

latitude and 75° 41′ 30″ longitude on the USGS Boyertown 7.5-minute topographic quadrangle map (Appendix 1, fig. 32, location D-5). The quarry exposes about 325 feet of upper Brunswick (Passaic) Formation hornfels (Olsen and Schlische, 1989a, p. 78). The quarry passed through a series of owners over the years and has been known as the Tilli quarry, the Mensch quarry, the Stowe Trap Rock quarry, and the Douglassvile quarry of the Pottstown Trap Rock Company. It is currently (2016) owned by the H&K Group and is known as the Pottstown Trap Rock Douglassville Quarry.

In 1900, the Philadelphia and Reading Railroad constructed a 2-mile siding to a new quarry on Rattlesnake Hill, which was opened on a farm owned by Squire Mauger. A crusher and plant were erected by James McQuade of Reading. In 1901, the quarry was operated by James Tilli, a Philadelphia banker, who was assassinated in 1905. The 114-acre farm containing the Tilli quarry was then sold to Caterina Romano for $10,000 to settle Tilli's estate (*Reading Eagle*, December 12, 1900; November 14, 1901; December 2, 1905; and July 24, 1914).

The quarry was next known as the Mensch quarry when it was owned by Samuel H. Mensch, who was a former warden of the Berks County prison. He also was a hotel keeper in Oley, Norristown, and Lebanon. Mensch operated one of the largest stone crushers in Pennsylvania. The crusher, known as the Champion No. 20, was installed at a cost of $75,000. It produced 35 carloads of crushed stone per day. Mensch quarried stone on the east side of the road, and the stone was moved from the quarry to the crusher on the west side of the road by 15 horses and carts that passed through a tunnel under the road. The stone was shipped by rail from the plant to the main line of the Reading Railroad. Large quantities of crushed stone were shipped to New York, New Jersey, and Philadelphia. The quarry employed 75 men, many of whom were Italian immigrants, in two shifts (*Reading Eagle*, January 21 and November 22, 1913, and July 12, 1914).

In November of 1913, Mensch was sued by William H. Pearce of Philadelphia for a 10 percent ($10,000) commission on the $100,000 sale of Mensch's quarry and 114-acre farm. According to Pearce, he and Mensch entered into a verbal agreement. Pearce allegedly sold the quarry to William J. Wilson of Philadelphia. Mensch said that he had never made such an agreement and refused to pay. The jury rendered a verdict in Mensch's favor (*Reading Eagle*, November 22, 1913, and May 20, 1914).

The quarry was next operated by the Storb Crushed Stone Company, which was owned by Horace Storb. Storb acquired the quarry around 1915. That same year, a 18 foot by 24 foot. two story building, used as a commissary by the company, was destroyed by fire. The loss was estimated at $300. In 1919, the plant caught fire. The cement and frame building containing five storage bins of 500-tons capacity each, the dynamos, crusher, and practically all of the equipment burned. The plant

Figure 192. The H&K Group Pottstown Trap Rock Douglassville Quarry, Douglass Township, 2015.

was completely gutted. Damage was estimated at $15,000 (*Reading Eagle*, March 30, 1915, and December 4, 1919).

By 1919, the horses and carts were replaced by a small narrow-gage rail system to move stone from the quarry to the crusher. The quarry employed 50 men, and the plant produced 30 car loads of crushed stone per day. In 1929, nine railroad cars broke loose. The runaway cars collided with an automobile where the railroad siding crossed the Benjamin Franklin Highway. In 1946, the Storb quarry was sold for back taxes (*Reading Eagle*, December 4, 1919; June 29, 1929; and March 13, 1946).

The quarry was operated by the Stowe Trap Rock Company from about 1946 to 1958. Pottstown Trap Rock Quarry, Inc. purchased the quarry in 1958 (*Reading Eagle*, August 30, 1999), and it became known as the Douglassvile quarry of the Pottstown Trap Rock Company. The Pottstown Trap Rock Company installed a new crusher and screening plant in 1963 (Kerr, 1963). In 1995, The quarry was purchased by Haines and Kibblehouse, Inc., now know as the H&K Group.

Little Oley Reibeckite Locality

The Little Oley reibeckite, var. crocidolite, locality was east of the intersection of Colebrookdale and Englesville Road, in the village of Colbrookdale, formerly called Little Oley, at about 40° 18′ 34″ latitude and 75° 38′ 57″ longitude on the USGS Boyertown 7.5-minute topographic quadrangle map (Appendix 1, fig. 34, location D-6). The locality was underlain by the Brunswick Formation.

The Little Oley locality was described by Bliss (1913) as locality number 11, which was 0.25 mile northeast of Little Oley. Bliss described the crocidolite as a blue amphibole in a Triassic sandstone. It commonly occurred disseminated throughout the rock as one of the chief constituents and as a thick coating on weathered surfaces, thereby giving the appearance of a weathering product. The coating, which varies in color from a deep blue or almost black to an ultramarine blue, was up to 5 mm thick. It was readily removed by scraping with a knife and crushed to an earthy powder. The luster was normally dull, but was occasionally silky. In addition to its occurrence as a coating, the amphibole occurred abundantly scattered throughout the mass of the rock in small, dark blue, irregular grains. These grains, which were less than 0.2 mm in diameter, had a dull luster, a hardness of 5 to 6, and showed neither crystal outline nor cleavage faces.

EARL TOWNSHIP

Iron Mines

Kauffman and Spang Iron Mine

The Kauffman and Spang iron mine was on the north side of Furnace Hill, northwest of the intersection of Woodchop-

Figure 193. Wavellite from Earl Township. Reading Public Museum collection 2005c-1-376.

pertown and Furnace Run Roads, 1.5 miles southwest of Shanesville at about 40° 21′ 17″ latitude and 75° 43′ 25″ longitude on the USGS Boyertown 7.5-minute topographic quadrangle map (Appendix 1, fig. 33, location E-1). It was across Furnace Road from the Rolling Hills (formerly Colebrookdale) landfill. The mine, in the Hardyston Formation, also was known as Spang's mine.

There were several open cuts on Furnace Hill, and an adit was driven in from the road to intersect the ore body. The Reading Eagle (July 19, 1912) reported that ore was mined from 1874 to 1876 and sent to the Bechtelsville Furnace; however, the Bechtelsville Furnace was not in blast until 1880. The 1880 Census (Pumpelly, 1886, p. 962) listed Spang and Kauffman as the operators of the mine. Production between July 1, 1879, and June 30, 1880, was 4,700 tons of limonite.

The ore was hard, compact, and crystallized with a bright metallic luster. In 1883, D'Invilliers (1883, p. 359) reported: "*The mine has been long abandoned, and the old workings being all closed up, nothing could be learned of the position of the ore.*" The mine was acquired by the Manatawny Mining Company, which was organized in 1906 (Poor's Manual Company, 1916), and was later acquired by the Manatawny Bessemer Ore Company.

The Manatawny Bessemer Ore Company was incorporated in June 1910, as the successor to the Manatawny Mining Company. The company was authorized to sell $5,000,000 in common stock (fig. 194) and $2,000,000 in seven percent preferred stock. Shares were sold for $10 each, and $440,440, raised from the sale of preferred stock, was used to purchase and install machinery, develop the property, and pay for options held on other properties. The company controlled 1,035 acres of land by ownership and lease (Poor's Manual Company, 1916).

The president of the Manatawny Bessemer Ore Company was Charles M. Allen of Bayonne, New Jersey. The vice president, general manager, and purchasing agent was M.J. Person. Stephen Robinson, Jr. of Audubon, New Jersey, was the secretary and treasurer. Howard Callingham of Philadelphia served

Figure 194. Stock certificate of the Manatawny Bessemer Ore Company, Earl Township, 1916.

as the auditor. Fred C. Simmons was the attorney for the company. Company directors included Allen, Person, Robinson, R.F. Huthmacher of Bethlehem, and D.J. Driscoll of Reading. Edwin R. Eaton was the chief engineer from 1912 to 1914. The company office was at 603 Morris Building in Philadelphia (Poor's Manual Company, 1916).

In 1911, the Reading Eagle (December 2, 1911) reported that the Manatawny Bessemer Ore Company spent $70,000 for mining equipment. They purchased a 70-ton steam shovel, two 60-ton steam shovels, four locomotives, 118 dump cars, steam drills, portable boilers, hoisting engines, and steel cables. About 1912, the adit was about 100 feet long, and a shaft 40 feet deep at the end of the adit was sunk into the ore body. Borings were made to determine the size of the ore body. Brown and Ehrenfield (1913, p. 79-81) noted that the ore body occurred at the contact of the Hardyston Formation and a highly serpentinized limestone. By July 1912, there were seven shafts on the property ranging from 23 to 70 feet deep, and the adit had been extended to 360 feet (*Reading Eagle*, July 19, 1912).

In 1914, Person and Simmons hired L.J. Campbell, a consultant specializing in iron-ore properties, to help them finance the company. After a meeting in Pittsburgh, Campbell brought in a mining expert to examine the property and produce a report that would induce investors to purchase stock. Campbell billed the company for $350, but they did not pay the bill. The company's position was that the by-laws did not permit the vice president or attorney to incur debt, and, therefore, they were unable to pay the bill. Campbell sued the company and was awarded $365.50. The company appealed the decision and lost (Schaffer and Weimer, 1916).

The Manatawny Bessemer Ore Company was bankrupt by 1917. In that year, Mayer Pollock, a Pottstown scrap dealer, removed the steam engine and other equipment. The company's properties were sold at sheriff's sale in February 1924 (*Reading Eagle*, July 27, 1917, and February 20, 1924).

The Manatawny Bessemer Ore Company formed a separate company, the Manatawny Railroad Company, to construct a railroad line from Stowe to the mine. Both companies shared an office at the West End Trust Building in Philadelphia, which was moved to 603 Norris Building in 1913 (*The Iron Age*, August 7, 1913, p. 305).

A Visit to Ore Mines in Earl Township

Reading Eagle, July 19, 1912

"John R. Kauuffman, of Earl Township, is in charge of mining. E.R. Eaton, mining engineer, of New York, directs the drilling and takes samples of the ore every two feet of excavation. Three eight-hour shifts are constantly at work, 25 men being employed.

Three steam shovels, four "dinkies," 120 small cars, three boilers, two steam engines, three gasoline engines, three rock drills and two large oil well drills are used in the operations. A pumping station has been erected at the western base of the mountain. The coal used in the engines on the top is hoisted by means of an aerial tramway. A telephone line connects at the exchange at Yellow House, reaching the office building, situated on the northeastern base of the mountain. A four-passenger automobile is used by the Superintendent and the engineer in making rounds from one side of the mountain to the other, a distance of about a mile; and in furnishing supplies.

Accompanied by the Superintendent and the engineer a recent visitor made a journey through the seven shafts thus far worked. First we went through what is called Drift No. 1, extending under ground a distance of 107 feet and branching out into three drifts 30 feet each. In one of the branching drifts we halted, having reached a 40-foot shaft. A rope lowered by a crank was the means of getting to the bottom. The engineer put his right foot into a loop and was gently lowered. Next was my time. It required some persuasion before I mustered enough courage to hang onto a rope with 40 feet of space below me. At the end of the 40 feet the only foothold was a 12-inch plank, underneath that an unknown depth of water. By the light of two tallow candles we chipped off ore specimens all along the gangways of the drifts as well as in the shaft. There appeared to be one solid mass of ore, said to be red hematite.

Our next trip was down a 60-foot shaft by means of a ladder. At the bottom of this shaft is a gasoline engine pumping water to the surface. Here, too, we chipped off iron rock.

Jere's tunnel was the next place visited. This is so named after the workman who started it. This tunnel goes underground 360 feet. Into it is laid a track and on cars the rock is being removed. Drilling on this tunnel is still in operation.

Shaft No. 1 is 70 feet deep and at the bottom extends two drifts. Close to this shaft is No. 2, 63 feet in depth. From the No. 4, which is 45 feet deep, five 56-feet [SIC] drifts branch out. The sixth shaft is only 23 feet deep and No. 7 is 68 feet deep. In the last named there is no red iron ore rock but a black spongy material said to be just as rich in ore."

 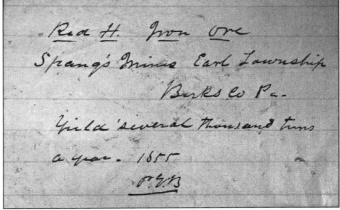

Figure 195. Hematite from the Kauffman and Spang mine, Earl Township, 9.2 cm. Carnegie Museum of Natural History collection ANSP 1513.

EARL TOWNSHIP

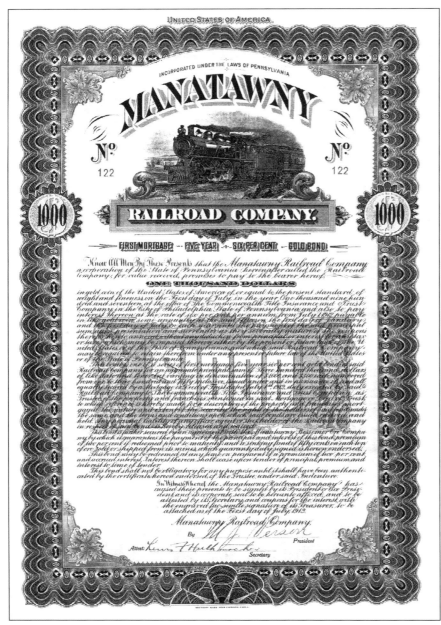

Figure 196. Bond issued by the Manatawny Railroad Company, Earl Township, 1912. The railroad was proposed to carry ore from the Manatawny Bessemer Ore Company mine to the main line of the Philadelphia and Reading Railroad.

An application was made by the Manatawny Railroad Company to the Pennsylvania Public Service Commission on February 17, 1914, (application docket No. 34-1914) for a certificate of public conveyance. The application was approved on July 13, 1914. The railroad company sold stock plus $300,000 worth of $1,000 coupon bonds at 6 percent interest payable in gold coin (fig. 196). Bonds issued in 1912 had a 1917 maturity date. The last coupon redeemed from the bond in figure 196 was dated January 1915; the railroad likely was bankrupt by then.

Contracts were awarded to the Highly Construction Company of Pottstown to build a connection to the Philadelphia and Reading Railroad at Stowe. The railroad was to run from Stowe 2 miles west to Douglassville and then 8 miles north through Amityville to the Manatawny Bessemer Ore Company mine. Grading was completed on 3 miles, but no track was laid. Two planned 150-foot bridges were never constructed (*Railway Age Gazzette*, September 19, 1913, p. 541). However, one section of track was laid, and an engine and several mine cars were conveyed overland to that section of track. The locomotive's boiler was stoked, and with smoke pouring from the stack, photos were taken. The pictures were used as evidence that a railroad was operating to promote the sale of stock. The line was never completed, and the investors lost their money. A 1.53-acre strip of land owned by the railroad was offered at sheriff's sale in 1949 (*Reading Eagle*, July 21, 1949, and July 1, 1956).

MINERALS

Hematite - specular with stalactitic crystallization (D'Invilliers, 1883, p. 390) (fig. 195)

Magnesite (D'Invilliers, 1883, p. 111); fibrous (Lininger, 1968)

Pyrite (Brown and Ehrenfield, 1913, p. 80)

Quartz (Brown and Ehrenfield, 1913, p. 80)

Dotterer Mine

The Dotterer mine was west of Woodside Road, just below the Earl-Pike Township border, about 1 mile east of Manatawny at about 40° 22′ 54″ latitude and 75° 42′ 45″ longitude on the USGS Manatawny 7.5-minute topographic quadrangle map (Appendix 1, fig. 17, location E-9). The mine was in the Hardyston Formation. The mine also was known as the H.N. Landis Red Oxide Iron Ore Mine and is labeled as such in the 1876 atlas. The Dotterer mine was located on the Dotterer property about 500 yards south of the dwellings of N.H. Landis and J. Dotterer. It was close to the summit of Saw Mill Hill. A plan of the mine (fig. 197) and a vertical section (fig. 199) were presented by D'Invilliers (1883).

Iron ore was discovered when large chunks of ore were found at the bottom of a spring on the property (*Reading Eagle*, December 3, 1896). Production from the Dotterer mine was as much as 100 tons of ore per day. In October 1881, the Reading Eagle (October 23, 1881) reported that the mine was idle because the ore could not be transported from the mine. The

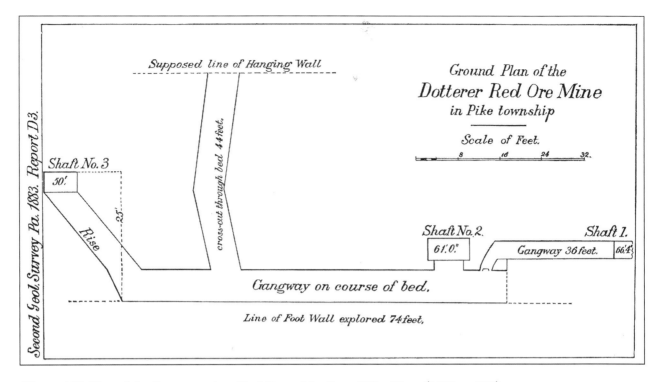

Figure 197. Plan of the Dotterer mine, Earl Township. From D'Invilliers (1883, p. 357).

mine was leased by the Pottstown Iron Company, and mining resumed late in the fall of 1882. The Pottstown Iron Company sank the No. 2 shaft to strike the foot wall of the ore body. In 1896, the mine was leased and worked by William Rowe (*Reading Eagle*, December 3, 1896).

There were three shafts on the property (fig. 197). In 1883, the No. 1 shaft, the most northern shaft, was 66 feet deep. The No. 2 shaft, the middle shaft, was 61 feet deep. The No. 3, or south shaft, was 50 feet deep (D'Invilliers, 1883, p. 354). The No. 2 shaft was sunk through slate, and at a depth of 61 feet, a drift was driven 12 feet northwest to intersect the foot wall of the ore body. A gangway was driven for about 60 feet south along the foot wall, and a rise was driven up at the end to meet shaft No. 3 at a depth of 60 feet. This gangway then was driven 15 feet north. From the top of a winze, a gangway was driven 36 feet in slate to intersect the No. 1 shaft. Between the No. 2 and No. 3 shafts and about 45 feet from the former, a cross cut was driven southeast through iron-bearing slates 44 feet from the foot wall (fig. 199) (D'Invilliers, 1883, p. 354-356).

The iron ore was disseminated through a chloritic slate that dipped conformably with the foot and hanging walls about S. 45° E. 70° to 80°. The foot wall was explored for 75 feet along the strike of the ore body. The ore was shipped to the Phoenix Iron Company and the Pottstown Iron Company, both of which found it unsuitable for smelting (D'Invilliers, 1883, p. 354-356). In 1883, the ore was shipped to the Bechtelsville Furnace, where it was mixed with ore from the Boyertown mines. An analysis of the ore by H.W. Hollenbush showed 33 per cent

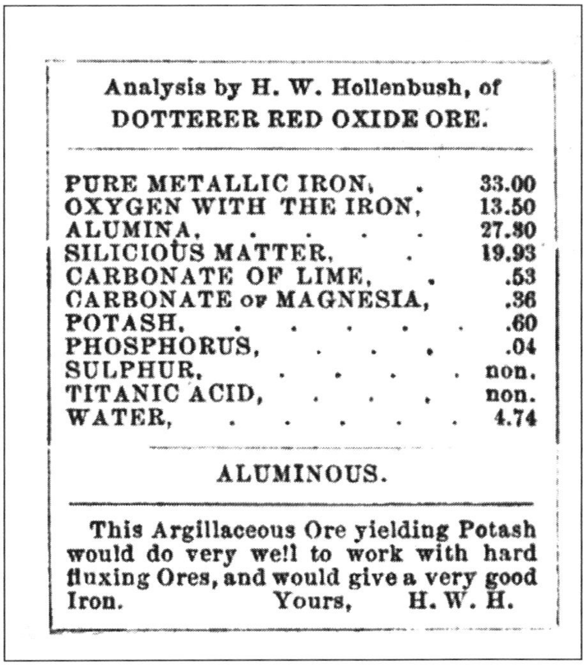

Analysis by H. W. Hollenbush, of DOTTERER RED OXIDE ORE.

PURE METALLIC IRON,	33.00
OXYGEN WITH THE IRON,	13.50
ALUMINA,	27.30
SILICIOUS MATTER,	19.93
CARBONATE OF LIME,	.53
CARBONATE OF MAGNESIA,	.36
POTASH,	.60
PHOSPHORUS,	.04
SULPHUR,	non.
TITANIC ACID,	non.
WATER,	4.74

ALUMINOUS.

This Argillaceous Ore yielding Potash would do very well to work with hard fluxing Ores, and would give a very good Iron.   Yours,   H. W. H.

Figure 198. Analysis of ore from the Doterer mine, Earl Township, by H.W. Hollenbush. From the Griffith Jones Papers Concerning Iron Ore Furnaces (3615), Historical Collections and Labor Archives, Special Collections Library, Pennsylvania State University.

metallic iron, 19.93 per cent silica; 0.04 percent phosphorus; and no sulfur (fig, 198).

Berks Development Company

The Berks Development Company was formed about 1904 by Wharton Barker to prospect for iron ore in the hills near Boyertown. Montgomery (1909, p. 602) reported that the company sank a number of shafts and found "*a good grade of ore*" on the Ehrst farm, which was occupied by Samuel S. Weis. The Engineering and Mining Journal (October 8, 1910) reported that the Berks Development Company had property on Long and Stone Cave Hills. The Journal reported that the ore was chiefly red hematite "*running well in iron and low in sulphur and phosphorus; there is also some magnetite.*" The company purchased a number of properties and the mineral rights on many properties in 1907-08. The company owned about 1,000 acres in Earl Township (*Reading Eagle*, July 16, 1913). In 1924, the directors and stockholders mutually dissolved the company because no additional ore had been discovered after years of prospecting (*Reading Eagle*, July 11, 1924).

Limestone Quarries

David Davidheiser Quarry

The David Davidheiser quarry was east of Camp Road, on the west side of Manatawny Creek, just east of the Earl-Oley Township border, about 0.9 mile north of Earlville at about 40° 19' 54" latitude and 75° 44' 15" longitude on the USGS Boyertown 7.5-minute topographic quadrangle map (Appendix 1, fig. 33, location E-10). The quarry was in the Leithsville Formation. Magnesian limestone was exposed in the quarry. The rock dipped N. 10° W. 50-60°. The limestone was about 80 feet thick and occurred in massive beds. It formed a bluff 30 feet high. D'Invilliers (1883, p. 162) published an analysis: 54.3 percent calcium carbonate, 42.6 percent magnesium carbonate, and 2.5 percent silica.

Fryermuth Quarry

The Fryermuth quarry was west of Linden Lane, southwest of the intersection of Linden lane and Pennsylvania State Route 562, 0.15 mile southwest of Worman at 40° 18' 49" latitude and 75° 41' 41" longitude on the USGS

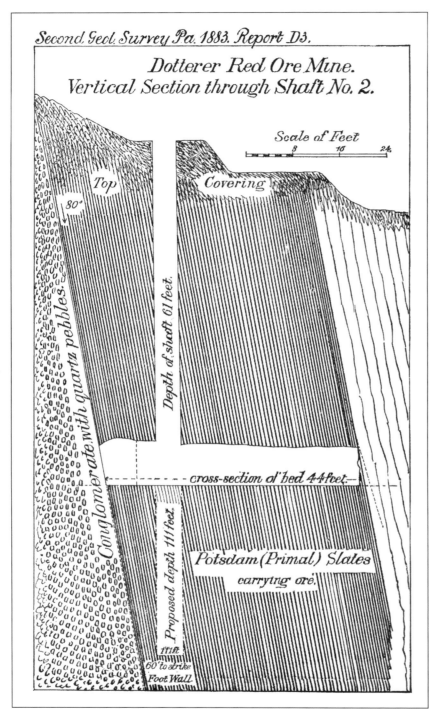

Figure 199. Geologic cross section of the Dotterer mine, Earl Township. From D'Invilliers (1883, p. 355).

Boyertown 7.5-minute topographic quadrangle map (Appendix 1, fig. 33, location E-2). It was about 400 feet south of Route 562. The quarry was in the Leithsville Formation. The quarry was owned by Nicholas Fryermuth. About 15 feet of slightly magnesian and very siliceous limestone was exposed in the quarry. The limestone dipped S. 20° E. 15°. The limestone was highly fractured with marked cleavage planes dipping S. 80° W. 80°, which caused it to split into small chunks. Most of the quarry output was burned in two kilns on the property to produce lime for private use (D'Invilliers 1883, p. 155). In 1891, the 146-acre Nicholas Fryermuth farm and quarry was sold at public sale (*Reading Eagle*, November 17, 1891).

Rapp Quarry

The Rapp quarry was south of Pennsylvania State Route 562, between Fancy Hill Road and Stauffer Lane, 0.4 mile southwest of Worman at about 40° 18′ 49″ latitude and 75° 42′ 02″ longitude on the USGS Boyertown 7.5-minute topographic quadrangle map (Appendix 1, fig. 33, location E-3). The quarry was in the Leithsville Formation. The quarry was on the Peter Rapp farm about 300 yards west of the Fryermuth quarry. The stone was burned in a kiln at the quarry to produce lime for use on Rapp's farm. About 12 feet of blue limestone and calcite mixed with quartz veins was exposed in the quarry. The dip was S. 23° E. 32° (D'Invilliers 1883, p. 155-156). D'Invilliers (1883, p. 156) presented an analysis of the limestone: 63.9 percent calcium carbonate, 12.9 percent magnesium carbonate, and 20.1 percent silica.

Mengle's Quarry

Mathias Mengle's quarry was east of Longview Road, southeast of the intersection of Longview and Manatawny Roads, 0.6 mile northeast of Earlville at about 40° 19′ 31″ latitude and 75° 43′ 43″ longitude on the USGS Boyertown 7.5-minute topographic quadrangle map (Appendix 1, fig. 33, location E-11). The quarry, in the Leithsville Formation, is shown on the 1882 topographic map. The rock dipped S. 80° W. 20°. The quarry exposed about 30 feet of blue magnesian limestone covered by about 10 feet of yellow clay. D'Invilliers (1883, p. 160-161) reported: *"This limestone is well liked by the farmers and is actively quarried, though as yet the opening is but slightly developed."* D'Invilliers published an analysis: 52.4 percent calcium carbonate, 40.4 percent magnesium carbonate, and 6.4 percent silica.

Boyer Quarry

The Boyer quarry was northwest of the intersection of Pennsylvania State Route 562 and Longview Road, between Longview Road and Manatawny Creek, 0.35 mile east of Earlville at 40° 19′ 06″ latitude and 75° 43′ 55″ longitude on the USGS Boyertown 7.5-minute topographic quadrangle map (Appendix 1, fig. 33, location E-4). It was located on the east bank of the Manatawny Creek. The quarry, in the Leithsville Formation, is shown in the 1876 atlas and on the 1882 topographic map. D'Invilliers (1883, p. 160) reported the *"old Boyer quarries"* were owned by H. Keefer.

Other Mines and Localities

Pennsylvania Uranium Mining Company Prospect

The Pennsylvania Uranium Mining Company prospect was located about 0.5 mile northwest of Shanesville. The prospect, in felsic to mafic gneiss, is believed to have been developed in the 1950s at the height of the uranium prospecting craze (Lininger, 1968). Radioactive minerals occur in a hornblende-plagioclase gneiss. Smith (1978, p. 44) reported that scattered boulders up to 40 cm across were about one-third thorite. A shipment from the mine contained mainly thorium (Barnes and Smith, 2001, p. 26).

Minerals -- from Smith (1978, p. 44)

Apatite-(OH)

Thorite (fig. 200)

Thorogummite - alteration product of thorite

Titanite - possibly the yttrium variety yttrotitanite or keilhauite

Mica Prospect Number 1

Mica prospect number 1 of Buckwalter (1954) was east of the intersection of Woodchoppertown Road and Ambrose Drive, on the southeastern side of Shenkel Hill, on both sides of Choppertown Road. The prospect was about 0.75 mile east of Spangsville at about 40° 21′ 25″ latitude and 75° 43′ 40″ longitude on the USGS Boyertown 7.5-minute topographic quadrangle map (Appendix 1, fig. 33, location E-5). The prospect was in hornblende gneiss. A shallow pit was dug about 1900 in search of iron ore. Some of the mica that was mined was used for isin-

Figure 200. Thorite from the Pennsylvania Uranium Company mine, Shanesville, Earl Township, 2.4 cm. Sloto collection 3332.

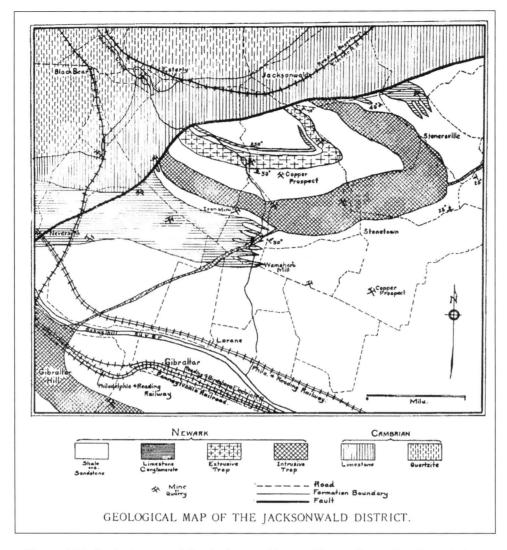

Figure 201. Geologic map of the Jacksonwald area, Exeter Township. From Wherry (1910, plate 1).

glass. The mica occurred in a zone about 20 feet wide, which contained several closely spaced pegmatite dikes. The main mica-bearing dike was about 350 feet long. Biotite occurred in books up to 7 inches long and 5 inches thick. Single crystals were less than 3 inches. The pegmatites were composed of quartz, orthoclase, microcline, biotite, and muscovite (Buckwalter, 1954, p. 17-19).

Riebeckite Localities

Four rebieckite, var. crocidolite, localities in Earl Township were described by Bliss (1913). Locality 1 is 1.25 miles southeast of Shanesville (Shanesville rebieckite locality described below); locality 2 is 1.75 miles northwest of Gabelsville; locality 3 is 1.75 miles northwest of Gabelsville (Gabelsville rebieckite localities described below); and locality 4 is 0.75 mile east of Shanesville in a cut of the Oley Valley Electric Railway (Oley Valley Electric Railway cut rebieckite locality described below). All four localities are in felsic to mafic gneiss.

Bliss described the crocidolite as a blue amphibole in Triassic sandstone. It commonly occurred disseminated throughout the rock as one of the chief constituents and as a thick coating on the weathered surfaces, thereby giving the appearance of a weathering product. The coating, which varies in color from a deep blue or almost black to an ultramarine blue, is up to 5 mm thick. It can readily be removed by scraping with a knife and crushed to an earthy powder. The luster is normally dull but is occasionally silky. In addition to its occurrence as a coating, the crocidolite occurs abundantly scattered throughout the mass of the rock in small, dark blue, irregular grains. The grains, which are less than 0.2 mm in diameter, have a dull lustre, a hardness between 5 and 6, and show neither crystal outline nor cleavage faces.

Gabelsville Riebeckite Localities

The Gabelsville riebeckite var. crocidolite localities are along Saw Mill Road, east of Pine Road, just east of the Earl-Colebrookdale Township border at about 40° 20′ 56″ latitude and 75° 40′ 52″ longitude on the USGS Boyertown 7.5-minute topographic quadrangle map (Appendix 1, fig. 34, location E-6). These are crocidolite localities 2 and 3 described by Bliss (1913, p. 519).

Shanesville Riebeckite Locality

The Shanesville riebeckite var. crocidolite locality is south of Pennsylvania State Route 73, southeast of its intersection with Pond Road at about 40° 21′ 36″ latitude and 75° 40′ 03″ longitude on the USGS Boyertown 7.5-minute topographic quadrangle map (Appendix 1, fig. 34, location E-7). It is crocidolite locality 1 described by Bliss (1913, p. 519).

Oley Valley Electric Railway Cut Riebeckite Locality

The Oley Valley Electric Railway riebeckite var. crocidolite locality is along Ironstone Drive, east of its intersection with Terrance Road at about 40° 21′ 48″ latitude and 75° 40′ 49″ longitude on the USGS Boyertown 7.5-minute topographic quadrangle map (Appendix 1, fig. 34, location E-8). It is crocidolite locality 4 described by Bliss (1913, p. 519). The Oley Valley Electric Railway was a trolley line that connected Boyertown and Reading and operated from 1902 to 1932.

Bliss (1913, p. 520) described locality 4: *"In a rich occurrence, discovered in a cutting of the Oley Valley Electric Railway about 3.5 miles northwest of Boyertown (Locality 4) the country rock which is a granite, has been thoroughly shattered and faulted, with the development of quartz veins along the lines of fracture. The blue mineral is formed over the surface of the quartz crystals to such a degree that in places fragments of practically pure amphibole two to three centimetres in length can be dug out."*

EXETER TOWNSHIP

Jacksonwald Occurrence

The Jacksonwald occurrence (fig. 201) is south of Jacksonwald along Shelbourne Road on the USGS Birdsboro 7.5-minute topographic quadrangle map (Appendix 1, fig. 29, location EX-1). The occurrence is in the Jacksonwald basalt. Most of the area is covered by residential housing developments.

The Jacksonwald basalt is the only area in Pennsylvania where magma flowed out onto the land surface during the early Jurassic. The Jacksonwald basalt crops out along Antietam Creek 1 mile south of Jacksonwald. Wherry (1910) provided evidence to show that the inner of the two igneous masses at Jacksonwald is not a sill, but an overflow sheet. He presented as

Figure 202. Amydular (top) and vesicular (bottom) basalt from the Jacksonwald basalt, Exeter Township. Sloto collection.

evidence the vesicular character of the rock (fig. 202) and the complete absence of thermal alteration in the enclosing shales.

A small quarry, located on the east side of Antietam Creek about 200 feet from the creek along the base of the hill, exposed the underlying shale, which dips 50° N. 20° E. To the west, the shale extends below the land surface with basalt conformably overlying it. On the eastern side, the basalt is exposed for 5 feet. The basalt is extremely compact, dense, and fine grained throughout the greater part of this exposure. About two feet above the contact with the shale, it contains occasional cylindrical amygdules filled with quartz averaging 0.5 inch in diameter by 3 inches long. Wherry (1910, p. 11-12) interpreted them as gas or steam cavities lengthened by the flow of the viscous lava; the direction of lengthening is N. 25° W.

A series of excavations made for road metal exposed the basalt through its entire thickness of 530 feet. The fine-grained character at the lower contact persists upwards for over 100 feet, with occasional slightly porphyritic areas. Then minute

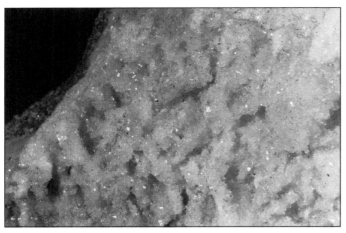

Figure 203. Prehnite from the Jacksonwald occurrence, Exeter Township. Bryn Mawr College collection Quickel 366.

cavities begin to appear, and these increase in number until the rock is highly vesicular at the top (fig. 202). The cavities are partly filled with secondary minerals, chiefly calcite, prehnite, and datolite. The datolite is confined to a narrow layer about 350 feet below the top of the basalt. The basalt near the base of the sheet is distinctly porphyritic, and its sections show large equidimensional feldspar crystals embedded in the groundmass containing blades of the same mineral (Wherry, 1910, p. 12).

Wherry (1910, p. 24) described the mineralogy of the basalt flow: "*At the very top of the mass the gas cavities are empty or, at best, filled with red mud, but ten feet down they are solidly filled with calcite, together with more or less chloritic material. About 100 feet below the top prehnite begins to appear, chiefly in seams solidly replacing the decomposed trap, but occasionally showing small globular clusters of crystals where a cavity has been occupied. This prehnite replacement is limited to a belt about 50 feet thick, although the same mineral occasionally accompanies the others in the cavities lower down. From 150 feet to 300 feet the calcite-chlorite filling is again the rule, and here the cavities are sometimes nearly an inch in diameter. Then datolite begins to appear, filling both bubble cavities and cracks, but never in large amount nor in distinct crystals, being instead coarsely granular and intimately mixed with calcite. The occurrence of this datolite is limited to perhaps fifty feet of thickness of the diabase, and no trace of the mineral has been noted in any other part of the sheet, nor at any other exposure. Below the datolite zone, as it may be called, gas cavities are much less prominent, and here the zeolites make their appearance along the joint planes. These comprise stilbite, heulandite and chabazite, in small but typical crystals, the first usually alone, the last two usually associated; they are sparingly present through the remainder of the thickness of the sheet. At the very base there are a few elongated gas bubbles, mentioned above, and these are filled solidly or nearly so with white crystalline quartz.*"

Minerals

In outcrops of the Jacksonwald basalt along Antietam Creek 1 mile south of Jacksonwald. From Wherry (1910, p. 12 and p. 24-25).

Calcite

Chabazite - small crystals associated with heulandite along joint planes

Chlorite group - flakes and concentric groupings

Datolite - coarsely granular; mixed with calcite

Feldspar

Heulandite - small crystals associated with chabazite along joint planes

Prehnite (fig. 203)

Quartz - in amygdaloidal cavities

Stilbite - small crystals along joint planes

Tourmaline group - minute, lenticular crystals in the shale above the basalt

X-ray diffraction analysis by the Pennsylvania Geological Survey of basalt samples collected by the author identified the following minerals: anorthite, albite, augite, clinopyroxene, chlorite group (probably clinochlore), quartz, muscovite, hematite, and possible dravite and schorl (John Barnes, Pennsylvania Geological Survey, written communication, 2005).

Kinsey Hill Locality

The Kinsey Hill locality is on Schoffers Road, between Fabers and Stonetown Roads about 0.4 mile north-northwest of Stonetown at about 40° 18′ 32″ latitude and 75° 49′ 49″ longitude on the USGS Birdsboro 7.5-minute topographic quadrangle map (Appendix 1, fig. 31, location EX-2). The hill, underlain by diabase, is labeled as Kintzi Mount on the 1860 map and Kinsey Mountain in the 1876 atlas. Minerals from the Kinsey Hill locality were collected by Allen Heyl and are preserved in the Bryn Mawr College mineral collection. The minerals include apophyllite (fig. 204), prehnite (fig. 205), and stellerite (fig. 206).

Bishop's Mill Locality

The Bishop's Mill locality was near Antietam Creek, west of Shelborne Road at about 40° 17′ 57″ latitude and 75° 50′ 55″ longitude on the USGS Birdsboro 7.5-minute topographic quadrangle map (Appendix 1, fig. 31, location EX-3). The locality is underlain by the Brunswick Formation. The Bishop's Mill locality also has been called Kinzi's Mill and Hartzog's Mill (D'Invilliers, 1883, p. 395-400). The 1854 Berks County map shows the mill as the Jones Mill. Bishop's grist mill produced flour from 1726 to 1926. It stood on the eastern edge of the Reading Country Club property diagonally across from a restaurant at the corner of U.S. Route 422 and Shelborne Road.

Figure 204. Apophyllite from the Kinsey Hill locality, Exeter Township. Bryn Mawr College collection Heyl 2404.

Figure 205. Prehnite from the Kinsey Hill locality, Exeter Township. Bryn Mawr College collection Heyl 1348.

Figure 206. Stellerite from the Kinsey Hill locality, Exeter Township. Bryn Mawr College collection Heyl 1991.

Minerals

Calcite - pink (D'Invilliers, 1883, p. 395)

Garnet group - single and twinned crystals (D'Invilliers, 1883, p. 397) (fig. 207)

Quartz, vars. jasper, chalcedonic jasper (D'Invilliers, 1883, p. 400)

Snydersville Malachite Occurrence

Eyerman (1889, p. 45) reported that malachite was found "at Snydersville." Robinson (1988) placed the location of the occurrence at 40°18' 25" latitude and 75° 49' 26" longitude on the USGS Birdsboro 7.5-minute topographic quadrangle map (Appendix 1, fig. 31, location EX-5). This location is north of Fabers Road, north of its intersection with Fenderson Road, about 0.4 mile northeast of Stonetown. The place name Syndersville does not appear on any map in this area.

Stonersville Magnetite Occurrence

Robinson (1988) placed the location of the Stonersville magnetite occurrence at 40° 19' 47" latitude and 75° 48' 39" longitude on the USGS Birdsboro 7.5-minute topographic quadrangle map (Appendix 1, fig. 30, location EX-6). This location is east of Devon Drive and west of Oley Line Road, **about 1 mile** north-northwest of Stonersville. The occurrence is in the Beekmantown Group.

Spencer (1908, p. 42) reported that "*specimens* [of magnetite] *have been plowed up in the fields near Spring Creek, between the northern arm* [of the diabase] *and Stonersville. Though some prospecting was done in this vicinity, no magnetite was found in bed rock. It is possible that the mineral may have been float from a deposit situated near the diabase wall on the hill slopes above the creek or it may have been derived from a pocket lying in the limestone conglomerate which covers a considerable area on the west side of Spring Creek south of the Reading turnpike.*"

EXETER TOWNSHIP

Figure 207. Garnets from Bishop's Mill, Exeter Township, 1.5 cm (top) and 1.1 cm (bottom). Carnegie Museum of Natural History collection CM 6961.

Bishop Mine

The Bishop iron mine was on the east side of Gibraltar Road, north-northeast of it's intersection with U.S. Route 422 at about 40° 18′ 28″ latitude and 75° 51′ 02″ longitude on the USGS Birdsboro 7.5-minute topographic quadrangle map (Appendix 1, fig. 31, location EX-8). The mine was in diabase. The mine is shown in the 1876 atlas as the Bishop magnetic ore mine, and the location of the Bishop mine shaft is shown on the 1882 topographic map. The Bishop mine was on the property of the Reading Country Club golf course. The area is developed, and the mine likely has been destroyed.

D'Invilliers (1883, p. 186) reported that a shaft 50 feet deep, known as Bishop's mine, was located close to the road. John M.F. Bishop was listed as *"Farmer and Ore Miner"* in the 1876 atlas. Spencer (1908, p. 41) reported that the 150-foot deep Bishop shaft was about 600 feet northwest of the Esterly mine. A crosscut was driven north from the shaft for 200 feet, ending in "garnet rock." A borehole from the bottom of the shaft reached limestone conglomerate at about 300 feet.

Monocacy Hill

Monocacy Hill is on the north side of the Schuylkill River, 2.5 miles northeast of Birdsboro on the USGS Birdsboro 7.5-minute topographic quadrangle map (Appendix 1, fig. 32, location EX-4). Kulp and Hughs (2001) described Brunswick Formation xenoliths in the Monocacy Hill diabase in outcrop and float. Minerals reported in and surrounding the xenoliths include apatite, chlorite, garnet group, crystallized olivine (see Kulp and Hughs, 1999), orthopyroxene, plagioclase, pyroxene, magnetite, quartz, titanite, fibrous wollastonite crystals, and zircon.

Esterly Mine

The Esterly mine was northwest of Lorane Road, along Antietam Creek, 0.3 mile northeast of Lorane at about 40° 17′ 30″ latitude and 75° 50′ 55″ longitude on the USGS Birdsboro 7.5-minute topographic quadrangle map (Appendix 1, fig. 31, location EX-7). The mine was in the Brunswick Formation.

About 1844, a group of prospectors searched for copper ore on the farm of Daniel and Jacob Esterly. Copper ore reportedly was found, but not in economic quantities (*Reading Eagle*, November 28, 1894). A small deposit of magnetite later was mined on the farm. The ore was mined through a 125-foot deep slope inclined at 58° toward the north. Drifts were run about 250 feet to the east (Spencer, 1908, p. 41). When the property was visited by Bever and Liddicoat (1954, p. 75) in 1953, they noted: *"At the property no trace of the shaft or rock dumps remain."* Rose (1970, p. 10) estimated total production at 4,000 to 5,000 tons of ore.

The Esterly mine ore body lies between a hanging wall of diabase and a foot wall of baked shale. The ore was formed by the replacement of shale under the mineralizing influence of the diabase. The diabase is an intrusive sill included in northward-dipping strata. In the neighborhood of the intruded rock, there was considerable baking. In the baked zone, metamorphic minerals, such as garnet, hornblende, and magnetite, were found. These minerals, along with some chlorite, were found on the mine dump. The silicate minerals occurred in close association with magnetite (Spencer, 1908, p. 41-42).

Guldin Hill Sandstone Quarry

The Guldin Hill sandstone quarry was on the southwestern side of Guldin Hill, east of Jacksonwald Road on the USGS Birdsboro 7.5-minute topographic quadrangle map (Appendix

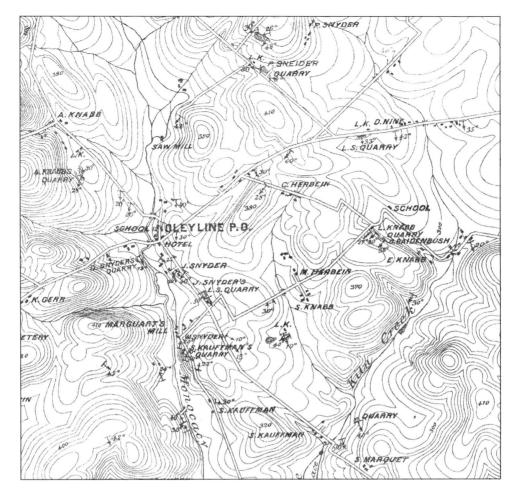

Figure 208. Quarries in the Oley Line area, Exeter Township, 1882. From Pennsylvania Geological Survey (1883).

1, fig. 29, location EX-17). The quarry, in the Hardyston Formation, is shown on the 1882 topographic map as "S.S. QUARRY" (fig. 210). A white, fine-grained sandstone was quarried (D'Invilliers, 1883, p. 380).

Limestone Quarries

D. Snyder Quarry

The D. Snyder quarry was southeast of the intersection of Limekiln and Oley Turnpike Roads in Oley Line at about 40° 20′ 35″ latitude and 75° 48′ 22″ longitude on the USGS Birdsboro 7.5-minute topographic quadrangle map (Appendix 1, fig. 30, location EX-11). The quarry is shown on the 1882 topographic map (fig. 208). The quarry was in the Beekmantown Group. The small quarry on the west bank of Monocacy Creek exposed a hard, blue, silicious limestone overlain by slate. The limestone dipped S. 14° E. 12° (D'Invilliers, 1883, p. 186). A lime kiln (fig. 209) still stands (2016) near the quarry.

S. Kauffman Quarry

The S. Kauffman quarry was south-southwest of the intersection of Limekiln and Oley Turnpike Roads, 0.25 mile southwest of Oley Line at 40° 20′ 24″ latitude and 75° 48′ 24″ longitude on the USGS Birdsboro 7.5-minute topographic quadrangle map (Appendix 1, fig. 30, location EX-9). The quarry is shown on the 1882 topographic map (fig. 208). The quarry was in the Beekmantown Group.

The quarry was located on the east bank of Monocacy Creek. According to D'Invilliers (1883, p. 186), the quarry contained "*both good and poor stone, metamorphosed and rendered hard and crystalline near the trap dyke that occurs in the south portion of the quarry, but soft and of firm, fine-grained, massive texture elsewhere.*" The dike was about 6 feet wide and composed of fine-grained, black diabase. The massive blue limestone was greatly deformed in the center of the quarry. The average dip was N. 30° E. 50°. At the north end of the quarry, the dip was N. 23° E. 20°. At the south end of the quarry near

Figure 209. Lime kiln on Limekiln Road near the D. Snyder quarry, Exeter Township. Photograph taken in March 2015.

the diabase dike, a bed of white, hard limestone dipped N. 40° E. 30°. There was 60 feet of limestone exposed in a face 30 feet high. D'Invilliers (1883, p. 186) collected a number of samples from the quarry. An analysis of the composite was 80.2 percent calcium carbonate, 9.5 percent magnesium carbonate, and 9.1 percent silica.

Albert Knabb Quarry

The Albert Knabb quarry was on the west side of Limekiln Road northwest of its intersection with Oley Turnpike Road, 0.45 mile northwest of Oley Line at about 40° 20′ 48″ latitude and 75° 48′ 39″ longitude on the USGS Birdsboro 7.5-minute topographic quadrangle map (Appendix 1, fig. 30, location EX-10). The quarry, in the Beekmantown Group, is shown on the 1882 topographic map (fig. 208).

The quarry was located 200 yards from the S. Kauffman quarry. The quarry exposed about 30 feet of a blue and white magnesian limestone in a small knoll. The limestone was cut by nearly vertical cleavage planes and exhibited a slight anticlinal roll at south end of quarry. The limestone dipped N. 30° W. 30° and S. 20° E. 20°. D'Invilliers (1883, p. 179-180) published an analysis of the limestone: 60.1 percent calcium carbonate, 38.2 percent magnesium carbonate, and 1.3 percent silica.

Benjamin Ritter Quarry

The Benjamin Ritter quarry was east of the intersection of and between Pennsylvania State Route 662 and Oley Turnpike Road, 0.9 mile east of Jacksonwald at about 40° 19′ 47″ latitude and 75° 50′ 04″ longitude on the USGS Birdsboro 7.5-minute topographic quadrangle map (Appendix 1, fig. 29, location EX-12). The quarry, in the Beekmantown Group, is shown on the 1882 topographic map (fig. 210). The area is now a residential development.

Benjamin Ritter operated a small quarry for his own use next to a private farm lane. The quarry exposed a 12-foot thick, massively bedded, blue limestone that dipped S. 87° W. 10°. D'Invilliers (1883, p. 186) published an analysis of the limestone: 87.9 percent calcium carbonate, 7.3 percent magnesium carbonate, and 4.6 percent silica.

Cornelius Tyson Quarry

The Cornelius Tyson quarry was east of the intersection of and between Pennsylvania State Route 662 and Oley Turnpike Road, 0.6 mile east of Jacksonwald at about 40° 19′ 35″ latitude and 75° 50′ 21″ longitude on the USGS Birdsboro 7.5-minute topographic quadrangle map (Appendix 1, fig. 29, location EX-13). The quarry, in the Beekmantown Group, is shown on the 1882 topographic map (fig. 210). The area is now a residential development.

The small quarry exposed the same limestone as the Benjamin Ritter quarry (D'Invilliers, 1883, p. 187). The quarried stone was used on the Tyson farm. At the east end of the quarry, there was a small anticlinal roll dipping S. 88° W. 15° and N. 86° E. 10° that exposed a cherty limestone. A better quality blue limestone was found in the western part of the quarry; the blue limestone dipped S. 85° W. 18°. D'Invilliers (1883, p. 187) published an analysis of the limestone: 86.9 percent calcium carbonate, 5.7 percent magnesium carbonate, and 7.2 percent silica.

Abandoned Quarry at Jacksonwald

An abandoned quarry was southeast of the intersection of Pennsylvania State Route 562 and Shelbourne Road in Jacksonwald at about 40° 19′ 25″ latitude and 75° 50′ 57″ longitude on the USGS Birdsboro 7.5-minute topographic quadrangle map (Appendix 1, fig. 29, location EX-14). The quarry, in Lower Cambrian rocks, is shown on the 1882 topographic map (fig. 210). D'Invilliers (1883, p. 187) described a small abandoned quarry close to a creek about 800 yards southeast of Jacksonwald that showed evidence of a compressed syncline. The quarry exposed 30 feet of blue limestone that dipped N. 20° W. 45°.

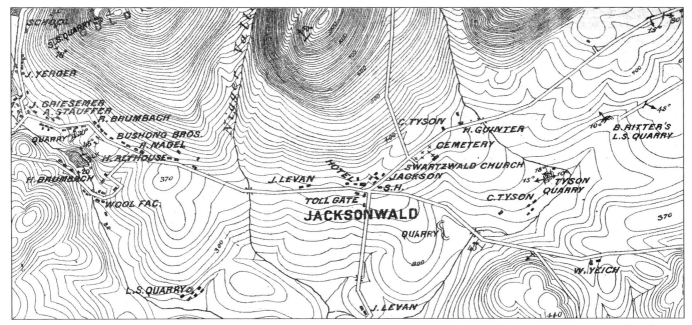

Figure 210. Quarries in the Jacksonwald area, Exeter Township, 1882. From Pennsylvania Geological Survey (1883).

Reifton Quarry

The Reifton quarry was west of the intersection of U.S. Route 422 and U.S. Business Route 422 at about 40° 18′ 44″ latitude and 75° 52′ 13″ longitude on the USGS Birdsboro 7.5-minute topographic quadrangle map (Appendix 1, fig. 31, location EX-15). The quarry is labeled "L.S. QUARRY" on the 1882 topographic map (fig. 210). The area has been completely developed. The quarry was in Lower Cambrian rocks. D'Invilliers (1883, p. 188) described a small quarry that exposed limestone dipping S. 12° W. 65°. The gray and white slaty limestone was about 10 feet thick. It contained considerable quantities of a dark black mineral with a dull luster similar to graphite. A dull, brown slate also was exposed in the quarry.

Jonas DeTurk Quarry

The Jonas DeTurk quarry was south of the intersection of Neversink Road and Heathstone Drive at about 40° 18′ 46″ latitude and 75° 53′ 02″ longitude on the USGS Reading 7.5-minute topographic quadrangle map (Appendix 1, fig. 27, location EX-16). The quarry, in Lower Cambrian rocks, is shown on the 1882 topographic map. The area is now completely urbanized. The 1876 atlas listed Frank B. DeTurk as the proprietor of two lime kilns 1 mile from the Black Bear Hotel.

The quarry was a large quarry that exposed a blue limestone below a 20-foot thick bed of slaty limestone. The blue limestone was deformed in places. The dip at the top of the quarry was about S. 20° W. 75°. The quarry was idle when visited by D'Invilliers in 1882. D'Invilliers (1883, p. 188) noted: "*It is remarkable for the large quantity of carbonate of manganese which it contains = 13.28 per cent.*"

Minerals

Calcite - pale pink or rose colored, small, flat rhombohedral crystals (Genth, 1876, p. 228)

Fluorite - occurs sparingly in coarse, granular and occasionally crystallized masses of a violet and purple color; associated with calcite (Genth, 1876, p. 210)

Klapperthal Park Oil Well

In 1919, an unnamed con man built an oil derrick near the old Klapperthal Park on the property of a Mr. Baum, near where the railroad extended into what is now Forest Hills Cemetery. He sold shares in what he promoted as an oil well. He even went so far as to bury a drum of oil poked full of holes in Klapperthal Creek to give the appearance of oil in the area (*Reading Times*, May 5, 1919; Druzba, 2003).

GREENWICH TOWNSHIP

Greenwich Manufacturing Company Quarry

The Greenwich Manufacturing Company slate quarry was east of Maiden Creek and north of old Pennsylvania State Route 22, east of Lenhartsville at about 40° 34′ 30″ latitude and 75° 52′ 42″ longitude on the USGS Hamburg 7.5-minute topographic quadrangle map (Appendix 1, fig. 2, location G-1). It is quarry number 5 of Behre (1933, p. 357). The quarry also was known as the Lenhartsville quarry.

The quarry is the southernmost quarry of the Greenawald group slate quarries. The Greenawald group quarries produced slate that was crushed and granulated for use as chips in roofing tar and as a filler with pigmenting qualities. Red and green slate beds occur interstratified with coarse sandy layers in the middle Matinsburg Formation.

The quarry was an open cut 120 feet along strike and 50 feet along the dip of the beds. The strata included purple, red, brown, and green beds. The brown layers were massive, sandy, and calcareous and showed cross bedding. In 1890, the quarry was worked for colored slate by James S. Focht, but operations probably date further back as local lore indicates that attempts were made to use the red rock as iron ore, and an old furnace once stood on the property. Quarrying was abandoned because of the large amount of waste.

In 1918, work was resumed and continued until about 1928 with minor interruptions. In 1928, it was the only slate quarry operating in Berks County. The main product was red and green "slate flour" (*Reading Eagle*, November 8, 1928). In 1918, a mill was built to process the rock. The Greenwich Manufacturing Company was the operator in 1927. The company used a train for hauling material from the quarry to the mill. Processing in the mill included drying in a low temperature kiln, transfer by bucket conveyor to a ball mill, then by a worm conveyor to a tube mill, and sizing by a Gates-type separator with the oversize material being returned to the ball mill. The crushed material was used as pigment and filler (Behre, 1933, p. 357).

Maiden Creek Furnace

Maiden Creek Forge was built about 1822, and the cold-blast charcoal Maiden Creek Furnace was built about 1854 near Lenhartsville, on Ontelaunee Creek near where it enters Maiden Creek. George Merkel built the furnace and operated both the furnace and forge until his death in 1875. Lesley (1859) stated that the forge was owned by George Merkel & Company and was managed by George Rimsel. The furnace produced 1,024 tons of iron in 48 weeks in 1857 using limonite from the Moselem, Coxtown, and Trexlertown mines. In 1876, the furnace was operated by heirs of George Merkel. The annual capacity was 1,600 tons (American Iron and Steel Association, 1876).

In 1878, Jacob K. Spang of Reading, Samuel Erb of Lebanon, and Joshua Hunsicker of Montgomery County purchased the furnace from the administrators of the Merkel estate and ran the furnace under the firm name of Spang, Erb & Company. Spang worked for Bushong & Company as superintendent of the Keystone Furnace in Reading for 15 years. He left Bushong & Company to purchase the Maiden Creek Furnace. By 1882, Spang was the sole owner. At that time, the capacity of the furnace was about 40 tons of iron per week. The furnace used limonite ore from the Moselem mine and produced car-wheel iron. In 1890, the furnace was converted to a hot-blast furnace. In 1894, Spang retired from the iron business and rented out the furnace. In 1896, the furnace used limonite ore from the Moselem mine and local magnetite ore to produce pig iron for car wheels and chilled rolls. The annual capacity was 3,500 tons, and the iron was sold under the brand name "Maiden Creek" (American Iron and Steel Association, 1896). The American Iron and Steel Association (1901) noted that the furnace had been idle for several years.

Figure 211. Pyrite nodule from Windsor Township, 3 cm (outside on top and inside on bottom). Carnegie Museum of Natural History collection ANSP 6881.

HEIDELBERG TOWNSHIP

The limestone quarries in Heidelberg Township were described by D'Invilliers (1886, p. 1554-1555). Quarry locations are shown in Pennsylvania Geological Survey (1891, reference map 11) (fig. 212).

William Moore Quarry

The William Moore quarry (fig. 213) was on the north side of Pennsylvania State Route 419, east of its intersection with Brickyard Road, about 0.9 mile south-southwest of the center of Womelsdorf at 40° 21′ 25″ latitude and 76° 11′ 31″ longitude on the USGS Womelsdorf 7.5-minute topographic quadrangle map (Appendix 1, fig. 21, location HB-1). It is quarry number 76 on reference map 11 in Pennsylvania Geological Survey (1891) (fig. 212). The quarry was cut into a hillside in the Richland Formation.

William G. Moore operated a limestone quarry on his farm in the 1880s. The quarry produced stone of *"medium quality"* from strata 6 to 18 inches thick. According to D'Invilliers (1886, p. 1554): *"There is a considerable amount of stripping and much earth occurs through the exposure, rendering the rock slaty and impure in places. The color of the rock is very dark-blue, and though the beds are thin, some of it would apparently do very well for building purposes."* The limestone dipped S. 10° W. 35°. A lime kiln (fig. 214) still stands (2016) east of the quarry on the north side of Pennsylvania State Route 419.

John Marshall Quarry

The John Marshall quarry was northeast of the intersection of Water Street and Ryeland Road, south of Womelsdorf at about 40° 21′ 25″ latitude and 76° 10′ 45″ longitude on the USGS Womelsdorf 7.5-minute topographic quadrangle map (Appendix 1, fig. 22, location HB-2). It is quarry number 77 on reference map 11 in Pennsylvania Geological Survey (1891) (fig. 212). The quarry was in the Millbach and Schaefferstown Formations, undivided. This small quarry was just northeast of the former Womelsdorf railroad station and was worked occasionally to supply railroad ballast. The rock dipped about 60° S. (D'Invilliers, 1886, p. 1554).

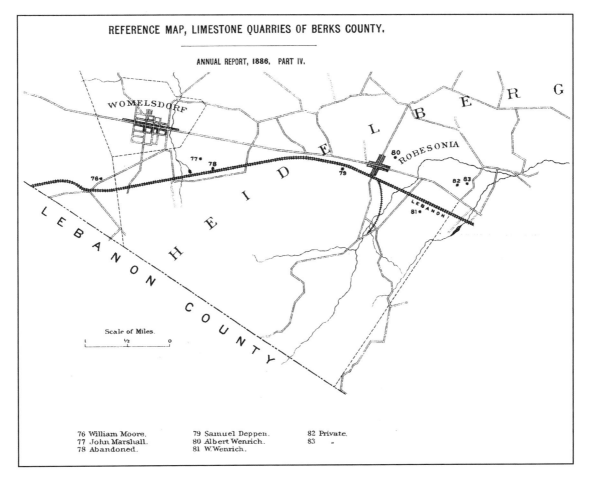

Figure 212. Limestone quarries in Heidelberg Township, 1891. From Pennsylvania Geological Survey (1891).

Figure 213. Moore quarry, Heidelberg Township, March 2015.

Samuel Deppen Quarry

The Samuel Deppen quarry was south of U.S. Route 422 and the railroad tracks, 0.7 mile west of the center of Robesonia at about 40° 21′ 17″ latitude and 76° 09′ 05″ longitude on the USGS Womelsdorf 7.5-minute topographic quadrangle map (Appendix 1, fig. 22, location HB-3). It is quarry number 79 on reference map 11 in Pennsylvania Geological Survey (1891) (fig. 212). The quarry was in the Ontelaunee Formation.

Stone was quarried for agricultural and building lime. There were two kilns on the property, each with a capacity of 600 bushels. The lime was sold to farmers and builders for 6.5 to 12 cents per bushel. A siding ran from the main railroad track to the quarry, and the plant at the quarry was equipped for shipping stone. D'Invilliers (1886, p. 1555) noted: "*None of the rock exposed here seemed suitable for building stone, the beds being too small and breaking into small blocks.*" The rocks dipped S. 20° W. 35°. Frear (1913) presented an analysis of the limestone from the quarry.

W. Wenrich Quarry

The Wellington W. Wenrich quarry was near the Robesonia Borough boundary, east-southeast of the intersection of Furnace and Freeman Streets at about 40° 20′ 29″ latitude and 76° 08′ 02″ longitude on the USGS Womelsdorf 7.5-minute topographic quadrangle map (Appendix 1, fig. 22, location HB-5). It is quarry number 81 on reference map 11 in Pennsylvania Geological Survey (1891) (fig. 212). The quarry was in the Richland Formation on a hill slope on the south side of the railroad, 0.75 mile southeast of the former Robesonia railroad station. The quarry was leased by the Progressive Mining Company during 1883. It was not being worked when visited by D'Invilliers around 1886. The rock dipped S. 15° W. 40° (D'Invilliers, 1886, p. 1555).

Reed Quarry

The Reed quarry was northeast of the intersection of U.S. Route 422 and Big Spring Road. It was directly across Big Spring Road from Big Spring at about 40° 20′ 56″ latitude and 76° 07′ 04″ longitude on the USGS Sinking Spring 7.5-minute topographic quadrangle map (Appendix 1, fig. 23, location HB-6). It is quarry number 83 on reference map 11 in Pennsylvania Geological Survey (1891) (fig. 212). The quarry, in the Epler Formation, exposed a finely laminated and very slaty rock, which dipped S. 35° E. 40° (D'Invilliers, 1886, p. 1555).

Big Spring Quarry

An unnamed quarry near Big Spring was immediately east of the Reed quarry. It was northeast of the intersection of U.S. Route 422 and Big Spring Road at about 40° 20′ 57″ latitude and 76° 07′ 10″ longitude on the USGS Sinking Spring 7.5-minute topographic quadrangle map (Appendix 1, fig. 23, location HB-7). It is quarry number 82 on reference map 11 in

Figure 214. Lime kilns at the Moore quarry, Heidelberg Township, March 2015.

Pennsylvania Geological Survey (1891) (fig. 212). The quarry was in the Epler Formation. D'Invilliers (1886, p. 1555) reported there was a lime kiln at the quarry and the rock dipped nearly 40° S.

Ryland Road Quarry

An abandoned quarry was north of the railroad tracks, northwest of the intersection of Hill and Ryland Roads at about 40° 21′ 21″ latitude and 76° 10′ 23″ longitude on the USGS Womelsdorf 7.5-minute topographic quadrangle map (Appendix 1, fig. 22, location HB-8). It is quarry number 78 on reference map 11 in Pennsylvania Geological Survey (1891) (fig. 212). The quarry was in the Millbach and Schaefferstown Formations, undivided. D'Invilliers (1886, p. 1554) described the abandoned quarry north of the railroad and east of the former Womelsdorf railroad station as *"quite unimportant in its bearing upon the commercial aspects of the region."* The limestone dipped about 85° SW.

Sheetz Sand Quarry

The Sheetz sand quarry was between U.S. Route 422 and the railroad tracks, southeast of the intersection of U.S. Route 422 and Hill Road at about 40° 21′ 24″ latitude and 76° 09′ 29″ longitude on the USGS Womelsdorf 7.5-minute topographic quadrangle map (Appendix 1, fig. 22, location HB-9). The quarry was in the Epler and Ontelaunee Formations. It is labeled as a sand pit on the 1862 map.

About 1815, surveyors for the Union Canal were seeking clay for construction of the canal. They dug a test pit on a farm owned by John Ruth and found a fine quality sand. When Ruth realized the sand was good enough to sell, he commenced digging and selling it. This developed into a profitable venture, and he soon hired several men to work in his sand quarry. Most of the sand was sold to builders. After several years in the sand business, he sold the property to George See, who afterwards sold it to John Seltzer. In 1858, Seltzer died, and his widow,

Figure 215. Aragonite from the Seisholtzville mines, Hereford Township. Reading Public Museum collection 2004C-007-646.

Mary B.R. Seltzer, operated the sand business until her death in 1877. Her daughter, Mary Sheetz, took over the farm and sand business. The sand sold for $1.20 per ton. Sand was hauled from the quarry to Reading by horse team nearly every day. In later years, sand was sold to iron furnaces and the Pennsylvania and Reading Railroad for 7 cents per bushel. During the last few years of the quarry operation, most of the output was shipped by rail to the Robesonia Furnace and rolling mills and foundries in Reading. Operations ceased in 1892 (*Reading Eagle*, December 4, 1892).

HEREFORD TOWNSHIP

Seisholtzville District Iron Mines

The Seisholtzville District (fig. 221) is about 1 mile east of Seisholtzville in the northern part of Hereford Township. The discovery of iron ore near Seisholtzville generally is credited to John Rush of Dale Forge. In 1838, he took some ore—from an outcrop he found on land then owned by D. Bittenbender—to his smith shop and successfully tested it for iron. About 1845, a few wagon loads of limonite mined from the surface were taken to the Hampton Furnace in Lehigh County, where the ore was successfully smelted. In 1850, the Trexler family of Reading began prospecting for the ore body; however, the surface ore was scattered over a considerable area, and they were unable to locate its source (D'Invilliers, 1883, p. 281).

Two types of iron ores were mined. The "hard ore" was a mixture of magnetite and limestone that rarely averaged over 22 percent iron. The "soft ore," which was found in the outcrops, resembled a fine-grained black to brown powder; it was about 45 percent iron (D'Invilliers, 1883, p. 279-280).

When the Seisholtzville District was visited by D'Invilliers on May 20, 1882, the mines were idle. Much of the mining machinery had been removed. The reasons given by D'Invilliers for abandonment were: (1) excessive royalties of 30 to 50 cents per ton demanded by property owners, (2) the better quality soft ore was exhausted, and (3) the hard ore had too little iron to be profitably worked. D'Invilliers (1883, p. 279-280) stated: *"I am informed on the best authority that a great quantity of this hard ore still remains; but as it rarely averages over 22 per cent, of iron, and is subject to a royalty of from 30 to 50 cents per ton, it is hardly fit to raise."*

Prior to 1880, all ore mined in the Seisholtzville District was delivered by horse teams to the Red Lion Station of the Catasauqua and Fogelsville Railroad for shipment. As many as 40 teams were engaged at one time. The cost of hauling ore ranged from 30 to 50 cents per ton. In 1880, the Pennsylvania Transportation and Improvement Company built a system of tramways from the mines to the railroad station. They charged 35 cents per ton. The tramway was not a success because of numerous breakages and delays. As soon as the problems were fixed and the tramway was in working order, the mines were abandoned. The tramway was later moved to New York (D'Invilliers, 1883, p. 290).

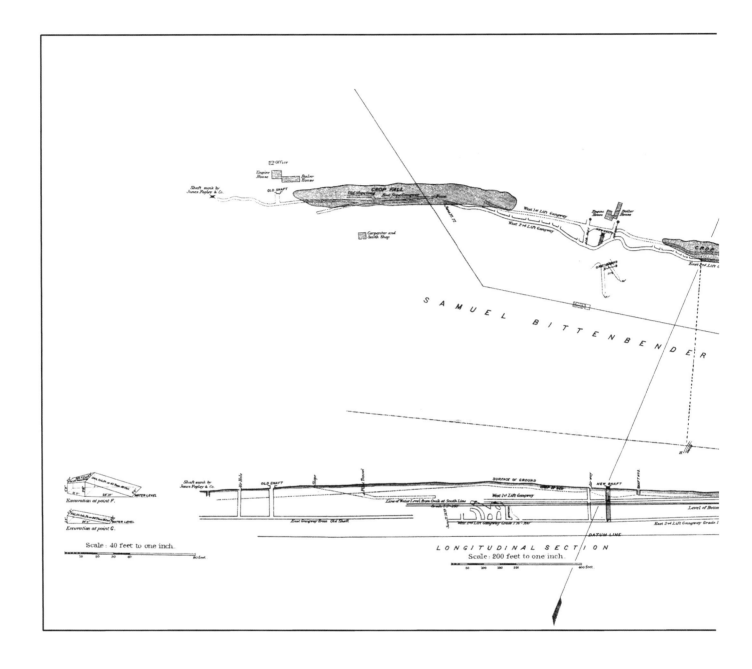

Figure 216. Second Pennsylvania Survey map of the Seisholtzville iron mines, Hereford Township. Modified from Pennsylvania Geological Survey (1883).

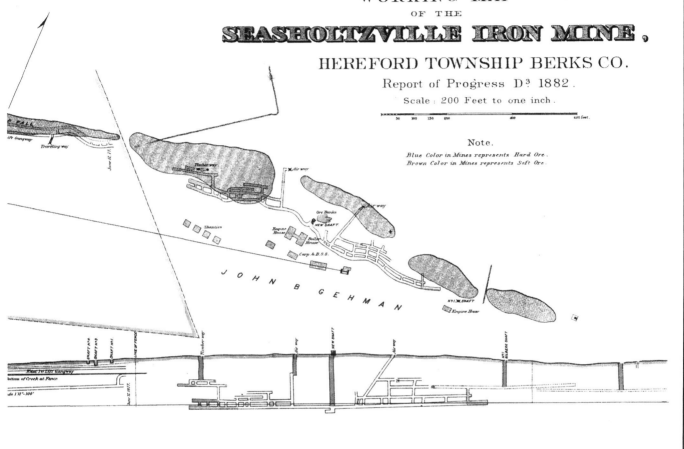

Figure 217. Incline at the Bittenbender mine, Hereford Township, 1901. Cars loaded with ore were hauled out of the mine on the incline. Carroll Delong is standing next to a walkway over the open pit. Courtesy of the Hereford Township Heritage Society.

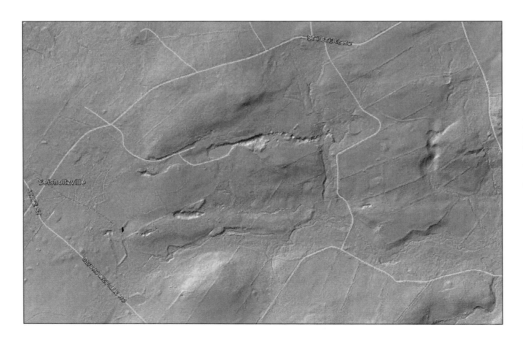

Figure 218. LIDAR image of the Seisholtzville mines, Hereford Township. Imagery provided by the Pennsylvania Spatial Data Access (PASDA).

Figure 219. Stilbite from Seisholtzville, Hereford Township, 11 cm. Ron Kendig collection.

Minerals

From D'Invilliers (1883, p. 396-399) unless noted.

Albite (Eyerman, 1889, p. 26)

Aragonite - sprays of acicular crystals associated with magnetite (fig. 215)

Augite (Eyerman, 1911, p. 4)

Calcite (Eyerman, 1889, p. 42)

Clinochlore

Graphite

Limonite (fig. 220)

Magnetite

Orthoclase (Prime, 1875, p. 6; analysis; Eyerman, 1911, p. 3, analysis)

Oligoclase

Pyrite

Pyroxene, var. sahlite (Eyerman, 1889, p. 15, analysis; Prime, 1875, p. 6, analysis)

Stilbite - sprays of crystals to 2 cm (fig. 219)

The main mines in the Seisholtzville District were on the Samuel Bittenbender and John B. Gehman properties. The mines were north of Perkiomen Creek in the area bounded by Seisholtzville, Saint Peters, Hollyberry, and Township Roads. D'Invilliers (1883, p. 288) estimated ore production from the Seisholtzville District to 1882 to be about 230,000 tons.

Bittenbender Mines

The William Bittenbender mine is shown on the 1882 topographic map (fig. 221). An open cut was located at about 40° 28' 07" latitude and 75° 35' 46" longitude; a pit or shaft to the east was at about 40° 28' 03" latitude and 75° 35' 54" longitude; and a pit or shaft east of the latter was on the south side of the Perkiomen Creek at about 40° 28' 02" latitude and 75° 35' 59" longitude on the USGS East Greenville 7.5-minute topographic quadrangle map (Appendix 1, fig. 19, location H-1). "Ore holes" shown on the 1882 topographic map (fig. 221) are visible on LIDAR imagery (fig. 218) on the east side of Hollyberry Road at about 40° 28' 08" latitude and 75° 35' 07" longitude; these are visible as an open pit on the 1937 aerial photography.

The Samuel Bittenbender property mines, shown as the Pennsylvania and Reading Coal and Iron Company iron ore mines on the 1882 topographic map (fig. 221), consisted of two open cuts and shafts. A flooded cut is at about 40° 28' 17" latitude and 75° 35' 43" longitude, and an open pit east of the flooded cut is at about 40° 28' 17" latitude and 75° 35' 56" longitude on the USGS East Greenville 7.5-minute topographic quadrangle map (Appendix 1, fig. 19, location H-2). The Bittenbender mine property, owned by Samuel Bittenbender in 1883, was jointly leased by the Philadelphia and Reading Coal and Iron Company and the Warwick Iron Company of Pottstown. The mine was leased by William S. Harvey in 1908 (Berks County deed book).

Samuel Bittenbender never mined the ore on his property, but rather leased his property to others for mining and collected a royalty on every ton of iron ore produced. As a result, he became one of the wealthiest men in the area. In the 1890s, he donated $3,000 in gold to Huff's Union Church to pay off their building fund (Huff's Union Church, 2000).

Early mining took place from open pits. In 1866, the Crane Iron Company leased the Bittenbender property, erected an ore washer, and began mining a limonite deposit. They mined the Samuel Bittenbender property for 3 months, producing only about 100 tons of ore, on which they paid 30 cents per ton royalty. The company mined only the hard ore, which they thought would be more profitable than the soft ore. They abandoned the mine on the Samuel Bittenbender property, but continued mining on the Christian Bittenbender property.

Figure 220. Limonite from the Gehman mine, Hereford Township, 8 cm. Reading Public Museum collection 2000C-027-840. Former D.B. Brunner specimen.

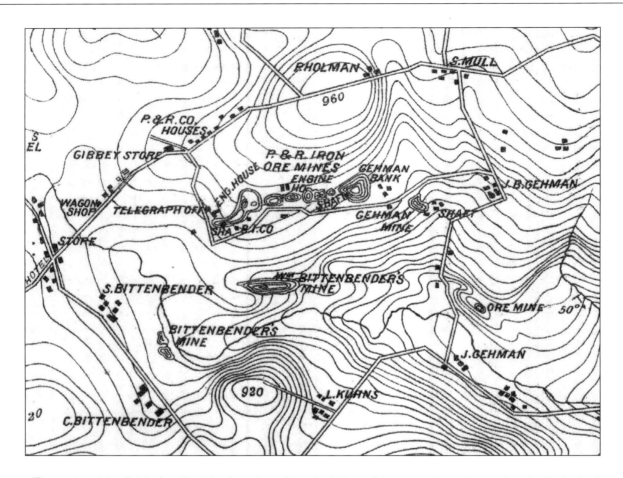

Figure 221. The Seisholtzville District mines, Hereford Township, 1882. From Pennsylvania Geological Survey (1883).

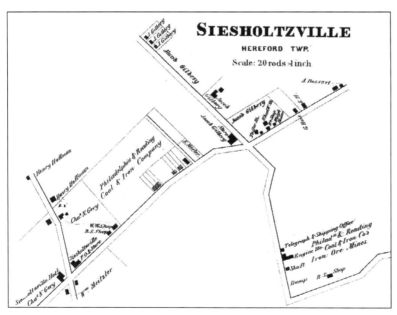

Figure 222. Location of the Philadelphia and Reading Coal and Iron Company mine at Seisholtzville, Hereford Township, 1876. From Davis and Kochersperger (1876).

The Samuel Bittenbender property was then leased by Frank Thomson for a short time. In 1872, Jacob Gilberg leased both the Bittenbender and Gehman properties. In March 1873, he sold the lease on the Bittenbender property for $64,000 to the Philadelphia and Reading Coal and Iron Company and Fegely, Jones & Gabel of Pottstown, who owned the Warwick Iron Company and also did business as the Steel Ore Company (fig. 225). The Philadelphia and Reading Coal and Iron Company (fig. 224) was the country's largest producer of anthracite coal from 1871 through the 1920s. At one time it controlled over 40 percent of the country's anthracite reserves. The two companies jointly began large-scale mining. Their shaft (fig. 222) at the extreme western end of the property was 40 feet deep.

Initially, the ore was taken by carts from open cuts and hauled 1.5 miles to the Red Lion railroad station for shipment. As many as 100 wagons were loaded at the mine in 10 hours. Hauling iron ore was a local industry, and some farmers kept five teams for the sole purpose of hauling ore. The mine operators later built a 2-mile rail line to the mine;

Figure 223. The Bittenbender Mine, Hereford Township, ca. 1890. Photograph courtesy of the Hereford Township Heritage Society.

the line was acquired by the Reading Railroad Company. Two trains carrying ore were run daily. Ore was hoisted up an incline using a steam engine. When the mine car reached the top, it was pushed along the platform and dumped directly into a rail car (*Reading Eagle*, June 4, 1898). The ore was smelted by the Warwick Iron Company, who later bought out the Reading Coal and Iron Company's share of the lease.

The Philadelphia and Reading Coal and Iron Company sank a slope east of the original shaft (fig. 216). The 100-foot slope produced a considerable quantity of soft ore. The thickness of the ore bed varied greatly; it swelled into large pockets and thinned to a bed only 3 feet thick (D'Invilliers, 1883, p. 282).

Mining continued without interruption from 1872 to 1877. In 1876, the mining operation employed 60 men and utilized two 18-horsepower steam engines. Production was 500 tons of ore per month. Ore was shipped to the Warwick Iron Company in Pottstown, the E & G Brooke Iron Company in Birdsboro, the Bethlehem Iron Company, the Keystone Furnace Company, and J. & J. Wister in Harrisburg. The No. 1 shaft (fig. 216) on the eastern side of the property was 112 feet deep with two gangways. The No. 2 shaft was 60 feet deep, and the No. 3 shaft was 102 feet deep (Commonwealth of Pennsylvania, 1877, p. 500).

From 1877 to 1879, the mine was idle; however, it continued to be dewatered. The main shaft on the Bittenbender property was furnished with a No. 10 Cameron pump, which pumped about 200 gallons of water per minute. The Reading Iron Company did some mining from 1879 to the spring of 1880. In 1880, the mine was abandoned and allowed to flood (D'Invilliers, 1883, p. 285-286). According to D'Invilliers, "*During the period of active working [1872 to 1877] 118,000 tons of ore were mined, upon which Mr. Bittenbender was paid a royalty of 50 cents per ton. This high royalty, together with the poor quality of the remaining hard ore, and the fact that further mining will necessitate the sinking of the shaft to procure ore, have all combined to entail idleness here. Nothing but a few pillars of ore, left to support the shaft and workings, remain above the present 100 foot level.*" Bittenbender was asked a number of times during 1878, when the mine was idle, to reduce his royalty to 35 cents per ton so that mining could resume, but he refused. In 1879, mining began again at a royalty of 50 cents per ton (*Reading Eagle*, September 24, 1879).

The 1880 Census (Pumpelly, 1886, p. 961) listed the Pennsylvania and Reading Coal and Iron Company as the oper-

Figure 224. Stock certificate of the Philadelphia and Reading Coal and Iron Company, 1939.

ator of the Seisholtzville mine. Production between July 1, 1879, and June 30, 1880, was 8,899 tons of magnetite. The ore was shipped to furnaces in the Schuylkill and Lehigh valleys.

A Mr. Hartzell leased a mine on the property from 1884 to 1898. In 1888, the Warwick Iron Company sank a new shaft to a depth of 180 feet. Gangways were driven from the shaft to the main vein. Bauen and Boyer were the mining contractors. About 1893, a Mr. Gotshall became a partner, and the mining was conducted under the name of the Bessemer Iron Company, Limited. Some of the ore was shipped to Birdsboro, but the bulk of it was shipped to Allentown (*Reading Eagle*, December 5, 1888, and June 4, 1898).

In 1896, the Bittenbender mine was in full operation. The Thomas Iron Company's Alburtis Furnace received 20 to 30 tons of ore per day. In 1898, there were two active shafts and one active slope, and 30 men were employed. Many of the older shafts had been abandoned. Two trains per day were run from the mine on the 2-mile long siding of the Reading Railroad. The mine was worked by a 230-foot deep slope, and all mining was done at a depth greater than 200 feet. The inflow of groundwater presented a great inconvenience; it required two 8-inch pumps running day and night to keep the mine dewatered. There was an additional 4-inch steam pump that was occasionally utilized. The deep cuts above the mine were kept dewatered to reduce infiltration. About 1,500 to 1,800 tons of ore were shipped monthly. In February 1898, 3,000 tons of ore were shipped (*Reading Eagle*, May 10, 1896, and June 4, 1898).

Mining continued until at least 1900. In that year, the *Reading Eagle* reported that a car loaded with iron ore from the Bittenbender mine derailed on the siding leading to the mine, and a dynamite explosion at the mine killed a miner named William Schindler (*Reading Eagle*, February 8 and April 14, 1900).

At the land surface, depressions in the ground marked deposits of soft ore, while the mounds and rolls indicated the position of underlying hard ore. The soft ore occurred as pockets of fine ore resembling black loam. The ore was never washed; it was so fine grained that it would wash away. The hard ore was more calcareous than the soft ore, which made for an excellent flux and was valuable for mixing with ores high in silica content. The hard ore was found at the bottom of the mine in continuous cone-shaped wedges, joined laterally together, with their apices pointing upwards. The soft ore was found in pockets between the wedges of hard ore (D'Invilliers, 1883, p. 283).

D'Invilliers (1883, p. 284) described the mining methods. Mining the soft ore required a great deal of expensive timbering. A shaft was first sunk from the surface to the hard or bottom ore, and sometimes through the latter to the foot wall of gneiss. The shaft was a double shaft, with one side for hoisting ore and the other for pumping water. From the bottom of the

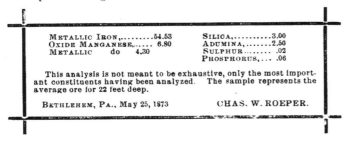

Figure 225. Steel Ore Company advertisement for ore from the Seisholtzville mines, Hereford Township. From the Griffith Jones Papers Concerning Iron Ore Furnaces (3615), Historical Collections and Labor Archives, Special Collections Library, Pennsylvania State University.

shaft, gangways were driven east and west, and pillars of ore were left standing for support. At regular intervals along the gangways, rises were mined up to the surface, sometimes along the hanging wall and sometimes along the foot wall. The rises were used to convey the mined ore down to the main gangway, where it was loaded on drift cars, taken to the bottom of the shaft, and hoisted to the surface in a cage.

After the rises reached the surface, counter gangways were driven, beginning at the top, towards each other from each rise until they met, and the top was allowed to fall in. If the ore bed was very wide, a second set of gangways was driven beside the first until the top of the bed fell. This plan was pursued down to the bottom of the mine. A new set of counter gangways were driven where the first had filled up, so that the bottom of the first set was the top of the second set and so on (see fig. 226). The soft ore was always thrown down the chutes to the main gangway.

Mining of the hard ore required blasting. Much larger openings could be made without fear of collapse. Cribbing was first started and refuse was used to fill up around it as it was mined. This was carried up to surface, keeping the bottom filled as the work progressed upward. Pillars of ore were left to protect the workings, some of which were afterwards removed while others were left standing.

Changes in the thickness of the ore bed were caused by the bulging of the hard ore. Just above the hard ore, and between it and the soft ore, there was a layer of ore with an open cellular structure from 3 inches to 1 foot thick forming a distinct bed. It was almost free from gangue and contained a high percentage of iron. This ore was considered a hard ore, but was different from the calcareous ore forming the lowest ore body in the mines. There was a distinct division between the hard and soft ores, and no gradation of one into the other. Large boulders of clay and limestone in the shape of

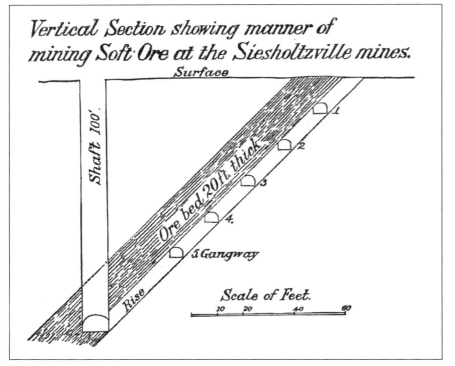

Figure 226. Section showing the method of mining soft ore at the Seisholtzville mines, Hereford Township. From D'Invilliers (1883, p. 285).

occasional partings of graphite. Some of the rocks consisted of massive hornblende, augite, some hypersthene, and graphite flakes. Some gneiss was composed of plagioclase, quartz, augite, and flakes of graphite 1/8-inch in diameter. Pegmatite, composed of orthoclase, quartz, and large graphite flakes, were observed, but were not common.

Analyses of ore from the Bittenbender tract were reported by D'Invilliers (1883, p. 286): 37.8 to 45.5 percent metallic iron, 2.16 percent manganese, and 9.5 to 16.5 percent silica.

Christian Bittenbender Mine

The Christian Bittenbender mine was located at about 40° 28′ 03″ latitude and 75° 35′ 56″ longitude on the USGS East Greenville 7.5-minute topographic quadrangle map (Appendix 1, fig. 19, location H-3). The mine consisted of two open cuts.

In 1866, the Crane Iron Company began mining on the Christopher Bittenbender property. They mined a 60-foot deep open cut until 1871 and produced 5,000 tons

balls from a few inches to 12 feet in diameter were found in the underlying calcareous ore (D'Invilliers, 1883, p. 283).

The cross section of the mine workings drawn on June 17, 1877, (fig. 216) shows six shafts sunk on the Bittenbender property from which a double set of east and west gangways were driven about 60 feet apart (D'Invilliers, 1883, p. 284).

Graphite occured in both the gneiss and limestone at the Bittenbender mine; however, the mine was never worked for graphite despite its abundance. Miller (1912a, p. 113-114) observed many specimens of white crystalline limestone with irregularly disseminated flakes of graphite 1/4-inch in diameter on the mine dumps. Some of the limestone was altered to a silicated rock with segregations of augite, graphite, and an asbestifom mineral. Other rocks consisted of dark red garnet, hornblende, augite, and graphite. Miller observed specimens of massive magnetite in which there were

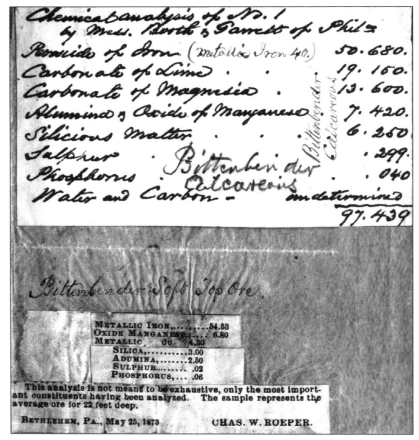

Figure 227. Analyses of ore from the Bittenbender mines, Hereford Township. From the Griffith Jones Papers Concerning Iron Ore Furnaces (3615), Historical Collections and Labor Archives, Special Collections Library, Pennsylvania State University.

of ore. The ore was sent to their furnaces in Catasauqua in Lehigh County. The mine was abandoned because of the large amount of quartz mixed with the ore at depth. The gangue was clay and limestone (D'Invilliers, 1883, p. 369-370).

In 1872, Henry Schankweiler briefly leased the mine. In 1879, R. Moll and Peter Worst leased the mine, but did very little mining. After passing through the hands of George Hess and C. Ziegenfuss, George Greis & Company obtained the lease. They erected a small washer and mined about 50 tons of ore, which was sent to the Crane Iron Company furnace in Catasauqua. The mine was abandoned after this last mining venture (D'Invilliers, 1883, p. 370).

J. B. Gehman Mine

Mines on the John B. Gehman property are shown as the Gehman Bank on the 1882 topographic map (fig. 226). The Gehman Bank was located at about 40°28' 19" latitude and 75° 35' 26" longitude, and the mine shown as the Gehman mine on the 1882 topographic map was located at about 40° 28' 22" latitude and 75° 35' 15" longitude on the USGS East Greenville 7.5-minute topographic quadrangle map (Appendix 1, fig. 19, location H-8). A small wedge of property between the Bittenbender property and the J.B. Gehmen property was known as the J.B. Gehman heirs' property.

Jacob Gilberg obtained a lease from John B. Gehman in 1873 (Berks County deed book) and began mining and washing the surface ore. However, the surface ore was soft ore and was so fine that much of it was lost in the washing process. Mining of the surface ore ceased after several shafts were sunk. Water was a considerable problem; 500 gallons per minute were pumped to keep the mine dry, which was difficult using only one No. 10 Cameron pump. Gilberg employed 20 to 40 men. A great quantity of ore was hoisted by horse power before steam power was introduced. About 12,000 tons of ore were mined on the eastern end of the property by 1882 (D'Invilliers, 1883, p. 286-288).

In 1877, Henry Guiterman, of Port Carbon in Schuylkill County, leased part of the western end of the property. By that year, ore had been mined from the western end to a depth of nearly 100 feet. In the spring of 1877, Guiterman sank a new shaft to a depth of 180 feet. The projected depth was 200 feet; however, hard gneiss was struck at 180 feet. A gangway was driven south on a 7-foot thick bed of soft ore. Drifts were then driven east and west on the ore bed for 300 feet on each side of the shaft. Hard ore similar to that on the Bittenbender property was encountered in the gangways; however, it had a low percentage of iron. The average thickness of the soft ore body was 13 feet. Guiteman worked as a contractor for the Bethlehem Iron Company in 1879 (Berks County deed book).

After Guiterman's death, the administrators of his estate, L.B. Morganroth and I.N. Hiteman of Shamokin, continued mining. About 50 to 70 men were employed, and daily production ranged from 30 to 100 tons (D'Invilliers, 1883, p. 287-288). The 1880 Census (Pumpelly, 1886, p. 961) listed Morganroth and Hiteman as the operator of the Guiterman mine.

Production between July 1, 1879, and June 30, 1880, was 11,200 tons of magnetite, which was shipped to the Bethlehem Iron Company furnaces.

Chalins & Company leased the eastern part of the property and sank a 100-foot shaft. After mining for a short time, they sold the lease to Schweinbinz & Company, who sold it to the Bethlehem Iron Company in 1881. The Bethlehem Iron Company deepened the shaft to 130 feet, but hard ore was not found on this part of the property, and the mine was abandoned in 1881. Other lessees of the property included F.S. Shimer and the Colerain Iron Company in 1876 and Max Schweibenz in 1878 (Berks County deed book). Production from this property was reported to be 100,000 tons of ore. The royalty was originally set at 65 cents per ton, but was eventually reduced to 35 cents per ton (D'Invilliers, 1883, p. 288).

A mine on the J.B. Gehman heirs' property was leased by Hartzel and F.S. Shimer in 1873. They mined about 1,000 tons of ore, which was never sold because of the high percentage of silica. In 1880, Smink and Morganroth sank a 60-foot deep shaft, but abandoned it because of the poor quality of the ore (D'Invilliers, 1883, p. 370). That same year, some mining also was done by George Hess and others on the property. A shaft was sunk, and 1,000 tons of ore were reported to have been mined from workings 100 feet deep and 40 feet long (D'Invilliers, 1883, p. 291 and 370).

Mining continued into the late 1890s. The Reading Eagle (June 4, 1898) reported that in 1888, the number of mine employees exceeded 100, but this number shrunk to 30 men by 1898. In the fall of 1897, there were 17,000 tons of ore stockpiled at the mine; it was all shipped by the spring of 1898.

D'Invilliers (1883, p. 291-292) published analyses of the ore from the Gehman property. The hard ore was 24.7 - 25.3 percent metallic iron, 0.312 percent sulfur, 0.024 - 0.029 percent phosphorus, and 9.24 - 10.1 percent silica. The soft ore was 41.9 - 46.7 percent metallic iron, 0.175 percent sulfur, 0.055 - 0.079 percent phosphorus, and 13.6 - 19.8 percent silica.

Other Iron Mines and Prospects

Olafson Mine

The Olafson mine was south-southwest of the intersection of Hunter Forge Road and Hert Lane at about 40° 27' 30" latitude and 75° 36' 49" longitude on the USGS East Greenville 7.5-minute topographic quadrangle map (Appendix 1, fig. 19, location H-5). The mine was in hornblende gneiss.

In 1974, the Olafson mine was located by Jay Lininger and Jim Quickel, who recovered mineral specimens from the nearby dumps. At the time, the mine shaft was being filled with trash and lawn clippings. The shaft appeared to be exploratory (Lininger, 1986). The mine was named by Smith (1976, p. 216) for the then current property owners, as no historical information on the mine seemed to be available.

Figure 228. Loellingite from the Olafson mine, Huffs Church, Hereford Township. Steve Carter Collection.

It is likely that this mine is the iron prospect briefly described by D'Invilliers and Miller. D'Invilliers (1886, p. 292-293) mentioned that iron ore was found at Huff Church, but the deposit was not "greatly developed." The ore was titaniferous iron. It is labeled as "ore hole" on the 1882 topographic map. Miller (1912a, p. 114) noted an iron prospect one mile south of Seisholtzville and stated: *"At this place there is a dark-colored banded gneiss composed of feldspar, quartz, hornblende, biotite, magnetite, garnet, pyrite and graphite. Some specimens of mica schist were seen on the rock heap and many pieces of massive magnetite."*

Minerals
From Smith (1978, p. 217) and Miller (1912a, p. 114)

Almandine - disseminated, massive, red; rarely crystals; common in gneiss

Biotite - flakes

Chalcopyrite - fine-grained, disseminated; associated with magnetite and pyrite

Fayalite - seams to 1 cm and larger with interlocking crystals in the magnetite-hornblende rock

Graphite - in gneiss from a trace to 10 percent of the rock mass

Hornblende

Loellingite - the safflorite-(Fe) of Smith (1978, p. 214); tin-white, metallic grains typically to 5 mm but as large as 1 cm; associated with magnetite (Smith, 1978, p. 214 and 217); identified by Gene Foord and Dick Erd as lollingite (Foord, 2000, p. 4)

Magnetite - massive

Pyrite - rare

Pyrrhotite - disseminated blebs in magnetite and gneiss

Quartz

Huff Church Locality

The Huff Church locality may be the Olafson mine. D'Invilliers (1883, p. 292-293) reported that iron ore was found at Huff Church, but was only a prospect. Brunner and Smith (1883, p. 280) reported titanite in magnetite from a mine close to Huff's Church.

Minerals

Biotite - near Huff Church (Eyerman, 1889, p. 21)

Orthoclase (D'Invilliers, 1883, p. 396)

Muscovite - at Huff Church in rhombohedral crystals (D'Invilliers, 1883, p. 399)

Magnetite, var. titaniferous (D'Invilliers, 1883, p. 398)

Titanite - clove brown crystals in magnetite (Brunner and Smith, 1883, p. 280)

Rauch Mine

The Rauch mine was between Bob White and Hunter Forge Roads, west of West Branch Perkiomen Creek, and northwest of the intersection of Hunter Forge and Old Mill Roads at 40° 27′ 27″ latitude and 75° 37′ 03″ longitude on the USGS East Greenville 7.5-minute topographic quadrangle map (Appendix 1, fig. 19, location H-6). The mine was in hornblende gneiss. Pyroxene crystals were reported by D'Invilliers (1883, p. 399).

Dale Mine

The open-pit Dale mine (fig. 229) is on the Hereford-Washington Township border, southeast of the intersection of Dale Road and Dairy Lane at 40° 25′ 33″ latitude and 75° 36′ 56″ longitude on the USGS East Greenville 7.5-minute topographic quadrangle map (Appendix 1, fig. 20, location H-7). The mine was in the Leithsville Formation. The Dale mine was an early iron mine and was operated in the late 1700s or early 1800s. Minerals reported from the Dale mine are titanite and magnetite (D'Invilliers, 1883, p. 401).

Rush's Ore Pit

The Rush's ore pit was next to Dale Road, west of its intersection with Dairy Lane at 40° 25′ 41″ latitude and 75° 37′ 03″ longitude on the USGS East Greenville 7.5-minute topographic quadrangle map (Appendix 1, fig. 20, location H-10). Limonite was mined from the open-pit mine in the Leithsville Formation. D'Invilliers estimated ore production at 4,000 to 5,000 tons. The ore was sold to the Thomas and Pottstown Iron Companies. D'Invilliers (1883, p. 369) noted: *"It has now been idle for 10 years and is entirely filled with water (the alleged cause of its abandonment.) The cut is lined with a decomposed limestone clay."* The mine was abandoned about 1872.

Figure 229. Dale mine, Hereford Township, March 2015.

Reitnauer Mine

Several old limonite mines were located just east of Seisholtzville. One of them was reopened by John D. Reitnauer of Alburtis, who sank a 45-foot deep shaft in the bottom of an old 25-foot deep open pit. At the bottom of the shaft he drifted about 18 feet in one direction and 20 feet in the other and obtained a considerable quantity of iron ore. In the spring of 1910, he mined and shipped about 22 tons of ocher from the bottom of the open pit where the washings from the old iron mine had been allowed to settle. The ocher was shipped to Lincoln, New Jersey, and was sold for $2.50 per ton. Some ocher also was shipped to Henry Erwin and Sons in Bethlehem (Miller, 1911, p. 40).

Limestone Quarries

Hampton Furnace Quarries

Six quarries (fig. 231), known as the Hampton Furnace quarries, were located along a limestone outcrop 150 to 200 feet thick. The quarries were on the north side of a hill, south of Sigmund Road, at the Berks-Lehigh County border on the USGS East Greenville quad. The quarries were in the Leithsville Formation. The quarries supplied limestone for flux to the Hampton Furnace as the limestone was a good flux for silicious iron ores. The quarries also produced stone for agricultural lime (D'Invilliers, 1883, p. 143-144). LIDAR imagery shows that the easternmost quarry (the David Benfield quarry) was in Lehigh County. The quarries are described below from east to west.

The next quarry to the west from the Benefield quarry was owned by James and Lewis Christman. The quarry was at about 40° 28′ 08″ latitude and 75° 33′ 55″ longitude (Appendix 1, fig. 19, location H-11). The quarry exposed an impure blue stone, which was unfit for making lime, and a dove-colored dolomitic stone, which was quarried and burned to produce lime. The rocks dipped about N. 21° W. 83° (D'Invilliers, 1883, p. 143).

The Jacob Christman owned the third quarry in the line. This small quarry was at about 40° 28′ 08″ latitude and 75° 33′ 58″ longitude (Appendix 1, fig. 19, location H-12). The rocks dipped N. 30° W. 87°, and the quarry exposed about 90 feet of mostly compact white dolomitic limestone. The quarry was idle when visited by D'Invilliers (1883, p. 143).

Henry Roth owned the fourth quarry in the line. The quarry was abandoned at the time of D'Invilliers visit (D'Invilliers, 1883, p. 143-144). The quarry was at about 40° 28′ 08″ latitude and 75° 34′ 00″ longitude (Appendix 1, fig. 19, location H-13). LIDAR imagery shows it to be the smallest of the six quarries.

Jonas Shaub worked the fifth quarry in the line. It was at about 40° 28′ 08″ latitude and 75° 34′ 04″ longitude (Appendix 1, fig. 19, location H-14). It was the only quarry of the six being actively worked at the time of D'Invilliers' visit. The quarry exposed 150 feet of limestone. An analysis of a composite sample was published by D'Invilliers (1883, p. 144): 52.9 percent calcium carbonate, 42.8 percent magnesium carbonate, and 3.9 percent silica. The composite sample did

Figure 230. Limonite from Dale Forge, Hereford Township, 9 cm. Reading Public Museum collection 2000c-27-157.

HEREFORD TOWNSHIP

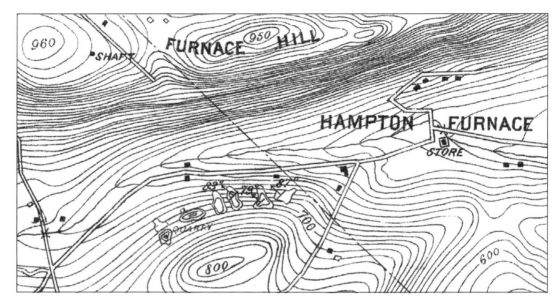

Figure 231. Hampton Furnace quarries, Hereford Township, 1882. From Pennsylvania Geological Survey (1883).

not include rock from a 6-foot thick bed of hard, siliceous limestone that was unfit for making lime. It was left untouched by the quarrymen, who called it "fire stone" because of its seeming immunity to fire. There was a kiln at the Shaub quarry with a capacity of 700 bushels. The lime was sold for agricultural use to neighboring farmers for 10 cents per bushel (D'Invilliers, 1883, p. 144).

A. Schantz owned the sixth quarry in the line and worked it only for his own use (D'Invilliers, 1883, p. 144). It was small quarry at about 40° 28′ 08″ latitude and 75° 34′ 06″ longitude (Appendix 1, fig. 19, location H-15).

Trollinger Quarry

The Trollinger quarry (fig. 234) was southeast of the intersection of Kemp and Airport Roads at 40° 25′ 48″ latitude and 75° 37′ 40″ longitude on the USGS Manatawny 7.5-minute

Figure 232. Lime kiln at one of the Hampton Furnace quarries, Hereford Township, March 2015. The photograph on the right shows the well-preserved loading hole at the top of the kiln.

topographic quadrangle map (Appendix 1, fig. 18, location H-16). The quarry was in the Leithsville Formation.

Cornelius and Peter R. Trollinger were listed as lime producers in the 1877 Berks County Business Directory (Phillips, 1877). In 1882, the quarry was operated by Cline and Weiler. The quarry exposed a massive, dark blue, quartzose limestone about 40 feet thick, conformably overlain by 2 to 3 feet of light grey decomposed slate. At the south end of the quarry, the arms of a shallow syncline dipped S. 20° E. 43° and N. 37° W. 6°. At the north end of the quarry, the limestone dipped S. 18° E. 29°. Lime was burned in a kiln on the property (fig. 235) and used by neighboring farmers (D'Invilliers, 1883, p. 146).

Clemmer Quarry

The A.G. Clemmer quarry was southwest of the intersection of Pennsylvania State Route 100 and Huffs Church Road

Figure 233. One of the Hampton Furnace quarries, Hereford Township, March 2015.

Figure 234. Trollinger quarry, Hereford Township, March 2015. The flooded quarry is covered with ice.

at 40° 25′ 23″ latitude and 75° 34′ 24″ longitude on the USGS East Greenville 7.5-minute topographic quadrangle map (Appendix 1, fig. 20, location H-17). The quarry was in the Leithsville Formation. Abraham G. Clemer [SIC] was listed as a lime producer in the 1877 Berks County Business Directory (Phillips, 1877).

The Clemmer quarry was located near the top of a hill. The quarry was about 50 feet long and exposed about 40 feet of white magnesian limestone. The limestone was generally massive, but in some places, it was heavily fractured. The dip was S. 65° E. 35° on the western side of the quarry (D'Invilliers, 1883, p. 148). About 50 feet further uphill, there was a smaller abandoned quarry exposing a cherty limestone that dipped S. 63° E. 40°. The limestone from both quarries was burned in two kilns near the small creek close to Clemmer's house. The kilns were operated about 3-1/2 months of the year and produced about 1,600 bushels of agricultural lime per week. The lime was sold for 11-1/2 cents per bushel. D'Invilliers (1883, p. 149) published an analysis of the limestone: 54.7 percent calcium carbonate, 43.4 percent magnesium carbonate, and 1.2 percent silica.

Seisholtzville Granite Quarry

The Seisholtzville granite quarry was east of Hunter Forge Road, north of its intersection with Old Mill Road near 40° 27′ 22″ latitude and 75° 36′ 41″ longitude on the USGS East Greenville 7.5-minute topographic quadrangle map (Appendix 1, fig. 19, location H-4). The quarry was in hornblende gneiss.

Figure 235. Lime kiln at the Trollinger quarry, Hereford Township, March 2015. (BE-396, LK-3, 7681)

Figure 236. Thomas Edison with a group of men at the Seisholtzville granite quarry, Hereford Township. Edison is at lower center standing on a rock next to David G. Seisholtz. From the collection of the Historical Society of Berks County Museum and Library, Reading, Pa.

Figure 237. Thomas Edison at the Seisholtzville granite quarry, Hereford Township. From left to right, Rev. James N. Blatt, Rev. William F. Bond, David G. Seisholtz, and Thomas A. Edison. Courtesy of the Hereford Township Heritage Society.

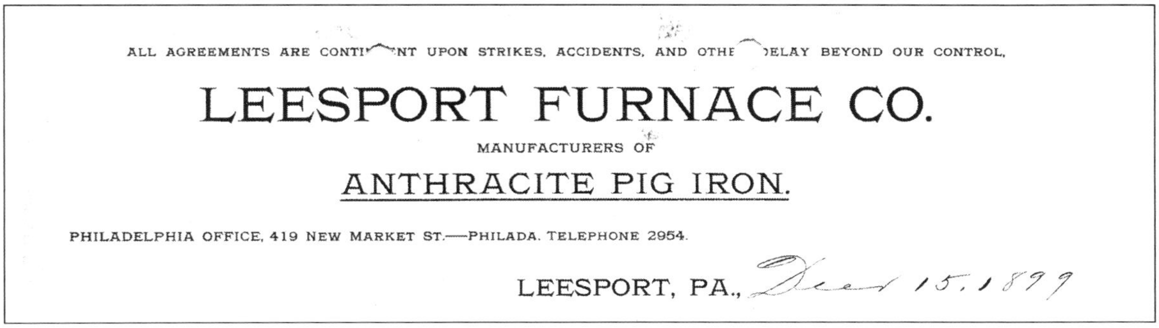
ALL AGREEMENTS ARE CONTI NT UPON STRIKES, ACCIDENTS, AND OTHE ELAY BEYOND OUR CONTROL.

LEESPORT FURNACE CO.

MANUFACTURERS OF

ANTHRACITE PIG IRON.

PHILADELPHIA OFFICE, 419 NEW MARKET ST.—PHILADA. TELEPHONE 2954.

LEESPORT, PA., Dec 15, 1899

Figure 238. Letterhead of the Leesport Furnace Company, Leesport, 1899. Courtesy of Dale Richards.

A "granite" quarry was located on the farm of David G. Seisholtz (1836-1918). The quarry supplied local building stone. Stone 4.5 feet square by 18 inches thick was used as the foundations for pillars in the P. Barbey and Son's brewery in Reading (*Reading Eagle*, December 16, 1896). Thomas Edison visited the Seisholtzville granite quarry (figs. 236 and 237). The date and the reason for the visit is unknown.

Gregory Graphite Prospect

About 1902, a prospect shaft about 15 feet deep was sunk on the farm of Nathaniel Gregory in Harlem in search of an economic deposit of graphite. Miller (1912a, p. 114) described the prospect: "*Considerable water was encountered and the project was abandoned. The host was graphitic gneiss, which contains much orthoclase and quartz with smaller amounts of graphite and biotite. The graphite occurred in small disseminated flakes and in sheeted masses. The feldspar was greatly decomposed and much of the graphite was exceedingly friable.*"

Hereford (Mayburry's) Furnace

Hereford Furnace, also called Mayburry's Furnace, was located on the West Branch Perkiomen Creek. The furnace was built by Thomas Mayburry and went into blast in 1745 or 1746. The furnace output, which was small, went to Mayburry's Green Lane Forge. In 1745, Mayburry rented the Mt. Pleasant Furnace from Thomas Potts & Company, but died intestate two years later, apparently not having made payments. At his death, his oldest son and heir, William, was only about twelve years old.

In 1755, William Mayburry, who was then 21 years old, petitioned the Orphans Court at Reading to value his father's estate so that it could be settled. The court valued the real estate at £1,100. By 1757, William had put the Hereford Furnace back in blast and also reopened Green Lane Forge. William operated both iron works until his death in 1764.

Because William's sons were underage at the time of his death, William's brother, Thomas Mayburry, a Quaker merchant from Philadelphia, rented and ran the furnace. William's widow, Anne, married Richard Tea, and they decided to rent out the furnace and forge in 1771. The furnace was closed for a period during the Revolutionary War. The furnace was out of blast in 1783 when Samuel Hermelin (1783) indicated that it was abandoned because of a lack of timber for fuel and nearby ore. In 1784, Thomas Mayberry's property was seized and sold by the sheriff. The 1,500-acre property included the Hereford Furnace and the Rock mine, located in the Rittenhouse Gap District in Longswamp Township.

KUTZTOWN BOROUGH

Kutztown Furnace

The Kutztown Furnace was built by the Kutztown Iron Company, which was incorporated in 1872. The furnace was constructed in 1873 by contractors Lee, Noble & Company. The annual capacity of the furnace was 8,300 tons. The furnace was first leased by Charles H. Nimson and Company. The furnace was acquired by the Philadelphia and Reading Coal and Iron Company, which leased it to different operators, of which William M. Kauffman & Company was the most successful. In July 1883, a boiler explosion toppled the smoke stack, which fell across the casting house and demolished it. The furnace was subsequently abandoned.

LEESPORT BOROUGH

Leesport Iron Company (Leeeport) Furnace

The Leesport Iron Company was formed on November 27, 1852. Issac Eckert served as president of the company from 1852 to 1873. Among the other founders of the company were George N. Eckert, John J. Kauffman, Samuel Kauffman, James Millholland, William M. Heister, W.H. Clymer, Fred S. Hunter, Edward M. Clymer, and Nicholas V.R. Hunter.

The company's anthracite furnace, on the Schuylkill Canal on the east side of the Schuylkill River, was built in 1852. The first iron was produced on September 18, 1853. Ore was

Mining and Washing Iron Ore

Iron ore found in carbonate rocks was known as limonite or brown hematite ore. The ore occurs massive, earthy, botryoidal, mammillary, concretionary, and occasionally stalactitic. It has a silky, often submetallic luster; sometimes, it is dull and earthy. When stalactitic, it forms pipe ore. When concretionary, it forms hollow spherical masses known as pot or bomb-shell ore. These hollow masses commonly contain water or masses of unctuous clay; their interior surface often presents a glazed appearance because of a very thin coating or incrustation of manganese oxide, which imparts a nearly black varnish-like surface. Sometimes, the bomb-shell ore is solid, and its interior has a honey-combed appearance. Most limonite ore occurs in small pieces, which have to be separated from the enclosing gangue by washers (Prime, 1875, p. 15).

Open-pit mining was the chief method of mining limonite ores in Berks County. When a pit was opened, horses and carts were used to carry the ore to the washer. As the pit became deeper, a road had was constructed for the horses to ascend. The overburden or stripping was washed if it contained sufficient ore to pay for this operation. The amount of stripping that needed to be removed before reaching ore varied greatly. In some places, a foot or two of overburden had to be removed, while in other localities 40 to 50 feet of overburden had to be removed. After the ore was reached, places in the pit with lean ore were left unless rich ore was found underneath (as was generally the case).

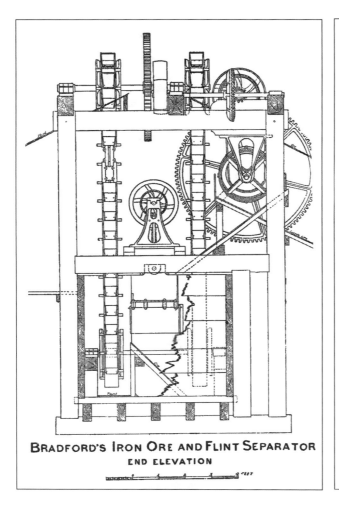
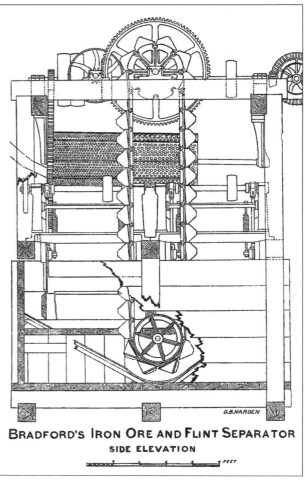

Figure 239. The Bradford ore separator. From Prime (1875, p. 53 and 55).

The miners took everything out of the pit, whether ore or barren clay. They used picks and shovels to extract the ore; blasting was never used. The ore was transported to the washer in wagons. At the washer, the ore was separated from the clay, slate, and sandstone associated with it. Three kinds of washers were used—single horizontal shaft, the Bradford washer with two shafts (fig. 239), or the Thomas washer with two shafts (fig. 240).

Bradford's washer was used by both the Thomas Iron Company and the Crane Iron Company. In this washer, the ore and rock, after being divided into two or more sizes by being passed through the rotary sieve, fell into a jig where the ore, quartz, and rock were separated according to their specific gravities; the ore being the heaviest fell to the bottom. In order to make the jig continuous in its action, self-acting rakes were arranged by which the rock and quartz were raked off the surface. These were then hoisted by a chain to be dumped on an inclined platform. This washer apparently worked very well for the separation of quartz and ore; it did not, however, separate the ore from the rock very well. This failure was apparently due in part to a lack of sufficient sieves for sizing and partly to the very light character of the ore mixed with the slate, so that the difference in the specific gravities of the two was small. Another drawback of this washer is the comparatively small quantity of ore washed per day.

The Thomas washer consisted of two shafts 20 to 24 feet long armed with teeth set at an angle as shown in figure 240. The shafts had an inclination of about 12 to 14 inches over their entire length. The washers were driven by steam power, which was connected to the lower end of one of the shafts by a chain. The other shaft revolved by geared wheels attached to the upper end of the shafts. The ore was carried to the upper end of the washer by the teeth, and after passing over a sieve, it fell into a wheelbarrow (Prime, 1875).

The ore washer was erected on a framework 10 to 20 feet above the ground in order to have enough distance to dump the ore, gravel, and clay. The ore washing process was as follows. The ore was dumped from a cart or car between the two shafts at the lowest end, care being taken to have an abundance of water to wash away the clay. To supply water, a wooden gutter, perforated with holes, ran the entire length of the washer. The flow of water was controlled by stopping up the holes. The heavier ore sank to the bottom of the trough underneath the shafts and was carried up to the top of the trough by the teeth. The lighter clay floated off at the lower end of the trough and was carried in gutters on a trestlework to the mud dam. The amount of ore washed in a day varied from 15 to 35 tons; the average was 20 to 25 tons (Prime, 1875).

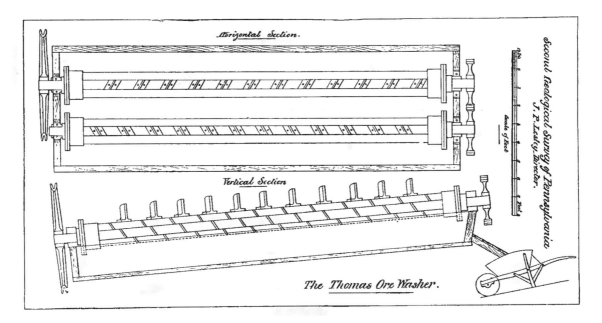

Figure 240. The Thomas ore washer. From Prime (1875, p. 49).

supplied by the Moselem mine, the Cornwall mine in Lebanon County, and several small mines in the vicinity of the furnace. The initial furnace capacity was 6,000 tons of iron per year. The pig iron produced was marketed under the brand name "Leesport." In 1855, the furnace produced 4,778 tons of iron. The furnace used a mixture of seven-eighths limonite and one-eighth magnetite (Lesley, 1859).

In 1871, the annual capacity was increased to 14,000 tons per year when the furnace was rebuilt. The new furnace used 75 percent anthracite coal and 25 percent coke. The ore used was limonite from the Moselem mine and magnetite from the Cornwall mine in Lebanon County. After the reconstruction, the furnace was operated by L.M. Kauffman.

After the furnace resumed operation, the company discovered that two of the three farms it owned were underlain by limestone suitable for furnace flux. The money saved by mining their own limestone was used to build a mansion for the general manager on one of the farms.

The depression of 1873 had a huge impact on the iron industry. From 1874 to 1885, the furnace operated only sporadically. In 1885, the Leesport Iron Company was reorganized with R.F. Leaf as president, Captain P.R. Stetson as secretary-treasurer, and M.P. Kenney as general manager. The company operated the furnace until 1895, at which time it was shut down. In 1899, the furnace was sold to the Leesport Furnace Company (fig. 238), which was owned by three Philadelphia men—O.A. Keim, T.W. Kiesaber, and a Mr. Swartley. The company produced its last iron on December 25, 1914. In 1928, the Kiesaber family sold the furnace to Garrett J. Rehr & Brothers of Reading for dismantling (Wittits and Dissinger, 1990).

LONGSWAMP TOWNSHIP

Mining was a very important industry in Longswamp Township from before the Revolutionary War to about 1920. Many more mines existed in Longswamp Township than are described in this book.

Jacob Lesher is credited with the discovery of iron ore in Longswamp Township sometime prior to 1800. It is known that a charcoal iron furnace was in operation on Little Lehigh Creek around 1797. This furnace later became known as the Mary Ann Furnace. The first stove used to burn anthracite coal, the Lehigh Coal Stove, was cast at the Mary Ann Furnace (Wagner and others, 1913, p. 183-184). John J. Thompson also owned tracts of land that contained iron ore deposits in Longswamp Township prior to 1820.

At one time, over 100 mines were in operation in Longswamp Township. According to a report in the June 4, 1890, Reading Eagle, a miner traveling from Macungie in Leigh County to Mertztown in Longswamp Township counted 120 iron mines in operation and estimated that another 50 were unseen. Estimated iron ore production in 1875 was 500,000 tons (Davis and Kochersperger, 1876). As late as 1910, the

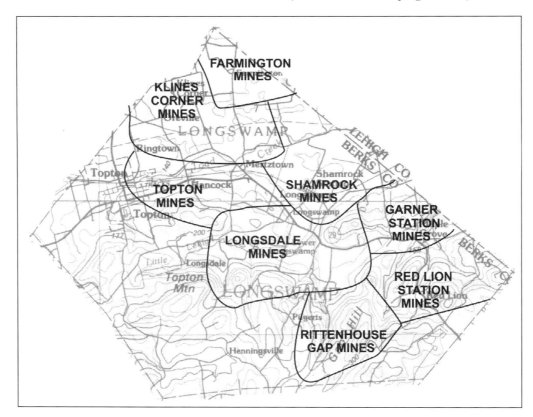

Figure 241. Mining areas in Longswamp Township.

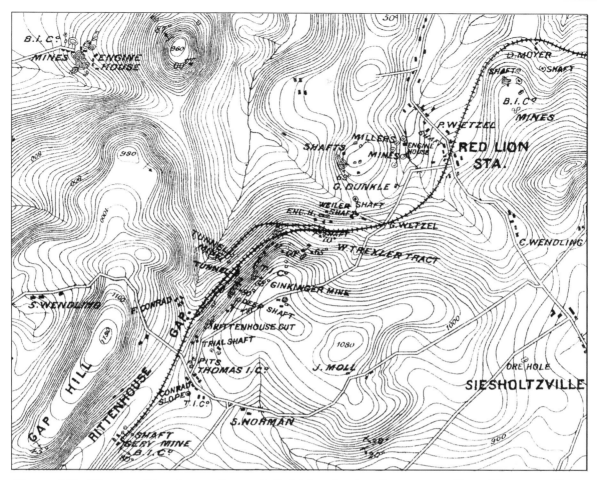

Figure 242. Mines at Rittenhouse Gap and Red Lion Station, Longswamp Township, 1882. From Pennsylvania Geological Survey (1883).

Census reported 161 full-time miners living and working in Longswamp Township, most of them living in frame shanties. Nearly all of them were Pennsylvania Germans ranging in age from 12 to 64, including one woman, 53 year old widow Sarah Stettler.

The iron mines in Longswamp Township are grouped into eight areas in this book (fig. 241). The areas are: Rittenhouse Gap District, Red Lion Station, Garner Station, Klines Corner, Farmington, Shamrock, Longsdale, and Topton areas.

Many of the iron mines in Longswamp Township shown on historical maps have numbers rather than names; the names of the mines are unknown. The numbers were on maps by Prime (1875), Prime (1878), and the 1882 topographic map. These maps show the mine locations. The origin of the numbers was described in a note from State Geologist J. Peter Lesley in Prime (1875, p. 57): *"The numbers on the map are those of the principal ore banks in the order in which they were visited; and the same numbers are given in the pages of the text, together with the names of the owners or lessees of the banks. The smaller and less important are not numbered, but are located accurately on the map. To have re-arranged the numbers in their geographical order would have involved the re-drawing of the map, and a delay in the publication; and any new openings to be hereafter inserted and reported, would disarrange the order of numbers, unless fractional numbers were employed, which would break the system in another sense."* Prime (1878, p. 2) stated: *"On the extreme west of the map are a few mines with numbers attached to which no reference is made in the key-list, as they are situated in Berks County, and will be referred to and described when the map of that portion of the State—now nearly completed—is issued."* However, the 1882 topographic map retained the numbers.

Rittenhouse Gap District

The Rittenhouse Gap District is located on the USGS Manatawny and East Greenville 7.5-minute topographic quadrangle maps. The district is underlain by felsic to mafic gneiss. The Rittenhouse Gap District mines also are known as the Rock mines, a name derived from the original Rock mine on the Rittenhouse property, as well as from the generally hard character of the gneiss gangue found in the mines (D'Invilliers, 1883, p. 246).

Figure 243. Stock certificate of the Southern Coal and Iron Company, 1924. The Southern Coal and Iron Company was the last mining company to do exploration work at Rittenhouse Gap.

The Rittenhouse Gap District (fig. 242) is one of the oldest iron ore-producing districts in Berks County with mining occurring prior to 1785. Iron ore was mined from 10- to 20-foot thick veins in open cuts up to 60 feet deep and 200 feet long (Smith, 2003). Mining continued intermittently until the 1920s. The mines were located with surveying instruments during the topographical survey made in 1879, and a new map of the district was made in 1882 (fig. 244) and published in 1883 by D'Invilliers (1883, p. 244).

D'Invilliers (1883, p. 248) took a dim view of the practice of contract mining in the Rittenhouse Gap District. He stated: *"The contract system of mining has also done its share towards ruining the property, for it acknowledged no future, and worked only for a certain end of taking out all the available cheap ore as quickly as possible, regardless of the eventual condition of the mine. The iron percentage of these mines will not average over 40 or 45, and they would not be worked at all (owing to their enormous quantity of silica) if it were not for the fact that they are almost entirely free from sulphur and phosphorus, and can be used, consequently, for Bessemer iron. They are largely mixed with the rich foreign ores, and some of the native earthy limonites, and so produce a good quality of iron."*

In 1889, Thomas Edison and Dr. H.K. Hartzell experimented with a small iron-concentrating plant near Bechtelsville (Ball, 1895). Shortly thereafter, Hartzell continued the experiments at Rittenhouse Gap. After experimenting with different methods and machines, he finally settled on the Ball Norton system in 1890. The Ball Norton separator used a combination of alternating magnetic poles and vibratory conveying works to separate out magnetic material. After being in successful operation for a while, the plant burned down in 1891, but was soon rebuilt, and operations resumed in 1892 (Ball, 1896).

The last known mining in the Rittenhouse Gap District was in 1920. The ore, principally from development work, was concentrated in a small mill (Charlton, 1921).

On January 18, 1922, the Rittenhouse Gap mines were acquired jointly by the Southern Coal and Iron Company of Virginia and the Rittenhouse Iron Company of Pennsylvania. John S. Birkinbine of Birkinbine Engineering of Philadelphia became general manager of the Southern Coal and Iron Company and

LONGSWAMP TOWNSHIP

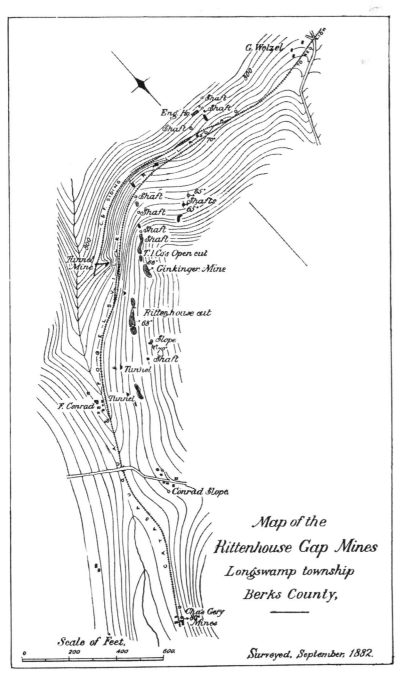

Figure 244. Map of the Rittenhouse Gap mines, Longswamp Township, 1882. From D'Invilliers (1883, p. 245).

The company sold stock to raise $1.5 million in capital (fig. 243). The company officers were E.E. Francy of Philadelphia, president; W.J. Jegen of New York, vice president; F.D. Holt, secretary and treasurer; and John S. Berkenbine, general manager. The company's office was located at 1500 Walnut Street in Philadelphia (Moody's Manual of Railroads and Corporation Securities, 1922, vol. 2, p. 417). According to the mortgage for the Rittenhouse Iron Company on file in the Berks County courthouse, Daniel E. Curran was the president and William E. Behan was the secretary and treasurer.

Mortgages taken out by the Southern Coal and Iron Company on record at the Berks County courthouse were from Adolph Pluemer ($29,000 in 1923), the Manayunk National Bank ($3,663.24 in 1924), the Rittenhouse Iron Company ($14,000 in 1924), and Frank E. Baker, ($19,000 in 1925).

It is not known if the Southern Coal and Iron Company or the Rittenhouse Iron Company did any actual mining. In 1924, the Reading Eagle (October 11, 1924) announced a deed transfer from the Thomas Iron Company to the Southern Coal and Iron Company. In 1926, the deed was transferred back to the Thomas Iron Company by order of the sheriff (recorded on November 15, 1926).

In 1928, Gorge D. Schmoyer acquired one of the Southern Coal and Iron Company properties at sheriff's sale. In 1929, Sheriff Thomas B. Kellon sold other properties of the Southern Coal and Iron Company totaling nearly 100 acres in Longswamp Township and Lower Macungie, Lehigh County, to Linn H. Schantz for $1. The sale was subject to a mortgage of $36,503.17 held by Adolph Pluemer. Sheriff Kellon was not able to locate an office or officer of the Southern Coal and Iron Company and served notice by posting the property (*Reading Eagle*, January 13, 1929). It is presumed that Schantz could not make the mortgage payments after the start of the Great Depression. In 1930, the bank of Catasauqua acquired one of the Southern Coal and Iron Company properties at sheriff's sale.

In 1967, the Industrial Valley Bank of Pottstown filed suit in Berks County court to recover $35,425 interest on the $13,000 mortgage taken out on January 18, 1922, by the Southern Coal and Iron and Rittenhouse Iron Companies. The mortgage, interest, and court costs totaled $48,815. The Rittenhouse Iron Company was no longer located at it's Philadelphia address, and the Southern Coal and Iron Company had its charter revoked in 1926. Both companies were believed to be defunct. The bank foreclosed on the property (*Reading Eagle*, August 23 and September 24, 1967). The property is currently (2016) owned by the Bear Creek Mountain Resort. Hiking trails transverse the area where the mines are located, and the open cuts are easily accessible.

was put in charge of its mines at Rittenhouse Gap (*Engineering World*, 1921, vol. 18, no.1, p. 441). The companies took out a $13,000 mortgage for 42 acres of land. The Southern Coal and Iron Company estimated the property contained 700,000 tons of ore. The company purchased the plants and machinery on the property, which included concentrating and separating plants for producing iron concentrates, sand, and crushed rock.

The Catasauqua and Fogelsville Railroad

The Catasauqua and Fogelsville Railroad was built in the 1850s to transport iron ore from local mines in Lehigh and Berks Counties to furnaces along the Lehigh River. The railroad was originally owned by the Crane and Thomas Iron Companies. The Catasauqua and Fogelsville Railroad later became part of the Reading Railroad system.

The initial application by the Crane Iron Company to the Pennsylvania General Assembly for a railroad charter, around 1853, was met with fierce resistance by local farmers, who feared that trains would frighten livestock, set fires, and destroy the local farming districts. The iron company was forced to compromise and charter the Catasauqua and Fogelsville Plank Road on July 2, 1853. While plank roads were a popular improvement in transportation at the time, the short stretch that was constructed was found inadequate for hauling ore. The heavy wagons rapidly damaged the road and rendered it dangerous for travel. On April 20, 1854, the plank road was issued a modified charter to operate as the Catasauqua and Fogelsville Railroad. The newly chartered Thomas Iron Company partnered with the Crane Iron Company to construct the railroad in March 1856, and construction began shortly thereafter. The Crane Iron Company owned 60 percent of the railroad stock, and the Thomas Iron Company owned 40 percent.

The eastern terminus was in West Catasauqua in Lehigh County. There, it connected with the Crane Iron Company furnaces at Catasauqua and the private railroads of the Thomas Iron Company, which ran a short distance north to its furnaces at Hokendauqua. It also connected with the Lehigh Valley Railroad.

Construction of the railroad continued southward, and the line entered Longswamp Township at Red Lion from the northeast and continued to the Thomas Iron Company's mines at Rittenhouse Gap. A turntable was built at the end of the line. The railroad could not run a spur to every mine, and a number of ore wharves were constructed along the right-of-way. At the wharves, ore could be dumped from wagons into piles and later transferred to rail cars to be shipped to the furnaces.

Important mines could afford the expense of constructing branch lines, which was authorized by an April 8, 1861, supplement to the charter. The longest of these lines continued to Farmington, the site of a large ore wharf, and ended at the mines in Klines Corner. In the late 1880s, a spur line was built from the main line between Red Lion and Rittenhouse Gap to serve the Siesholtzville iron mines. In 1890, the Reading Railroad purchased most of the stock of the Catasauqua and Fogelsville Railroad as cheap iron ore from the Mesabi Range began to cause local iron mines to close (Wikipedia contributors, 2014).

Figure 245. Catasauqua and Fogelsville Railroad bed at the Rittenhouse Gap mines, Longswamp Township, March 2015.

The iron ore deposits in the Rittenhouse Gap District occur as a series of 8- to 16-inch wide veins along a dipping N. 35° E. contact between granite and granite gneiss. Individual magnetite veins are over 350 feet long along strike. The host rocks are intensely albitized and lack biotite and accessory magnetite. Magnetite mineralization is associated with quartz, biotite, and pyroxene, both within the veins and disseminated within 3 feet of the veins. Several metadiabase and felsic dikes crosscut the magnetite veins. The metadiabase dikes are intensely chloritized and contain minor fluorite, molybdenite, and specular hematite veinlets that suggest continued hydrothermal activity (Orner and Friehauf, 2000). The felsic dikes were described in detail by Smith (2003), who presented analyses of rock samples.

The Rittenhouse Gap mines had foot and hanging walls of gneiss composed of small crystals of pink and white feldspar, grains of white quartz, and occasionally a few grey mica scales. The general appearance of the rock is quartzose, varying in color from a grey to a dark pink. The ore veins were intimately associated with the rock; they imperceptibly grade into one another. The prevailing outcrop dips are very steep, from 70° to 80°, though declining sharply to perhaps 30° in some of the deeper mines. The steep dips at the surface made the outcrop a prominent feature, rising up from the hillside in nearly vertical columns of ore 20 feet thick and high (D'Invilliers, 1883, p. 247-248). Individual mines in the Rittenhouse Gap District are described below.

Figure 246. Rock mine open cut, Rittenhouse Gap, Longswamp Township, March 2012.

Rock Mine

The Rock mine (figs. 246 and 252) was located at about 40° 28′ 23″ latitude and 75° 37′ 31″ longitude on the USGS Manatawny 7.5-minute topographic quadrangle map (Appendix 1, fig. 16, location L-1). The rock mine also was known as the Rittenhouse cut (D'Invilliers, 1883, p. 251).

Prior to 1785, the Rock mine was worked by the Mayburry family, who owned the Hereford Furnace in Hereford Township. From 1785 to 1809, Jacob Lesher, one of the pioneer ironmasters of Berks County, worked the mines and smelted the ore in the Mary Ann Furnace. Reuben Trexler, Jacob Lesher's son-in-

Figure 247. Thomas Iron Company open cut at Rittenhouse Gap, Longswamp Township, March 2012.

Figure 248. Tunnel mine, Rittenhouse Gap, Longswamp Township, March 2012.

law, purchased the mines and the furnace and renamed the furnace the Trexler Furnace. He operated the mine from 1809 to 1846. In 1861, Lucinda Rittenhouse, who inherited the mine and about 44 acres of land from her father, Reuben Trexler, leased the mine to the Thomas Iron Company for 20 years until the lease expired in 1882. This was the first lease on the property. The Thomas Iron Company paid a royalty of 26 cents per ton and, in return for the low royalty, agreed to extend the Catasauqua and Fogelsville Railroad to Rittenhouse Gap. The Thomas Iron Company also sank several shafts and opened several open cuts in conjunction with their own property to the east. When D'Invilliers visited the mine in the summer of 1882, it was idle (D'Invilliers, 1883, p. 246).

The Thomas Iron Company open cut (fig. 247) was 200 feet long; it extended for 150 feet long on the Rittenhouse property and continued for an additional 50 feet on the Thomas Iron Company property. A small felsic dike 4 feet thick divided the properties. The mine was originally worked as an open cut to a depth of about 60 feet. Shafts were sunk from the bottom of the open cut to a depth of 150 feet on the Rittenhouse property and to a depth of 100 feet on the Thomas Iron Company property (D'Invilliers, 1883, p. 246-247).

The ore occurred as a lenticular body, averaging 10 to 15 feet thick, which swelled in the center. To the east and west, the ore body came to an abrupt end in rock. The foot and hanging walls were a fine-grained granitoid gneiss with flesh-colored feldspar. The ore vein dipped about S. 40° E. 68°, but the dip decreased lower in the mine (D'Invilliers, 1883, p. 247). Two analyses of the ore were published by D'Invilliers (1883, p. 247): (1) 39.3 percent metallic iron, 43.2 percent silica, and 0.016 percent phosphorus, and (2) 41.4 percent metallic iron, 39.7 percent silica, and 0.048 percent phosphorus. The high percentage of silica (quartz) in the ore is notable.

Tunnel Mine

The Tunnel mine (fig. 248) was located at about 40° 28′ 24″ latitude and 75° 37′ 36″ longitude on the USGS Manatawny 7.5-minute topographic quadrangle map (Appendix 1, fig. 16, location L-2). The Tunnel mine was opened about 1830 to furnish ore to area iron furnaces. The mine was re-opened in 1879 by the Thomas Iron Company. When D'Invilliers visited the Rittenhouse Gap District on September 22, 1882, the Tunnel mine was the only mine in operation (D'Invilliers, 1883, p. 248-249). Twenty-two men were employed at the mine under William Donsen, the mine superintendent. Production was about 25 tons of ore per day.

The ore body was reached by a tunnel 900 feet long, which was driven through hard quartzose gneiss into the north flank of the hill at the end of the Catasauqua and Fogelsville Railroad siding. Two ore beds were struck in the tunnel — one at 900 feet that was 8 to 20 feet thick, and one at 14 feet beyond the first ore bed that averaged 10 to 12 feet thick. Both beds dipped 50° to 60° SE. Drifts were driven for 150 feet east and 125 west along strike. At the foot wall of both beds, there was about a 1-foot thick layer of stratified gray ferruginous micaceous gneiss with scales of gray and black mica and thin seams of ore. The occurrence of this rock, called soft ore by the miners, was very marked in the mine. The hard ore had to be blasted because of its great hardness. The hard ore could not be mined economically when mixed with rock. The foot and hanging walls were gneiss with crystals of pink feldspar (D'Invilliers, 1883, p. 248-249). A metadiabase dike cut by the tunnel mine was described by Smith (2003), who also presented analyses of samples.

Figure 249. Magnetite from the Gap (Moll and Gery) mine, Rittenhouse Gap, Longswamp Township, 7 cm. Reading Public Museum collection 2004c-002-193.

Figure 250 Mine building ruins, Rittenhouse Gap, Longswamp Township, March 2012.

Figure 251. Ore wharf on the Catasauqua and Fogelsville Railroad at Rittenhouse Gap, Longswamp Township, March 2015.

Figure 252. Rittenhouse open cut, Rittenhouse Gap, Longswamp Township, March 2012.

Figure 253. Open shaft at Rittenhouse Gap, Longswamp Township, March 2012.

After picking and sizing, the ore was transported on a small track to the Catasauqua and Fogelsville Railroad siding, where it was dumped directly into the rail cars of the Thomas Iron Company and taken to their furnaces at Hokendauqua. It cost 15 cents per ton to load the ore, which was included in the $1.50 per ton cost of mining (D'Invilliers, 1883, p. 249). An analysis of the ore was published by D'Invilliers (1883, p. 250): 38.2 percent metallic iron, 40.8 percent silica, and 0.011 percent phosphorus.

Conrad's Slope

Conrad's slope was located at about 40° 28′ 14″ latitude and 75° 37′ 39″ longitude on the USGS Manatawny 7.5-minute topographic quadrangle map (Appendix 1, fig. 16, location L-3). Conrad's slope was driven 75 feet downward on a 4-foot thick ore bed. The mine was abandoned when a hard mass of quartz was struck. An abandoned (in 1882) open cut was located about 800 feet northeast of Conrad's slope. It was on the Rittenhouse property opposite E. Conrad's house. The open cut was 150 feet long and 100 feet deep. There also was a 125-foot long tunnel driven into the hillside that intersected the open cut (D'Invilliers, 1883, p. 251).

Figure 254. Magnetite and quartz from the Ginkinker mine, Rittenhouse Gap, Longswamp Township, 7 cm. Reading Public Museum collection 2004c-002-186.

Gap Mine

The Gap mine was located at about 40° 27′ 58″ latitude and 75° 37′ 58″ longitude on the USGS Manatawny 7.5-minute topographic quadrangle map (Appendix 1, fig. 16, location L-4). The Gap mine also was called Geary's Gap mine. It was on property known as the Moll and Geary Tract on the north flank of a hill at the end of the Catasauqua and Fogelsville Railroad.

The Thomas Iron Company leased the Gap mine in 1864. A 70-foot deep shaft was sunk on an 85° SE. dip to follow the ore. Water was pumped out of the mine by a pole pump. About 3,000 tons of ore were mined and shipped to the Thomas Iron Company furnaces. The mine was abandoned in 1869 (D'Invilliers, 1883, p. 250).

In 1876, Moll and Spinner leased the Gap mine. They mined about 40 tons of ore from the outcrop and shipped it to the Kutztown Furnace. The mine was abandoned because of financial problems. In 1878, the mine was leased by the Bethlehem Iron Company, which installed new pumping and hoisting machinery and deepened the shaft an additional 30 feet. A gangway was driven 20 feet west in rock at the bottom of the shaft. About 1,100 tons of ore were mined and shipped to the Bethlehem Iron Company furnaces. Mining was expensive because of the hardness of the tough, close-grained gneiss and the intimate mixture of the ore with the rock, which required careful sorting. The 1880 Census (Pumpelly, 1886, p. 960) listed Moll and Geary as the mine operators. They were contractors for the Bethlehem Iron Company. Production between July 1, 1879, and June 30, 1880, was 2,425 tons of magnetite (fig. 249). The ore was shipped to the Bethlehem Iron Company furnaces. The mine was abandoned in 1881 (D'Invilliers, 1883, p. 250-251). D'Invilliers published an analysis of the ore: 30.3 percent metallic iron, 0.175 percent sulfur, 0.020 percent phosphorus, and 51.2 percent silica.

Figure 255. Thomas Iron Company open cut, Rittenhouse Gap, Longswamp Township, March 2012.

Figure 256. Open cut of the Thomas Iron Company, Rittenhouse Gap, Longswamp Township, March 2012.

Figure 257. Magnetite from the Thomas Iron Company open cut, Rittenhouse Gap, Longswamp Township, 8.8 cm. The included quartz grains caused the ore from this mine to be very high in silica content. Sloto collection 3323.

Ginkinker Mine

The Ginkinker mine was located at about 40° 28′ 22″ latitude and 75° 37′ 31″ longitude on the USGS East Greenville 7.5-minute topographic quadrangle map (Appendix 1, fig. 16, location L-5). The Ginkinger mine consisted of an open cut and a shaft. The ore was mined at its outcrop in a crescent-shaped open cut with the two ends of the crescent pointing east. The rock dipped about S. 55° E. 66°. The open cut was about 100 feet long, 40 feet wide in the center, and tapered to 8 to 10 feet at both ends. It was 50 feet deep. The shaft, located at the bottom of the cut, was 100 feet deep. The magnetite (fig. 254) ore was intermixed with rock. The gangue rock was a blue, very hard, stratified gneiss consisting of quartz, feldspar, and mica. An analysis of the ore was published by D'Invilliers (1883, p. 251-252): 46 percent metallic iron, 0.010 percent phosphorus, and 24.8 percent silica.

Thomas Iron Company Mines

The Thomas Iron Company mines were located at about 40° 28′ 23″ latitude and 75° 37′ 31″ longitude on the USGS East Greenville 7.5-minute topographic quadrangle map (Appendix 1, fig. 16, location L-6). Magnetite (fig. 257) was mined between the Ginkinker mine and the Catasauqua and Fogelsville Railroad from several small open cuts (figs. 255 and 256) and shafts from 50 to 100 feet deep. The mines were idle when visited by D'Invilliers in 1882 (D'Invilliers, 1883, p. 251-252).

Other Mines

There were several mines west of the Ginkinger mine. All were abandoned at the time of D'Invilliers visit in 1882. They were described by D'Invilliers (1883, p. 252-254) and included the following.

(1) West of the Ginkinger mine, there was a slope about 235 deep, which was abandoned in 1881. A large quantity of ore was reported to have been mined. The ore dipped about S. 50 E. 70°. It was one of the oldest mines in the Rittenhouse Gap District.

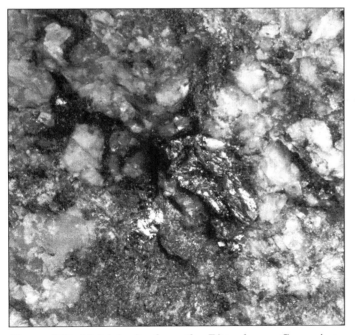

Figure 258. Molybdenite from the Rittenhouse Gap mines, Longswamp Township. Field of view is 2.5 cm. Sloto collection 3324.

THOMAS IRON COMPANY

The Thomas Iron Company was a major iron producer in the Lehigh Valley from its organization in 1854 until its dismantling in the early 20th century. The company operated a number of mines in Berks County.

On February 14, 1854, a meeting was held at White's Tavern in Easton to discuss plans for the organization of an iron company. A resolution was adopted calling the new enterprise the Thomas Iron Company in honor of David Thomas and in recognition of his work as a pioneer in the successful manufacture of iron using anthracite coal. The Lehigh Crane Iron Company brought David Thomas, a Welsh ironmaster, to America in 1839 to introduce the hot blast manufacture of anthracite iron. The Thomas Iron Company was organized on February 14, 1854. A special charter from the Commonwealth of Pennsylvania was granted and approved on April 4, 1854. The capital stock was fixed at $200,000 with 4,000 shares issued at fifty dollars per share. David Thomas' son, Samuel, was selected as the superintendent to oversee construction of two blast furnaces.

Thomas was authorized to purchase the Thomas Butz farm situated on the west bank of the Lehigh River near the Hokendauqua Dam as the most promising site for the furnaces. The farm was 185 acres, and the purchase price was $37,112.50. Other land was acquired, bringing the total to 294 acres at a cost of $120,502.

On June 8, 1854, the name of the furnace site was selected at a meeting of the Board of Directors. David Thomas' suggestion of Hokendauqua, derived from the name of the nearby creek, was adopted. On November 9, 1854, the town of Hokendauqua was laid out by the company, and lots were donated for a schoolhouse and a Presbyterian Church. Over two hundred company homes were later erected.

The new company built two furnaces on the site. Furnace No. 1 was put in blast on June 1, 1855, and furnace No. 2 on October 23, 1855. The furnaces were substantially built, each 60 feet high, with an eighteen-foot bosh. They were a success from the first blast.

The Thomas Iron Company joined the Crane Iron Company to construct the Catasauqua and Fogelsville Railroad in 1856. The rail line reduced difficult and inefficient wagon haulage to supply local ore to both companies. The Thomas Iron Company obtained magnetite from mines at Rittenhouse Gap, at the southern end of the Catasauqua and Fogelsville Railroad.

The Thomas Iron Company produced record quantities of iron; during 1857, the No. 1 furnace produced 9,731 tons of iron, and the No. 2 furnace produced 8,366 tons. New furnaces were built at Hokendauqua. Furnace No. 3 was blown in on July 18, 1862, and furnace No. 4 on April 29, 1863. Samuel Thomas, who earlier had been appointed a director, was elected president of the company on August 31, 1864. After resigning from the Lehigh Crane Iron Works, John Thomas became superintendent of the Thomas Iron Company in 1867.

Figure 259. Thomas Iron Company Furnace, Hokendauqua, Pa., 1903.

The Lock Ridge Iron Company of Alburtis acquired a charter on December 26, 1866, with Samuel Thomas as president and J.H. Knight as secretary-treasurer, the same positions they held with the Thomas Iron Company. Furnace construction began in 1867, and the first furnace was put in blast on March 18, 1868; the second furnace was blown in on July 9, 1869. On May 1, 1869, the entire capital stock was acquired by the Thomas Iron Company

Furnace No. 5 at the Hokendauqua plant was blown in September 15, 1873, and furnace No. 6 was put in blast January 19, 1874. All six furnaces were of the traditional masonry stack construction and used iron pipe stoves. In 1882, the Thomas Iron Company secured complete ownership of the Ironton Railroad for hauling the ore, coal, limestone, and iron ore for their furnaces at Hokendauqua. On April 1, 1882, the Keystone Furnace, located in the Borough of Glendon, was purchased, and on December 13, 1884, the Thomas Iron Company purchased the Saucon Iron Company near Hellertown for $300,000. With these two purchases, the Thomas Iron Company operated eleven furnaces. They were designated as No. 1 through No. 6 at the Hokendauqua plant, Nos. 7 and 8 at Lock Ridge, No. 9 at Keystone, and Nos. 10 and 11 at Saucon. From March 22, 1886, until December 15, 1887, the Thomas Iron Company leased the Lucy Furnace at Glendon.

During the 1890s, as the local limonite industry declined, and railroad transportation improved, Thomas Iron switched from local ore to hematite from the Lake Superior Region. The declining importance of local ore also prompted Crane and Thomas to divest themselves of the Catasauqua and Fogelsville Railroad; most of the stock in the railroad was sold to the Reading Railroad in 1890.

The year 1891 marked the beginning of a round of furnace upgrades, which added Durham-style regenerative heating stoves to the No. 6 and 7 stacks. In 1893, the No. 1 and 2 stacks were abandoned; No. 1 was demolished and rebuilt with the new stoves, which also were added to the No. 10 and 11 furnaces. In 1893, Benjamin Franklin Fackenthal, Jr., became president of the company.

By the beginning of the 20th century, many changes had come to the iron industry. In addition to the shift from local to foreign ores, coke had largely replaced anthracite as the principal furnace fuel. The shift away from local ores and fuels eliminated much of the original competitive advantage of the Lehigh Valley furnaces. Against this backdrop, President Fackenthal resigned on May 1, 1913, after recommending a program of retrenchment and abandonment of the old furnaces at Alburtis. His successor, chosen on July 1, 1913, was Ralph H. Sweetser, who held opposite views. Furnaces No. 7 and 8 at Alburtis were reportedly the last furnaces in the country to use anthracite, converting to coke in 1914. Sweetser also attempted to restart local limonite mining, an effort which proved a costly failure. William A. Barrows, Jr., succeeded Sweetser as president in 1916. However, the company faced terminal decline.

Dismantling of the company began. On June 28, 1917, the Thomas Iron Company sold the Keystone Furnace to the Northern Ore Company of Philadelphia. Operations at the Saucon furnaces ceased in late 1921, and soon thereafter, the plant was sold and dismantled for scrap. The Lock Ridge furnaces ceased operations in December 1921, and the property was sold to William Butz of Alburtis for scrap. On June 30, 1922, the company's stock was sold to Drexel & Company, which disposed of its assets over the next few years. The railroad stocks were sold. The furnaces and other assets were sold to the Reading Coal and Iron Company, which dismantled the plants for scrap. The last Hokendauqua furnace was abandoned in 1927, and the Hokendauqua plant was sold to Bethlehem Steel, which scrapped it in the mid 1930s. Samuel R. Thomas surrendered the company's charter to the State in June 1942, fifteen years after the last furnace was blown out.

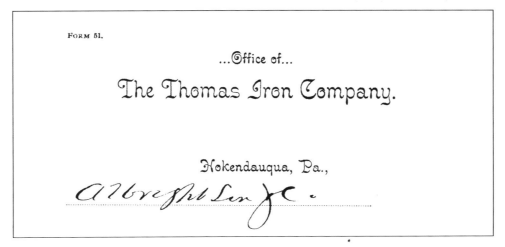

Figure 260. Billhead of the Thomas Iron Company.

(2) A small shaft 75 feet deep was located about 100 feet southwest of mine (1) described above. The ore was only 3 feet thick. At 75 feet, the ore vein was offset to the north by a fault.

(3) There were two small shafts and an open cut in a cove. The rock in the open cut dipped 65° SE.

(4) There was a small shaft to the northeast of mine (3) described above between the main track and siding of the Catasauqua and Fogelsville Railroad. The rock in the shaft dipped S. 40° E. 70°.

Minerals from the Rittenhouse Gap Mines

Fluorite - in metadiabase dikes (Orner and Friehauf, 2000)

Hematite, var. specular - in metadiabase dikes (Orner and Friehauf, 2000)

Magnetite - massive, often containing considerable quartz (figs. 254 and 257)

Molybdenite -- small silvery flakes (fig. 258); molybdenite at Rittenhouse Gap was found and verified by Richard E. Myers of Moravian College (Montgomery, 1969, p. 77)

Muscovite - silvery flakes

Sphalerite - lemon yellow, in felsic dikes (Smith, 2003)

Zircon - in felsic dikes (Smith, 2003)

Red Lion Station Mines

The Red Lion Station mines were centered around the Red Lion railroad station on the Catasauqua and Fogelsville Railroad (fig. 242). The mines were in felsic to mafic gneiss.

Trexler Mine

The Trexler mine was about 0.5 mile southwest of Red Lion at about 40° 28′ 28″ latitude and 75° 37′ 25″ longitude on the USGS East Greenville 7.5-minute topographic quadrangle map (Appendix 1, fig. 19, location L-7). An open cut on the William Trexler property adjoining the Ginkinger mine tract was mined by the Thomas Iron Company. D'Invilliers (1883, p. 252) published an analysis of the ore: 52.1 percent metallic iron, 0.022 percent phosphorus, and 24.3 percent silica.

Weiler Mine

The Weiler mine was about 0.4 mile southwest of Red Lion at about 40° 28′ 30″ latitude and 75° 37′ 17″ longitude on the USGS East Greenville 7.5-minute topographic quadrangle map (Appendix 1, fig. 19, location L-8). The Weiler mine was on the property of Calvin Weiler, which was north of the Catasauqua and Fogelsville Railroad. There were three shafts; one of them was reported to be sunk to a *"considerable depth"* (D'Invilliers, 1883, p. 253-254).

The Thomas Iron Company operated the mine until the panic of 1873, at which time it was abandoned. The mine was later worked for 3 years by the Bethlehem Iron Company. About 10 tons of ore per day were mined from a 125-foot deep shaft. Drifts were run 30 feet east and west from the shaft. Although the mine was abandoned when D'Invilliers visited in the summer of 1882, an engine house with a Barber & Son hoisting and pumping engine remained. Mining was expensive because of the hardness of the gneiss and the great quantity of groundwater inflow.

The 1880 Census (Pumpelly, 1886, p. 960) listed the Bethlehem Iron Company as the operator of the Weiler mine. Production between July 1, 1879, and June 30, 1880, was 2,006 tons of magnetite, which was shipped to the Bethlehem Iron Company furnaces. An analysis of the ore was published by D'Invilliers (1883, p. 253-254): 52.7 percent metallic Iron, 24.3 percent silica, and 0.018 percent phosphorus.

Wetzel Mine

The Wetzel mine was about 0.25 mile south of the Red Lion station at about 40° 28′ 36″ latitude and 75° 36′ 55″ longitude on the USGS East Greenville 7.5-minute topographic quadrangle map (Appendix 1, fig. 19, location L-9). The Northampton Iron Company leased the mine from 1873 to 1878. The Crane Iron Company leased the mine in 1888. George Greiss and Dr. H.K. Hartzel were mining contractors for Crane. The mine was in operation as late as 1904 (*Reading Eagle*, July 11, 1888, and January 13, 1904).

The mine on the George Wetzel property consisted of an open cut about 30 feet deep and several *"insignificant shafts."* They were on the west flank of a small hill about 0.5 mile west of the Red Lion railroad station. A sandy, feldspathic gneiss dipping about S. 3° W. 63° was exposed in the open cut. The ore was shipped from a wharf on the Wetzel property. The ore was a fine-grained magnetite with a bright metallic luster. When D'Invilliers visited the mine in the summer of 1882, he noted: *"There is no ore in sight, and the mines have been abandoned for some time"* (D'Invilliers, 1883, p. 254-255). An analysis of the ore was published by D'Invilliers: 38.4 percent metallic iron, 0.006 percent sulfur, 0.037 percent phosphorus, and 40.1 percent silica.

Dunkle Mines

The Dunkle mines were about 0.3 mile southwest of the Red Lion station at about 40° 28′ 39″ latitude and 75° 37′ 16″ longitude on the USGS East Greenville 7.5-minute topographic quadrangle map (Appendix 1, fig. 19, location L-10). The Dunkle property mines consisted of a group of open cuts and shallow shafts on the north flank of a hill about 500 yards east of the Red Lion railroad station. The mines were actively worked in 1879 by contractors for the Bethlehem Iron Company. Contractors Schroeder, Weiss, and Brensinger sunk numerous shafts 30 to 50 feet deep. No regular ore bed was found. The ore, a poor-quality magnetite, was mined from a light grey, decomposed micaceous gneiss (D'Invilliers, 1883, p. 256).

The 1880 Census (Pumpelly, 1886, p. 960) listed the Bethlehem Iron Company as the operator of the Dunkle mine. Production between July 1, 1879, and June 30, 1880, was 6,822 tons of magnetite.

Miller Farm Mines

The Miller farm mines were about 0.1 mile southwest of the Red Lion station at about 40° 28' 45" latitude and 75° 37' 07" longitude on the USGS East Greenville 7.5-minute topographic quadrangle map (Appendix 1, fig. 19, location L-11). Iron mines on the farm of Charles Miller were located near Doe Mountain Road about 200 yards west of the Red Lion railroad station. The mine was opened in 1879 when the 2-acre property was leased to the Bethlehem Iron Company by the Crane Iron Company. During the summer and fall of 1879, 100 tons of ore per day were mined, but production fell to 40 to 50 tons per day in 1882 (D'Invilliers, 1883, p. 255).

In May 1882, ore was mined from five shafts. Numerous shafts were needed because of the faulted character of the ore bed and the *"pernicious contract system of mining practiced throughout the region."* The shafts averaged 70 feet deep. Each shaft was worked by a different contractor, all under the management of A.S. Miller, the representative of the Bethlehem Iron Company. Forty men were employed — 20 miners, 2 engineers, and 18 laborers. The ore was raised by windlass and horse power and hauled to the Red Lion station wharf. An inflow of surface drainage water caused the foot and hanging walls to decompose, which required extensive timbering. Two steam pumps, a No. 7 Cameron and a No. 4 Knowles, were in constant use. They required 16 to 18 tons of coal per month at a cost of $3.60 per ton to operate (D'Invilliers, 1883, p. 255-256).

The ore bed dipped 60° to 70° SE. The ore was an earthy magnetite that occurred between layers of a soft brown micaceous and feldspathic gneiss. Scales of black and grey mica occurred in great profusion in the gangue (D'Invilliers, 1883, p. 255-256).

Garner Station Mines

The Garner Station mines were centered around the Garner railroad station on the Catasauqua and Fogelsville Railroad (fig. 261). Unless noted, the mines were in felsic to mafic gneiss.

Bethlehem Iron Company Gardner Station Mines

The Bethlehem Iron Company Gardner Station mines were a group of mines oriented in a north-south direction just southwest of the intersection of Hensingersville, Walker, and Wetzel Roads at about 40° 29' 09" latitude and 75° 36' 04" longitude on the USGS East Greenville 7.5-minute topographic quadrangle map (Appendix 1, fig. 19, location L-12). The mines were close to the Lehigh County border.

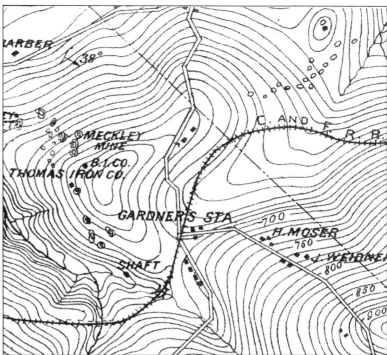

Figure 261. Garner Station mines, Longswamp Township, 1882. From Pennsylvania Geological Survey (1883).

In 1879, the mines were leased from George Greis and Benjamin Wendling and worked by James Moser, a contractor for the Bethlehem Iron Company. The mine employed 10 men—7 miners, 1 engineer, and 2 laborers. The ore was raised by windlass and horse power. The mine had a large inflow of groundwater, which made mining expensive. The mine was dewatered by a No. 7 Cameron steam pump, which pumped water every 10 minutes through a 3.5-inch diameter pipe. The pump consumed 20 tons of coal per month at a cost of $3.25 per ton (D'Invilliers, 1883, p. 256).

An old shaft, sunk prior to 1879, was about 200 feet west of the Walker Road bridge. The shaft was 50 feet deep with a 60-foot long drift driven to the northeast. A second shaft, known as the new shaft, was sunk in July 1881 closer to the Catasauqua and Fogelsville Railroad tracks. In 1882, mining took place from the new shaft, which was 125 feet deep with a drift driven on ore bed. When D'Invilliers visited the mine in May of 1882, production was 10 to 12 tons of ore per day (D'Invilliers, 1883, p. 257).

The ore was magnetite, and the ore bed ranged from 4 to 8 feet thick and dipped into the hill at about S. 40° E. 45°. The foot and hanging walls were feldspathic gneiss, and little timbering was required in the lower workings. An analysis of the ore was published by D'Invilliers (1883, p. 257): 35.5 percent metallic iron, 0.023 percent sulfur, 0.025 percent phosphorus, and 42.3 percent silica.

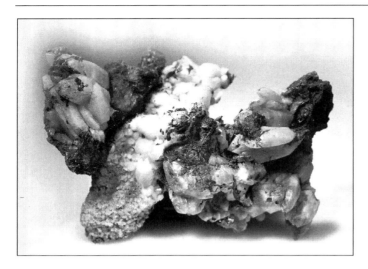

Figure 262. Quartz and Hematite from Maple Grove, Longswamp Township, 5 cm. Ron Kendig collection.

Thomas Iron Company Garner Station Mines

The Thomas Iron Company Garner station mines were a group of mines oriented in a north-south direction about 0.2 mile west of the intersection of Hensingersville, Walker, and Wetzel Roads at about 40° 29′ 12″ latitude and 75° 36′ 16″ longitude on the USGS East Greenville 7.5-minute topographic quadrangle map (Appendix 1, fig. 19, location L-13).

The mines were west and northwest of the Garner railroad station. The property was mostly owned by the Thomas Iron Company, and the mines were worked by contractors. The mines were active in 1879 and idle in 1882. When visited by D'Invilliers in 1882, he noted: "*It is riddled with mine holes of all descriptions, small shafts and open cuts too numerous and insignificant to enumerate. It looks as if a traveling mining camp had located in this hill, moving camp every few weeks, and leaving as landmarks some shafts or open cuts to mark its progress*" (D'Invilliers, 1883, p. 260-261).

Finley Mine

The Finley (also spelled Findley) mine was about 0.4 mile northwest of the intersection of Hensingersville, Walker, and Wetzel Roads at about 40° 29′ 24″ latitude and 75° 36′ 18″ longitude on the USGS East Greenville 7.5-minute topographic quadrangle map (Appendix 1, fig. 19, location L-14). The James Findley residence is shown just north of the mine in the 1876 atlas. James Findley, Sr. was an iron ore and coal mining contractor (*Reading Eagle*, August 4, 1911). The Finley mine, called Finley's shaft by D'Invilliers, was 60 feet deep and located on the western slope of the hill. D'Invilliers considered it "*one of the best*." The ore bed was over 5 feet thick, and the ore was "*very much broken up*" (D'Invilliers, 1883, p. 261).

Old Mickley Mine

The old Mickley mine, also known as the Meckley mine, was about 0.25 mile northwest of the intersection of Hensingersville, Walker, and Wetzel Roads at about 40° 29′ 25″ latitude and 75° 36′ 12″ longitude on the USGS East Greenville 7.5-minute topographic quadrangle map (Appendix 1, fig. 19, location L-15). It is mine number 46 on Prime's 1875 map. The old Mickley mine was worked by the Thomas Iron Company. It was higher up on the same hill and southeast of the Finley mine. The mine produced a considerable amount of ore before it was abandoned. An analysis of the ore was published by D'Invilliers (1883, p. 261): 39.6 percent metallic iron, 0.033 percent sulfur, 0.028 percent phosphorus, and 41.1 percent silica.

Smoyer Mine

The Smoyer mine was about 0.3 mile west of the intersection of Hensingersville, Walker, and Wetzel Roads at about 40° 29′ 09″ latitude and 75° 36′ 16″ longitude on the USGS East Greenville 7.5-minute topographic quadrangle map (Appendix 1, fig. 19, location L-16).

The mine was on the south side of a hill 50 feet above a tributary to Swabia Creek. There was a series of excavations, many of which were worked in 1879 by Smoyer, who was a contractor for the Bethlehem Iron Company. Smoyer worked an ore bed 5 to 10 feet thick; however, the bed was irregular and faulted, and it was unprofitable to mine (D'Invilliers, 1883, p. 261).

Peter Kline Mine

The Peter Kline mine was somewhere east of the intersection of State Street and Mountain Road at about 40° 29′ 29″ latitude and 75° 36′ 37″ longitude on the USGS East Greenville 7.5-minute topographic quadrangle map (Appendix 1, fig. 19, location L-20). The mine was in the Hardyston Formation. The mine is shown in the 1876 atlas as the Kline mine and as mine number 45 on Prime's 1875 map. The mine was on the south side of and close to Mountain Road, west of Maple Grove. The mine was abandoned when visited by Prime in 1874. At the mine, slate dipped 35° S. The mine was worked to a depth of about 25 feet (Prime, 1875, p. 22).

Klines Corner Mines

There were many more mines in the Klines Corner area than are described here. Several are shown on the 1882 topographic map (fig. 263) that are neither named nor numbered. The history of the mines is complicated by changing mine owners, lessees, and contractors, who often operated different mines at different times. LIDAR imagery shows an extensively mined area at Klines Corner (fig. 264).

Klines Corner was named for Jonas L. Klein. He opened a store on an 82-acre farm that he purchased in 1842. He later

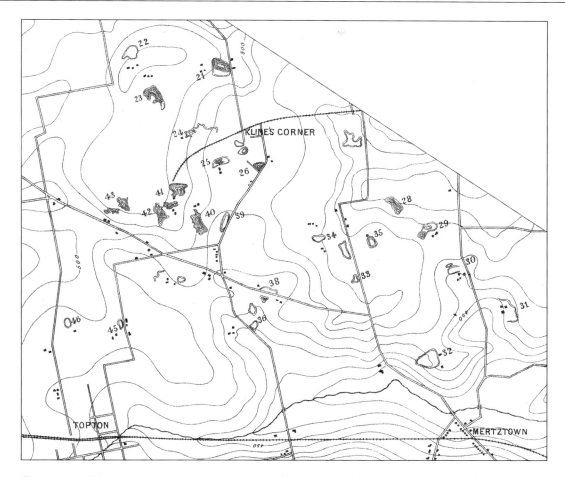

Figure 263. Klines Corner mines, Longswamp Township, 1882. From Pennsylvania Geological Survey (1883).

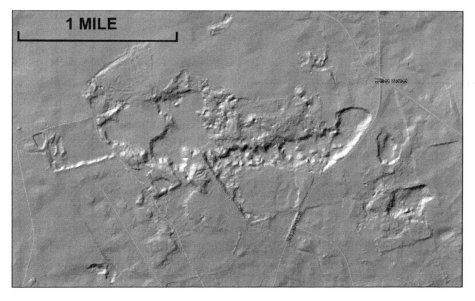

Figure 264. LIDAR image of the Klines Corner area, Longswamp Township. The image shows the extensive open-pit mining that took place. Imagery provided by the Pennsylvania Spatial Data Access (PASDA).

Figure 265. Mines and ore washeries at Klines Corner, Longswamp Township, 1897. Photograph taken by William Smith. Used with permission from Suzanne Smith (William Smith's granddaughter).

Figure 266. Mines and ore washeries at Klines Corner, Longswamp Township, 1897. Photograph taken by William Smith. Used with permission from Suzanne Smith (William Smith's granddaughter).

Figure 267. Ore wharf at Klines Corner, Longswamp Township, 1897. A rail car is at the loading dock. Photograph taken by William Smith. Used with permission from Suzanne Smith (William Smith's granddaughter).

Figure 268. Mining and ore washing at Klines Corner, Longswamp Township, 1897. Photograph taken by William Smith. Used with permission from Suzanne Smith (William Smith's granddaughter).

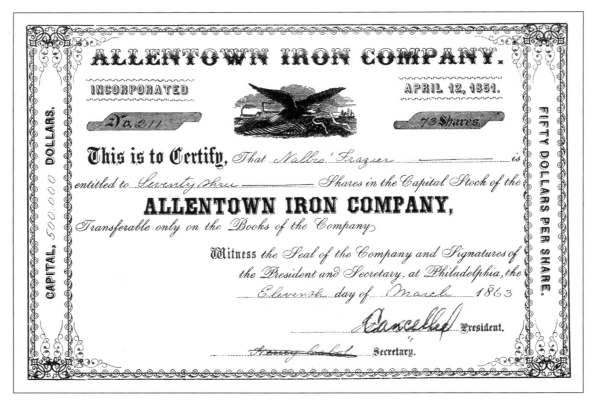

Figure 269. Stock certificate of the Allentown Iron Company, 1863. The Allentown Iron Company operated a mine at Klines Corner adjacent to the Catasauqua and Fogelsville Railroad.

purchased much land in the vicinity and was the owner of Klein's mine. Klein died in March 1860. Afterwards, Allen Schweyer and his son Alfred made major improvements in Klines Corner. Allen opened a hotel, and Alfred opened a store. The Schweyers post office was established at the store in April 1884 with James Schweyer as postmaster (*Reading Eagle*, September 2, 1981). Because of the establishment of the post office, Klines Corner is often called Schweyer or Schweyers.

All of the Klines Corner mines are located on the USGS Topton 7.5-minute topographic quadrangle map. The mines in the Klines Corner area (fig. 263) are numbered 21 through 46 (with 27 and 37 missing or unnumbered) on the 1882 topographic map. The area includes Oreville, which is directly south of Klines Corner.

D.K. Kline Mine

The D.K. Kline mine was leased by the Temple Iron Company. The mine produced limonite ore that was compact and dark brown and yellowish-brown and also cellular ore with the cells usually filled by ocher. An analysis was presented by McCreath (1879, p. 210-211): 45.5 percent metallic iron, 0.028 percent sulfur, and 0.148 phosphorus.

D.L. Trexler Mine

The D.L. Trexler mine was located at about 40° 31' 28" latitude and 75° 41' 25" longitude on the USGS Topton topographic map (Appendix 1, fig. 4, location L-46). The mine was in the Allentown Formation. The D.L. Trexler ore beds are shown on the 1860 and 1862 maps. The mine was worked by Edwin Trexler (ca. 1876) and Jonas Trexler (ca. 1879). The 1880 Census (Pumpelly, 1886, p. 960) listed E.H. Trexler as the operator of the Trexler mine. Production between July 1, 1879, and June 30, 1880, was 5,600 tons.

The mine produced limonite ore, which was generally dark brown to brownish-black, compact, and fine grained with an admixture of fibrous iron ore. An analysis was presented by McCreath (1879, p. 211): 52.2 percent metallic iron, 0.005 percent sulfur, and 0.177 percent phosphorus.

Thomas Iron Company Klines Corner Mines

The Thomas Iron Company mines at Klines Corner were located at about 40° 31' 21" latitude and 75° 42' 21" longitude on the USGS Topton 7.5-minute topographic quadrangle map (Appendix 1, fig. 4, location L-24). The Thomas Iron Company

Figure 270. Thomas Iron Company ore washer at Klines Corner, Longswamp Township, 1897. Uriah Biery was the mine superintendent. Photograph taken by William Smith. Used with permission from Suzanne Smith (William Smith's granddaughter).

Figure 271. Incline at iron mine number 41 between Klines Corner and Oreville, Longswamp Township, 1897. A large crowd observes damage from a recent accident on the incline. Photograph taken by William Smith. Used with permission from Suzanne Smith (William Smith's granddaughter).

Figure 272. Magnetite from Fritch and Brother's mine, Longswamp Township, 2.8 cm. Reading Public Museum collection 2004c-002-192.

operated three mines on a 72-acre tract that were both open-pit and underground mines. Uriah Biery was the mine superintendent.

In 1888, all three mines on the property were in operation, 100 men were employed, and 26,000 tons of ore were mined. In the same year, the boiler of the steam engine at the mine exploded, demolishing the boiler house (*New York Times*, March 9, 1888). In 1898, only two of the three mines were in operation. Production was 100 to 150 tons of ore per day, and 60 men were employed. One of the shafts was 80 feet deep (*Reading Eagle*, June 4, 1898).

White clay was exposed in two shafts. In one shaft, the clay was reported to be 30 feet thick with an overburden of 12 to 25 feet of yellow clay and sand (Hopkins, 1900, p. 24).

Allentown Iron Company Mine

The Allentown Iron Company mine was located at about 40° 31′ 35″ latitude and 75° 41′ 21″ longitude on the USGS Topton 7.5-minute topographic quadrangle map (Appendix 1, fig. 4, location L-23). The mine is shown in the 1876 atlas as the Allentown mine. The Pittsburgh Commercial (July 12, 1873) reported that a boiler exploded on June 30, 1873, at the Allentown Iron Company mine.

The Allentown Iron Company (fig. 269) was located at Front and Furnace Streets in Allentown. Their first furnace was constructed in 1846 for the production of pig iron. By 1855, the company was operating five furnaces.

Moatz and Schrader Mine

The Moatz and Schrader mine was formerly operated by Isaac Eckert of the Henry Clay Furnace in Reading. In 1895, Philip Moatz and Jonas M. Schrader, experienced mine operators from Breinigsville, bought the 16-acre mine property for $30,000 from George B. Eckert, administrator of the Henry Eckert estate. There were two deep shafts on the property. Several years prior to the sale, the Thomas Iron Company, which operated mines on the adjoining property, offered Eckert $42,000 for the property, but the offer was not accepted because Eckert wanted more money (*Reading Eagle*, September 13, 1895). The ore was shipped to the Crane Iron Company furnaces in Catasauqua.

In April 1896, the mine buildings and equipment burned. The loss was estimated at $2,000 to $3,000; insurance covered only $1,200. The buildings were rebuilt, the machinery was restored, and the mine was put back in operation (*Reading Eagle*, April 28, 1896).

By 1899, Moatz and Schrader were operating three mines. Production was 50 to 60 tons of ore per day. The ore was shipped to the Topton Furnace. The mines were operated after the peak years of iron mining in Berks County, and there was a shortage of miners (*Kutztown Patriot*, May 6, 1899). The mine was in operation in 1906 (*Reading Eagle*, November 22, 1906), and mining may have continued to 1912. In 1912, the Reading Eagle (June 16, 1912) reported that 65 horses and mules were used to haul ore from the late Isaac Eckert mine at Klines Corners to Topton.

Fritch and Brother Mine

The Fritch and Brother mine was on a 26.5-acre tract that adjoined the Thomas Iron Company and Moatz and Schrader mining properties. Tilghman Fritch and his brother, Manoah L. Fritch, were ocher manufacturers. Their ocher mill was destroyed by fire at a loss of $25,000. They went into receivership in 1902 and filed for bankruptcy in 1903 (*Reading Eagle*, September 28, 1904).

Both surface and underground mining methods were used. In 1899, Fritch was mining an iron ore vein 53 feet thick. In 1900, a drift 79 feet underground in the Fritch mine, operated by contractor Shankweller, was flooded by a thunderstorm. By 1912, the Fritch open cut was abandoned and water-filled (*Reading Eagle*, October 27, 1899; July 23, 1900; and August 17, 1912).

Fenstermacher Mine

In 1873, one of the principle mines at Klines Corner was the Fenstermacher & Company mine (*Reading Eagle*, July 12, 1873). That same year, the Reading Eagle (September 8, 1873) reported that a man was killed when he fell into an ore washer

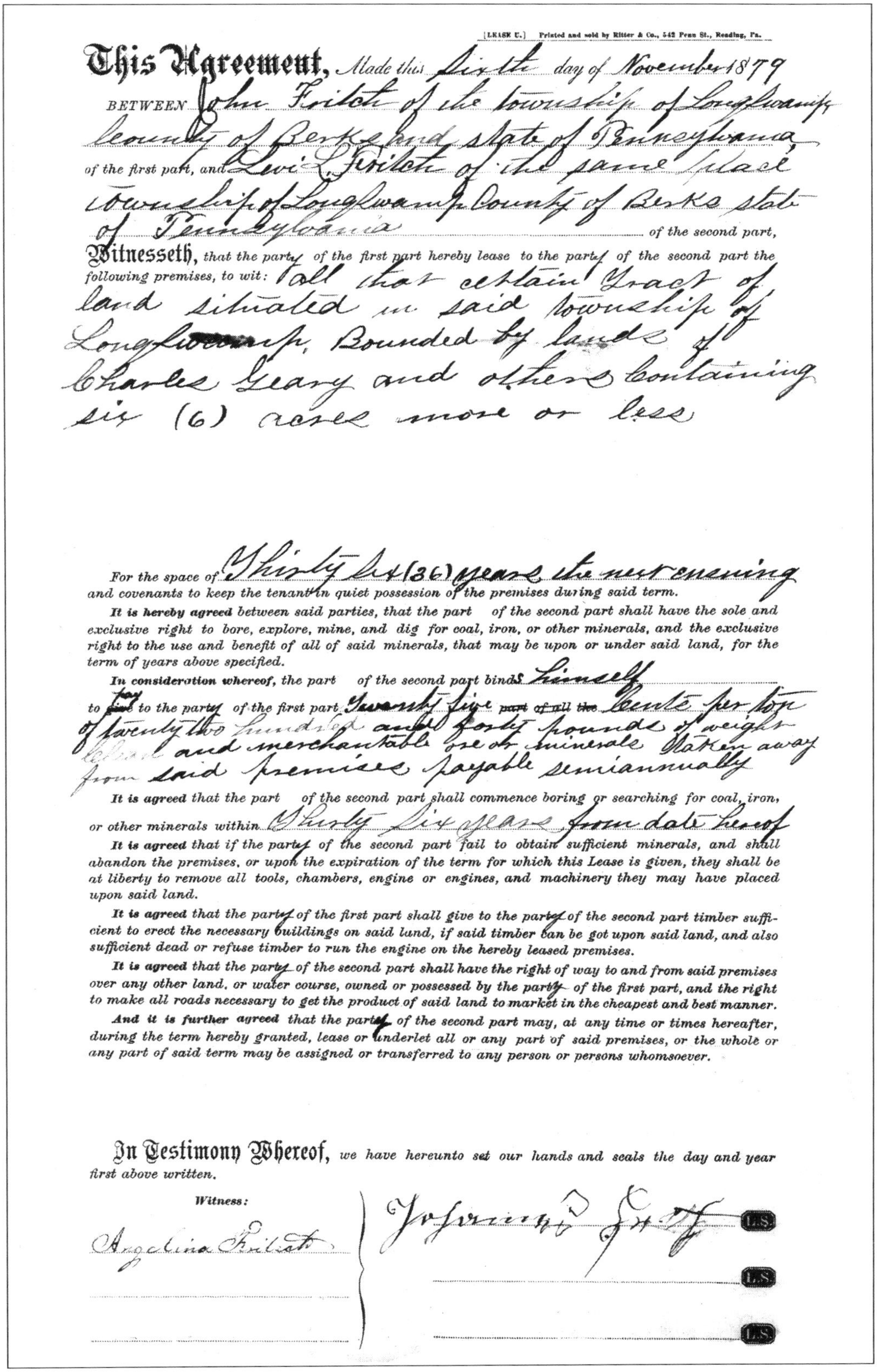

[LEASE U.] Printed and sold by Ritter & Co., 542 Penn St., Reading, Pa.

This Agreement, *Made this* Sixth *day of* November 1879

BETWEEN John Fritch of the township of Longswamp County of Berks and state of Pennsylvania

of the first part, and Levi L. Fritch of the same place township of Longswamp County of Berks state of Pennsylvania *of the second part,*

Witnesseth, *that the party of the first part hereby lease to the party of the second part the following premises, to wit:* All that certain Tract of land situated in said township of Longswamp, Bounded by lands of Charles Geary and others containing six (6) acres more or less

For the space of Thirty Six (36) years the next ensuing *and covenants to keep the tenant in quiet possession of the premises during said term.*

It is hereby agreed *between said parties, that the part of the second part shall have the sole and exclusive right to bore, explore, mine, and dig for coal, iron, or other minerals, and the exclusive right to the use and benefit of all of said minerals, that may be upon or under said land, for the term of years above specified.*

In consideration whereof, *the part of the second part binds* himself *to pay to the party of the first part* Twenty five ~~part of all the~~ Cents per ton of twenty two hundred and forty pounds of weight Clean and merchantable ore or minerals taken away from said premises payable semiannually

It is agreed *that the part of the second part shall commence boring or searching for coal, iron, or other minerals within* Thirty Six years from date hereof

It is agreed *that if the party of the second part fail to obtain sufficient minerals, and shall abandon the premises, or upon the expiration of the term for which this Lease is given, they shall be at liberty to remove all tools, chambers, engine or engines, and machinery they may have placed upon said land.*

It is agreed *that the party of the first part shall give to the party of the second part timber sufficient to erect the necessary buildings on said land, if said timber can be got upon said land, and also sufficient dead or refuse timber to run the engine on the hereby leased premises.*

It is agreed *that the party of the second part shall have the right of way to and from said premises over any other land, or water course, owned or possessed by the party of the first part, and the right to make all roads necessary to get the product of said land to market in the cheapest and best manner.*

And it is further agreed *that the party of the second part may, at any time or times hereafter, during the term hereby granted, lease or underlet all or any part of said premises, or the whole or any part of said term may be assigned or transferred to any person or persons whomsoever.*

In Testimony Whereof, *we have hereunto set our hands and seals the day and year first above written.*

Witness:

Angelina Fritch

Johannes Fritch [L.S.]

[L.S.]

[L.S.]

Figure 273. Mining lease from John Fritch to Levi Fritch, November 19, 1879. The lease was for a 6-acre mining tract for 36 years. The royalty payment was 25 cents per long ton (2,240 pounds).

Mining in Klines Corner in 1898

From the Reading Eagle, June 4, 1898

"On the floor of the gangway is laid a railroad on which the cars are conveyed. The ore from the highest level is run through a chute to the second level. From the second it is conveyed to the next chute, through which it is sent to the lowest level. Here it is loaded on cars and taken to the shaft, from where it is hoisted to the surface.

The gangways vary in height from 5 1/2 to 7 1/3 feet. They are lined with heavy timber. This is put in as the drift is extended. The ore is removed about 5 feet ahead of the last supports, when 2 uprights, with a cross-piece at the top are again placed in position. These uprights are called legs and the cross-pieces [are called] caps. The roof is prevented from caving in by the planks, which run across the top of the caps. Short planks are placed along the sides to prevent the earth from falling into the gangway. Fresh air is passed into the workings by the air shaft. This is a second opening connected with the gangway some distance from the main shaft. When this has once been opened, a strong draft is created.

As a rule, the joists and shovel will suffice to mine the ore, but now and then rocks are struck and blasting must be resorted to. When the miners reach the end of a vein they begin robbing it. This is done by drilling holes into the timber which forms the support for the drift. These are loaded with dynamite and torn to pieces. This work is begun at the far end of the mine. As much of the loose ore is then removed as is possible, when the next supports are withdrawn in a similar manner. It is a sight to see the many lanterns and candles flicker in one of these long gangways.

That an enormous weight rests upon the timbers in a mine is shown by the fact that pieces 12 and even 18 inches in thickness are often snapped in the middle. Sometimes the legs are bent and cracked, as well as the cap. They are replaced by new supports. The mine is always cool and damp and water is continually dripping from the roof of the gangway. The ore, after it is separated from the [quartz], clay and sand, is ready to be transferred to the wharf. Heavy teams are employed to do this."

---

at Fenstermacher's ore mine. The 1880 Census (Pumpelly, 1886, p. 960) listed William M. Kauffman & Company as the operator of the Fenstermacher mine. Production between July 1, 1879, and June 30, 1880, was 2,547 tons of limonite, which was shipped to the Topton and Kutztown Furnaces.

Amos Fisher Mine

The Amos Fisher mine was located at about 40° 30' 57" latitude and 75° 41' 15" longitude on the USGS Topton 7.5-minute topographic quadrangle map (Appendix 1, fig. 5, location L-27). It is mine number 36 on the 1882 topographic map. The Amos Fisher mine was leased by the Northampton Iron Company for 50 years in 1823; the lease was renewed in 1873. The 1880 Census (Pumpelly, 1886, p. 960) listed Edwin Darman as the operator of the Fisher mine. Production between July 1, 1879, and June 30, 1880, was 4,480 tons of limonite.

Fleetwood Iron Company Mine

The Fleetwood Iron Company mine was located approximately 1 mine northwest of Mertztown. The ore was dark brown limonite, generally compact and fine grained; sometimes it was cellular. An analysis was presented by McCreath (1879, p. 210-211): 47.5 percent metallic iron, 0.015 percent sulfur, and 0.23 percent phosphorus.

Klines Corner Mine

The 1880 Census (Pumpelly, 1886, p. 960) listed Allen Schweyer as the operator of the Klines Corner mine. Production between July 1, 1879, and June 30, 1880, was 3,360 tons of limonite, which was shipped to the Keystone Furnace in Reading.

Other Iron Mines

Iron mines at Klines Corner with numbers shown on the 1882 topographic map (fig. 263) and no available information include the following mines.

Mine number 27 is numbered on the index to the 1882 topographic map; it was shown but not numbered on the 1882 topographic map. The mine was located at about 40° 31' 37" latitude and 75° 40' 45" longitude on the USGS Topton SE 7.5-minute topographic quadrangle map (Appendix 1, fig. 5, location L-32).

Mine number 30 was located at about 40° 31' 11" latitude and 75° 40' 21" longitude on the USGS Topton 7.5-minute topographic quadrangle map (Appendix 1, fig. 5, location L-34).

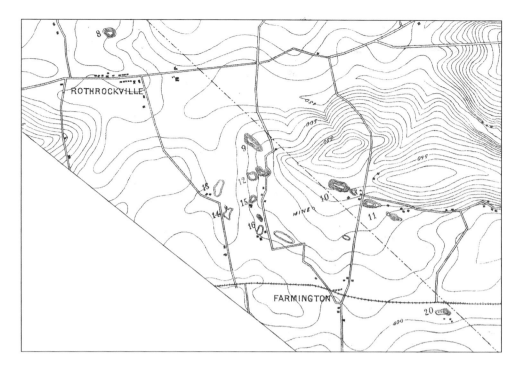

Figure 274. Mines in the Farmington area, Longswamp Township, 1882. From Pennsylvania Geological Survey (1883).

Mine number 31 was located at about 40° 30′ 35″ latitude and 75° 39′ 53″ longitude on the USGS Topton 7.5-minute topographic quadrangle map (Appendix 1, fig. 5, location L-35).

Mine number 32 was located at about 40° 30′ 47″ latitude and 75° 40′ 30″ longitude on the USGS Topton 7.5-minute topographic quadrangle map (Appendix 1, fig. 5, location L-36).

Mine number 33 was located at about 40° 31′ 04″ latitude and 75° 40′ 47″ longitude on the USGS Topton 7.5-minute topographic quadrangle map (Appendix 1, fig.5, location L-37).

Mine number 34 was located at about 40° 31′ 09″ latitude and 75° 40′ 53″ longitude on the USGS Topton 7.5-minute topographic quadrangle map (Appendix 1, fig. 5, location L-38). The 1876 atlas shows this mine located near the J. Dunkle residence.

Mine number 35 was located at about 40° 30′ 15″ latitude and 75° 40′ 41″ longitude on the USGS Topton 7.5-minute topographic quadrangle map (Appendix 1, fig. 5, location L-26). The mine is shown on or close to the Henry Barber property in the 1876 atlas.

Mine number 38 was located at about 40° 30′ 56″ latitude and 75° 41′ 06″ longitude on the USGS Topton 7.5-minute topographic quadrangle map (Appendix 1, fig. 5, location L-40). This mine was likely a Fenstermacher mine. The 1876 atlas shows the mine located near the Fenstermacher estate and properties.

Mine number 40 was located at about 40° 31′ 22″ latitude and 75° 41′ 15″ longitude on the USGS Topton 7.5-minute topographic quadrangle map (Appendix 1, fig. 5, location L-29).

Mine number 41 was located at about 40° 31′ 20″ latitude and 75° 41′ 25″ longitude on the USGS Topton 7.5-minute topographic quadrangle map (Appendix 1, fig. 4, location L-41). This mine likely was a Thomas Iron Company mine as the railroad tracks, which were financed in part by the Thomas Iron Company, terminated at this mine.

Mine number 42 was located at about was located at about 40° 31′ 22″ latitude and 75° 41′ 38″ longitude on the USGS Topton 7.5-minute topographic quadrangle map (Appendix 1, fig. 4, location L-42).

Mine number 46 was located at about 40° 30′ 53″ latitude and 75° 42′ 11″ longitude on the USGS Topton 7.5-minute topographic quadrangle map (Appendix 1, fig. 4, location L-45).

Figure 275. Ziegler's ore washer, Farmington, Longswamp Township, 1897. Photograph taken by William Smith. Used with permission from Suzanne Smith (William Smith's granddaughter).

Farmington Area Mines

The Farmington area iron mines were located north of Farmington (fig. 274) and are numbered 9 and 13 through 16 on the 1882 topographic map. Mines shown on figure 274 numbered 9, 12, 13, 14, and 15 are in Maxatawny Township and are described in that section.

Zeigler Mine

The Jonathan Zeigler mine was northwest of the intersection of Valley Road and Pine Street. It is mine number 10 on the 1882 topographic map, which shows it in Lehigh County. However, LIDAR imagery indicates it also is in Berks County with more extensive workings to the east in Berks County than is shown on the 1882 topographic map. The eastern open cut is at 40° 32′ 10″ latitude and 75° 40′ 22″ longitude, and the western open cut is at 40° 32′ 10″ latitude and 75° 40′ 15″ longitude on the USGS Topton 7.5-minute topographic quadrangle map (Appendix 1, fig. 5, location L-47A and L-47B respectively). In 1876, the operator of the mine was J. Zeigler & Company (1876 atlas). Other owners included Jonathan Zeigler and David Zeigler. Gideon Zeigler's ore washery (fig. 275) was in operation as late as 1895 (*Reading Eagle*, September 2, 1981).

The 1880 Census (Pumpelly, 1886, p. 960) listed the Temple Iron Company as the operators of the Zeigler mine. Production between July 1, 1879, and June 30, 1880, was 5,831 tons of limonite, which was shipped to the Temple Iron Company furnace.

The ore was a dark brown and reddish brown, very sandy, cellular limonite, which also was compact and fine grained. It contained a considerable quantity of quartz. An analysis was presented by McCreath (1879, p. 211-212): 36.55 percent metallic iron, 0.034 percent sulfur, 0.275 percent phosphorus, and 0.129 percent metallic manganese.

Merkel Mine

Mrs. John Merkel's mine is mine number 16 on the 1882 topographic map (fig. 274). It was located at about 40° 32′ 01″ latitude and 75° 40′ 51″ longitude on the USGS Topton 7.5-minute topographic quadrangle map (Appendix 1, fig. 5, location L-25). The mine is shown in the 1876 atlas. The ore was a light and dark brown, cellular limonite. The cells were filled with clay and contained masses of quartz. An analysis was presented by McCreath (1879, p. 211-212): 47.2 percent metallic iron, 0.034 percent sulfur, and 0.113 percent phosphorus.

Fegley and Walbert Mine

The Fegley and Walbert mine was southeast of the intersection of Valley Road and Magnolia Drive at about 40° 31′ 42″ latitude and 75° 40′ 08″ longitude on the USGS Topton 7.5-minute topographic quadrangle map (Appendix 1, fig. 5, location L-51). The Fegley and Walbert mine is shown in the 1876 atlas. The 1880 Census (Pumpelly, 1886, p. 960) listed the Thomas Iron Company as the operator of the Walbert mine.

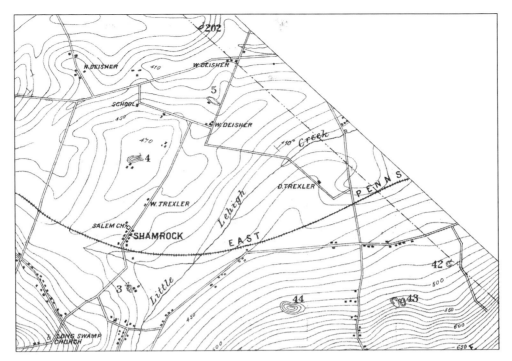

Figure 276. Mines in the Shamrock area, Longswamp Township, 1882. From Pennsylvania Geological Survey (1883).

Other Iron Mines

Iron mines in the Farmington area with numbers on the 1882 topographic map (fig. 274) or the index to the 1882 topographic map and no available information include the following mines.

Mine number 17 is numbered on the index to the 1882 topographic map; it was shown but not numbered on the 1882 topographic map. The mine was located at about 40° 31′ 57″ latitude and 75° 40′ 39″ longitude on the USGS Topton 7.5-minute topographic quadrangle map (Appendix 1, fig. 5, location L-49). LIDAR imagery indicates that it was a shallow, and likely short-lived, operation.

Mine number 18 is numbered on the index to the 1882 topographic map; it was shown but not numbered on the 1882 topographic map. The mine was located at about 40° 31′ 27″ latitude and 75° 40′ 02″ longitude on the USGS Topton 7.5-minute topographic quadrangle map (Appendix 1, fig. 5, location L-64).

Mine number 19 is numbered on the index to the 1882 topographic map; it was shown but not numbered on the 1882 topographic map. The mine was located at about 40° 31′ 29″ latitude and 75° 39′ 41″ longitude on the USGS Topton Topton S 7.5-minute topographic quadrangle map (Appendix 1, fig. 5, location L-66).

Mine number 20 was located at about 40° 31′ 42″ latitude and 75° 39′ 37″ longitude on the USGS Topton SE 7.5-minute topographic quadrangle map (Appendix 1, fig. 5, location L-50).

Mine number 37 is numbered on the index to the 1882 topographic map; it was shown but not numbered on the 1882 topographic map. The mine was located at about 40° 30′ 58″ latitude and 75° 41′ 23″ longitude on the USGS Topton SW 7.5-minute topographic quadrangle map (Appendix 1, fig. 4, location L-67).

Shamrock Area Mines

The mines near Shamrock are numbered 3, 4, 5, 42, 43, and 44 on the 1882 topographic map (fig. 276) and Prime's 1878 map. Mine names are from Prime (1878).

John Fegley Mine

The John Fegley mine was north of Mertztown Road, very close to the Lehigh County border, at about 40° 31′ 12″ latitude and 75° 38′ 49″ longitude on the USGS Topton 7.5-minute topographic quadrangle map (Appendix 1, fig. 5, location L-58). The mine was in the Allentown Formation. The John Fegley iron ore beds are shown on the 1862 map.

Solomon Boyer & Company Mine

The Solomon Boyer & Company mine was on the west side of Swabia Creek at about 40° 29′ 12″ latitude and 75° 37′ 23″ longitude on the USGS East Greenville 7.5-minute topographic quadrangle map (Appendix 1, fig. 19, location L-59).

D'Invilliers visited what he called the "*Fegley mine or Sol. Boyer & Co.'s mine*" in 1879. He stated: "*I met no one there who could speak English, and all the information I could obtain through my limited knowledge of Dutch [Deitsch or Pennsylvania German] was that Nathan Ziegenfuss was the contractor and the B. I. Co. [Bethlehem Iron Company] the consumers. The shaft was then 50 feet deep, employed 12 men, and turned out 12 tons a day.*" The mine was idle in 1882 (D'Invilliers, 1883, p. 261-262).

The gangue was gneiss mixed with hornblende that dipped conformably with the ore bed S. 54° E. 60°. D'Invilliers (1883, p. 262) published an analysis of the ore: 43.5 percent metallic iron, 0.005 percent sulfur, 0.057 percent phosphorus, and 29.46 percent silica. Smith (1885, 414; analysis) reported stilbite in masses of white radiating needles from the Fegley mine.

Jesse Laros' Mine No. 3

Jesse Laros' mine No. 3 was east of the intersection of Longswamp Road and Kennedy Avenue at about 40° 29′ 57″ latitude and 75° 38′ 54″ longitude on the USGS Manatawny 7.5-minute topographic quadrangle map (Appendix 1, fig. 16, location L-22). It is mine number 3 on the 1882 topographic map (fig. 276) and Prime's 1875 map. The mine was in the Lethsville Formation. Prime (1875, p. 41) stated: "*Jesse Laros's Mine, No. 3, abandoned. This excavation has not been worked for a long time. The ore, from the fragments on the side of the mine, must have been associated with damourite slate.*"

Jesse Laros' Mine No. 4

Jesse Laros' Mine No. 4 was west of Kennedy Avenue, 0.45 mile north of Shamrock at about 40° 30′ 32″ latitude and 75° 38′ 45″ longitude on the USGS Topton 7.5-minute topographic quadrangle map (Appendix 1, fig. 5, location L-52). It is mine number 4 on the 1882 topographic map (fig. 276) and Prime's 1875 map. The mine was in the Allentown Formation.

The mine was leased by the Crane Iron Company. When Prime (1875, p. 41) visited the mine, it was idle and flooded. Prime noted: "*The depth of the mine to water is 38 feet. On the west side of the mine there is a mixture of clay, quartz, and damourite slate, all in very small pieces, for a depth of 17 feet from the surface. Larger pieces of the slate occur on the dump, showing that it occurs at a greater depth. The ore where visible, at a few points, occurs in clay.*" The ore was a compact, sandy limonite of a very dark color. An analysis was presented by McCreath (1875, p. 53-54): 43.7 percent metallic iron, 0.763 percent manganese, 0.005 percent sulfur, 0.869 percent phosphorus, and 18.58 percent insoluble residue.

Aaron Hertzog Mine

The Aaron Hertzog mine was south of Longswamp Road close to the Berks-Lehigh County boundary. The northern open cut was at about 40° 30′ 08″ latitude and 75° 37′ 08″ longitude, and the southern open cut was at about 40° 30′ 06″ latitude and 75° 37′ 07″ latitude on the USGS Allentown West 7.5-minute topographic quadrangle map (Appendix 1, fig. 5, location L-53). It is mine number 42 on the 1882 topographic map (fig. 276) and and Prime's 1875 map. The mine was in the Leithsville Formation. When Prime (1875, p. 17) visited the mine, it was abandoned, but the machinery was still present.

Wescoe Mine

The Wescoe mine was northwest of the intersection of Fairchild and Hemphill Roads. The eastern open cut was at about 40° 29′ 53″ latitude and 75° 37′ 14″ longitude (location L-54A), and the western open cut was at about 40° 29′ 53″ latitude and 75° 37′ 22″ longitude (location L-54B) on the USGS East Greenville 7.5-minute topographic quadrangle map (Appendix 1, fig. 19). It is mine number 43 on the 1882 topographic map (fig. 276) and Prime's 1875 map. The mine was in the Leithsville Formation. Iron ore is shown on the 1862 map at this location. Prime (1875, p. 17) noted: "*This consists of several pits, none of them very deep; work has long since been abandoned, due to exhaustion of the ore. A large body of white clay was observed in the bottom of the mine.*"

Wagenhorst Mine

The Wagenhorst mine was west of Timber Drive at about 40° 29' 55″ latitude and 75° 37′ 58″ longitude on the USGS Topton 7.5-minute topographic quadrangle map (Appendix 1, fig. 16, location L-55). It is mine number 44 on the 1882 topographic map (fig. 276) and Prime's 1875 map. The mine was in the Leithsville Formation. Prime (1875, p. 17) stated: "*This mine was not being worked when visited; it consists of a single excavation about 30 feet deep. Nothing could be seen there.*"

Henry Stein Mine

The Henry Stein mine was east of Kennedy Avenue southeast of the intersection of Kennedy Avenue and Mertztown Road at about 40° 30′ 26″ latitude and 75° 37′ 47″ longitude on the USGS Topton 7.5-minute topographic quadrangle map (Appendix 1, fig. 5, location L-56). It is mine number 5 on the 1882 topographic map (fig. 276) and Prime's 1875 map. The mine was in the Leithsville Formation.

Henry Stein's mine was leased by the Thomas Iron Company. Prime (1878, p. 40-41) described the mine: *"In this mine no damourite slate was apparent, but the ore occurs in and over white and pink clay, resulting from the decomposition of the slate. The mine is 47 feet deep. At a depth of 40 feet limestone was struck in one portion of the mine which dips 12° S. 41° E. The top limestone is slaty and drab colored. It is 4 feet thick and overlies the ordinary blue waterworn limestone. The clay overlying the limestone and containing the ore, dips 42° in the same direction. In the limestone at the bottom of the mine there is an aperture about 10 inches square into which all the water of the mine pours and disappearing avoids the necessity of any pump to keep the mine dry."*

The ore was a compact limonite with a considerable coating of white clay; some of the pieces were brick red, and others a reddish-brown color. An analysis was presented by McCreath (1875, p. 53-54): 49.6 percent metallic iron, 0.583 percent manganese, 0.007 percent sulfur, 1.29 percent phosphorus, and 9.44 percent insoluble residue.

Dresher Mine

The Dresher mine was north of the intersection of Mertztown and Mertz Roads. The northern mine was at about 40° 30′

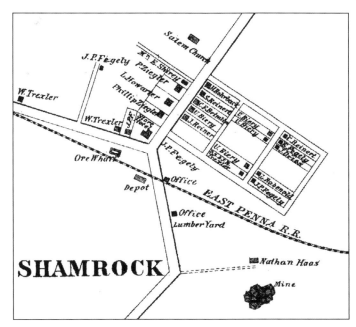

Figure 277. Location of the Nathan Haas mine, Shamrock, Longswamp Township. From Davis and Kochersperger (1876).

46″ latitude and 75° 38′ 32″ longitude, and the southern mine was at about 40° 30′ 38″ latitude and 75° 38′ 48″ longitude on the USGS Topton 7.5-minute topographic quadrangle map (Appendix 1, fig. 5, location L-57). The mine was in the Allentown Formation. The Stephen Dresher iron ore beds are shown on the 1862 map.

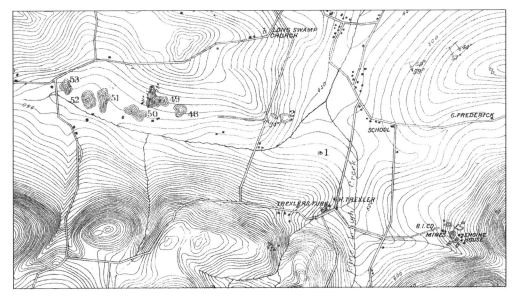

Figure 278. Mines in the Longsdale area, Longswamp Township, 1882. From Pennsylvania Geological Survey (1883).

Nathaniel Dresher leased his property to the Emaus Furnace for mining iron ore beginning on October 19, 1871 (*Reading Eagle*, November 26, 1879). The Reading Eagle (February 13, 1874) reported that two workmen were killed in a blast in the mine. The 1880 Census (Pumpelly, 1886, p. 960) listed H.B. Benfield as the operator of the Nathan Dresher mine. Production between July 1, 1879, and June 30, 1880, was 405 tons of limonite, which was shipped to the East Penn Furnace in Lyons.

Haas Mine

The Nathan Haas mine was northeast of the intersection of Kennedy Avenue and Stutzman Lane at about 40° 30' 04" latitude and 75° 38' 48" longitude on the USGS Topton 7.5-minute topographic quadrangle map (Appendix 1, fig. 5, location L-62). The mine was in the Leithsville Formation. The mine is shown in the 1876 atlas (fig. 277). The Reading Eagle (May 24, 1893) reported that a man drowned while fishing in an abandoned iron mine on the Nathan Haas farm.

Longsdale Area Mines

The mines in the Longsdale area are numbers 1, 2 and 48 through 53 on the 1882 topographic map (fig, 278). The Frederick mine is included in this area.

Litzenberger Mine

The H. Litzenberger mine was southeast of the intersection of Centennial Road and Mary Ann Drive at about 40° 29' 11" latitude and 75° 38' 58" longitude on the USGS Manatawny 7.5-minute topographic quadrangle map (Appendix 1, fig. 16, location L-60). It is mine number 1 on the 1882 topographic map (fig. 278) and Prime's 1875 map. The mine was in the Leithsville Formation.

The Thomas Iron Company sank a number of exploratory shafts on the Litzenberger farm. Ore was found to be distributed throughout clay in all of them. About 20 feet of iron ore was found in three of the shafts. Prime (1875, p. 42-44) stated: *"It seems somewhat doubtful whether the ore in sight will justify the erection of machinery. Probably no very large quantity of ore will be found."*

Lichenwallner Mine

The Levi Lichenwallner mine was west of Centennial Road and southwest of the intersection of Centennial Road and State Street. The eastern open cut was located at about 40° 29' 23" latitude and 75° 39' 09" longitude (location L-61A), and the western open cut was located at about at 40° 29' 24" latitude and 75° 39' 14" longitude (location L-61B) on the USGS Manatawny 7.5-minute topographic quadrangle map (Appendix 1, fig. 16). It is mine number 2 on the 1882 topographic map (fig. 278) and Prime's 1875 map. The mine was in the Leithsville Formation. The mine was leased by the Crane Iron Company.

Prime (1875, p. 42) observed: *"At this mine there are several excavations. The most southerly one is full of water at the bottom, and is 26 feet deep to the water. There is a plane in this pit. The small one to the north of this is only 10 feet deep; only stripping having been taken from it. The largest excavation of all lies still more to the north; this is 40 feet deep, and is no longer worked. To the west of this is a small pit 15 feet deep. The mine has not been worked since the Fall of 1873. The ore occurs in white clay, and overlying damourite-slate. In some places the clay apparently overlies the ore, but clay or slate underlies it. It is said that the well, which has been sunk to a considerable distance, struck blue slaty limestone at a depth of 130 feet. As the mine is not being worked, it was impossible to ascertain the dip."*

The ore was a hard, sandy limonite with considerable ocherous iron ore. An analysis was presented by McCreath (1875, p. 54): 42.8 percent metallic iron, 0.252 percent manganese, 0.036 percent sulfur, 0.222 percent phosphorus, and 25 percent insoluble residue.

Long Mines

The Long mines were within the area bounded by Tower Road, Hidden Valley Road, and Longsdale Drive. Several iron and clay mines and one graphite mine were located in this area.

Figure 279. Mine number 53, Longsdale, Longswamp Township, March 2015.

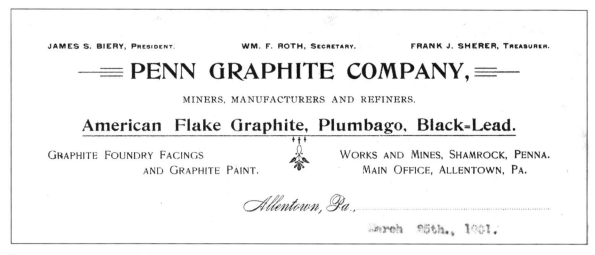

Figure 280. Letterhead of the Penn Graphite Company, 1901. Courtesy of Dale Richards.

The mines were worked by members of the Long family and a number of lessees. This area includes mines numbered 49 through 53 on the 1882 topographic map (fig. 278). The mines are centered at 40° 29′ 28″ latitude and 75° 40′ 10″ longitude on the USGS Manatawny 7.5-minute topographic quadrangle map (Appendix 1, fig. 16, location L-48).

William Long leased mining rights to Alfred H. Newhart (1893), Bryan O'Neil (1895), Reuben H. Kramm (1897 and 1898), James S. Biery (1899), M.J. Person & Company (1906), and Theodore Leas (1906) among others.

Several iron mines were located on the Long property. The 1880 Census (Pumpelly, 1886, p. 961) listed Dr. M.S. Long as the operator of the Long mine. Production between July 1, 1879, and June 30, 1880, was 8,512 tons of limonite. The ore was shipped to furnaces in the Lehigh Valley and Reading. The 1880 Census (Pumpelly, 1886, p. 960) also listed the Thomas Iron Company as the operator of the M. Long limonite mine.

The Reading Eagle (August 23, 1900) reported that Nicholas Long, of S. Long & Son, was engaged in mining and preparing ocher from about 1880 until 1889 when the mill burned down. The steam-powered ocher mill had a capacity of 10 tons per day. In 1897, the mine employed 12 people and produced 3,000 tons of yellow ocher per year. The mine operated 200 days per year. (*Mine and Quarry News Bureau*, 1897). In 1910, the owner of the ocher mine was Dr. Wilson P. Long.

In 1910, Stoddard and Callen (1910, p. 432-433) reported: *"Long's property is a quarter of a mile south of Hancock. It shows signs of long abandonment, as no shafts or workings remain open. A considerable quantity of ocher is stocked in piles, and it seems to be of good color and weight. The surrounding ground is honeycombed with test pits, in which water has risen very high. It has been found impossible to construct shafts that will last beyond one working season, as the wet clay breaks the timbering of shafts and gangways, and it was partly because of this trouble that the place was abandoned. The veins on the property dip southeast. They were worked to a depth of about 80 feet below the surface, but at greater depths they were valueless. There are now no remains of hoisting machinery, and washing apparatus was never installed."*

Ocher occurred in association with iron ore nodules, chert fragments, and clay in irregular pockets of variable size. During mining, ocher was separated from the clay as well as possible. Most of the shafts were 30 to 40 feet deep, although some were said to have been sunk to a depth of 130 feet. When visited by Benjamin Miller during August 1910, only one man was engaged in mining. He sunk several shallow test pits and found ocher in some of them. The shaft that was worked during the preceding winter had been abandoned. The entire output of the mine was sold to the C.K. Williams Company of Easton (Miller, 1911, p, 33).

About 1873, white clay was discovered on the Long farm. A steam engine was erected, and mining began. No profit was realized for the first year of mining. At depth, a deposit of better quality clay was found. It was shipped to Philadelphia and used principally for the manufacture of white china ware (Davis and Kochersperger, 1876) The mine was known as the Berks County China Clay Works.

The East Penn Graphite Company, described below, was located on the Long property.

East Penn Graphite Company Mine

About 1878, William Reily of Philadelphia began mining graphite and constructed a mill on the farm of William Long, Sr. Shortly after it was put into operation, the mill burned down and was not rebuilt. Reily died soon afterward (*Reading Eagle*, April 19, 1893). Little mining was done until 1897 when the East Penn Graphite Company was organized to work the graphite deposit. The East Penn Graphite Company (fig. 280) erected a well-equipped refining mill and extended the workings.

In 1898, the East Penn Graphite Company fired several employees. The work day was increased from 10 to 11 hours with the same pay of $1 a day. Several employees did not want to work an extra hour for the same pay and were fired (*Reading Eagle*, December 6, 1898).

Two shafts, one 90 feet deep, were sunk to the graphite vein, and drifts were extended from the shafts at different levels. The ore was mined with pick and shovel and hoisted to the surface using horse power. The ore was crushed, washed, further refined by air currents, and packed for shipment. Seventy-five percent of the product was flake graphite, and the remaining twenty-five percent was a lower grade used for producing graphite paint. Production in July 1899 was 2 tons per day. Quartz sand, a by-product from processing, was sold to foundries for moulding sand.

The East Penn Graphite Company never made a profit. In 1901, the company was reported to have produced 14 tons of flake graphite and 3 tons of lower grade graphite per day. The mine closed in 1901. The machinery was removed, and the mill was in ruins by 1911 (Miller, 1912a, p. 112-113).

The graphite occurred in a fine-grained conglomerate or a coarse-grained sandstone, which was weathered to a considerable depth. The deposit was lens shaped and was nearly vertical. The maximum thickness of the vein was reported to be 39 feet (Rothwell, 1899, p. 350). The ore was reported to contain between 25 and 80 percent graphite, with an average of 28 percent. Miller (1912a, p. 112-113) disputed those figures and stated: "*Certainly the average ore did not contain that amount of graphite, judging from the ore specimens found about the mine and mill. It is probable that the ore contained considerably less than 7%, although no figures to verify such a conclusion are available.*"

The graphite formed a prominent constituent of a highly weathered gneiss in which the feldspar was thoroughly kaolinized. Quartz was abundant. Some of the kaolin was pink, probably due to iron staining. A great amount of iron rust suggested the former presence of pyrite, which was oxidized. Mica was not observed. The graphite flakes were matted together in irregular curved sheets. The rock had been sheared, and, in the shearing process, the graphite flakes were compressed into irregular sheets. The graphite was unusually friable (Miller, 1912a, p. 112-113).

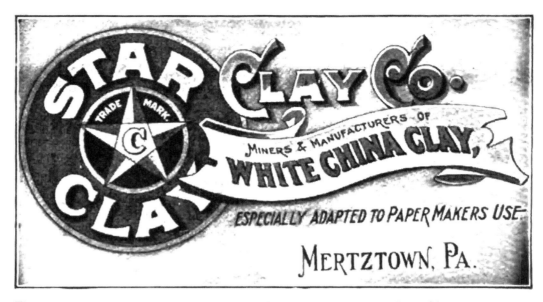

Figure 281. Advertisement for the Star Clay Company, Longswamp Township.

Star Clay Company

A clay pit was opened, and a clay refining plant was built by William Reily on the property of David DeLong in 1876. The company grew into the Star Clay Company managed by J.B. Wilson. The Star Clay company leased property from William Riley (1878), Davis DeLong (1889), the Davis DeLong estate (1900 and 1912), Eliza Schweyer (1907 and 1912), Thomas J. Koch (1912), Cyrenius J. Mabry (1917), Mary E. Denkel (1919 and 1920) and Beneville Eck (1920).

In 1886, the plant used steam power, and 10 men were employed (Montgomery, 1886, p. 1050). During 1899, the Star Clay Company employed 15 men and shipped refined clay at the rate of one car load per day. The open pit covered an area of about 1 to 2 acres and was 50 to 90 feet deep. The overburden of yellow sand and clay varied from 5 to 25 feet thick. On one side of the pit, there were bands of pink and purplish clay interspersed with white clay to a considerable depth. The mass of clay was blue to blue gray when mined, but white when refined. Mining by the Star Clay Company was described by Fegley (1915) (see opposite page). The product was shipped to potteries (for china ware) and paper factories (fig. 281). The best grade sold for $11 per ton (Hopkins, 1900, p. 24).

Star Clay Company supplied clay to a number of companies for the manufacture of paper. The corporate office was at 220 South 2nd Street in Philadelphia (*American Stationer*, September 4, 1879). In 1919, fire destroyed the boiler room and engine house. The loss amounted to several thousand dollars and was only partly covered by insurance (*Reading Eagle*, July 14, 1919). The Star Clay Company is known to have operated as late as 1922 (Pennsylvania Department of Labor and Industry, 1922).

Kaolin Mining In Longswamp Township

by H. Winslow Fegley

Brick and Clay Record, May 20, 1915

"The kaolin deposits were first discovered in 1876 on the lands of David DeLong, where the beds were quite numerous and the product almost pure white, making it especially adapted for the uses of paper manufacturers, the finer clay products, and for the use of chemists and ceramists.

The kaolin that is mined in Longswamp township is taken from pits from 25 to 100 feet in depth. The clay, as it is dug from the earth, is carried to the surface by an endless bucket system, which works automatically, and as soon as it reaches its destination, workmen shovel it from the dumping wharves upon wheel-barrows, in which it is wheeled to the extensive pits, lined with heavy planking, that are hollowed out of the earth near by. There are six or eight of these pits at each of the more important mines, measuring from 200 to 300 feet in length by half as wide, and about 12 feet deep.

When one of the pits is filled with the rough kaolin it is flooded with clear water, which is left in the pit for three or four months. Occasionally these pits are flooded in the fall of the year and left until the following spring, but in summer two floodings of a pit may be made. The water will evaporate to a certain extent, and a close watch is kept upon every pit, so that there is always enough water above the surface of the deposited kaolin. After this soaking process is thoroughly completed, the kaolin is again lifted from the pits, and spaded out in small blocks, which are again placed on the wheel-barrows and hauled to the drying houses, where the blocks are placed on shelves, one tier above another. for the purpose of thoroughly drying. The kaolin being full of pores, has, during its stay in the pits, absorbed a good deal of water, and this has all to be disposed of by evaporation. It usually takes several months' drying on these racks before the sun and air has finally drawn away the water.

After this drying process, the kaolin is ready to be ground into powder. The blocks, now slightly smaller and much lighter in weight, are again placed on the wheel-barrows in the smaller plants and upon trucks in the more extensive ones, and conveyed to the grinding pits, which resemble to a certain extent those used by potters to grind their clay, horse or steam power being used to do this work. After the kaolin is ground so fine that it almost resembles dust, it is deposited into large bags holding about 200 pounds or in some cases packed in barrels, and shipped to the concerns that at present operate the largest of these kaolin mines, and from this central point it finds its ways into the trade circles, finally reaching the manufacturing plant, where its use may be for half-a-dozen different purposes.

It is estimated that fifty per cent of the Longswamp kaolin is used in the manufacture of clay products and paper commodities. A good deal is used to make alum, the valuable component of which for dyeing and printing, is the aluminum sulphate. In the late use of kaolin in paper mills, the fine particles are simply cemented to the fibre by means of aluminum hydrate, and serve to give body to the paper. Especially is this the case in the manufacture of wall paper.

Kaolin deposits are of various kinds, and are composed of three species of minerals—quartz, feldspar and mica. The kaolin of the mines in Longswamp township shows a whitish color, generally, and when touched by the hand it resembles soap; but is soft and brittle and has an adherent nature and a peculiar odor. Kneaded together with water, it absorbs the latter in considerable quantities and becomes plastic. A good deal of it has been used to make the cheaper grades of chinaware."

Figure 282. Star Clay Company kaolin pits, Longswamp Township, 1915. The pits in which kaolin is soaked for 3 to 4 months are in the foreground. Photograph by H, Winslow Fegley. Courtesy of the Swenkfelder Heritage Center.

Figure 283. Digging kaolin from the pits, Star Clay Company, Longswamp Township, 1915. Drying sheds are in the background. Photograph by H. Winslow Fegley. Courtesy of the Swenkfelder Heritage Center.

Figure 284. Kaolin drying sheds, Star Clay Company, Longswamp Township, 1915. Photograph by H, Winslow Fegley. Courtesy of the Swenkfelder Heritage Center.

Figure 285. Kaolin pits and drying sheds, Star Clay Company, Longswamp Township, 1915. Photo-graph by H, Winslow Fegley. From Fegley (1915).

Figure 286. Fritch and Brother's mine, Longsdale, Longswamp Township, March 2015.

Other clay companies that operated in the area were the National Clay Company of Shamrock (ca. 1920), the Long Valley Clay Works operating under the supervision of Levi L. Fritch (ca. 1920), and a clay mine operated by F.H. Yoder of the Reading White Clay Company (ca. 1911).

Hancock Mud-Dam Deposit

The Hancock mud-dam ocher deposit was southwest of the intersection of Tower and Home Roads. It was southwest of the Long mines. LIDAR imagery shows it as a large flat area at about 40° 29′ 22″ latitude and 75° 40′ 28″ longitude on the USGS Manatawny 7.5-minute topographic quadrangle map (Appendix 1, fig. 16, location L-21). The deposit was owned by the Thomas Iron Company.

The mud-dam deposit covered about two acres and was several feet deep. It represented the washings from the limonite ore extracted from a large iron mine a short distance farther up the hill. The ocher deposit was prospected, and fairly good material was found, but none had been shipped as of 1910 (Miller, 1911, p. 34).

Fritch and Brother Iron and Clay Mines

Fritch and Brother's iron and clay mines (fig. 286) were southwest of the intersection of Hidden Valley Road and Longsdale Drive. They were centered around 40° 29′ 36″ latitude and 75° 39′ 27″ longitude on the USGS Manatawny 7.5-minute topographic quadrangle map (Appendix 1, fig. 16, location L-18). The mines included a number of shafts and pits in the Hardyston Formation.

D'Invilliers (1883, p. 105) noted: "*A few hundred yards S. S. W. of Longswamp church Fritch Bros.' bank of brown hema-tite is located, which, though now abandoned, has furnished some good ore in the past.*" The Clymer Iron Company leased one of the mines from Nathan Fritch in 1882. D'Invilliers (1883, p. 264) published an analysis of the ore: 22.1 percent metallic iron, 0.72 percent metallic manganese, 0.028 percent sulfur, 0.018 percent phosphorus, and 30.6 percent silica.

A clay mine on the Fritch property was operated by Benjamin Moore & Company, which had an office in Brooklyn, New York. The mine was worked in underground drifts, and the clay was hoisted from the mine by steam power through a shaft reported to be 80 feet deep. The clay was dumped into a log washer similar to the iron ore washers, where it was separated from the coarser materials and run through a series of troughs into large settling vats. After the clay settled, the water was decanted into a separate reservoir and pumped back to the washer. The clay was removed from the vats to the drying sheds where it was air dried and then hauled to the refining plant at the Hancock station, where it was further processed to prepare it for market. The dry clay was finely ground on a burr mill, mixed with the proper pigments and other materials, and packed for market (Hopkins, 1900, p. 24). The Reading Eagle (August 17, 1895) reported that William Christman and W.O. Lichtenwalner were hauling white clay for Benjamin Moore & Company from the mine on the Edwin Fritch property near Longswamp.

In 1899, the Reading Eagle (November 1, 1899) reported that Job V. Folk of Reading sold 1,000 tons of white clay to various parties. He had a number of teams hauling clay from the Edwin Fritch property to the Shamrock railroad station. Folk paid his teamsters 40 cents per ton.

Figure 287. Goethite from Hancock, Longswamp Township, 8.6 cm. Carnegie Museum of Natural History collection ANSP 24888. Former E.T. Wherry specimen.

Figure 288. Stalactitic pipe ore (limonite) from the Topton area, Longswamp Township, 5.3 cm. Reading Public Museum collection 2000C-027-118-C.

Frederick Mines

The Frederick mines were a group of mines east of Dogwood Drive at about 40° 28′ 58″ latitude and 75° 38′ 11″ longitude on the USGS Manatawny 7.5-minute topographic quadrangle map (Appendix 1, fig. 16, location L-17). The mines, in felsic to mafic gneiss, were on the property of G. Frederick (D'Invilliers, 1883, p. 263).

The Frederick mines were worked by the Crane Iron Company from 1875 to 1878, and later by the Bethlehem Iron Company. The Bethlehem Iron Company leased the mines to individual operators through the company's agent, Stiles Levan of Alburtis. In 1879, P. and J. Razor, Thomas James, George Fenstermacher, I. Reinhardt, and others mined iron ore from six shafts from 15 to 120 feet deep. The deepest shaft was equipped with a hoisting engine and Cameron and Knowles pumps; the others were worked by horse and windlass. Production was about 15 tons of ore per day. The ore was shipped from the Shamrock railroad station. It cost 30 to 40 cents per ton to haul the ore from the mines to the station (D'Invilliers, 1883, p. 263). The 1880 Census (Pumpelly, 1886, p. 960) listed the Bethlehem Iron Company as the operator of the Frederick mine. Production between July 1, 1879, and June 30, 1880, was 3,820 tons of magnetite.

The Frederick mines, with one exception, were abandoned by 1882. According to D'Invilliers (1883, p. 263): "*The causes of abandonment were natural — too much water and too little ore.*" The only mine worked in 1882 was at the extreme north edge of the property. It was leased in March 1882 by I. Reinhardt. The shaft was about 40 feet deep and rigged with a windlass. The mine produced about 20 tons of ore per week. The ore was sold to Horatio Trexler at the Shamrock wharf for $1.60 per ton, which included a royalty of 35 cents per ton.

The ore was disseminated in a soft, micaceous, quartzose gneiss. The ore bed was about 2.5 feet thick and dipped 70° SE. into the hill. An analysis of the ore was published by D'Invilliers (1883, p. 264): 35.5 percent metallic iron, 0.003 percent sulfur, 0.042 percent phosphorus, and 43.8 percent silicious matter (chiefly quartz).

Tatham Mine

The Tatham mine was west of the intersection of Tower Road and Park Avenue at about 40° 29′ 47″ latitude and 75° 40′ 41″ longitude on the USGS Manatawny 7.5-minute topographic quadrangle map (Appendix 1, fig. 16, location L-19). The mine was in hornblende gneiss.

James and William Tatham of Tatham Brothers of Philadelphia were the owners of the mine. The mine was originally opened by the Bethlehem Iron Company, which leased the property from George N. Tatham in 1864. Iron ore was mined by an open cut on the outcrop until 1878. The cut was about 30 feet deep and 40 feet wide. From a narrow opening at the north end, it extended about 70 feet to the southeast, ending in a perpendicular face of gneiss. In 1878, a 46-foot deep shaft was sunk at the bottom of the open cut. The ore bed thinned, and the shaft was abandoned. A slope was then started further to the west. The slope started at the top of the open cut and extended on a 60° dip for about 80 feet with about 50 feet of the slope underground. The slope was furnished with a double track for hoisting ore cars. From the bottom of the slope, a 25-foot wide drift was driven east in the ore-bearing strata for 50 feet. From the bottom of the slope, a stope was carried up 25 feet in ore. A tunnel was driven east into the perpendicular face of the rock, but ore was not encountered. D'Invilliers (1883, p. 265) described the tunnel: "*For some inexplicable reason a tunnel was driven east into the perpendicular face of rock before mentioned. From its direction no success could have been expected, as it ran nearly parallel with and back of the ore bed.*" An adit was afterwards driven south from the underground workings and was reported to have struck a 14-foot thick ore bed (D'Invilliers, 1883, p. 264-265). Mining ceased about 1884.

In 1899, after being idle for 15 years, James Tatham contracted with Simon P. Romig to mine 100 tons of iron ore. The mine was flooded and had to be dewatered (*Reading Eagle*,

Figure 289. Bombshell limonite ore from the Topton area, Longswamp Township, 9 cm. Reading Public Museum collection 2000c-27-112.

August 1, 1899). The *Reading Eagle* (December 9, 1912; May 6, 1913) reported that Romig, superintendent of the Tatham Brothers mine, had 12 miners at work and was shipping magnetite iron ore. Romig (1847-1917) was one of the pioneers in the ore, clay, and ocher mining businesses in Longswamp Township. He was operating Tatham's mine at at the time of his death in 1917 (*Reading Eagle*, February 19, 1917).

The ore was magnetite disseminated in gneiss in a band 20 to 25 feet thick between walls of dark green pyroxene and black hornblende rock. After mining, the ore had to be carefully sorted. The ore was shipped to the Topton Furnace. The ore and gneiss dipped conformably 40° SE. The strike of the ore-bearing strata was nearly east and west. An analysis of the ore was published by D'Invilliers (1883, p. 266): 29 percent metallic iron, 0.002 percent sulfur, 0.006 percent phosphorus, and 26 percent silica.

Topton Area Mines

Henry Erwin and Sons Ocher Mine

The Henry Erwin and Sons ocher mine was about 0.5 mile south of Topton. LIDAR imagery shows a mine east of the intersection of Woodside Avenue and Ramp Lane at about 40° 29' 20" latitude and 75° 41' 59" longitude on the USGS Manatawny 7.5-minute topographic quadrangle map (Appendix 1, fig. 15, location L-65). The mine was in felsic to mafic gneiss.

The Henry Erwin and Sons mine was reported to be one of the largest ocher mines in the area. It was opened about 1880 and operated only during the summers. In 1910-11, mining was from an open cut. However, there were several shallow shafts through which ocher was once mined. The timbering of one of the shafts was exposed in the open cut. In 1911, the open cut was about 45 feet deep at the deepest point, approximately 300 feet long, and about 100 feet wide. In some parts of the pit, the ocher was within 2 or 3 feet of the surface, but in other parts, the overburden was 5 to 8 feet thick. The ocher was not uniform in character, and some of it was too light in color to be sold. No regular arrangement of the variously colored materials was apparent (Miller, 1911, p. 34). When visited by Stoddard and Callen (1910, p. 432-433), the workings were flooded.

The mined ocher was hauled to the nearby washing plant where it first passed through a 16-foot long log washer to separate the larger particles of iron ore and chert from the ocher. The ocher, held in suspension in the water, passed through three sets of sand troughs. The first trough was about 32 feet long with two compartments; the other two troughs were about 16 feet long, each with three compartments. The ocher then went to one of the eight mud dams or settling ponds, each of which held from 30 to over 100 tons (fig. 290). Enough material was sent to each pond to produce 2 to 3 feet of ocher after drying. The ocher required from 2 to 2-1/2 months to be sufficiently dry before it could be removed. Next, it was taken to one of the six drying sheds, which ranged from 50 to 75 feet long and 15 to 20 feet wide. In one shed, there were steam coils beneath the floor to assist in drying. Another shed next to the boiler house was enclosed on all sides and fitted with steam coils. This drying shed held about 15 tons and was used in winter when air drying was impractical (Miller, 1911, p. 35-36).

The greatest obstacle at the plant was the small amount of water available for washing the ocher. The 145-foot deep well, fitted with a Cornish pump, frequently failed to furnish sufficient water. At the time of Miller's visit, it was possible to pump only enough water to wash ocher for 2 to 2-1/2 hours at a time (Miller, 1911, p. 36)

The ocher was hauled to Topton and shipped to the Irwin paint mills at Bethlehem, where it was ground and either sold as raw ocher or mixed with oil to form paint. The yearly output of the plant averaged 800 to 1,000 tons (Miller, 1911, p. 36).

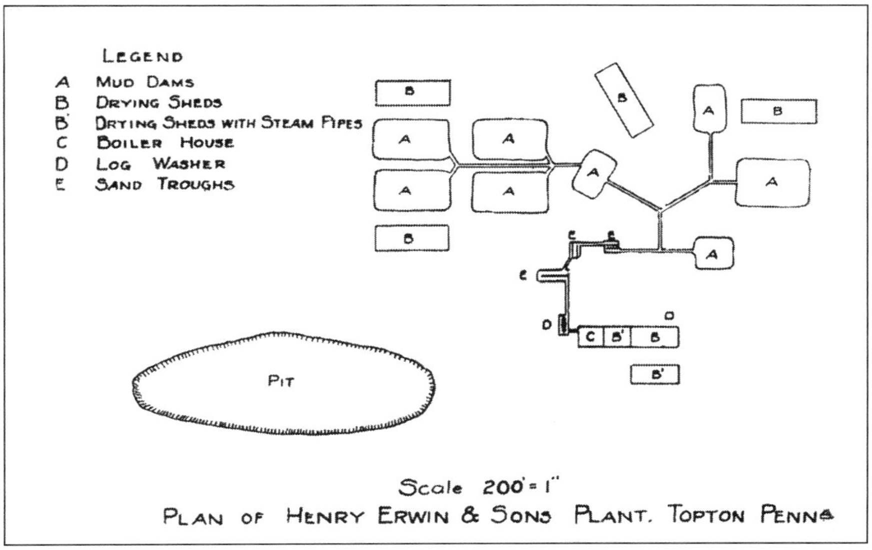

In the 1930s, the mine and plant were operated by Reichard-Coulston, Inc. (Miller, 1933, p. 148). Mining was done by open cut. In 1933, the pit was 300 feet long, 175 feet wide, and 45 feet deep. The ocher was processed onsite and shipped to the Reichard-Coulston paint mills in Bethlehem, where it was ground in an air mill and sold as raw ocher or mixed with oil to make paint. The company sold about 150 tons of product per year for $30 per ton in carload lots.

Many nodules of limonite, partly filled with white clay and ranging in size from 1/8 inch to 1 foot in diameter, were mixed with the clay and ocher. They were not shaped like the ordinary limonite geodes generally found in association with ocher. A large percentage had a neck or tapering projection half as long as the main nodule. In addition to the limonite geodes, there were many

Figure 290. Plan of the Henry Erwin and Sons ocher mine, Topton, Longswamp Township. From Miller (1911, p. 35).

Figure 291. Henry Erwin and Sons open-pit ocher mine, Topton, Longswamp Township. From Miller (1911, plate 3).

Figure 292. Henry Erwin and Sons boiler house and ocher drying sheds, Topton, Longswamp Township. From Miller (1911, plate 4).

Figure 293. Henry Erwin and Sons ocher settling troughs, Topton, Longswamp Township. From Miller (1911, plate 5).

Figure 294. Walker quarry, Longswamp Township, ca. 1932. From Stone (1932, p. 58). Courtesy of the Pennsylvania Geological Survey.

Walker Granite Company Quarry

The Walker Granite Company quarry (figs. 294 and 295) was north of Walker Road at about 40° 28′ 15″ latitude and 75° 37′ 02″ longitude on the USGS East Greenville 7.5-minute topographic quadrangle map (Appendix 1, fig. 19, location L-63). The quarry was in felsic to mafic gneiss.

The Walker Granite Company quarry was owned and operated by Willis F. Walker, who was born in Somerset County, Pa. about 1862. Walker settled in northeastern Berks County in 1906 and opened the the granite quarry. In addition to his position as manager and treasurer of the Walker Granite Company, Walker was active in railroad design. He held a number of patents, which he contended for many years held the key to a new era in rail transportation. His patents covered railroad ties, rail hatches, rail joint bridge bars, rail joint insulation, track rail insulation, and reinforced composition slabs for floors and other purposes.

Walker discovered the granite deposit near Siesholtzville in 1906 while employed by the York Bridge Building Company. His interest in the origin of specimens of hard stone used for the construction of piers and foundations for bridges led him to the area. Within a few weeks of locating the deposit, he leased 200 acres of land and proceeded with the construction of a crusher plant and opening of the quarry. The stone from this quarry was particularly popular with stone cutters and contractors because of the ease with which it could be worked into suitable sizes and shapes. In later year, railroad construction

angular chert particles, some of which were 8 to 10 inches in diameter. They were derived from the limestone and left behind, together with the other insoluble materials forming the clay and ocher when the calcium carbonate was removed in solution. In some places in the ocher and clay, there were unusual aggregations of chalcedony resembling bunches of berries. They generally were less than 1 inch in diameter, although a few were somewhat larger. They were unlike any siliceous nodules observed in place in the limestone, and their origin was unknown (Miller, 1911, p. 34). Miller (1911, p. 35) provided two analyses of the ocher.

Atlas Ocher Mine

The Atlas ocher mine and plant were on a farm formerly owned by Rev. M.H. Brensinger. The plant was adjacent to the Henry Irwin and Sons plant. The plant was abandoned in 1908. Ocher was mined by the Atlas Paint Company of Mertztown for several years and later by Reitnauer & Strohl. The first operations were by open cut and later by a shaft about 35 feet deep. There was one main hoisting shaft and a number of test pits, but no extensive workings. The ocher was washed in a manner similar to that used by Henry Erwin and Sons, described above (Miller, 1911, p. 36; Stoddard and Callen, 1910, p. 432-433).

Other Localities

D'Invilliers (1883, p. 400) reported chalcedonic jasper and jasper near Mertztown. Eyerman (1899, p. 20) reported massive epidote from Hancock.

Figure 295. Walker quarry, Longswamp Township, April 2016.

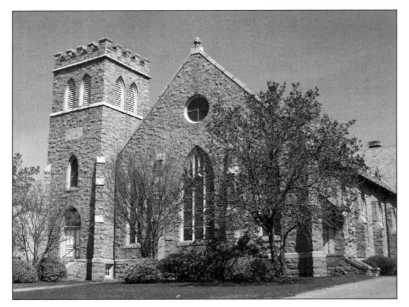

Figure 296. Palm Schwenkfelder Church, Palm, Pa., constructed with Seisholtzville granite from the Walker quarry. Photograph taken April 2016.

plans occupied most of Walker's efforts, and quarrying operations were subsequently discontinued (*Reading Eagle*, November 7, 1950). The Walker quarry is know to have operated as late as 1940 (Meyers, 1940).

Seisholtzville granite is a light buff to light pink gneiss with fine- to medium-grained texture, poorly defined banded structure, composed predominant of quartz and microcline, and containing 5 to 10 percent hornblende and biotite (Geyer, 1977, p. 43).

Russell (1941) described a metadiabase dike that cut through the Walker quarry. Minerals associated with unaltered granitic gneiss included quartz, microcline, orthoclase, and sodic plagioclase plus small amounts of hornblende, apatite, and biotite. Minerals associated with the dike included quartz, augite, epidote in bladed aggregates, anhedral magnetite, chlorite, muscovite var. sericite, and pyrite grains.

Stone from the quarry was used in the construction of numerous buildings in southeastern Pennsylvania, including the Reading Railroad stations in Pottstown (ca. 1928), Wernersville, and at Seventh and Franklin Streets in Reading (1930). Many churches were constructed from Seisholtzville granite, including Bethany Lutheran Church in West Reading, many churches in Allentown, Scranton Nativity School, Swedenborgian Cathedral in Bryn Athyn, Palm Schwenkfelder Church (1911)(fig. 296), St. John's Church in Emaus (1924), St. Gabriel's Roman Catholic Church in Hazelton (1925), and Grace Lutheran Church in Macungie. In addition to the production of building stone, the quarry also produced grit for chicken feed, which was manufactured from the chips too small for building purposes.

E.H. Trexler Quarry

The E.H. Trexler limestone quarry was located on the west Side of Valley Road north of its intersection with Mertztown Road at about 40° 30′ 48″ latitude and 75° 40′ 06″ longitude on the USGS Topton 7.5-minute topographic quadrangle map (Appendix 1, fig. 5, location L-31). The 1860 and 1862 maps and the 1876 atlas show a lime kiln on the property. Frear (1913, p. 78) presented an analysis of the limestone from the T.L. Trexler quarry in Mertztown.

Graphite Prospects

Franklin DeLong Graphite Prospect

A graphite prospect was located on the Franklin DeLong farm, two miles southwest of Longswamp. A prospect shaft about 50 feet deep was sunk in search of graphite. The feldspar in the gneiss was decomposed. Pyrite, probably originally present, decomposed leaving iron stains. Biotite was present in small amounts. Graphite was present in rather large flakes arranged parallel to the gneiss bands (Miller, 1912a, p. 115).

Charles Brensinger Graphite Prospect

A graphite prospect was located on the Charles Brensinger farm, two miles south of Longswamp. Several pits were dug in graphitic gneiss about 1901. The rock was similar to that described above at the DeLong prospect, except that it wassomewhat finer grained (Miller, 1912a, p. 115).

Schmeck's Farm Graphite Prospect

D'Invilliers (1883, p. 105) reported that graphite was mined on Smeck's farm, which was located south of Longswamp. D'Invilliers (1883, p. 397-398) also reported the occurrence of epidote and kaolin on Schmeck's farm.

LOWER ALSACE TOWNSHIP

Big Dam Quarry

The Big Dam Quarry was near the Schuylkill River, northwest of the Forest Hills Cemetery at about 40° 18′ 48″ latitude and 75° 53′ 45″ longitude on the USGS Reading 7.5-minute topographic quadrangle map (Appendix 1, fig. 27, location LA-1). The location of the Big Dam quarry is shown on the 1882 topographic map.

The Big Dam quarry was so named because of its proximity to Big Reading Dam No. 24 on the Schuylkill River, which was built as part of the slack water canal navigation system. Rogers (1858, vol. 2, p. 681) noted that the quarry was in limestone fanglomerate in a cleft in Lower Cambrian rocks (fig. 297). According to D'Invilliers (1883, p. 189), the limestone fanglomerate rested conformably on the Hardyston Formation, which dipped 60° SE. under the overlying fanglomerate. The fanglomerate is very calcareous and was used for making lime in kilns near the railroad. The face of the quarry exposed blue and pink beds of compact, fine-grained stone. Wherry (1913, p. 117) noted that the quarry exposed conglomerate composed of rounded limestone pebbles cemented by red mud resting on a limestone breccia.

Figure 298. Calcite from the Big Dam quarry, Lower Alsace Township. Crystals to 1.9 cm. Skip Colflesh collection.

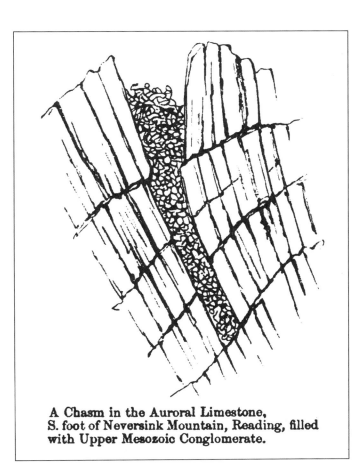

Figure 297. Geological setting of the Big Dam quarry, Lower Alsace Township. From Rogers (1858, vol. 2, p. 682, fig. 568).

A cave was found by the E & G Brooke Iron Company while quarrying limestone in the 1880s. According to the Reading Eagle (May 6, 1888), the company found a cave the size of "an ordinary room." It was filled with "ground." The fill had an abundance of stalagmites at the base and stalactites throughout the body of the fill. An analysis of the limestone was published by D'Invilliers (1883, p. 189): 67.8 percent calcium carbonate, 24 percent magnesium carbonate, and 7.1 percent silica.

Minerals

Calcite - fine scalenohedral crystals (D'Invilliers, 1883, p. 189) (figs. 298 and 299)

Fluorite (Eyerman, 1889, p. 7); deep blue cubic crystals

Malachite (Eyerman, 1889, p. 45)

Antietam Reservoir Locality

The Antietam Reservoir locality (fig. 300) is west of the intersection of Antietam and Lewis Roads, across Antietam Road from the Antietam Reservoir dam at 40° 21′ 20″ latitude and 75° 52′ 10″ longitude on the USGS Birdsboro 7.5-minute topographic quadrangle map (Appendix 1, fig. 29, location LA-2). This locality, in hornblende gneiss, also is known as Antietam Lake (Eyerman, 1889) and Ohlinger Dam (D'Invilliers, 1883).

In 1874, the City of Reading purchased the property for $7,000. Four years later, a dam was built, and the lake served as a water supply for Reading. The city bought land around the

Figure 299. Calcite from the big Dam quarry, Lower Alsace Township, 11.7 cm. West Chester University collection.

lake to protect water quality. The site became part of Lower Alsace Township in 1888 (*Reading Eagle*, July 2, 2001).

Smith (1978, p. 194-196) reported several occurrences of molybdenite in flakes to 3/4 inch in pegmatite dikes in this area. Smith postulated that the source of the molybdenite was hydrothermal solutions from an igneous intrusion. Some of the molybdenite is associated with powellite, pyrrhotite, pyrite, and jarosite.

A powellite-bearing dike is found in the roadcut (fig. 300) on the east side of Antietam Road, 295 feet south-southeast of the dam breast. The dike is 2 to 4 inches wide, trends roughly N. 45° E., and dips 70° NW. The dike is of probable Precambrian age and is composed mostly of coarse-grained, greenish-gray plagioclase and quartz with minor biotite, hornblende, and pyrrhotite. Molybdenite is sparsely distributed along the dike margins. The molybdenite is of the 2H polytype and alters to sparse pseudomorphs of powellite. The powellite is highly fluorescent (cream to yellow) under short-wave ultraviolet light. This is the first verified occurrence of powellite in Pennsylvania.

A larger, irregular molybdenite-bearing dike of gray granite occurs from 450 to 500 feet south-southwest of the dam breast; molybdenite occurs along parts of the dike margins. Molybdenite flakes at this location are up to 3/4 inch across.

A third, finely granitic dike, which is approximately 600 feet south-southwest of the dam breast and about 82 feet from the downhill end of the roadcut, contains disseminated molybdenite and pyrite (Smith, 1975, p. 16; Smith, 1978, p. 195).

D'Invilliers (1883, p. 75) described a small quarry in the roadcut (fig. 302), which he called the Ohlinger dam quarry: *"The east flank of hill shows generally a distinctly granulite rock composed of quartz and feldspar and everywhere giving rise to a sandy soil. From the summit however down to the reservoir the rocks are generally darker becoming mixed with hornblende epidote and pyroxene and corresponding with that peculiar face of rock exposed in the road cut at Ohlinger Dam."*

Minerals

Roadcut on the east side of the Antietam Reservoir. From Smith (1978, p. 194-195)

Biotite

Jarosite - orange-brown weathering product

Molybdenite - flakes to 3/4 inch

Plagioclase

Powellite - thin, transparent, gray or straw yellow scales, pseudomorphs after molybdenite; fluoresces cream yellow under short-wave ultraviolet light

Pyrite

Pyrrhotite

Quartz

Outcrops on the west side of the Antietam Reservoir opposite the dam; includes the Ohlinger Dam locality of Eyerman (1889)

Actinolite, var. byssolite (Eyerman, 1889, p. 17)

Actinolite, var. asbestos (D'Invilliers, 1883, p. 394)

Augite (D'Invilliers, 1883, p. 399)

Biotite (Smith, 1976, p. 195)

Epidote - massive and crystals (Eyerman, 1889, p. 20)

Garnet (Eyerman, 1889, p. 18)

Graphite (Eyerman, 1889, p. 3)

Figure 300. Antietam Reservoir locality, Lower Alsace Township, April 2015.

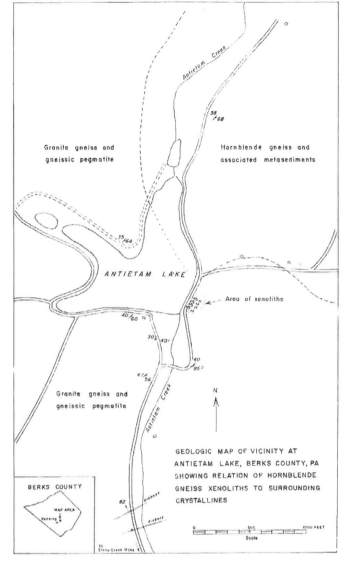

Figure 301. Geologic map of the Antietam Reservoir area, Lower Alsace Township. From Buckwalter (1958).

muscovite var. sericite, plagioclase, magnetite, apatite, diopside var. diallage, quartz, albite var. oligoclase, microcline, perthite, and orthoclase in the area west of the dam along Antietam Road.

Railroad Cut Opposite Poplar Neck

The railroad cut opposite Poplar Neck is along the Schuylkill River, northwest of the Forest Hills Cemetery at about 40° 19′ 00″ latitude and 75° 54′ 06″ longitude on the USGS Reading 7.5-minute topographic quadrangle map (Appendix 1, fig. 27, location LA-3). Poplar Neck is the area inside a tight loop of the Schuylkill River southeast of Reading. The cut is in the Hardyston Formation. Eyerman (1889, p. 45) reported malachite from this locality.

Neversink Mountain Sienna Mine

The Neversink Mountain sienna mine was in the Neversink Mountain Reserve, south of Reservoir Road at about 40° 19′ 26″ latitude and 75° 54′ 31″ longitude on the USGS Reading 7.5-minute topographic quadrangle map (Appendix 1, fig. 27, location LA-4). The mine was in the Hardyston Formation.

According to Miller (1911, p. 46): *"There is no sharp line between ocher and sienna as materials occur in nature showing all gradations between the two. The term sienna is properly applied, however, only to those pulverulent substances whose composition is practically that of high grade limonite iron ore."*

Sienna was mined for several years on the north slope of Neversink Mountain, directly south of the eastern part of Reading, about 100 feet from the crest of the mountain. A considerable thickness of quartzite decomposed to sand occurs on the slopes, and many sand pits were located there. Some of the quartzite is decomposed to such an extent that it can be readily crumbled in the hand, particularly those beds in which arkose is a prominent constituent. All of the quartzite is weathered enough to be easily crushed in a rock crusher (Miller, 1911, p. 46).

Hornblende (D'Invilliers, 1883, p. 394)

Labradorite (?) (D'Invilliers, 1883, p. 396)

Molybdenite (D'Invilliers, 1883, p. 399)

Muscovite (D'Invilliers, 1883, p. 399)

Orthoclase (D'Invilliers, 1883, p. 396)

Pyroxene - abundant (Eyerman, 1889, p. 15)

Pyrrhotite (D'Invilliers, 1883, p. 400)

Rutile (?) - threadlike inclusions in gneiss (D'Invilliers, 1883, p. 401)

Stilbite - radiating crystalline masses (Eyerman, 1889, p. 32)

Buckwalter (1958) described xenoliths (fig. 301) with greenish-brown hornblende, chlorite, clinozoisite, epidote,

Figure 302. Photograph of the Ohlinger Dam quarry, Lower Alsace Township, 1882. From D'Invilliers (1883, p. 81).

Figure 303. Molybdenite from the Antietam Reservoir locality, Lower Alsace Township. Bryn Mawr College collection Rand 2836.

About 1903, a layer of sienna interbedded with weathered quartzite was encountered in one of the sand pits on land owned by the Michael Haak estate. At first, the material was not recognized as valuable. On learning that the material was a high-grade sienna, drifts were run on the bed, and intermittent mining began.

At one time, C.K. Williams and Company operated a mine on the hill, but for several years prior to 1910, the only mining was carried on by Mrs. John P. Lance of Reading, who was one of the heirs to the Haak property. Most of the time, only a few men were employed, and the annual production ranged from 125 to 200 tons. When the mine was visited by Miller in August 1910, only two men were engaged in mining, and it was reported that C.K. Williams and Company recently leased an adjoining property and planned to open a mine (Miller, 1911, p. 46-47).

The sienna possessed a rich yellow color with thin streaks of a somewhat darker material running through it and was remarkably free from impurities. Miller (1911, p. 47) presented an analysis. The ore had the same dip as the enclosing beds of weathered quartzite, 25° to 30° N., which is about the same as the slope of the mountain. It represents a replacement of quartz and arkose of certain beds and shows the stratification lines of the original rock. These lines show even more distinctly in the rich, yellow, fine-grained sienna than in the unaltered coarser quartzite.

The sienna bed was up to 5 feet thick, but rapidly thinned to a few inches or entirely disappeared within 10 to 15 feet. The strata of the quartzite was remarkably regular, showing that the sienna did not represent a single stratum of the original rock. The thickening and thinning of the sienna occurred both in the direction of the dip as well as along the strike of the quartzite. However, certain layers seem to have been replaced to a greater degree than others. When the sienna was found, mining followed the same strata, even though the sienna was absent in places. Two sets of sienna beds were mined (Miller, 1911, p. 47).

The workings consisted of a adit driven into the mountain about 35 feet to the sienna bed. The drift then turned along strike about 20 feet, part of which was in sienna. At the end of this drift, a pocket of good sienna was found, and a stope was opened along the dip. The stope was about 10 feet deep. The sienna was taken from the mine in wheelbarrows and was partially dried by being placed on a sheet iron platform above a wood fire. It was stored in a covered shed and subsequently hauled to the Reading railroad station for shipment. The sienna sold for about $20 per ton (Miller, 1911, p. 47).

Stony Creek (Ohlinger) Mills Locality

The Stony Creek (Ohlinger) Mills locality is near the boundary between Alsace and Lower Alsace Townships on the USGS Birdsboro 7.5-minute topographic quadrangle map. D'Invilliers (1883, p. 396 and 399) reported augite and red orthoclase from this locality. A specimen of hornblende from the Ohlinger Mills locality is in the Carnegie Museum of Natural History Jefferis collection (fig. 304).

D'Invilliers (1883, p. 396) stated that the locality was 1/8 mile east of the Stony Creek Mills post office. The modern USGS Topographic map shows Stony Creek Mills in Exeter Township; however, the 1882 topographic map shows Ohlinger Mills and the Stony Creek post office in Lower Alsace Township at the intersection of Friedensburg, Casonia Avenue, and Antietam Roads. One-eight mile east of this intersection places the locality in Lower Alsace Township.

David Knabb Mine

The David Knabb mine was east of the intersection of Friedensburg and old Friedensburg Roads at about 40° 21′ 14″ latitude and 75° 51′ 35″ longitude on the USGS Birdsboro 7.5-minute topographic quadrangle map (Appendix 1, fig. 29, location LA-6A). An "ore hole" is shown on the 1882 topographic map to the northeast almost at the top of the hill at about 40° 21′ 23″ latitude and 75° 51′ 30″ longitude (Appendix 1, fig. 29, location LA-6B). According to D'Invilliers (1883, p. 276), an abandoned mine on the property of David Knabb produced *"many tons of good magnetite ore."*

Fischer Prospect

The Fischer prospect was northwest of the intersection of Friedensburg Road and Spook Lane at about 40° 20′ 28″ latitude and 75° 52′ 55″ longitude on the USGS Reading 7.5-minute topographic quadrangle map (Appendix 1, fig. 27, location LA-7). The prospect was in felsic to mafic gneiss. D'Invilliers (1883, p. 276) described the prospect: *"There is a shaft which seems to have proved unsuccessful, located about 100 feet up from the road just west of D. Fisher's house, on the Reading-*

Figure 304. Hornblende from the Ohlinger Mills locality, Lower Alsace Township, 9 cm. Carnegie Museum of Natural History collection CM 6215. This specimen was collected in 1877.

Ohlinger dam road. The country rock here is black hornblendic syenite, and no ore is visible in the pile of refuse stuff at the mine dump. The numerous tributaries to Antietam creek on the side of this hill expose small chunks of magnetite ore in their beds mixed with sediment, and this probably led to the sinking of the trial shaft."

LOWER HEIDELBERG TOWNSHIP

Berkshire Furnace

The Berkshire Furnace was built by William Bird in 1755-56 on a branch of Spring Creek, a tributary to Tulpehocken Creek, about 2 miles southwest of Wernersville. Its original name was Roxborough Furnace, and it also was known locally as Roxberry Furnace. The furnace initially used iron ore from South Mountain. After Bird's death in 1762, the furnace was run by John Patton and Bird's son, Mark, under the firm name Patton & Bird. Patton married Bird's widow, and Mark Bird transferred his share of the furnace to his mother and John Patton. Patton acquired sole ownership of the property in 1764 for £2,550 and renamed the furnace Berkshire about 1767. Patton was a Colonel in the Berks County Militia, and it is unclear if he ran the furnace during the entire Revolutionary War, as George Ege is credited with assisting in its management during that time. Hessian prisoners were used at the furnace, and the furnace workman were exempt from military duty. Montgomery (1884) stated that in November 1780, George Ege, the lessee of Berkshire Furnace, supplied the American Government with 2,894 pounds of shot and shell. The firm name later was changed to Patton & Ege.

Both Hermelin (1783) and Schoepf (1788) noted the furnace during their travels in Berks County in the 1780s. Hermelin indicated it produced 600 tons of pig iron annually, and Schoepf stated that there was not a sufficient quantity of local ore to run the furnace, and additional ore was obtained from Peter Grubb's Cornwall mine. In 1786, Ege purchased an ore right to the Cornwall mine and obtained his ore from there. In 1790, John Patton died, and Ege purchased the furnace for £2,500 from Bridget Patton, who was then living in Fairfax, Virginia. Montgomery (1884) indicated that Ege operated the furnace for several years but abandoned it in 1794 for lack of water.

Berkshire Furnace Mine

Samuel Hermelin, a Swedish industrialist who served as an ambassador to the U.S., visited the mine that supplied ore for the Berkshire Furnace and provided the following description: *"The top layer consists of sandy earth of different depths; the next layer is red clay, in certain places of a thickness of from one to two feet, then yellow clay, in which stratum ore is found in large stones and in small grains; which layer has not been gone through. The ore quarry has a large air shaft on the slope of a hill, from 4 to 5 feet deep at the lower end and a few and twenty feet at the upper end. The varieties here are partly reddish-brown clay, partly reddish-brown iron ore"* (Hermelin, 1783).

Benjamin Hull Quarries

The Benjamin Hull quarries were northeast of the intersection of U.S. Route 422 and Sportsman Road. The northernmost quarry (location LH-1A) was at 40° 20′ 17″ latitude and 76° 06′ 03″ longitude, and the southernmost quarry (location LH-1B) was at 40° 20′ 15″ latitude and 76° 06′ 01″ longitude on the USGS Sinking Spring 7.5-minute topographic quadrangle map (Appendix 1, fig. 23). The quarries were in the Hershey and Myerstown Formations, undivided. The location is number 85 (fig. 305) on reference map 11 in Pennsylvania Geological Survey (1891).

Benjamin Hull owned two small quarries and one lime kiln. The quarries were in a conglomeritic limestone that dipped about 20° S. D'Invilliers (1886, p. 1556) described the limestone: *"The conglomerate is not coarse, but contains a number of different silicious limestone and sandstone pebbles, only slightly rounded, and all firmly cemented together."* D'Invilliers considered these quarries *"experimental openings, abandoned when the poor quality of the limestone was detected."*

Hospital Creek Quarry

A limestone quarry was on the east bank of Hospital Creek, in a steep bluff a short distance east of the Benjamin Hull quarries. The quarry was northwest of the intersection of

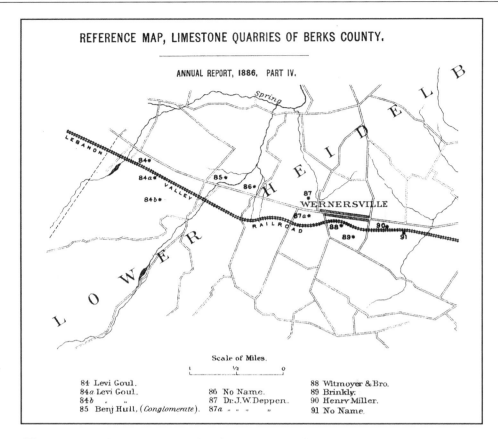

Figure 305. Limestone quarries in Lower Heidelberg Township, 1886. From Pennsylvania Geological Survey (1891, reference map 11).

U.S. Route 422 and Point Road at about 40° 20′ 08″ latitude and 76° 05′ 49″ longitude on the USGS Sinking Spring 7.5-minute topographic quadrangle map (Appendix 1, fig. 23, location LH-2). The quarry was in the Martinsburg Formation. It is quarry location number 86 (fig. 305) on reference map 11 in Pennsylvania Geological Survey (1891). D'Invilliers (1886, p. 1556-1557) noted: *"The limestone exposed here is lean, considerably washed out and cavernous and generally quite shaly. It has a pale blue color."* D'Invilliers considered this quarry and the nearby Benjamin Hull quarries *"experimental openings."*

Glen Gery Quarry

The Glen Gery quarry (fig. 306) was north of the intersection of State Hill and Sweitzer Roads at 40° 21′ 35″ latitude and 76° 01′ 03″ longitude on the USGS Sinking Spring 7.5-minute topographic quadrangle map (Appendix 1, fig. 24, location LH-3). Clay for manufacturing bricks was mined from an open pit in Hamburg sequence shale (MacLachlan and others, 1975, p. 205-206).

Old Quarry No. 1

Old quarry No. 1 of Miller (1934, p. 218) was north of U.S. Route 422 at 40° 19′ 46″ latitude and 76° 03′ 34″ longitude on the USGS Sinking Spring 7.5-minute topographic quadrangle map (Appendix 1, fig. 24, location LH-4). An old lime kiln and a somewhat irregular quarry in the Epler Formation were located close to Little Cacoosing Creek at the base of a hill. About 10 feet of bluish-black beds were exposed. Miller (1934, p. 218) presented an analysis of the rock.

Old Quarry No. 2

Old quarry No. 2 of Miller (1934, p. 218) was north of U.S. Route 422 at 40° 19′ 58″ latitude and 76° 03′ 31″ longitude on the USGS Sinking Spring 7.5-minute topographic quadrangle map (Appendix 1, fig. 24, location LH-5). Old quarry No. 2 was in the Epler Formation 100 yards north of old quarry No. 1 in the same hill on the south bank of Little Cacoosing Creek. In this quarry, a 3.5-foot thick basal bed of massive limestone was overlain by a massive blue-gray bed 10 feet thick. The beds dipped south (Miller, 1934, p. 218).

LYONS BOROUGH

East Penn Furnace

Lyons was founded as Lyon Station in 1860 when the railroad was extended to that point The East Penn Furnace was located at Railroad and Hunter Streets. The two stack, hot blast anthracite furnace was constructed in 1871 by John T. Noble of Pottsville, Pa., for the East Penn Iron Company at a cost exceeding $200,000. Franklin Brownback was the first furnace manager. In 1875, the furnace was acquired by the Philadelphia and Reading Coal and Iron Company. The furnace used limonite from Berks and Lehigh Counties and magnetite from New Jersey to produce gray forge and foundry pig iron under the brand name "East Penn." The annual capacity was 17,000 tons (American Iron and Steel Association, 1882, p. 22). In 1881, a fire destroyed the engine house and damaged the furnace to such an extent that it was never used again (Montgomery, 1886). The furnace was dismantled in 1890 (American Iron and Steel Association, 1892).

MAIDENCREEK TOWNSHIP

A post office named Calcium was established in Maidencreek Township on July 4, 1885. It was so named because it was an important shipping point for lime and limestone. Six kilns were owned and operated by Kline and Hoffman near the station. This business began on the Bushong farm in 1873 (Historical Committee of the Blandon Bicentennial Committee, 1976).

Limestone Quarries

The Jacksonburg Formation cement rock occurs in Maiden Creek Township. This formation was quarried in several locations to obtain rock for manufacturing cement. The current (2016) Evansville cement plant had several predecessors in the cement business, including the Molltown, Newport, and Ajax Cement Companies. In 1899, the roaster for the Molltown Cement Company was hauled from Evansville to Molltown by Samuel Bobst using a team of 4 horses and 2 mules. The roaster was 60 feet long and weighed 12.5 tons. (*Reading Eagle*, February 17, 1900). In 1901, the Molltown Cement Company changed its name to the Ajax Cement Company (Berks County deed book). The company went out of business in 1906 (Bollfras, 2013).

Figure 306. Glen-Gery quarry, Lower Heidelberg Township. From MacLachlan and others (1975, p. 206). Courtesy of the Pennsylvania Geologic Survey

Reading Cement Company Quarry

The flooded Reading Cement Company quarry was south of the intersection of Maiden Creek and Pleasant Hill Road at 40° 28′ 19″ latitude and 75° 52′ 46″ longitude on the USGS Temple 7.5-minute topographic quadrangle map (Appendix 1, fig. 8, location MC-5).

In 1899, the Reading Cement Company was incorporated in New Jersey with $100,000 in capital. The company was organized by John P. Lance, president; J.C. Illig, secretary and treasurer; M.C. Aulenbach of Reading; and George H.M. Martin of New Jersey. The company office was at 536 Penn Street in Reading. J.L. Repplier was the general sales agent. The company purchased 69 acres that included an existing cement rock quarry. The company constructed a plant on the Heffner farm near the Evansville railroad station on the Philadelphia and Reading Railroad (*Reading Eagle*, September 2 and November 14, 1899); the plant capacity was 600 barrels of cement per day. In 1902, the company constructed a 1.5 mile spur from the rail line to the plant (*Stone*, May 1901, p. 476).

The first cement was produced in September 1900. It was sold under the brand names "Reading Portland Cement" and "Improved Rosedale Natural Hydraulic Cement." Italian immigrant workers from Virginia were employed in the cement plant. In 1901, the president of the company was M.C. Aulenbach, and the secretary and treasurer was F.W. Harold, both of Reading. Directors included J.C. Illig of Reading, George W. Beard of Reading, J.L. Repplier of Reading, and George H. Dunsford and George H.N. Martin of Camden, New Jersey. The general manager was R.G. Bush, and the plant chemist was Henry Muller, both of Fleetwood (Brown, 1901, p. 294). In 1901, Albert Heffner sold his farm to the Reading Cement Company for $300 per acre (*Reading Eagle*, July 15, 1901). The farm then became known as the cement farm. Chemical analyses of limestone from the Reading Cement Company quarry were presented by Day (1904, p. 162).

Figure 307. Calcite from Maidencreek Township. Carnegie Museum of Natural History collection ANSP 25673.

Business was excellent for the Reading Cement Company in 1902; they declared a stock dividend of 6 percent. Two years later, the price of cement dropped, and the company stockpile was large, so work at the plant was suspended. The company started three new roasters on July 22, 1905. However, a few months later, the company filed for bankruptcy (*Reading Eagle*, July 26 and November 2, 1905). In 1906, a bankruptcy sale took place, and the property, quarry, and plant were sold to the Vindex Portland Cement Company.

Figure 309. Calcite from Maidencreek Township, 7 cm. Carnegie Museum of Natural History collection ANSP 25674.

Vindex Portland Cement Company

On November 19, 1906, John M. Frame of Reading applied to the Commonwealth of Pennsylvania for a charter to incorporate the Vindex Portland Cement Company (fig. 331), a new company that was to be capitalized at $25,000 (*Kutztown Patriot*, November 3, 1906). A meeting of the stockholders created a $150,000 bond issue in 1907 to finance the company (*Reading Eagle*, February 19, 1907). N.J. Ritter was the general manager of the company. He graduated from Millersville State Normal School with a degree in mechanical engineering and had 15 years experience in the cement manufacturing business prior to his position with Vindex (*Cement World*, February 15, 1909).

Figure 308. Zircon from Maidencreek Township, 1.6 cm. Sloto collection 3322.

Figure 310. Limonite from Maidencreek Township, 6.5 cm. Bryn Mawr College collection Rand 928.

Figure 311. Stock certificate of the Vindex Portland Cement Company, 1907.

The Vindex Portland Cement Company quarry also produced building stone. In 1910, the quarry furnished stone for the construction of a new hotel at Schlemsville (*Kutztown Patriot*, June 18, 1910). Miller (1934, p. 214) indicated that the cement plant was abandoned because of the poor quality of the rock. The Vindex Portland Cement Company filed articles of dissolution in 1914 (*Reading Eagle*, July 25, 1914). In 1915, the property and plant were offered for sale by the Bondholders Protective Committee (*Cement and Engineering News*, October 1915); however, the property did not sell. It was offered for sale again in 1916 (fig. 312). According to the sale advertisement in the Reading Eagle (September 23, 1916), the public sale included 67 acres with dwellings, house, barn, farm buildings, and a quarry. Again, the property was not sold. In 1917, the property finally was sold in three parcels to Morris H. Brensinger, Mahon G. Hummel, and Daniel E.S. Dries (Berks County deed book).

Maidencreek Portland Cement Company

In 1902, the Maidencreek Portland Cement Company was incorporated in New Jersey to "*manufacture and deal in cement, lime, limestone, calcined and other plasters and artificial stone.*" The company was capitalized with $100,000 in stock and $60,000 in bonds. It was incorporated by Kenneth K. McLaren, Raymond Newman, and Horace S. Gould (*Clay Record*, December 29, 1902). H.M. Hawkesworth of the Carpenter Steel Works in Reading was president, and John Barbey was treasurer. The Maidencreek Portland Cement Company purchased five farms totaling 198 acres adjacent to the South Evansville station of the Berks and Lehigh Railroad and held options on five more farms totaling an additional 600 acres. The Berks and Lehigh Railroad passed through the property. The company planned to build a cement plant with an output of 1,200 barrels per day (*Reading Eagle*, December 20, 1902).

The Maidencreek Portland Cement Company took out a mortgage for $250,000 from the Guardian Trust Company of New York City dated February 1, 1904, and a second mortgage for $600,000 dated May 7, 1903, from the Fidelity Trust Company of New Jersey (*Reading Eagle*, February 2, 1906; *Kutztown Patriot*, March 19, 1904). On September 15, 1904, a public sale of the bonds of the Maidencreek Portland Cement Company was held at the Commercial Exchange at 25 North Sixth Street in Reading. Bonds with a face value of $50,000, held by the First National Bank of Reading as collateral security for an indebtedness to the bank, were offered for sale (*Reading Times*, September 15, 1904). The company declared bankruptcy in 1906.

MAIDENCREEK TOWNSHIP

FOR SALE—BUSINESS PLACES

Cement Plant FOR SALE

All machinery, material and buildings of Vindex Portland Cement Plant, at Molltown, Berks County, are offered for sale, either as a whole or in part. This plant contains modern machinery, such as engines, boilers, air compressors, rotary burring kilns, ball mills, tube mills, kominuters, dryers, crushers, Fuller pulverizing mills, storage tanks, elevators, conveyors, shafting, pulleys, hangers, dump cars, light and heavy rail, all sizes pipes, valves, fittings, etc. Also all buildings, lumber, brick and laboratory equipment, all of which will be sold at bargain prices. For information apply to

Sterling Iron & Steel Co.
10th and Hamilton Sts.
PHILADELPHIA, PA.
Or Representatives at Plant.
Bell Telephone at Plant, Leesport 44-B4.

Figure 312. Advertisement offering the Vindex Portland Cement Company plant for sale, 1916 (Reading Eagle, June 10, 1916).

Evansville Quarry and Cement Plant

The Evansville quarry is a large active (2016) quarry (fig. 313) also known as Richmond quarry No. 1. It is west of the intersection of Maidencreek and Evansville Roads at 40° 28′ 13″ latitude and 75° 53′ 57″ longitude on the USGS Temple 7.5-minute topographic quadrangle map (Appendix 1, fig. 8, location MC-1).

In 1906, the Allentown Portland Cement Company was incorporated in New Jersey with $2,000,000 in capital. In 1908, the company began buying farms in the Evansville area, acquiring a total of 452 acres. Charles A. Matcham (fig. 314) of Allentown was the founder and general manager of the company. The Evansville cement plant was built by Fuller Engineering of Allentown, of which Matcham was the president. Col. J.W. Fuller of Catasauqua was the inventor of the Fuller-Kinyon pump, which simplified piping cement and other powdered materials (*Reading Eagle*, November 19, 1961). Prior to his involvement with the Allentown Portland Cement Company, Matcham was superintendent of the Alpha Portland Cement Company for 6-1/2 years and the general manager of the Lehigh Portland Cement Company for 9-1/2 years.

A steam "thumper" drill bored holes for dynamite in the quarry. After the rock was blasted, workers pounded the rock into pieces small enough for one man to load into a horse drawn cart (fig. 317). Later, 5-1/2 ton quarry cars (figs. 319 and 325) operating on a single track narrow gage railroad (figs. 321 and 322) replaced the horse-drawn carts, and rock was loaded into cars with a steam shovel (fig. 320). The cars were hauled to a No. 10 McCully crusher with an electric hoisting engine. Between 1910 and 1920, the crushing operation improved so that a rock 6 feet in diameter could be crushed. Dormitories and a mess hall for workers were located at the site.

The material from the crusher discharged to a large elevator that distributed the material to three No. 6-1/2 Lehigh gyratory crushers. The crushed material was conveyed to large bins over the dryers directly in back of the kilns. The three dryers utilized waste heat from the kilns. After drying for about 12 to 24 hours, the material was moved to a 5,000-ton storage area. From the storage area, the material was fed into two No. 8 Krupp ball mills and then into seven 42-inch Fuller-Lehigh pulverizers. From the pulverizers, the material was transferred to storage bins above the kilns. The material was then fed by a screw conveyer to an inclined feed pipe directly into the kilns. The four coal-fired kilns, built by Robert Wetherill & Company of Chester, were 8 feet in diameter and 120 feet long. Each kiln had a capacity of 700 to 800 barrels per day. After leaving the kilns, the material moved through rotary coolers and then to the gypsum house, where gypsum was mixed in. The mixture then passed through 11 Fuller-Lehigh finishing mills, traveled to the storage house, and then to the bagging house where Bates bag machines were used to bag the finished cement. Electricity used to power the machinery was produced at an

Figure 313. Aeriel view of the Evansville Quarry, Maidencreek Township, September, 2015.

Figure 314. Portrait of Charles A. Matcham (1862-1911).

on-site power plant located 1,000 feet from the cement plant. Power was produced by three large Wetherill condensing engines (fig. 316) connected to alternating current generators (*Reading Eagle*, October 31, 1909). On April 1, 1910, the first carload of cement left the plant.

In 1909, the Allentown Portland Cement Company leased the Vindex Portland Cement Company plant in Molltown. The Reading Eagle (October 31, 1909) reported that they were repairing it and hoped to have it operational by early 1910. The company expected to use the Vindex plant if the capacity of the new Evansville plant was exceeded (*Reading Eagle*, May 30, 1910).

By 1925, the quarry was 1,500 feet long, 700 feet wide, and 120 feet deep. The rock resembled a black shale or slate, but was about 74 percent calcium carbonate. The beds were considerably deformed but had a general strike of N. 44° E. and a dip of 22° SE. (Miller, 1934, p. 214). Analyses of the rock were published by Miller (1934, p. 215).

In 1930, W.R.R. Weaver was the president of the Allentown Portland Cement Company, and C.H. Breerwood was the vice president and general manager. F.A. Weibel was the secretary and treasurer. D.C. Morgan was the plant superintendent. The plant capacity was 1,000,000 bushels of portland cement per year. The plant had four rotary kilns, 8 feet by 120 feet, each with a 960-bushel capacity. The company had 225 employees, including five chemists and helpers (Shaw, 1930, p, 161-162).

In 1932, stripping was done by a dragline and shovel. Stone was moved from the quarry to the cement plant by a gasoline-powered locomotive and cars. Primary crushing was done by a No. 10 gyratory crusher; secondary crushing was done by a No. 6 gyratory crusher and hammer mills. The annual production of portland cement was 1,400,000 barrels (Pit and Quarry, 1932).

Almost from the beginning, the Allentown Portland Cement Company used the swastika, a symbol of good luck dating back to ancient times, as its trademark. When Adolph Hitler rose to power and adopted the swastika, the company quickly dropped the symbol. To aid the war effort during World War II, the company converted phosphate rock into fertilizer and waste sludge into a high-grade iron concentrate for iron furnaces.

In 1951, the quarry was 160 feet deep. Sixty percent of the cement was sold in bulk, and 40 percent was sold in paper bags. Silos provided storage for 250,000 barrels of cement (*Reading Eagle*, November 30, 1952). By 1961, the plant capacity was 2,500,000 barrels or 10 million bags of cement per year (*Reading Eagle*, November 19, 1961). One barrel of cement equals 4 cubic feet, which equals 4 bags. One bag of cement weighs 94 pounds.

Almost a years' worth of cement production from the Evansville plant, 2.25 million bags, was used in the construction of the Pentagon. Cement from the Evansville plant also was used in the construction of the Panama Canal, the Jefferson Memorial in Washington, D.C., and the Penn Street bridge

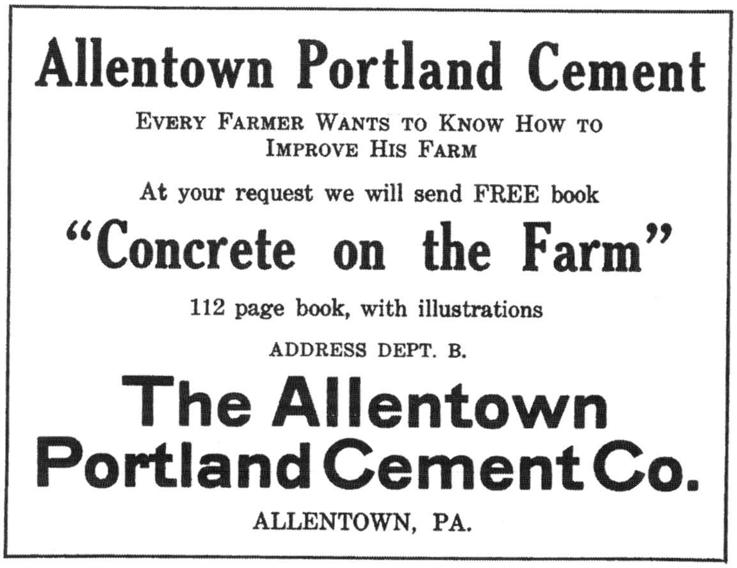

Allentown Portland Cement

EVERY FARMER WANTS TO KNOW HOW TO
IMPROVE HIS FARM

At your request we will send FREE book

"Concrete on the Farm"

112 page book, with illustrations

ADDRESS DEPT. B.

The Allentown Portland Cement Co.

ALLENTOWN, PA.

Figure 315. Advertisement for the Allentown Portland Cement Company, 1914.

Figure 316. Power plant at the Allentown Cement Company Evansville plant, Maidencreek Township, ca. 1910.

Figure 317. Horse-drawn carts in the Evansville quarry, Maidencreek Township, early 1900s. Courtesy of Lehigh Cement Company LLC.

Figure 318. Evansville quarry, Maidencreek Township, early 1900s. Courtesy of Lehigh Cement Company LLC.

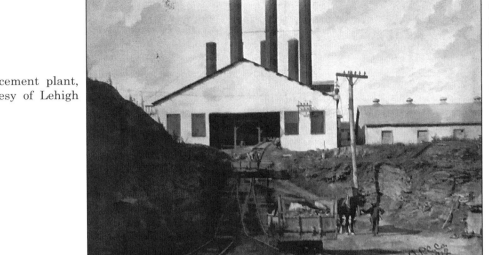

Figure 319. Evansville quarry and cement plant, Maidencreek Township, 1912. Courtesy of Lehigh Cement Company LLC.

Figure 320. Steam shovel loading stone at the Evansville quarry, Maidencreek Township, early 1900s. Courtesy of Lehigh Cement Company LLC.

Figure 321. Evansville quarry, Maidencreek Township, early 1900s. Courtesy of Lehigh Cement Company LLC.

Figure 322. Stone carts at the Evansville quarry, Maidencreek Township. Courtesy of Lehigh Cement Company LLC.

Figure 323. Stone car dumping stone inside the Evansville cement plant, Maidencreek Township. Courtesy of Lehigh Cement Company LLC.

Figure 324. Evansville cement plant, Maidencreek Township, 1915.

Figure 325. Restored stone cart at the Evansville cement plant, Maidencreek Township. The cart was unearthed during mining in 2009 and restored by plant employees.

supplied building stone for the Blue Mountain Dam and many of the bridges on the Berks and Lehigh Railroad. Stone was taken from the quarry in five-ton blocks. Stone also was used for manufacturing lime by J.M. Meredith and Thomas Lightfoot (Historical Committee of the Blandon Bicentennial Committee, 1976).

Pleasant Hill Road Quarry

The flooded Pleasant Hill Road quarry was northwest of the intersection of U.S. Route 222 and Pleasant Hill Road at about 40° 28′ 01″ latitude and 75° 52′ 20″ longitude on the USGS Fleetwood 7.5-minute topographic quadrangle map (Appendix 1, fig. 11, location MC-6). This quarry is locality 8 of Gray (1951, p. 33). The quarry, in the Epler Formation, was a former source of rock for producing agricultural lime. The quarry exposed blue laminated limestone with one thin dolomite bed in an 8-foot high face. Gray (1951, p. 33) presented analyses of rock from the quarry.

in Reading. The Allentown Portland Cement Company was the principal cement supplier for the construction of the eastern end of the Pennsylvania turnpike and the Verranzo-Narrows Bridge in New York (*Reading Eagle*, November 19, 1961). Allentown Portland Cement was used in the construction of the Three Mile Island and Limerick nuclear power plants (*Reading Eagle*, October 2, 1974).

In June 1960, the Allentown Portland Cement Company was acquired by the National Gypsum Company (fig. 327) through an exchange of stock valued at $30 million and became a subsidiary of the National Gypsum Company. As a result of the purchase, National Gypsum gained control over one-third of the cement distribution in the United States. In 2001, the Allentown Cement company was absorbed by Lehigh Cement LLC. Lehigh Cement LLC is a subsidiary of Lehigh Hanson, a part of the Heidelberg Cement Group, which acquired the Lehigh Cement Company in 1977.

Maidencreek Quarry

The Maidencreek quarry was northwest of Maiden Creek Road, on a peninsula jutting out into Lake Ontelaunee at about 40° 27′ 35″ latitude and 75° 54′ 21″ longitude on the USGS Temple 7.5-minute topographic quadrangle map (Appendix 1, fig. 8, location MC-7). The quarry is locality 9 of Gray (1951, p. 34). The quarry was in the Epler Formation.

The quarry originally produced stone for agricultural lime, which was made in a kiln at the quarry. The Allentown Portland Cement Company obtained some limestone from the

J.M. Meredith Quarry

The likely location of the J.M. Meredith quarry was along Maidencreek Road, northeast of its intersection with Calcium Road at about 40° 27′ 33″ latitude and 75° 54′ 07″ longitude on the USGS Temple 7.5-minute topographic quadrangle map (Appendix 1, fig. 8, location MC-2). A limestone quarry at that location is shown in the 1876 atlas. LIDAR imagery shows several old quarries within a small area on both sides of the railroad tracks.

The J.M. Meredith quarry was 0.25 mile from the Calcium railroad station. The quarry

Figure 326. Evansville cement plant, Maidencreek Township, March 2016.

Figure 327. Unissued stock certificate of the National Gypsum Company.

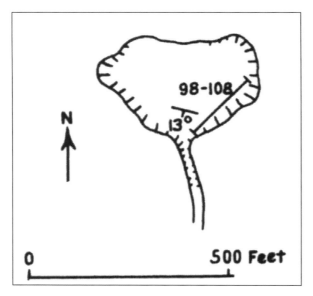

Figure 328. Sketch of the Maidencreek quarry, Maidencreek Township, 1951. Modified from from Gray (1951, p. 34). Numbers indicate sampling locations.

quarry, but abandoned it because of the high magnesium content (Miller, 1934, p. 215). The quarry was connected to the main rail line by a 1,000-foot spur. The quarry exposed limestone in beds 5 to 8 feet thick separated by dolomite beds 1 to 6 feet thick. The limestone was blue massive to laminated, weathering to light to dark gray. The dolomite was gray and massive. Beds in the floor of the quarry were composed of angular dolomite fragments cemented with quartz and calcite. Locally, pyrite replaced dolomite in the breccia. Gray (1951, p. 34) presented a sketch of the quarry (fig. 328) and analyses of rock from the quarry. Miller (1934, p. 215) reported numerous fossils in the weathered beds.

Iron Mines

Crane Iron Company Mine

The Crane Iron Company mine was on the Kosmerl farm, between Ridge and Shoemakersville Roads, 2.5 miles northeast of Leesport. There were two open pit mines; the northern flooded pit (location MC-4A) (fig. 330) is at 40° 28′ 42″ latitude and 75° 56′ 22″ longitude, and the southern flooded pit (location MC-4B) (fig. 333) is at 40° 28′ 38″ latitude and 75° 56′ 21″ longitude on the USGS Temple 7.5-minute topographic quad-

MAIDENCREEK TOWNSHIP

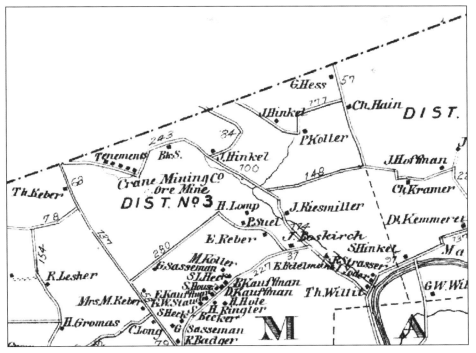

Figure 329. Crane Iron Company mine, Maidencreek Township. 1876. From Davis and Kochersperger (1876). The map shows a row of tenant houses belonging to the Crane Iron Company.

rangle map (Appendix 1, fig. 7). The mine was in Hamburg Sequence rocks and was mine number 258 of Sanders (1883).

According to the Maidencreek Township tax assessment records, the Kosmerl farm was owned by John Grett in 1846 (Wayne T. Kosmerl, written communication, 2015). The 1864 Maidencreek Township tax assessment records list the Grett Ore Company, so the mine likely was in operation by then.

In 1874, the Reading Times (October 19, 1874) reported that the Crane Iron Company stopped shipping ore because of a depression in the iron industry. Annual production was about 10,000 tons, and the company had stockpiled about 5,000 tons of ore, which was awaiting shipment. The mine was abandoned and flooded before 1882. Sanders (1883, p. 133), in his report on slate, observed: *"The surface is covered with loose slate, and pieces of slate coated with hematite. From the looks of the dump I should say that the mine had a great deal of slate in it."*

Wayne Kosmerl reported that there was a horizontal tunnel about 100 feet long in the southern mine about midway between the top of the water surface and the land surface The tunnel was about 3 to 4 feet high (Wayne T. Kosmerl, personal communication, January 16, 2014).

In 1877, the Reading Times (June 14, 1887) reported that a 22-year old local man named Sylvester Dries departed for Philadelphia carrying a considerable amount of money to answer an advertisement for a business partner. Finding the arrangement unsuitable, he left Philadelphia the following day. He took the evening train through Reading to the Mohrsville station, where he got off the train intending to walk several miles home. The next day, a boy named Martin Koller was fishing in the flooded mine pit when he found a brand new hat floating on the water. He fished it out and took it home. He told no one about it and wore it to Sunday School the following Sunday. The hat was identified by Sylvester's younger brother, and the name Sylvester Dries was found on the inside hat band. Sylvester's father organized a search party, and, after 2 days of dragging the mine pit, Sylvester's body was recovered in 17 feet of water. His money and watch were still on his person. His watch had stopped at 9:03 PM, and the path near the mine hole was caved, so presumably he took a shortcut through the field, the path gave way in the dark, and he fell into the water. Being unable to swim, he drowned.

Figure 330. Northern Crane Iron Company mine, Maidencreek Township, April 2015.

THE Crane Iron Company

The Crane Iron Company was a major iron producer in the Lehigh Valley from its founding in 1839 until its sale in 1899. It was founded under the patronage of the Lehigh Coal and Navigation Company, which hoped to promote the then novel technique of smelting iron ore with anthracite coal. The new company was named for George Crane, a British ironmaster whose superintendent, David Thomas, was hired to come to America and set up an ironworks using the new technique.

The Lehigh Coal and Navigation Company, which was organized in 1818, led the exploitation of anthracite coal in Pennsylvania. Seeking to expand its sales in 1838, the company offered valuable water-power privileges to any firm that would spend $30,000 to build and operate an iron furnace using anthracite coal on the Lehigh River. This offer led to the informal organization of a company, which included members of the Lehigh Coal and Navigation Company. In the November 1838, Erkskine Hazard traveled to Wales to hire a competent person to come to the United States and oversee the construction of anthracite iron furnaces. There, he met George Crane, proprietor of the Crane Iron Works at Yniscedwin, who recommended David Thomas, an expert employee.

David Thomas, regarded as the father of anthracite iron manufacture in America, was born in South Wales on November 3, 1794. He entered the iron business in 1812, and, after working in various places, he went to the Yniscedwin Works in Brecknockshire, Wales in 1817. As early as 1820, Thomas and George Crane began to experiment with anthracite; however, the experiments were unsuccessful. In 1834, Neilson, manager of the Glasgow Gas Works, invented the hot-blast method. Thomas read a pamphlet on the hot-blast method written by Neilson. In September, 1836, Thomas built ovens for heating the blast. On February 5th, 1837, the new process was applied, and the result was a complete success. Yniscedwyn Works became the first ironworks in Great Britain to produce anthracite iron in commercial quantities by use of the hot-blast method. Influenced by a liberal offer and the consideration that his sons would have better opportunities in America than they could hope for in Wales or Great Britain, Thomas sailed for the United States from Liverpool in May 1839. Thomas brought his wife and five children with him.

The Lehigh Crane Iron Company was formally organized on April 23, 1839. Robert Earp was elected president. A charter for 25 years was granted on May 16, 1839, under a general act of the Pennsylvania Legislature. The Lehigh Coal and Navigation Company supported the new company and granted them land and rights to water power for its furnace.

Before leaving England, David Thomas had the blowing machinery and castings for the hot blast made, and all were shipped except the two blowing cylinders, which were too large for the hatches of the ship. At the time, no foundry in the United States was large enough to cast the blowing cylinders. Thomas went to Philadelphia to the Southwark Foundry of S.V. Merrick and J.H. Towne, who enlarged their boring machinery and made the five-foot cylinders. The firebrick was imported from Wales.

In August 1839, ground was broken at Craneville, later renamed Catasauqua, for the first furnace. The stack was 45 feet high from the base of the hearth to the trunnel head, with a chimney extending about 12 feet above the trunnel head. It was constructed of limestone, 30 feet square at the base and tapering to about 23 feet square at the top. The ovens for the hot blast were coal fired, and the blowing engine was driven by a waterwheel tapping the canal at Lock 36. The furnace was blown in on July 3, 1840, and the first four tons of iron were produced on July 4, 1840. The ore was two-thirds local limonite and one-third New Jersey magnetite. The new furnace produced 1,088 tons of pig iron in its first six months, with a peak production of 52 tons in one week, much more than could typically be made by a cold blast charcoal furnace. The company soon became the leading producer of anthracite iron in Pennsylvania, manufacturing 14,272 tons of iron in 1849, which was 13 percent of all anthracite iron made in Pennsylvania. The success of the furnace and the growing demand for iron led the company to construct five more anthracite furnaces between 1842 and 1868.

Figure 331. Crane Iron Company furnaces, Catasauqua, Pa.

Over the next several decades, the Crane Iron Company developed an extensive portfolio of assets, buying mines in Berks County, the Lehigh Valley, and northern New Jersey and taking over many smaller iron furnaces in the region. Crane Iron also financed the building of railroads in the area to haul limestone and iron ore to its furnaces.

The success of Crane Iron and the many other iron companies that sprang up in the Lehigh Valley led to a major mining boom in Berks County. Some mines were worked by independent operators, and the ore was sold on the open market. Other mines were leased, and some were owned outright by the companies. The teams that brought iron ore from local mines were sometimes lined up for more than two miles. The teams and iron ore wagons frequently made the roads impassable to farmers. To remedy this, the Lehigh Crane Iron Company, assisted by the Thomas Iron Company, secured a charter from the Commonwealth of Pennsylvania for a railroad. Construction of the Catasauqua and Fogelsville Railroad began in the spring of 1856, and the rail line was opened during the summer of 1857.

Magnetite brought from New Jersey was hoisted on an inclined plane by horse power and piled up 60 feet high in front of the furnaces. The limestone flux for the furnaces was largely shipped from local quarries. A mixture of about 75 percent limonite and 25 percent magnetite ore was used. Most of the limonite was mined locally.

In 1855, David Thomas left his post as superintendent of the company and was succeeded by his son, John Thomas. During 1872, when an increase was being made in the capital stock, the corporate name of the Lehigh Crane Iron Company was changed to the Crane Iron Company. The company survived the Panic of 1873 and the subsequent poor iron market. In 1880, the capacity of the five furnaces was increased to 100,000 tons per year.

Leonard Peckitt became associated with the Crane Iron Company in 1887 as chief chemist. In 1891, Peckitt was appointed superintendent, succeeding William R. Thomas. During the Panic of 1893, Crane went into the hands of receivers, one of whom was Peckitt. In 1894, Peckitt was elected president of the firm. On January 30, 1895, all property, rights, franchises, and privileges of the Crane Iron Company were transferred to the Crane Iron Works. In 1899, the company was sold to the Empire Steel and Iron Company. In 1899, Peckitt, who took an active part in the formation of the Empire Steel and Iron Company, was elected president of the company.

Commencing with the United States involvement in World War I, America's iron and steel industry did a thriving business, and the Crane Iron Works was no exception. With the war's end, the company's fortunes began a slow decline from which there would be no return. On April 10, 1922, the Replogle Steel Company acquired control of the Empire Steel and Iron Company, with Leonard Peckitt being elected president of that company. In 1930, the last Crane Iron Works furnace was dismantled, and in 1938, the remaining Catasauqua property was sold.

Figure 332. Stock Certificate of the Lehigh Crane Iron Company, 1872. Courtesy of David Beach (dbeach@cfl.rr.com).

Figure 333. Southern Crane Iron Company mine, Maidencreek Township, April 2015.

Shaeffer's Old Mine

Adam Shaeffer's "old" iron mine was northeast of Blandon. The area is now a residential development. Schaeffer's old mine was abandoned in 1882 when he started a new iron mine 1 mile southeast of the old mine (D'Invilliers, 1883, p. 372).

Shaeffer's New Mine

Adam Shaeffer's "new" iron mine was northeast of Pennsylvania State Route 73, 0.8 mile east of Blandon at about 40° 26′ 15″ latitude and 75° 52′ 25″ longitude on the USGS Fleetwood 7.5-minute topographic quadrangle map (Appendix 1, fig. 11, location MC-3). The mine was in the Epler Formation. Shaeffer's new mine is shown on the index map for the 1882 topographic map. The area is now a residential development.

Figure 335. Dishong quarry and kiln, Marion Township, April 2015.

Shaeffer's new mine was started late in the fall of 1882. When the limonite mine was visited by D'Invilliers (1883, p. 372) that same year, he noted: "*They had just commenced operations here, washing the surface ore from a 6-foot pit, the ore going to the Temple Iron Company at Temple. The mine is entirely in Potsdam and the ore very silicious.*"

Wade Mine

Stoddard and Callen (1910, p. 430) described the Wade mine at Blandon: "*E. B. Wade's property is one-half mile due south of Blandon and is at present being worked only for clay, which is shipped and refined mostly for use in iron works. The only evidences of ocher are a few very shallow pits immediately south of the clay pits. Some ocher was formerly extracted from these pits and is said to have been of very high grade, but from all accounts none has been shipped for five years or more, and no considerable quantities were ever shipped. The clay was also formerly refined for certain kinds of paints and is said to have*

Figure 334. Goethite from Schaeffer's mine, Maidencreek Township, 9.5 cm. Bryn Mawr College collection Rand 940.

brought $7.50 a ton. The deposit here lies upon the quartzite, and clays are found from 30 to 50 feet thick overlain by 8 to 10 feet of soil. The clay varies widely in color and texture, the pure-white layers being found on top in moderate quantities. It is said to be useful as an oilcloth base, as a wall-paper base, and for paint. The locality is only in the first stages of development, however, and the extent of its resources, if it has any, is unknown."

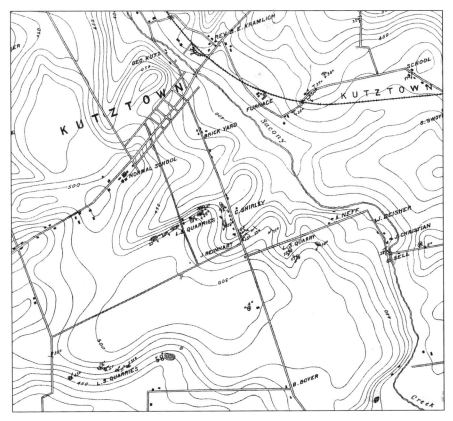

Figure 336. Limestone quarries in Maxatawny Township south and southeast of Kutztown, 1882. From Pennsylvania Geological Survey (1883).

MARION TOWNSHIP

Dishong (Stouchsburg) Quarry

The Dishong quarry (fig. 335) was between Pennsylvania State Route 422 and Main Street, just east of their intersection, 0.6 mile northwest of Stouchburg at 40° 23′ 00″ latitude and 76° 14′ 59″ longitude on the USGS Strausstown 7.5-minute topographic quadrangle map (Appendix 1, fig. 21, location MA-1). The quarry, in the Annville and Ontelaunee Formations, is locality 43 of Gray (1951, p. 61). The quarry is shown on the Levi King property in the 1876 atlas. The quarry produced stone for lime production until 1950. A kiln still (2016) stands at the quarry (fig. 335).

The Dishong quarry exposed a thick-bedded, blue, finely crystalline limestone strongly veined with calcite. The crest of an anticline striking about N. 80° W. and plunging west was exposed in the quarry. Gray (1951, p. 61) provided chemical analyses of the rock.

Smaltz Road Quarry

The Smaltz Road quarry was northeast of the intersection of Pennsylvania State Route 422 and Smaltz Road, 1 mile northwest of Stouchburg at 40° 23′ 07″ latitude and 76° 14′ 59″ longitude on the USGS Strausstown 7.5-minute topographic quadrangle map (Appendix 1, fig. 21, location MA-2). The quarry, in the Annville Formation, is locality 45 of Gray (1951, p. 62). The quarry exposed a blue, finely crystalline, laminated limestone. The rocks strike N. 85° W. to N. 90° W. and 45° S. Gray (1951, p. 62) provided chemical analyses of the rock.

Filbert Ocher Mine

The Filbert ocher mine was 0.5 mile southwest of Stouchsburg on the farm of Hiester Filbert of Robesonia. There was a large open-cut limonite mine on the adjoining farm of Thomas W. Reed. About 1886, an iron mine was opened on the Filbert farm, and the material from the washery was run into the Reed open pit. Around 1899, the ocher in the Reed pit was found to have considerable value and was mined for use in paint manufacturing. The mine was reopened for ocher, and limonite was considered the by-product. A plant for washing, drying, and grinding the ocher was built, and several car loads of finished product were shipped to Lebanon, Reading, and Philadelphia (Miller, 1911, p. 41).

Figure 337. Eastern Industries Hinterleiter quarry, Maxatawny Township, March 2016.

MAXATAWNY TOWNSHIP

Limestone Quarries

Rogers (1858, vol. 1, p. 266) noted: *"On the ridge S. of Kutztown there are extensive quarries of limestone from which much stone is transported into the slate country on the N., and there burned into lime for manure. The dip of the strata at these quarries is about 20° S. A white clay is also found at the iron mine."*

Hinterlieter Quarry

The Eastern Industries Hinterleiter quarry (figs. 327 and 328) is a large active (2016) quarry. The original and older quarry (location M-1A) is northwest of the intersection of Hinterleiter and Quarry Roads at 40° 30′ 59″ latitude and 75° 43′ 56″ longitude, and a newer quarry (location M-1B) is southeast of the intersection of Hinterleiter and Quarry Roads at 40° 30′ 50″ latitude and 75° 43′ 23″ longitude the USGS Topton 7.5-minute topographic quadrangle map (Appendix 1, fig. 4). The newer quarry is in the Allentown Formation; the older quarry is in the Stonehenge Formation. The quarry has been known as the Hinterleiter quarry, the quarry at Hinterleiter Crossing, nd the Topton Furnace quarry of Miller (1925 and 1934). The quarry currently (2016) is called as the Eastern Industries Kutztown quarry.

The Hinterleiter quarry was originally opened by local German farmers to produce agricultural lime for their farms. The quarry was owned and operated by many individuals and companies over the years. Sell D. Kutz of Kutztown operated the quarry ca. 1889-1903. In 1892, the powder and storage house at the quarry caught fire. Sixty pounds of dynamite and 100 blasting caps exploded. The fire was caused by an overheated stove (*Reading Eagle*, January 1892). In 1900, the quarry was owned by William D. Hess of Lebanon, who leased the quarry to Kutz (*Reading Eagle*, July 23, 1900). Kutz supplied limestone for furnace flux to the Topton Furnace and the Macungie Furnace in Lehigh County. In 1901, the Reading Eagle (September 4,

Figure 338. Aeriel view of the Hinterleiter quarry, Maxatawny Township, September 2015.

Figure 339. Portrait of Dr. U.S.G. Bieber. From Historical Committee of the Kutztown Centennial Association (1915, p. 213).

1901) reported that Kutz was shipping several car loads of limestone to Reading daily.

The quarry was next operated by the Penn Limestone Company of Reading. The company was incorporated in New Jersey on October 11, 1905, with $15,000 in capital for the purpose of "*quarrying limestone, selling and dealing In limestone, and the burning, selling and dealing in lime.*" The company office was located at 536 North 8th Street in Reading. In 1908, the Penn Limestone Company suspended operations at the quarry because of a depression in the iron business (*Stone*, January, 1908). In 1909, 25 men were employed in the quarry, which furnished limestone for flux to the Topton Furnace (*Reading Eagle*, March 9, 1909).

In 1926, Dr. U.S.G. Bieber (fig. 339), proprietor of the Kutztown Stone Company, purchased the Penn Limestone Company and the Hinterleiter quarry. He then incorporated the Penn Limestone Company, with a capital of $100,000, of which $50,000 was preferred shares and $50,000 was common stock. The stock was offered locally (*Kutztown Patriot*, October 7, 1926).

The quarry was closed for several years before it was purchased by the Eastern Lime Corporation in 1941. The Eastern Lime Corporation was incorporated in Delaware in 1941 to produce chemical-grade limestone for cement companies, crushed stone for ready-mix concrete and highway construction, and agricultural limestone. Its office was located in Kutztown. In 1952, the quarry produced agricultural limestone and aggregate; the combined capacity was 150 tons per hour. H.H. Snyder was the plant superintendent (Pit and Quarry, 1952). In 1965, the company's stockholders voted to change the name to Eastern Industries, Inc. On March 31, 1976, Protection Services, Inc. of Harrisburg bought 98 percent of the outstanding common stock of Eastern Industries. The quarry was next acquired by Stabler Companies, Inc., which operated 16 aggregate and sand quarries in eastern Pennsylvania. In 2008, the New Enterprise Stone and Lime Company acquired Stabler Companies Inc.

The Eastern Lime Corporation opened a new quarry southeast of the old quarry. In 2016, stone was quarried from the new quarry, and the old quarry was used to stockpile stone.

The old quarry is on the crest of an anticline whose axis strikes N. 55° E. and plunges gently southwest. The beds quarried for calcium content were massive, blue, fine-grained, laminated limestone. Overlying those beds is a blue and brownish, mottled, finely-crystalline magnesian limestone (Gray, 1951, p. 24). Chemical analyses of the limestone were provided by Day (1904), Frear (1913, p. 78), and Gray (1951, p. 24).

Figure 340. Portrait of Edward Hottenstein. From Historical Committee of the Kutztown Centennial Association (1915, p. 188).

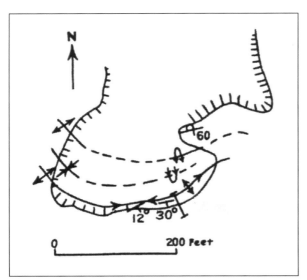

Figure 341. Sketch of the Hottenstein quarry, Maxatawny Township, 1951. Modified from Gray (1951, p. 43).

Hottenstein Quarry

The Hottenstein quarry was southeast of the intersection of Willow Street and Normal Avenue at about 40° 30′ 56″ latitude and 75° 48′ 37″ longitude on the USGS Kutztown 7.5-minute topographic quadrangle map (Appendix 1, fig. 3, location M-4). A rail spur once ran to the quarry. The quarry was in the Jacksonburg Formation.

The Hottenstein quarry was owned by Dr. Edward Hottenstein (fig. 340). In 1879, Hottenstein secured a contract to furnish limestone for flux to the Topton and Kutztown Furnaces. In that year, he constructed a railroad siding to the quarry to expedite shipping (*Reading Eagle*, May 3, 1879).

In 1891, Dr. William Gross, proprietor of the Pennsylvania House Hotel in Kutztown, purchased the equipment and the lease on Hottenstein's quarry from a Mr. McCellan, the previous lessee. The quarry shipped 1,200 to 1,500 tons of stone per month to the Topton Furnace for flux stone (*Kutztown Patriot*, February 7, 1891).

In 1901, Maxatawny Township residents voted to procure a stone crusher to produce aggregate for resurfacing roads. A committee of about a dozen residents visited different stone crushers to see which was the best. They also decided to purchase a gasoline engine to furnish the power. The crusher was installed in the Hottenstein quarry (*Kutztown Patriot*, March 16, 1901).

The quarry exposed dense, blue, laminated, massive limestone in 10- to 20-foot zones separated by 1- to 4-foot thick beds of gray dolomite. Two asymmetrical anticlines were exposed in the quarry. Gray (1951, p. 43) provided a sketch of the quarry (fig. 341) and analyses of the stone.

The Pumping Station Cave was located in the quarry. A small entrance at the base of the quarry led to a 12-foot crawlway that sloped downward, ending in deep water. The entrance later was buried by tons of rock (Dave and Jon Adam, 1990, *Bucks County Diviner*, vol. 11, no. 8).

Keystone Quarry

The Keystone quarry is a flooded quarry, southwest of the intersection of Baldy and Commons Roads at about 40° 30′ 24″ latitude and 75° 46′ 39″ longitude on the USGS Kutztown 7.5-minute topographic quadrangle map (Appendix 1, fig. 3, location M-25). It is on the property of Kutztown University. The quarry, in the Stonehenge Formation, has been known as the Baldy Street quarry and the Bieber quarry.

In 1891, Ephraim Sharadin of Kutztown purchased Charles Berck's lime and building stone quarry. The purchase included several lime kilns on the property (*Kutztown Patriot*, April 18, 1891). In 1911, the quarry was purchased by Dr. U.S.C. Bieber, who operated the quarry as the Kutztown Crushed Stone and Lime Company. He also operated a brickyard in connection with the quarry (*Reading Eagle*, July 9, 1913).

In 1911, Bieber purchased a new stone crusher from Stocker Brothers in Harrisburg. The crusher was moved from the train depot to the quarry by Jerry Wessner of Wessnersville using a large traction engine. The crusher, which had a capacity of 150 tons per day, was run by electricity furnished by the trolley company (*Kutztown Patriot*, July 15, 1911).

In 1912, Bieber installed a new crusher motor, which received power from the Kutztown and Fleetwood power plant.

Figure 342. Crusher at the Bieber (Keystone) quarry, Maxatawny Township. From Historical Committee of the Kutztown Centennial Association (1915, p. 226).

Figure 343. Aeriel view of the Berks Products (left) and Koller (right) quarries, Maxatawny Township, September 2015.

Bieber received a contract from the Greenwich Board of Supervisors for 800 tons of crushed stone (*Reading Eagle*, April 24, 1912). By 1913, the Kutztown Lime and Crushed Stone Company employed 20 men and had begun to manufacture hydrated lime. In 1924, Edgar Bieber was the general manager of the company (*Reading Eagle*, November 5, 1924). The quarry was offered for sale by Bieber in 1925 (*Reading Eagle*, April 24, 1925), but was not sold.

In October 1926, the largest blast ever detonated in the quarry (up to that time) dislodged an estimated 30,000 tons of rock. The charge of three tons of dynamite was prepared and set off by Andrew Saroson of Wilkes-Barre, a representative of an explosive manufacturing company (*Kutztown Patriot*, October 7, 1926). In 1934, the quarry was about 400 feet long along strike and 175 feet wide with a 60-foot high face. The plant produced 500 to 600 tons of crushed stone per day, which was used for road construction (Miller, 1934, p. 213).

In March of 1935, the court allowed the executor of the Maria E. Beiber estate to lease the quarry and machinery to the Wernersville Lime and Stone Company for a term of 5 years (*Reading Eagle*, March 7, 1935). The Wernersville Lime and Stone Company formed a subsidiary called the Keystone Quarry Company, which purchased the equipment and leased the property. It began operations on March 15, 1935 (*Kutztown Patriot*, July 25, 1935). Eugene E. Uhler became the plant manager in 1937. By 1942, the daily output was 300 tons of crushed stone (*Kutztown Patriot*, April 16, 1942).

Berks Products Quarry

The Berks Products Kutztown quarry (fig. 343) is an inactive (2016) quarry northwest of the intersection of Baldy and Bastian Roads at about 40° 29′ 50″ latitude and 75° 46′ 54″ longitude on the USGS Fleetwood 7.5-minute topographic quadrangle map (Appendix 1, fig. 12, location M-2). The quarry is in the Stonehenge Formation. The quarry is known as the Keystone Quarry Division of Berks Products. In 1995, the Berks Products Corporation received approval to deepen the quarry to 250 feet and to discharge water to a mine hole east of Baldy Road (*Reading Eagle*, January 2, 1993; January 25, 1995). The quarry produced raw materials for cement, concrete aggregate, road metal, and screenings. The quarry closed in 2015.

Bowers Quarry

The Bowers quarry was west of the intersection of Bowers and Bastian Roads at about 40° 29′ 31″ latitude and 75° 44′ 51″ longitude on the USGS Manatawny 7.5-minute topographic quadrangle map (Appendix 1, fig. 15, location M-3). The quarry was in the Allentown Formation.

The Bowers quarry was owned by Jonathan Bower (ca. 1869), who sold it to Peter W. Fisher of Topton. Fisher sold the farm and quarry to the Clymer Iron Company for $20,000, but later bought the farm back, leaving Clymer with the quarry. The 1877 Berks County Business Directory listed William H. Clymer & Company as the owner of the quarry (Phillips, 1877). The quarry supplied flux stone to three iron furnaces. In 1899, the quarry was approaching the property line, and the adjoining property owner, Mrs. Charles Trexler, refused to sell her property to Clymer. Fearing the loss of their flux stone supply, the Topton and Macungie (Lehigh County) Furnaces began looking for another source of supply and found it at the Hinterleiter quarry (*Kutztown Patriot*, September 23, 1899).

After the quarry had lain idle for several months, it was leased by Abraham (also spelled Abram) Sweitzer (also spelled Schweitz, Schweitzer, and Sweitz in different newspaper accounts) of Reading (*Kutztown Patriot*, July 1, 1905). The quarry reached the property line in 1905, and quarrying ceased. Abraham Sweitzer died in 1905, and his son Edward moved the quarry operations to the nearby Angstadt quarry.

Angstadt Quarry

The Angstadt quarry was north of the Bowers quarry at about 40° 29′ 34″ latitude and 75° 44′ 49″ longitude on the

Figure 344. Aragonite from the Baldy Street (Keystone) quarry, Maxatawny Township, 11.5 cm. Skip Colflesh collection.

Figure 345. Lime kiln near the Angstadt quarry, Maxatawny Township, April 2015.

USGS Manatawny 7.5-minute topographic quadrangle map (Appendix 1, fig. 15, location M-5). The quarry was in the Allentown Formation. The quarry was about one-fourth of a mile northwest from Bowers railroad station, and a siding was constructed from the East Pennsylvania Railroad to the quarry (*Reading Eagle*, June 27, 1908).

The Angstadt quarry was owned by Rudolph H. Angstadt, a merchant in Dryville, Rockland Township. Three lime kilns, each having an average capacity of 600 bushels, were located at the quarry. The first kiln was built around 1868 by George S. Sell. The quarry was sold at public sale on February 4, 1897, by Mary A. Gabel, sole heir of George Sell (*Kutztown Patriot*, January 23, 1897). The quarry was purchased by Obadiah B. Angstadt, who operated the quarry for several years. His brother Rudolf purchased the property and built two more lime kilns. Rudolf then operated the kilns from about 1903 to 1906. He allowed farmers to pay him for the privilege of quarring stone and burning lime (*Reading Eagle*, June 27, 1908).

After the Bower quarry ceased operation in the spring of 1905, Edward E. Sweitzer moved his quarry operations from the Bowers quarry to the Angstadt quarry, paying Rudolph Angstadt a royalty on each ton of stone mined. Angstadt received over $1,000 in royalty payments in the first three years (1905-08). As many as 2,870 tons of limestone were quarried in one month (*Kutztown Patriot*, July 18, 1908). The quarry employed 36 men, and stone was shipped to the Emmaus Furnace and two pipe mills (*Reading Eagle*, January 12, 1908). Charles Young was the quarry superintendent. The quarry was in operation as late as 1919, when it was operated by the Reading Iron Company (*Reading Eagle*, March 6, 1919).

A.G. Smith Quarry

The A.G. Smith quarry was northeast of the intersection of Hinterlieter and Quarry Roads at about 40° 31′ 22″ latitude and 75° 43′ 38″ longitude on the USGS Topton 7.5-minute topographic quadrangle map (Appendix 1, fig. 4, location M-23). The quarry, in the Stonehenge Formation, is locality 3 of Gray (1951, p. 26). In 1951, the Smith quarry was leased by the Eastern Lime Corporation.

The A.G. Smith quarry was opened around 1905 (*Reading Eagle*, May 1, 1905). The Kutztown Patriot (May 6, 1906) reported: "*The supervisors of Maxatawny township are busy crushing stones for the Topton Orphans' Home. They receive the stones from A. G. Smith's stone quarry, on H. A. Miller's land.*" Limestone and lime producers listed by Hice (1911) included A.G. Smith. In 1924, Joseph Hamsher purchased the 135-acre Wilson Smith farm. The farm contained several large limestone quarries and kilns, which were in operation until about 1918 (*Reading Eagle*, September 27, 1924).

The quarry exposes thick limestone beds with occasional limestone layers. The limestone is blue and dense with brown, silty laminae on weathered surfaces. The dolomite is gray and finely crystalline in beds 6 to 12 inches thick. Calcite-filled gash veins are common in the dolomite. Gray (1951, p. 26) provided a sketch of the quarry (fig. 346) and analyses of the stone.

Kohler Quarry

The flooded Kohler quarry (fig. 343) was southwest of the intersection of Noble and Foch Streets at 40° 30′ 05″ latitude

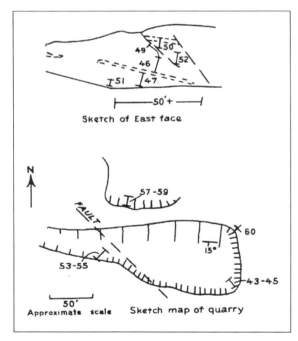

Figure 346. Sketch of the A.G. Smith quarry, Maxatawny Township, 1951. From Gray (1951, p. 26). Numbers represent sampling locations.

Figure 347. Kohler Road quarry, Maxatawny Township, March 2015.

Figure 348. Lime kiln at the Kohler Road quarry, Maxatawny Township, March 2015.

Figure 349. Platform for loading rail cars at the Kohler Road quarry, Maxatawny Township, March 2015.

and 75° 46′ 06″ longitude on the USGS Kutztown 7.5-minute topographic quadrangle map (Appendix 1, fig. 3, location M-24). The quarry, in the Stonehenge Formation, is locality 6 of Gray (1951, p. 30).

Limestone and lime producers listed by Hice (1911) included G.J.S. Kohler of Allentown. Miller (1934, p. 213-214) reported that the quarry was 350 feet long, 100 feet wide, and had a working face 35 feet high. The plant produced 150 tons of crushed stone per day. In 1937, the Met Ed Electric Company sued the Martin W. Kohler estate to collect $462.75 for electric charges at the quarry (*Reading Eagle*, April 5, 1937). In 1938, the John Kohler quarry produced building stone, crushed stone, two kinds of sand, and dust for finishing (*Reading Eagle*, June 12, 1938).

The quarry exposed dolomite with interbedded limestone. The dolomite was gray, even bedded, finely crystalline and graded into mottled magnesian limestone. The limestone beds were 1 to 3 feet thick, and varied from blue laminated to gray with faint mottling. Beds in the quarry was essentially horizontal. On the south side of the quarry, the strike is N 90° E with a gentle dip to the north. Gray (1951, p. 30) provided analyses of the stone.

Kohler Road Quarry

The Kohler Road quarry (fig. 347) was southwest of the intersection of Kohler Road and South Laurel Street at 40° 31′ 07″ latitude and 75° 45′ 53″ longitude on the USGS Kutztown 7.5-minute topographic quadrangle map (Appendix 1, fig. 3, location M-7). The quarry, in the Epler and Ontelaunee Formation, is locality 21 of Gray (1951, p. 46). The quarry produced crushed stone and stone for manufacturing lime. A kiln is located west of the quarry (fig. 348), and a platform for loading rail cars is south of the quarry (fig. 349).

The lowest beds were a thin-bedded, light gray calcarenite. Above them were interbedded, blue, fine-grained, laminated limestone beds 3 to 8 feet thick and gray, buff-weathering dolomite beds 1 to 3 feet thick. Some black chert was present. The strike of the beds was N. 35° E. to N. 45° E., and the dip was 35° to 45° SE. Gray (1951, p. 46) provided analyses of the stone.

Bailey Quarry

The Bailey quarry was west of the intersection of Long Lane and High Roads at 40° 31′ 36″ latitude and 75° 43′ 19″ longitude on the USGS Topton 7.5-minute topographic quadrangle map (Appendix 1, fig. 4, location M-12). The quarry, in the Stonehenge Formation, is locality 4 of Gray (1951, p. 28). The quarry was worked for crushed stone.

The quarry exposed interbedded limestone and dolomite. The limestone was gray with thick, black, siliceous laminae or bands that stood out strongly in weathered surfaces. The dolomite beds were gray and finely crystalline with only faint lamination. Gray (1951, p. 28) provided analyses of the stone.

Iron Mines

A 20-year lease dated May 20, 1799, allowed Jacob Winey and Jacob L. Wyler to mine iron ore on Valentine Skelkap's land in Maxatawny Township. They paid 6 pence per load of ore and, if the ore produced good iron, they were to pay 20 pounds gratis at the end of the year.

Rogers (1858, vol. 1, p. 266) noted: "*About a mile S. of Kutztown, good iron-ore was at one time obtained in some quantity, though subsequently the works were neglected, on account, it is said, of the influx of water. This is in the low ground near the South side of a limestone ridge, which lies between it and Kutztown. The surface-soil between this locality and Kutztown is abundantly strewed with blocks of chert of various colours, but generally dark bluish or black, in masses of considerable size.*"

Many of the iron mines in Maxatawny Township shown on maps produced by the Second Pennsylvania Geological Survey have numbers rather than names; the names of the mines are unknown. The numbers are shown on maps by Prime (1875), Prime (1878), and the 1882 topographic map. The origin of the numbers was described in a note from State Geologist J. Peter Lesley (Prime, 1875, p. 57), which stated, "*The numbers on the map are those of the principal ore banks in the order in which they were visited; and the same numbers are given in the pages of the text, together with the names of the owners or lessees of the banks. The smaller and less important are not numbered, but are located accurately on the map. To have re-arranged the numbers in their geographical order would have involved the re-drawing of the map, and a delay in the publication; and any new openings to be hereafter inserted and reported, would disarrange the order of numbers, unless fractional numbers were employed, which would break the system in another sense.*" Prime (1878, p. 2) stated, "*On the extreme west of the map are a few mines with numbers attached to which no reference is made in the key-list, as they are situated in Berks County, and will be referred to and described when the map of that portion of the State-now nearly completed-is issued.*" However, the 1882 topographic map retained the numbers.

Figure 350. Samuel Lewis mine near Klines Corner, Maxatawny Township, March 2015.

Klines Corner-Farmington Area

Samuel Lewis Mine

The Samuel Lewis mine (fig. 350) was west of Klines Corner Road, between Fisher Lane and Farmington Road at 40° 31′ 53″ latitude and 75° 41′ 28″ longitude on the USGS Topton 7.5-minute topographic quadrangle map (Appendix 1, fig. 4, location M-16). It is shown in the 1876 atlas and is mine number 21 on the 1882 topographic map (fig. 263). The mine was in the Allentown Formation. It is now a large, tree-filled open pit partially filled with trash.

The 1880 Census (Pumpelly, 1886, p. 960) listed Samuel Lewis as the operator of the mine. Production between July 1, 1879, and June 30, 1880, was 11,299 tons of limonite. The ore was shipped to the Topton, Kutztown, and Keystone Furnaces and to furnaces in Pottstown. The Samuel Lewis mine produced limonite, which was dark brown to yellowish brown, compact, cellular, and full of ocher seams. The ore contained a considerable quantity of quartz. An analysis was presented by McCreath (1879, p. 210-211): 44 percent metallic iron, 0.021 percent sulfur, and 0.553 percent phosphorus.

Charles Miller Mine

The Charles Miller mine was northwest of the intersection of High and Long Lane Roads at about 40° 31′ 45″ latitude and 75° 43′ 34″ longitude on the USGS Topton 7.5-minute topographic quadrangle map (Appendix 1, fig. 4, location M-21). It is shown in the 1876 atlas and is mine number 47 on the 1882 topographic map. The mine was in the Epler Formation. The mine produced limonite, which generally was dark brown, cellular, and full of ocher seams with spangles of quartz. An analysis was presented by McCreath (1879, p. 210-211): 53.1 percent metallic iron, 0.062 percent sulfur, and 0.038 percent phosphorous.

Schweyer and Liess Mine

The Schweyer and Liess mine was located near the Bowers railroad station. The mine was active from the 1870s to about 1900. In 1879, it produced about 350 tons of ore per day (*Reading Times*, August 8, 1879). In 1897, Schweyer and Liess were contractors for the Thomas Iron Company (Mine and Quarry News Bureau, 1897).

Other Iron Mines

S. Smith's mine was at 40° 33′ 04″ latitude and 75° 41′ 50″ longitude on the USGS Topton 7.5-minute topographic quadrangle map (Appendix 1, fig. 4, location M-15). The mine is shown in the 1876 atlas.

Iron mines with numbers and no available information include the following mines.

Mine number 7 is shown on the 1882 topographic map as two large open pits. The northern pit was in Lehigh County. The southern pit was mostly in Lehigh county, but the western part extended into Berks County at 40° 32′ 58″ latitude and 75° 41′ 16″ longitude on the USGS Topton 7.5-minute topographic quadrangle map (Appendix 1, fig. 4, location M-13).

Mine number 8 on the 1882 topographic map (fig. 274) was at 40° 32′ 39″ latitude and 75° 41′ 16″ longitude on the USGS Topton 7.5-minute topographic quadrangle map (Appendix 1, fig. 4, location M-14).

Mine number 9 on the 1882 topographic map (fig. 274) was at 40° 32′ 20″ latitude and 75° 40′ 43″ longitude on the USGS Topton 7.5-minute topographic quadrangle map (Appendix 1, fig. 5, location M-6).

Mine number 12 on the 1882 topographic map (fig. 274) was at 40° 32′ 14″ latitude and 75° 40′ 39″ longitude on the USGS Topton 7.5-minute topographic quadrangle map (Appendix 1, fig. 5, location M-8).

Mine number 13 on the 1882 topographic map (fig. 274) was at 40° 32′ 07″ latitude and 75° 41′ 04″ longitude on the USGS Topton 7.5-minute topographic quadrangle map (Appendix 1, fig. 5, location M-9).

Mine number 14 on the 1882 topographic map (fig. 274) was at 40° 32′ 02″ latitude and 75° 40′ 51″ longitude on the USGS Topton 7.5-minute topographic quadrangle map (Appendix 1, fig. 5, location M-10).

Mine number 15 on the 1882 topographic map (fig. 274) was at 40° 32′ 06″ latitude and 75° 40′ 39″ longitude on the

Figure 351. Limonite from near Lyons, Maxatawny Township, 7.5 cm. Reading Public Museum collection 2000c-27-115.

USGS Topton 7.5-minute topographic quadrangle map (Appendix 1, fig. 5, location M-11).

Mine number 22 on the 1882 topographic map (fig. 263) was at 40° 32′ 06″ latitude and 75° 41′ 58″ longitude on the USGS Topton 7.5-minute topographic quadrangle map (Appendix 1, fig. 4, location M-17).

Mine number 23 on the 1882 topographic map (fig. 263) was at 40° 31′ 41″ latitude and 75° 41′ 49″ longitude on the USGS Topton 7.5-minute topographic quadrangle map (Appendix 1, fig. 4, location M-18).

Mine number 24 on the 1882 topographic map (fig. 263) was at 40° 31′ 36″ latitude and 75° 41′ 39″ longitude on the USGS Topton 7.5-minute topographic quadrangle map (Appendix 1, fig. 4, location M-19).

Mine number 43 on the 1882 topographic map (fig. 263) was somewhere around 40° 31′ 14″ latitude and 75° 42′ 06″ longitude on the USGS Topton 7.5-minute topographic quadrangle map (Appendix 1, fig. 4, location M-20).

Mine number 58 on the 1882 topographic map was at 40° 29′ 55″ latitude and 75° 46′ 32″ longitude on the USGS Fleetwood 7.5-minute topographic quadrangle map (Appendix 1, fig. 12, location M-22).

Lyons Area

Lyons Borough is on the USGS Fleetwood 7.5-minute topographic quadrangle map. D'Invilliers (1883, p. 402) reported limonite, var. xanthosiderite, and wad (manganese oxide) from the Lyons area.

MUHLENBERG TOWNSHIP

W. Hartman Farm Locality

The W. Hartman farm was east of Bernhart and south of the Bernhart Reservoir on the USGS Temple 7.5-minute topographic quadrangle map (Appendix 1, fig. 10, location MU-1). Genth (1876, p. 220) reported brown dravite associated with epidote from the W. Hartman farm.

Barnhart's Dam locality

Barnhart's dam lies east of the intersection of Spring Valley and Crystal Rock Roads at about 40° 22′ 45″ latitude and 75° 54′ 28″ longitude on the USGS Temple 7.5-minute topographic quadrangle map (Appendix 1, fig. 10, location MU-2). D'Invilliers (1883, p. 273) reported that zircons were found in magnetite near Barnhart's dam.

Figure 352. Limonite from Muhlenberg Township. Reading Public Museum collection 2000c-27-065.

Brook's Quarry

The exact location of Brook's quarry is unknown. Gordon (1922, p. 156) placed the quarry in Muhlenberg Township. D'Invilliers (1883, p. 397) reported fluorite crystals "of an amethystine hue" from Brook's quarry north of Reading. There are several calcite crystal specimens (fig. 353) from Brook's quarry in the Reading Public Museum collection. The calcite crystals are up to 2.6 cm.

Laureldale Quarry

The Reading Quarry Company's Laureldale quarry was southeast of the intersection of Pennsylvania State Route 61 and Water Street at about 40° 24′ 04″ latitude and 75° 55′ 53″ longitude on the USGS Temple 7.5-minute topographic quad-

Figure 353. Calcite from Brook's quarry, Muhlenberg Township. Reading Public Museum collection 2004C-7-452.

Figure 354. Aerial photograph of the limestone quarries in Muhlenberg Township, September 14, 1937. Courtesy of the Bureau of Topographic and Geological Survey, PennPilot Historical Aerial Photo Library. 1, Reading Quarry Company Laureldale quarry; 2, South Temple quarry; 3, G.W. Focht Stone Company quarry.

Figure 355. Advertisement for the Reading Quarry Company Laureldale quarry, 1914. From Boyd's Reading Directory.

Figure 356. Reading Quarry Company's Laureldale quarry lime kilns, Muhlenberg Township, 1913.

Figure 357. Reading Quarry Company's Laureldale quarry rail car loading area, Muhlenberg Township, ca. 1915.

Figure 358. Reading Quarry Company's Laureldale quarry and lime kilns, Muhlenberg Township, 1915.

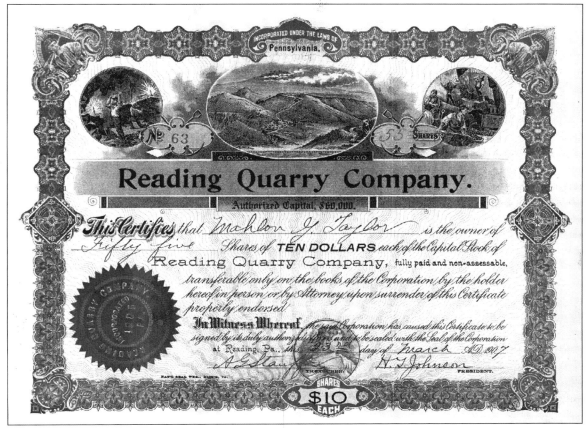

Figure 359. Stock certificate of the Reading Quarry Company, 1907. The certificate was signed by Harry T. Johnson.

rangle map (Appendix 1, fig. 10, location MU-3). There were three quarries close together in this area (fig. 354); the northernmost quarry was the Reading Quarry Company Laureldale quarry. The quarry also has been known as the Shalters quarry, Dreibelbis quarry, and Dietrick quarry. The quarry was in the Allentown Formation.

In 1896, John S. Dreibelbis purchased the 120-acre property of Jonas Shalters for $16,000. The property contained a limestone quarry and two coal yards, and three railroad lines ran through the property. In 1895, enough stone was quarried to produce 160,000 bushels of lime (*Reading Eagle*, January 18, 1896).

Sometime around 1902, Harry T. Johnson purchased the quarry from John Dreibelbis. Johnson produced lime and also shipped flux stone to the E & G Brooke Iron Company furnaces in Birdsboro. The Reading Eagle (September 4, 1902) reported that Johnson shipped several car loads of flux stone to the Temple Furnace daily. In 1903, 30 men were employed at the quarry (*Reading Eagle*, August 8, 1903). Unfortunately, the quarry was noted for having numerous accidents (*Reading Eagle*, May 8, 1904).

In 1906, the Reading Quarry Company was organized by Johnson, who served as the president and general manager (fig. 359). During its peak, the company employed about 40 men, operated nine lime kilns, and shipped several rail car loads of lime daily (Hartman, 1976). In 1909, the Reading Quarry Company installed five new modern kilns for the production of agricultural lime and two kilns for the production of building lime. The company also leased the Gehret quarry (*Reading Eagle*, March 12, 1909).

The Reading Quarry Company went bankrupt in 1922. Pursuant to a court order, the real and personal property of the company was sold at a public sale on April 29, 1922. The property was purchased by N.E. Dietrick and August Dietrich (Berks County deed book), who operated the quarry for a period of time.

The Laureldale quarry was a large quarry with a 70-foot high quarry face. Miller (1934, p. 216) reported that in 1933 the quarry was idle, and much of the machinery had been removed. There were seven lime kilns present. The stone was high in magnesia and was mostly gray with some dark blue. The beds were thick, the strike was N. 73° E., and the dip 18° SE. Styolites were prominent along the bedding planes. Miller also noted the presence of oolites and limestone conglomerate in the quarry.

Figure 360. Aragonite from Nolan's (Tuckerton) Cave, Muhlenberg Township. Reading Public Museum collection 2000C-027-799.

Figure 361. Calcite from Nolan's (Tuckerton) Cave, Muhlenberg Township, 8.5 cm. Reading Public Museum collection 2000C-027-829.

Figure 362. Calcite from Nolan's (Tuckerton) Cave, Muhlenberg Township, 5.2 cm. Reading Public Museum collection 2000C-027-829.

Figure 363. Calcite (cave pearls) from Nolan's (Tuckerton) Cave, Muhlenberg Township, 10.1 cm. Reading Public Museum collection 2000C-027-800.

Figure 364. Aragonite from Nolan's (Tuckerton) Cave, Muhlenberg Township, 5.5 cm. Sloto collection 3573.

Figure 365. Calcite from Nolan's (Tuckerton) Cave, Muhlenberg Township, 8.5 cm. Reading Public Museum collection 2000C-027-794.

Nolan's (Tuckerton) Cave

A cave, called Nolan's cave or Tuckerton Cave, was discovered about 1887, when the Reading Railroad was constructing its line through Muhlenberg Township. The cave was in a quarry about 0.5 mile north of the Muhlenberg railroad station. Tuckerton cave was sealed by Muhlenberg Township in the 1980s (*Reading Eagle*, August 6, 1934, and March 12, 2009).

South Temple Quarry

The South Temple quarry (fig. 370) was between U.S. Business Route 222 and the former railroad tracks at about 40° 23′ 33″ latitude and 75° 55′ 45″ longitude on the USGS Temple 7.5-minute topographic quadrangle map (Appendix 1, fig. 10, location MU-4). There were three quarries close together in this area (fig. 354); the South Temple quarry was a short distance southeast of the northernmost quarry (the Reading Quarry Company Laureldale quarry). The quarry was in the Allentown Formation. The quarry was labeled as a stone quarry on the 1860 and 1862 maps, and the quarry is shown on the 1882 topographic map. The quarry also was known as the Berks Cast Stone quarry and the Berks Products South Temple quarry.

The Berks Cast Stone Company was a subsidiary of the Berks Products Corporation, which consolidated all the crushed stone producers in the Reading area (fig. 367). The consolidation resulted in the closing of the G.W. Focht Stone Company quarry, which was west of the South Temple quarry, as well as the Mays Brothers quarries in West Reading and Glenside. Crushed stone production was concentrated in the Berks Cast Stone quarry and the former Fehr and O'Rourke

Figure 366. Aragonite from the South Temple quarry, Muhlenberg Township. Ron Kendig collection.

quarry in Shillington. In 1929, the Berks Cast Stone Company supplied stone to local contractors and manufactured concrete blocks. Harry Muth was the company manager. In 1930 during prohibition, Pennsylvania State Police and Federal Bureau of Prohibition agents raided the quarry and seized a car load of beer hidden under the crusher (*Reading Eagle*, March 10, 1929, and July 15, 1930).

When Miller (1934, p. 216) visited the quarry in 1933, it was 400 feet long and 200 feet wide with an 80-foot face. The beds were fairly massive; the strike was N. 66° E. and the dip was 14° SE. The stone was light gray to bluish gray. The plant had a capacity of 80 tons of crushed stone per hour. Some of the fines were sold for sand. The quarry ceased operation in the 1980s. In 2000, the Muhlenberg Township Authority purchased the 240-foot deep quarry for a future water supply. Prior to abandon-

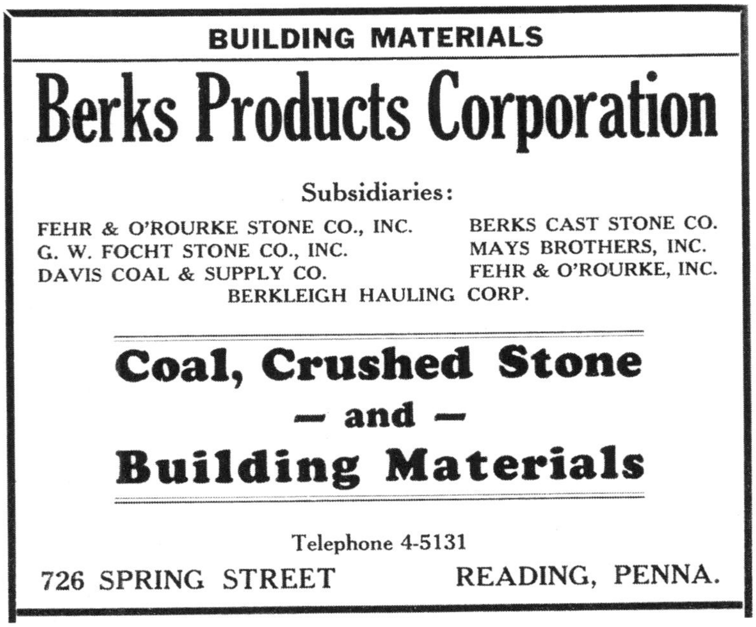

Figure 367. Advertisement for the Berks Products Corporation, 1933. From Boyd's Reading Directory.

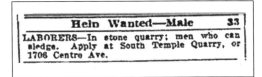

Figure 368. Help wanted advertisement for laborers at the South Temple quarry, 1923 (*Reading Eagle*, April 4, 1923)

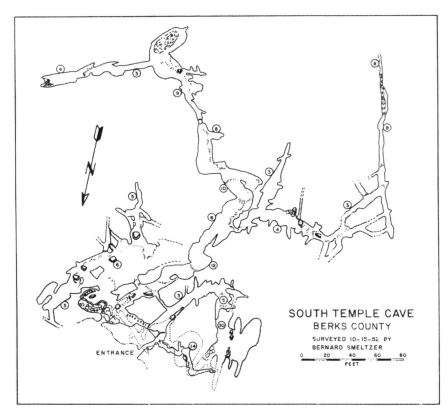

Figure 369. Map of the South Temple Cave, Muhlenberg Township. The cave was located in the South Temple quarry. From MacLachlan (1979, p. 55). Courtesy of the Pennsylvania Geological Survey.

ment, Berks Products pumped 5 million gallons of water a day to keep the quarry dewatered (*Reading Eagle*, December 10, 1999, and March 21, 2010).

A cave, known as the South Temple cave (fig. 369), was located in the South Temple quarry. The cave was described by Stone (1932, p. 33-34), who provided a map of the cave. The entrance to the cave was 40 feet above the quarry floor in the southwest corner of the quarry. The entrance was buried under fill when that part of the property was developed into an industrial park in 1978 (Mostardi and Durant, 1991, p. 75). Thomas (1961, p. 7-8) reported fluorescent cave calcite from the South Temple quarry. Specimens of aragonite (fig. 366) from the South Temple quarry are in the Ron Kendig collection.

G.W. Focht Stone Company Quarry

The G.W. Focht Stone Company quarry was between Pennsylvania State Route 61 and the former railroad tracks at about 40° 23' 40" latitude and 75° 55' 59" longitude on the USGS Temple 7.5-minute topographic quadrangle map (Appendix 1, fig. 10, location MU-5). There were three quarries close together in this area (fig. 354); the Focht quarry was west of the South Temple quarry. The quarry was in the Allentown Formation. The 1882 topographic map shows a railroad siding to the quarry (fig. 371). The quarry also has been known as the Reading Railroad quarry and the Gehret quarry.

The Gehret quarry was opened about 1830 (Hawes, 1884, p. 70-71). In 1871, the Reading Railroad Company (fig. 372) purchased the 84-acre Daniel Gehret farm for $20,200 ($240 an acre) for its limestone quarry (*Reading Eagle*, February 6, 1871). In 1884, the quarry was 50 feet deep and could not be deepened because of the lack of drainage. The quarry was worked to obtain stone for bridge construction and other railroad construction work. The Reading Railroad Company operated the quarry until about the late 1880s (*New York Times*, February 16, 1886).

The quarry was later acquired by the G.W. Focht Stone Company. Frank P. Pehlman became the superintendent of the quarry in 1919. Prior to that year, he was superintendent of the Laureldale quarry (*Reading Eagle*, August 10, 1924). The Reading Eagle (July 24 and August 11 and 19, 1922) reported that the G.W. Focht Stone Company set off a huge blast in the Gehret quarry. Five tons of dynamite were placed in 11 holes between 80 and 85 feet deep drilled by a well driller. The blast yielded an estimated 40,000 tons of stone. In 1924, The G.W. Focht Stone Company produced crushed stone for the Pennsylvania Highway Department, which was shipped by rail to Hamburg (*Reading Eagle,* July 18, 1924).

Figure 370. South Temple quarry, Muhlenberg Township, April 2015.

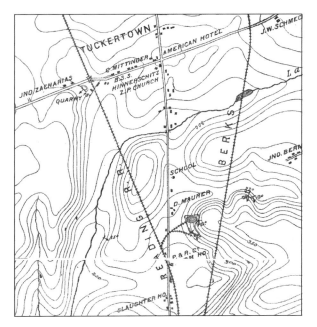

Figure 371. G.W. Focht quarry, Muhlenberg Township, 1882. The map shows a rail spur from the Reading Rail Road line to the Focht quarry. From Pennsylvania Geological Survey (1883).

The quarry later became a subsidiary of Berks Products, Inc., and Walter H. Fehr became the quarry superintendent. The quarry produced 75,000 tons of crushed limestone per year. An Osgood steam shovel was used to load quarry cars, and a small industrial railroad was used to deliver the stone from the quarry to the crusher. The crushing plant was equipped with a Traylor jaw crusher and two Traylor gyratory crushers (Pit and Quarry, 1932). The Focht stone quarry was abandoned before 1953. Water from the adjacent South Temple quarry was pumped into it (*Reading Eagle*, February 18, 1953).

Shaler (1883, p. 151) described the limestone at the quarry as thin- to thick-bedded, fine, blue limestone and gray, massive, calcareous dolomite. The joints were spaced 3 to 20 feet apart, and the strata dipped 4° to 5°.

Temple Sand Company Quarry

The Temple Sand Company quarry is on Mountain Road, north of its intersection with Mount Laurel Avenue. LIDAR imagery shows three quarries in the Hardyston Formation on Irish Mountain. The largest quarry is at 40° 24′ 57″ latitude and 75° 54′ 29″ longitude on the USGS Temple 7.5-minute topographic quadrangle map (Appendix 1, fig. 10, location MU-6). Two smaller quarries are to the northeast and northwest of the largest quarry.

Figure 372. Stock certificate of the Reading Rail Road Company, 1899. The Reading Rail Road Company operated the G.W. Focht (Gehret) quarry in Muhlenberg Township for construction stone in the 1870s and 1880s.

MUHLENBERG TOWNSHIP

HIGH GRADE
Temple Silica Sand
Used in the Construction of This Modern Building

Temple Silica Sand Co.
511 PENN ST.
Bids Gladly Furnished

Figure 373. Advertisement for the Temple Silica Sand Company, 1925 (*Reading Eagle*, February 15, 1925).

The Temple Sand Company, which was organized in June 1894, purchased 131 acres of land on Irish Mountain west of Temple. The sand deposit was discovered by William and Moses Rothermel in 1890. They attempted to purchase the land, but the owners initially asked too much money. They finally were able to purchase land from six different owners. They purchased a crusher from Davis, Printz, & Company of Reading, which was erected by G.A. Francis of Temple. A pair of rollers were purchased from Chester Bertolet & Company of Norristown and were added to the crusher. There were three steam engines; the largest was a 50-horsepower engine. It was purchased from the Temple Iron Company, which used it for an unsuccessful mining venture near Harmonyville in Chester County (see Sloto, 2009, p. 250-251). There was a small hoisting engine and an engine used for pumping water from a spring below the engine house. The company was awarded a contract for all sand used by the City of Reading for paving, which was about 6,000 tons annually. The company employed 8 to 10 men, and the capacity of the plant was 60 tons per day. Loose sand was sold for plastering. All other sand was produced by crushing quartzite. Five grades of sand were produced. A wharf was erected 0.5 mile from the quarry for loading sand into rail cars.

The property of the Temple Sand Company was sold by the sheriff to Charles H. Hunter in 1905 (Berks County deed book). Wellington M. Bertolet became Hunter's business partner. In 1906, Bertolet and Hunter sued to restrain Dr. Horace Schlemm from removing a switch, which was part of a siding connecting his quarry with the rail line on which he shipped sand (*Reading Eagle*, October 15, 1906). Schlemm owned the Temple Cement Stone Company and operated a nearby quarry (*Reading Eagle*, August 31, 1902).

Bertolet died in 1923, and his half-interest in the quarry passed to his wife Esther M. Bertolet and his grandson Wellington Bertolet Hunter. Hunter, Bertolet, and Bertolet then incorporated as the Temple Silica Sand Company (fig. 373). That year, the Temple Silica Sand Company produced and shipped about 100 tons of sand per day (*Reading Eagle*, September 11, 1923, and October 25, 1923).

Gossler Quarry

The Henry Gossler quarry was between Pennsylvania State Route 61 and U.S. Business Route 222 at about 40° 22′ 45″ latitude and 75° 55′ 45″ longitude on the USGS Temple 7.5-minute topographic quadrangle map (Appendix 1, fig. 10, location MU-7). The Henry Gossler quarry and lime kiln are shown on the 1882 topographic map. The quarry was in the Allentown Formation. The area is now completely urbanized.

Becker Quarry

The location of the Becker quarry or quarries is unknown. Newspaper reports list several Beckers operating a quarry. In 1892, Edward Becker was crushed to death when an embankment gave way, and 25 tons of stone and earth fell on him (*Baltimore American*, August 31, 1892). John Becker employed four men in his quarry and used two teams to haul stone to Reading (*Reading Eagle*, May 24, 1893). Two women were killed in an explosion at the home of ex-sheriff Elias Becker when he put several sticks of dynamite in the kitchen stove to thaw and left the house. A few minutes later, an explosion occurred (*Reading Eagle*, December 31, 1896). Day (1904, p. 162) provided an analysis of limestone from the Elias Becker quarry in Tuckerton.

Blue Bell Lime Company

In 1930, a state charter was granted to the Blue Bell Lime Company of South Temple to quarry, crush and burn lime and manufacture articles of stone and clay. The company was capi-

Figure 374. Temple Iron Company Furnace, ca. 1860s.

Shares of the Capital Stock of

The Clymer Iron Company,

OF TEMPLE, BERKS COUNTY, PA.,

On each of which Shares .... Dollars have been paid.

Witness .... hand and Seal this .... day of 18

SEALED AND DELIVERED IN PRESENCE OF

Figure 375. Receipt for the purchase of Clymer Iron Company stock.

talized with $10,000 in stock. Norman B. Ringler was the treasurer. Other company officers included Frank A. Kremser and Elizabeth M. Kremser of Philadelphia (*Reading Eagle*, January 22, 1930). The company produced 25,000 tons of crushed limestone, 30,000 tons of hydrated lime, and 12,000 tons of lime per year. John Cordone was the plant superintendent. The plant equipment consisted of an Atlas-Chalmers crusher, Champion jaw crusher, Steacy-Schmidt kiln, Clyde hydrator, and a Raymond pulverizer (Pit and Quarry, 1932).

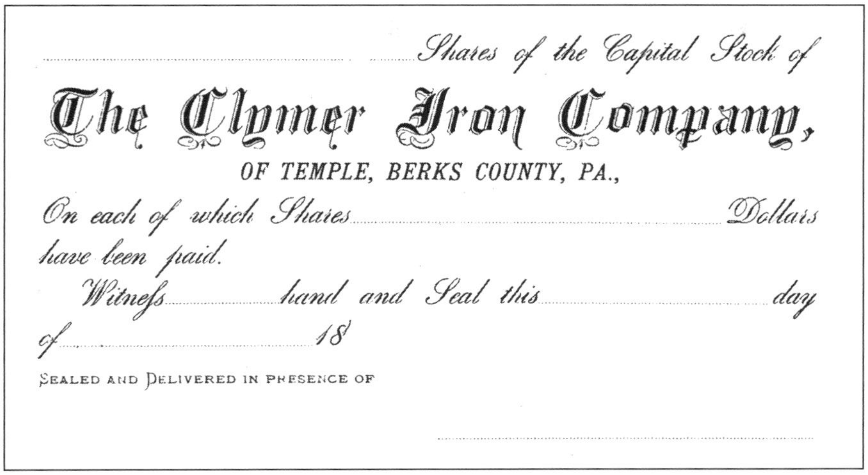

Temple Furnace

The Temple Furnace (fig. 374) was an anthracite iron furnace built by The Clymer Iron Company, commonly known as William H. Clymer & Company (fig. 375) in 1867 adjoining the East Pennsylvania Railroad in Temple. The furnace was run by Clymer & Company until 1873, when the Temple Iron Company (fig. 376) was organized with William H. Clymer as president. The furnace was rebuilt in 1875.

In 1880, the furnace used ore from Lehigh, Berks, and Lebanon counties. Its specialty product was foundry pig iron, and the annual capacity was 11,000 tons. In 1892, the furnace used ore from the Lake Superior region and magnetite from New Jersey. The annual capacity was 27,000 tons, and iron was sold under the brand name "Temple." By 1902, the annual capacity had increased to 50,00 tons. The specialty products were foundry and forge pig iron (American Iron and Steel Association, 1880, 1892, and 1902).

George F. Baer became president of the Temple Iron Company about 1881. Baer studied law and was admitted to the bar in 1864. He moved to Reading in 1868 and established a law practice. His connection with the Reading companies began in 1870 when he prosecuted a case for damages against the rail-

Capital $2,500,000

No. 15

1 Shares

The Temple Iron Company

BERKS COUNTY, PENNSYLVANIA

Shares $100 Each

THIS CERTIFIES that R. M. Olyphant is entitled to One Shares in the Capital Stock of The Temple Iron Company, transferable only on the Books of the Company in person or by Attorney on surrender of this Certificate.

Witness the Signature of the President and Treasurer and the Corporate Seal of the Company at Reading, Berks County, Pennsylvania, this 15th day of March A.D. 1902.

TREASURER

Geo. F. Baer

PRESIDENT

Cancelled

INCORPORATED UNDER THE LAWS OF THE STATE OF PENNA., MAR. 22, 1873

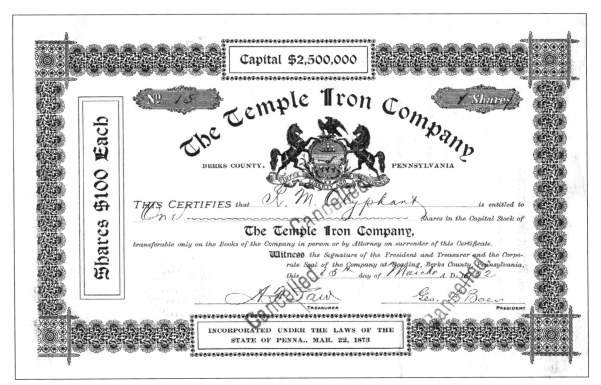

Figure 376. Stock certificate of the Temple Iron Company, 1902. The certificated was signed by George F. Baer.

Figure 377. Limonite from Muhlenberg Township, 8.7 cm. Reading Public Museum collection 2000c-27-054.

road so successfully that he was hired as counsel for the Philadelphia and Reading Railroad. He was the confidential legal adviser to financier J.P. Morgan. When Morgan acquired and reorganized the Philadelphia and Reading Railroad in 1901, he named Baer president of three Reading companies—the Philadelphia and Reading Railway Company, the Philadelphia and Reading Coal and Iron Company, and the Central Railroad Company of New Jersey. Baer became president of several other companies, including the Lehigh and Wilkes Barre Coal and Iron Company, Reading Paper Mills, and Reading Iron Company, and served as a director for several other companies.

Pennsylvania law did not allow coal-carrying railroads to own coal mines, but it did allow a furnace company to do so. The Temple Iron Company had a broad charter that allowed it to act as a holding company. In 1899, Baer was one of the largest stockholders. The balance of the stock was held by personal friends of Baer, and they were willing to sell their shares to him. On January 26, 1899, the capital stock of the Temple Iron Company was increased to $2,500,000, and at the same time $3,500,000 in bonds were issued. By a series of complicated transactions, the Temple Iron Company became the owner of seven coal companies. The stock of the Temple Iron Company was then purchased by the Reading company in partnership with other coal-carrying railroads. Ownership of the Temple Iron Company stock was determined by the annual percentage of anthracite coal each railroad transported, with the Reading owning 29.96 percent; the Lehigh Valley Railroad 22.88 percent; the Delaware, Lackawanna and Western Railroad 19.52 percent; the Central Railroad Company of New Jersey 17.12 percent; the Erie Railroad 5.84 percent; and the New York, Susquehanna and Western Railroad 4.65 percent (Jones, 1914). By December 31, 1906, practically the entire capital stock of the Temple Iron Company was owned by the railroads. The partnership, known as the Hard Coal Trust, controlled 90 percent of the supply and transportation of anthracite coal in Pennsylvania and fixed its price.

A long series of antitrust investigations and lawsuits followed. On June 12, 1907, the Department of Justice filed suit in the Circuit Court for the Eastern District of Pennsylvania to dissolve the Hard Coal Trust. In the petition, the Government charged that the defendants had entered into a combination or conspiracy, by which they restrained and monopolized the anthracite coal trade. The Circuit Court rendered its decision on December 8, 1910. The charge that the defendants had entered into a general combination or conspiracy was unanimously dismissed. The only contention of the Government that was sustained was the charge that the railroads had unlawfully combined through the Temple Iron Company to prevent the building of the proposed independent coal-carrying New York, Wyoming and Western Railroad. The Government appealed to the Supreme Court. The Supreme Court decision on December 16, 1912, dissolved the Temple Iron Company and declared void the railroads' perpetual contracts with independent coal producers, which paid the producers 65 percent of the market price for their coal.

On May 1, 1914, the directors of the Temple Iron Company announced that the company had sold the stock of its eight coal companies to Mr. S.B. Thorne, who was at one time general manager of the Temple Iron Company. In November 1915, the furnace was sold to H.H. Adams of New York to complete compliance with the court's dissolution order. In January 1916, Adams sold the furnace to the Delaware Steel and Ordnance Company (*Iron Age*, January 6, 1916, p. 113). After having been inactive for months, it was started up in October 1915; however, it operated only a short time before it was shut down for repairs (*Iron Age*, December 2, 1915, p. 1337). The furnace was dismantled in 1926-27.

Figure 378. Quartz from Muhlenberg Township, 9.5 cm. Reading Public Museum collection 2005C-2-191.

NEW MORGAN BOROUGH

Grace Mine

The Grace mine was west of Pennsylvania State Route 10, about 1.1 miles northeast of the center of Morgantown at 40° 10′ 16″ latitude and 75° 52′ 49″ longitude on the USGS Morgantown 7.5-minute topographic quadrangle map (Appendix 1, fig. 35, location N-1).

The Grace mine and mill was owned by the Bethlehem-Cuba Iron Mines Company (also known as the Bethlehem Steel Corporation) and was operated by the Bethlehem Cornwall Corporation. The mine was named in honor of Eugene C. Grace (fig. 379), the chairman of Bethlehem Steel at the time of the discovery of the ore body. Grace was an enthusiastic supporter of the company's exploration program (Wharton, 2003, p. 6).

The Grace mine ore body was discovered as a the result of an airborne magnetometer survey. It was the first instance of an ore body that was discovered by an airborne magnetometer survey In 1948, Bethlehem Steel's geology department contracted with the Aero Service Corporation of Philadelphia to conduct aeromagnetic surveys of the Mesozoic Basins in Pennsylvania as they were known hosts for several important Cornwall-type iron deposits (Sims, 1968, p. 109). The survey consisted of 7,000 linear miles of flight lines spaced 0.25 mile apart and extended from the Delaware River on the east to Gettysburg on the west (Bingham, 1957). In 1948, residents reported seeing small airplanes flying low over the hills around Morgantown towing a cylindrical object at the end of a long wire. William Agocs, an Aero Service geophysicist, was assigned to Bethlehem Steel to help evaluate the magnetic anomalies on the new aeromagnetic maps.

Figure 379. Portrait of Eugene C. Grace, chairman of the Bethlehem Steel Corporation at the time of the discovery of the ore body.

Figure 380. Sign at the entrance to the Grace mine on Pennsylvania State Route 10 north of Morgantown.

One of the most promising anomalies was near Morgantown, which is shown on figure 381. After examining the aeromagnetic maps, Bethlehem Steel's geologists made limited surface traverses over the area to check the location and found no surface expression of an ore body.

In May 1949, a secret buyer quietly began buying large tracts of land in Caernarvon Township and across the Berks County line in Chester County. Rumors about the buyer and the reason for the purchases ranged from Midwestern cattle dealers who planned to fatten steers for eastern markets to uranium mining. Other rumors included a huge tire plant, a chemical plant, and a coffee bean plantation. As part of an agreement between Bethlehem Steel and the Berks County Trust Company, the Trust Company held the titles to all the properties (Frankhouser, 1962). Officers of the trust company refused to reveal the identity of the buyer, citing client confidentiality. Local newspapers were unable to learn what was occurring. Headlines in the newspapers read "Berks County Buzzing with Uranium Talk" (*Gettysburg Times*, September 30, 1949) and "Nine More Farms Go at Fabulous Prices" (*Reading Eagle*, September 26, 1949). A newspaper from Philadelphia even sent reporters equipped with Geiger counters to the area (*Reading Eagle*, September 26, 1949).

By September 26, 1949, 27 farms had been purchased. The last two farms sold for many times their assessed value,

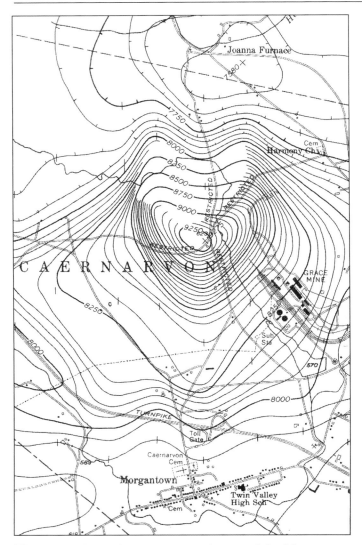

Figure 381. Part of an aeromagnetic map of the Morgantown area showing the anomaly at the Grace mine iron ore body. From Bromery and others (1959).

encountered at a depth of 1,524 feet when the drill cuttings suddenly changed color from red to black. Heyward M. Wharton, the site geologist, inspected the drill cores and then drove to Reading in order to assure privacy while phoning Bethlehem Steel management. Donald Fraser, George Adair (a senior geologist), and two Bethlehem Steel vice presidents drove to the drill site, arriving about midnight with a celebratory bottle of bourbon (Wharton, 2002). The No. 1 drill hole (fig. 383) penetrated over 400 feet of iron ore, followed by about 25 feet of weakly mineralized limestone, and then about 20 feet of nearly pure tremolite in contact with the thick diabase sheet.

The No. 2 drill hole was about 1,000 feet south of the No. 1 drill hole (fig. 383) on the former farm of Fred Trunk. The rig was a 30-foot tall wooden tripod. Drilling was carried on around the clock by a crew of six men from Sprague and Henwood of Scranton, Pa. Water to cool the drill bit was pumped from a nearby creek. The drillers did not reveal their purpose or their employer (*Reading Eagle*, November 22, 1949). The No. 2 drill hole did not encounter ore because of the abrupt termination of the ore body in that direction (Bingham, 1957).

Rumors were put to rest on January 27, 1950, when Bethlehem Steel announced that it was the company behind the land purchases and test drilling. The company acknowledged that it was exploring for iron ore and issued the following statement: "*We are interested in prospecting for possible deposits of*

bringing the total purchase to 1,500 acres at a cost of more than $350,000 (*Reading Eagle*, August 5, 2007). Fred Trunk believed he had sold his farm to cattlemen who wanted grazing land near eastern markets (*Reading Eagle*, September 30, 1949).

The first core drilling rig (fig. 382) arrived at the site on September 1, 1949 (Bingham, 1957, p. 45). There was a lot of speculation by local residents after visits to the drill site by newspaper reporters and others who were trying to find out what the drillers were doing. The drillers and on-site geologists did not divulge any information about the purpose of drilling (Wharton, 2002).

The drilling crew consisted of four men. When drilling began, there was a single daylight shift, and progress was slow, often due to mishaps. Managers in the Bethlehem Steel main office soon became nervous, wondering if the anomaly might not actually be anything important. A second shift was added to speed things up. Finally, on December 19, 1949, iron ore was

Figure 382. Core drilling rig exploring for the Grace mine ore body, January 12, 1954. Photograph by Richard Angelo. Bethlehem Steel Corporation Mining Collection of Photographs, Archives Center, National Museum of American History, Smithsonian Institution.

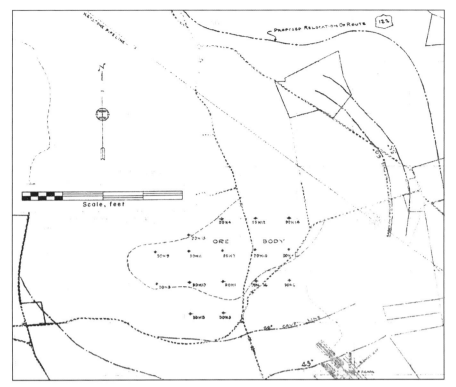

Figure 383. Location of core holes used to delineate the Grace mine ore body. Courtesy of the Office of Surface Mining Reclamation and Enforcement National Mine Map Repository.

ore in the general area of Morgantown, Pennsylvania, and have acquired some land on which exploratory work is being done. This is a continuation of a general program of prospecting in the vicinity of former workings in this area for ore occurrences" (*Reading Eagle*, January 27, 1950).

By April 6, 1950, Bethlehem Steel had five core drill rigs working. The work was overseen by Dewey Bernard from the Cornwall mine in Lebanon County. After 6 months of drilling, the ore body had not been encountered. The rigs were only capable of drilling to a depth of 2,000 feet. Fearing that the ore body was deeper than the rigs capability, Bethlehem Steel brought in a larger rig capable of drilling to 5,000 feet. Water to cool the bit was piped in from the No. 1 drill hole, which had been gushing water since October 1949 (*Reading Eagle*, April 5, 1950). By January 1951, 17 core holes (fig. 383) had been completed, spaced on 600-foot coordinates and averaging 2,200 feet deep. The extent of the ore body had been defined, and mine development was then initiated.

The sinking of two shafts was begun in 1952. In 1953, shaft A, the ore hoisting shaft, was sunk to nearly 1,300 feet. Shaft B, the man and material shaft, was greater than 700 feet deep. Many of the surface buildings were constructed in 1953, as well as the railroad spur and loading tracks (Kaufman, 1956). By 1954, shaft A was 1,600 feet deep, and shaft B was 1,200 feet deep (Thomson, 1957). In 1955, shaft A reached 2,208 feet, and shaft B, reached 3,079 feet. Initial development included driving and concreting 6,000 feet of main haulage drift on the first level of shaft A, constructing a power station on the second level, concreting two main sumps for mine drainage water on the third level, and constructing a main pump room on the fourth level (Tomson, 1958).

In 1955, shaft A reached a final depth of 2,208 feet. The concrete-lined, circular shaft was 21.5 feet in diameter and contained two skipways and a cage compartment (fig. 387). Ladders, pipe columns, and electrical lines occupied the remaining space. In conjunction with the shaft sinking, an ore pocket, car dump, and 250 feet of drifting were completed on each of the first four levels. The shaft A head frame (fig. 384) was constructed of steel and stood 180 feet high. It included an elevator to transport personnel, a 700-ton ore storage bin, and an 800-ton rock storage bin (Bingham, 1957, p. 46).

In 1955, shaft B, located 280 feet south of shaft A, reached a final depth of 3,079 feet. Shaft B was a circular, concrete-lined shaft 17.5 feet in diameter with one cage compartment and a ladderway (fig. 387). Starting at 1,812 feet below collar and at 100-foot intervals thereafter, landings were cut as the shafts were advanced. Twelve landings were constructed in shaft B. A pump room and

Figure 384. Grace mine shaft A headframe during construction, June 14, 1956. Photograph by Richard Angelo. Bethlehem Steel Corporation Mining Collection of Photographs, Archives Center, National Museum of American History, Smithsonian Institution.

Figure 385. Aerial view of the Grace mine, April 13, 1970. Photograph by Richard Angelo. Photograph by Richard Angelo. Bethlehem Steel Corporation Mining Collection of Photographs, Archives Center, National Museum of American History, Smithsonian Institution.

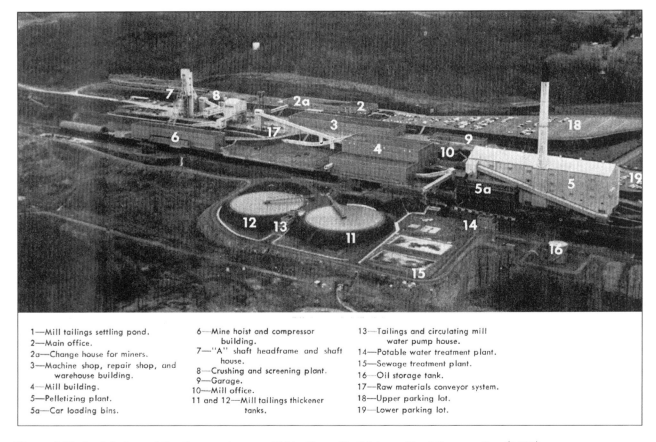

1—Mill tailings settling pond.
2—Main office.
2a—Change house for miners.
3—Machine shop, repair shop, and warehouse building.
4—Mill building.
5—Pelletizing plant.
5a—Car loading bins.
6—Mine hoist and compressor building.
7—"A" shaft headframe and shaft house.
8—Crushing and screening plant.
9—Garage.
10—Mill office.
11 and 12—Mill tailings thickener tanks.
13—Tailings and circulating mill water pump house.
14—Potable water treatment plant.
15—Sewage treatment plant.
16—Oil storage tank.
17—Raw materials conveyor system.
18—Upper parking lot.
19—Lower parking lot.

Figure 386. Aerial view of the Grace mine, ca. 1961. .From Bethlehem Steel Corporation (1961).

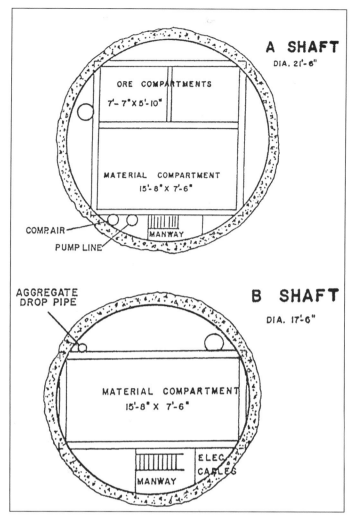

Figure 387. Cross sections of the Grace mine shafts.

Figure 388. Grace mine shaft A hoists and compressors. The two hoists are shown in the background; air compressors are shown in the foreground. From Bingham (1957, p. 46). Used with permission from the American Institute of Mining, Metallurgical, and Petroleum Engineers.

Figure 389. Grace mine shaft A hoists. From Bethlehem Steel Corporation (1961).

sump drifts were constructed at the 1,548-foot level. The sump drifts were driven out to intersect the bottom of the core holes in order to dewater the area around the two shafts (Bingham, 1957, p. 46).

The shaft A ore hoist (fig. 389) had double cylindrical drums 15 feet in diameter. The hoist was powered by two 2,250 horsepower DC motors and was rated at a hoisting speed of 2,400 feet per minute. The hoisting rope was 2-1/4 inches in diameter. Ore and waste rock were hoisted in balance with 305-cubic-foot bottom dump skips (Bingham, 1957, p. 46). The shaft A man and material hoist had double cylindrical drums 10 feet in diameter. The hoist was powered by a 1,000 horsepower DC motor and was rated at a hoisting speed of 1,450 feet per minute.

The shaft B hoist had double cylindrical drums 12 feet in diameter. The hoist was powered by an 800 horsepower AC motor and was rated at a hoisting speed of 1,500 feet per minute. Both cage hoists were operated in balance with one drum for the counterweight rope and the other for the cage rope. The rope diameter on both hoists was 1-3/4 inches. The man and

UNDERGROUND AT THE GRACE MINE

Figure 390. Grace mine 513B ramp drift, January 5, 1967. Drift is 11 feet high and 12.5 feet wide. Courtesy of the Tri-County Heritage Society, Morgantown.

Figure 391. Grace mine entry drifts to panel 51, December 6, 1966. Courtesy of the Tri-County Heritage Society, Morgantown.

Figure 392. Grace mine sixth level, August 12, 1964. Photograph by Richard Angelo. Bethlehem Steel Corporation Mining Collection of Photographs, Archives Center, National Museum of American History, Smithsonian Institution.

Figure 393. Grace mine sixth level, January 6, 1966. Photograph by Richard Angelo. Bethlehem Steel Corporation Mining Collection of Photographs, Archives Center, National Museum of American History, Smithsonian Institution.

Figure 394. Grace mine 31 north production drift on the seventh level of the ore body, January 5, 1967. Courtesy of the Tri-County Heritage Society, Morgantown.

Figure 395. Scoop tram in the Grace mine, August 3, 1968. Courtesy of the Tri-County Heritage Society, Morgantown.

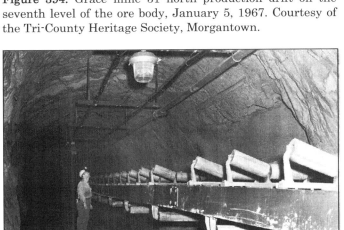

Figure 396. Grace mine sixth level conveyor drift, June 6, 1969. Courtesy of the Tri-County Heritage Society, Morgantown.

Figure 397. Electric locomotive hauling ore cars in the Grace mine. From Bethlehem Steel Corporation (1961).

Figure 398. Grace mine ore cars, July 20, 1962. Each car held 20 tons of ore. Photograph by Richard Angelo. Bethlehem Steel Corporation Mining Collection of Photographs, Archives Center, National Museum of American History, Smithsonian Institution.

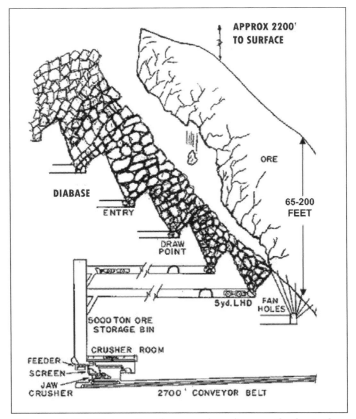

Figure 399. Mining by the panel-caving method at the Grace mine.

to shaft A where it was hoisted to the surface in 20-ton skips. While block caving is very economical, it's chief drawback is surface subsidence, which occurred on the Grace mine property (fig. 400).

Ore was rasised from the mine at the rate of 800 tons per hour and discharged from 20-ton bottom dump skips into the 700-ton storage bin located in the headframe structure. Figure 401 shows the ore processing process. A roll feeder fed the ore into a 42-inch gyratory crusher. Ore less than 5 inches in diameter was fed to two parallel 5-foot by 14-foot double deck vibrating screens. The top deck screens had 2-inch openings, and the bottom deck screens had 3/4-inch openings.

Ore less than 3/4-inch in diameter that passed through the screens was conveyed to the mill ore bins or ore storage area. Ore that was greater than 3/4-inch in diameter passed over two 42-inch by 50-inch magnetic cobbers. Rejected material went to a waste rock bin, and the cobbed ore went to two 7-foot cone crushers. The discharge from the crushers was conveyed to two 6 by 14-foot double deck vibrating screens with a 1-1/4-inch top deck screen and a 3/4-inch bottom deck screen. The ore less than 3/4 inch in diameter was conveyed to the mill ore storage area. Ore greater than 3/4 inch in diameter was sent to two 30-inch by 48-inch magnetic cobbers. Rejected material went to the rock bin; the concentrated ore went back through the cone crushers (Bingham, 1957, p. 47-48).

Ore from the crushing plant was conveyed to four 800-ton feed bins. The ore moved from the bins on conveyor belts to four 4-foot by 12-foot wet vibrating screens. Ore greater than 1/4 inch in diameter passed over four 30-inch by 60-inch mag-

material single deck cage interiors were 15 feet long by 6 feet 4 inches wide (Bingham, 1957, p. 46).

Ore was mined by the panel caving method (fig. 399). Funnel-shaped openings were driven into the ore body from drifts below. The undercutting caused the ore to collapse under its own weight in a controlled fashion into drawpoints. The ore was removed through a system of transfer, slushing, productIon, and haulage drifts and was hoisted to the surface through shaft A. All development work was done in diabase so that, except for exploratIon drIfts, access to the ore body was limited to the footwall contact (Sims, 1968, p. 114). The caving action started at the center of the panel and advanced toward the extremities. This method was chosen for the following reasons: (1) preliminary development could be advanced more rapidly, (2) full-scale production is reached at an earlier date, and (3) boundary scraping stopes could be worked in conjunction with the caving without the necessity of a protective pillar (Bingham, 1957, p. 47). The ore initially was transferred by an electric railway (fig. 397) and later by a conveyor belt system (fig. 396)

Figure 400. Ariel view of subsidence at the Grace mine. Photograph by Richard Angelo. Bethlehem Steel Corporation Mining Collection of Photographs, Archives Center, National Museum of American History, Smithsonian Institution.

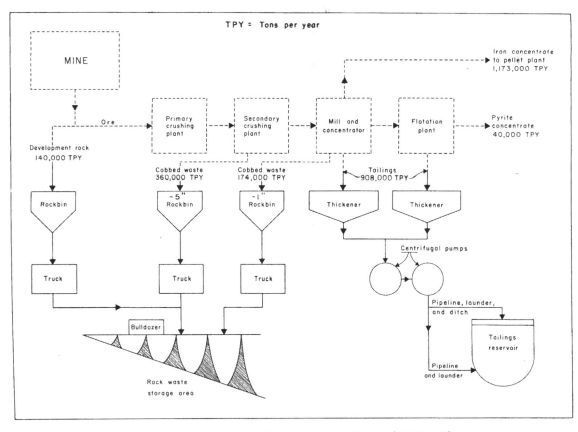

Figure 401. Grace mine ore-processing flow diagram. From Cocran (1969, p. 5).

netic cobbers. Rejected material went to a fine rock bin, and the concentrated ore went to four 9-foot by 13-foot rod mills (fig. 402).

Ore less than 1/4 inch in diameter that passed through the screens went to four 3-foot by 21-foot screw classifiers. Ore greater than 1/4 inch went to the rod mills. The overflow from the classifiers joined the ball mill discharge and was pumped to eight 24-inch cyclones in a closed circuit with the four 11.5-foot by 14-foot ball mills (fig. 402). The rod mill discharge passed by gravity directly to the ball mills. Overflow less than 100 mesh from the eight cyclones passed to 32 three-drum magnetic separators (fig. 403). The 65-percent iron concentrate went to the agglomeration plant. The tailings went to sixteen 12-inch cyclones in the pyrite flotation plant (Bingham, 1957, p. 48).

Figure 402. Grace mine rod and ball mills. From Bethlehem Steel Corporation (1961).

Figure 403. Grace mine magnetic separators. From Bethlehem Steel Corporation (1961).

Figure 404. Grace mine tailings pond and dam. From Cocran (1969, p. 8).

shipped to the sulfuric acid plant at Sparrows Point, Maryland (Bingham, 1957, p. 48).

Fines from the 12-inch cyclones and the pyrite flotation tailings joined in a common launder and flowed to two 200-foot diameter tailings thickeners. Tailings from the thickeners flowed by gravity to two centrifugal pumps in series, which delivered the 30 percent solids slurry to the tailings dam at 1,600 gallons per minute. The pumping distance was approximately two miles. The tailings were impounded behind an earth fill dam 125 feet high (fig. 404), which was constructed between two ridges to form a 325-acre disposal area (Bingham, 1957, p. 48).

The 65-percent iron concentrate was pumped from the mill to a thickener located in the agglomeration plant. A small amount of culm coal was added. The thickened concentrate was filtered, and, to improve the balling quality, a small amount

Underflow from the 12-inch cyclones passed to conditioners and then to flotation cells consisting of roughers, cleaners, and recleaners. The pyrite flotation concentrate was thickened in a 30-foot thickener from which the underflow was pumped to two 6-foot disk filters. The filtered pyrite concentrate was

Figure 406. Grace mine pellet furnaces. From Bethlehem Steel Corporation (1961).

Figure 405. Grace mine revolving balling cones. From Bethlehem Steel Corporation (1961).

of bentonite was added before the concentrate was mechanically rolled into pellets averaging 3/8 to 5/8 inch in diameter (fig. 405). The pellets were fed into an oil-fired vertical shaft furnace (fig. 406) on an oscillating belt conveyor. In passing down through the furnace, the pellets (fig. 407) were baked to render them hard enough to withstand breakage during shipment and handling. Much of the heat required was provided by the exothermic transformation of magnetite to hematite. Discharged pellets were screened and conveyed to railroad cars (fig. 409) for shipment to Bethlehem, Pa., or other Bethlehem Steel plants. Each car held 5,760 tons of pellets. A 3/4-mile spur was built from the railroad main line to the plant to accommodate shipping of the finished pellets (Bingham, 1957, p. 48).

The Grace mine was designed to produce 9,600 tons of ore per day, which averaged 42.5 percent iron. The plant was designed to produce 5,760 tons of iron ore pellets per day, which averaged 65 percent iron. The design annual capacity of the mine and plant was 3 million tons of ore and about 1.5 million tons of iron pellets, respectively (Bingham, 1957, p. 47-48). Mine production began in 1958. Between 1958 and 1968, 18 million tons of ore were mined. There were seven miles of track, 17 electric locomotives, and 133 cars in the mine, and 940 people were employed.

Figure 407. Grace mine magnetite pellets.

A modernization program was undertaken in 1969-70. The Grace mine mobile mining system, which attained full operation during 1970, included 8-ton-capacity, diesel powered, load-haul-dump vehicles; a 2,500-foot-long conveyor belt; an ore crusher; and a complete equipment repair shop 2,200 feet underground. The conveyor belt discharged into a hoist-loading pocket. All main headings for the block-caving method used in the Grace mine were drilled by two, three-boom universal jumbo drills (Cooper, 1975).

The Grace mine produced crude magnetite containing recoverable pyrite, copper, gold, silver, and cobalt. Pyrite (containing cobalt and copper) was recovered by flotation. Cobalt concentrates were produced as early as 1961. The pyrite was shipped to a sulfuric acid plant in Sparrows Point, Maryland, for use in sulfuric acid production. After roasting, the cinder was leached to recover a liquid sulphate concentrate containing cobalt and copper. The concentrate was shipped to Pyrites Company, Inc. in Wilmington, Delaware, for refining. The concentrate was processed into metal oxide and hydrate and electrolytic copper or copper chemicals. The Grace and Cornwall mines were the only domestic producers of cobalt in the 1960s. No cobalt was produced in the 1970s.

Figure 408. Toppling of the Grace mine smoke stack, November 16, 1984. The smokestack was the tallest structure in Berks County at the time. Courtesy of the Tri-County Heritage Society, Morgantown.

Figure 409. Magnetite pellets being loaded into rail cars at the Grace mine. From Bethlehem Steel Corporation (1961).

The Grace mine was a major employer and an important contributor to the regional economy. Employment peaked at 1,100 workers in the early 1970s. When the mine closed on July 30, 1977, 850 employees lost their jobs. Grace mine's closing was due to the following factors: (1) cheap steel from abroad that depressed the U.S. steel industry, (2) underground mining was much more expensive than the increasingly important open-pit operations, (3) top-heavy management, and (4) a generous labor contract for the miners, who were members of the steelworkers union (*Reading Eagle*, August 5, 2007).

Charley Taylor, superintendent of the Grace mine, kept the pumps running for 4 years after the mine closed. Barnes and Smith (2001, p. 19 and 24) reported that at the time of clo-

GRACE MINE
DECEMBER 2012

Figure 410. Scenes around the Grace mine property, December 2012.

Figure 411. Mineralogical Society of Pennsylvania mineral-collecting field trip to the Grace mine, April 13, 1958.

sure, a zone of ore 480 feet thick and rich in copper had been outlined by drilling to the northeast. A total of 45 million tons of crude ore was produced at the mine (Wharton, 2002). The ore body was not exhausted.

On November 16, 1984, the smokestack, which was the highest structure in Berks County, was toppled because it was considered a hazard to aviation (fig. 408). The property was sold to Raymond Carr. In 1987, Carr, the sole owner of Morgantown Properties, petitioned the Berks County Court to create a new borough from 3,500 acres in Caernarvon and Robeson Townships. The fate of the proposal was determined by the ten residents living within the boundaries of the proposed borough, most of whom were Carr's tenants. In an April 1988 election, those residents voted 9-1 in favor of forming a new borough, which is now New Morgan Borough. The state's procedures regarding the formation of a new borough have since been revised to require new boroughs to have at least 500 residents.

Final reclamation of the property by Morgantown Properties began sometime after 2004. A landfill and a large Victorian style residential development to be named Bryn Eyre was proposed to be built on the site. Bryn Eyre was proposed as an entire town with 12,393 residential homes and commercial, school, and public areas (Robinson, 2007). The development was to include a 1,000 room hotel and an 18-hole golf course. Bryn Eyre was never built; the Conestoga landfill occupies 454 acres of New Morgan Borough.

The Grace mine dumps were a popular mineral collecting locality from the 1950s through the 1980s (fig. 411). Most of the mine dumps (fig. 412) were crushed and used for aggregate and in highway construction. Beginning in 1960, the dumps were crushed and sold for aggregate by the Bradford Hills Quarry, Inc. of East Petersburg (O'Neil, 1965, p. 12). Bradford Hills Quarry, Inc. later was acquired by General Crushed Stone, Inc. Little of the dumps remain (fig. 410).

The Grace mine deposit is a Cornwall-type iron ore deposit occuring at the contact between diabase and Cambrian carbonate rock (fig. 414) along the southern border of the Mesozoic Basin. The bedrock in the vicinity of the mine includes Triassic shales, sandstones, quartzites, and conglomerates, and Cambrian carbonate rocks. These rocks have been intruded by diabase, which varies in thickness from a few feet to 1,200 feet. The diabase was intruded at a depth of about 20,000 feet. The diabase dips northward and forms the footwall of the ore body. Mineralizing solutions that followed the intrusion of the diabase replaced the carbonate rocks with an iron ore body consisting of magnetite, pyrite, pyrrhotite, and other minerals, such as chlorite, serpentine, tremolite, and garnet (Bingham, 1957).

The ore body is a replacement of Cambrian carbonate rocks of the Buffalo Springs Formation (formerly called the Elbrook Formation). The 1,200-foot thick Morgantown diabase sheet strikes about N. 60° W. and dips 20° NE., cross-cutting both the Cambrian carbonate rocks and the overlying Triassic sandstones and conglomerates.

In general, the Grace mine ore body has a footwall of early Jurassic diabase and a hanging wall of the late Triassic Stockton Formation. Locally, the walls may be unreplaced Cambrian carbonate rock. The ore body is surrounded laterally by Cambrian carbonate rock. Contacts between the ore and surrounding

Figure 412. Ariel view of the Grace mine dumps, April 13, 1970. Photograph by Richard Angelo. Bethlehem Steel Corporation Mining Collection of Photographs, Archives Center, National Museum of American History, Smithsonian Institution.

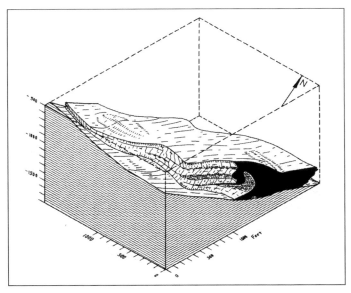

Figure 413. Block diagram of the Grace mine ore body. From Sims (1968). Used with permission from the American Institute of Mining, Metallurgical, and Petroleum Engineers.

rocks typically are sharp. The footwall contact, which was most frequently exposed in the mine, is very sharp and well defined between the chilled margin facies of diabase and the ore. In places, hydrous calcium-magnesium silicate contact minerals separate the ore from diabase. Contacts with the limestone are relatively abrupt, commonly with a transitional zone less than 10 feet thick. Contacts between the ore and the hanging wall, based on drill hole data, also appear to be sharp (Sims, 1968, p. 114). In a few places, the boundary between the deposit and the diabase footwall is a fault with a strike of N. 65° W. and a dip of 20° to 35° NE.

The Grace mine magnetite deposit is a single ore body with no outcrop. It is roughly tabular in shape (fig. 413), strikes about N. 60° W, dips 20° to 30° NE., and plunges about 20° N. 80° E. The ore body is approximately 3,500 feet long by 700 feet to 1,500 feet wide and ranges from about 22 feet to about 425 feet thick (fig. 413). It lies between 600 feet and 2,200 feet below sea level (Bingham, 1957; Sims, 1968). Eben (1996) estimated the ore body contained about 118 million tons of ore.

The ore body swells, pinches, and interfingers into the surrounding limestone host rock. It consists predominantly of magnetite and irregularly distributed zones of hydrous calcium-magnesium silicates, scattered lenses of limestone, and a few veins of milky quartz. The ore differs in its physical properties and is typically moderately friable and easily broken when struck with a hammer. It can, however, range from very crumbly to very hard. The ore is granular, fine to medium grained, and typically consists of unevenly distributed magnetite grains in a matrix of light green to white gangue minerals. In places, magnetite is concentrated along parallel layers, giving the ore a banded aspect. Elsewhere, magnetite is concentrated along fractures that intersect the layers; or it may be evenly distributed in the gangue. On a small scale, the ore texture differs from place to place, but on the large scale of the ore body, it is quite uniform (Sims, 1968, p. 113-114).

The limestone was invaded by mineralizing solutions. Along with the introduction of magnetite and its accessory minerals, the silicate minerals in the limestone were hydrated and converted to serpentine, talc, chlorite, and tremolite, the main gangue minerals. The main accessory minerals are pyrite, chalcopyrite, and, locally, pyrrhotite. Less common are sphalerite, marcasite, galena, hematite, digenite, and goethite. Gangue minerals also include diopside, phlogopite, serpentine, calcite, and dolomite (Sims, 1968).

Serpentine aggregates and evidence of talc replacing tremolite suggest that the first mineral assemblage developed after the diabase intrusion was calcite-dolomite-diopside-forsterite-tremolite-mica. Further hydrothermal or retrograde alteration converted diopside, forsterite, and some tremolite and mica to serpentine, talc, and chlorite. This alteration was probably the beginning of the phase that resulted in ore mineralization. The assemblage that was replaced by ore minerals was serpentine-talc-chlorite-tremolite-calcite-dolomite. Magnetite was the earliest ore mineral to form as indicated by the great predominance of inclusions of gangue over those of sulfides and by the occurrence of sulfides inter-granular to magnetite. Pyrite and pyrrhotite then grew between and around magnetite (including euhedral crystals of magnetite) and locally replaced magnetite. Chalcopyrite followed the last-formed pyrite and pyrrhotite, replacing pyrrhotite margins and filling fractures in pyrite. Sphalerite, galena, and digenite were probably contemporaneous with chalcopyrite. The secondary minerals were formed last. Pyrrhotite first was oxidized to goethite plus marcasite, and marcasite then inverted to pyrite. Where pyrrhotite was oxidized, an aureole of iron staining was produced in the gangue minerals around the pyrrhotite grains. Overgrowths of marcasite and/or pyrite on primary pyrite grains and hematite on magnetite are interpreted as secondary mineralization. Hematite along fractures in magnetite is considered as secondary oxidation of magnetite (Sims, 1968, p. 118-119).

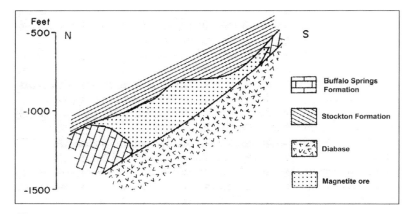

Figure 414. Cross section of the Grace mine ore body. Modified from Tsusue (1964, p. 3). Courtesy of the Pennsylvania Geologic Survey.

Figure 415. Apophyllite on prehnite from the Grace mine, New Morgan Borough, 6 cm. Joseph Varady collection.

Minerals

Actinolite - massive, acicular (Lapham and Geyer 1969, p. 38); common on the dumps

Actinolite, var. byssolite (Lapham and Geyer 1969, p. 38)

Albite (Smith, 1978, p. 232)

Anatase - reported in microcrystals

Andradite - small yellow-orange crystals

Anhydrite - light purple cleavages

Antigorite (Sims, 1968, p. 117); reported as serpentine (Lapham and Geyer 1969, p. 38)

Apatite - accessory mineral in diabase (Sims, 1968, p. 115)

Apophyllite - often associated with zeolites (Lapham and Geyer 1969, p. 38); well-formed, tabular crystals to 2.6 cm (figs. 415, 416, and 418)

Aragonite - reported

Augite - small, dark green crystals (fig. 417)

Aurichalcite - mats of tiny acicular blue-green crystals

Azurite - reported in microcrystals

Barite - reported in microcrystals

Biotite - associated with pyroxene (Sims, 1968, p. 115)

Bornite (?)

Brochantite (?) - reported as green coatings

Brucite - identified on one specimen of limestone by Sims (1968, p. 116)

Brookite (?)

Calcite - in vugs associated with zeolites (Lapham and Geyer 1969, p. 38); fine crystals to 2.1 cm (fig. 419)

Chabazite - reported in microcrystals

Chalcanthite (?) - reported

Chalopyrite - massive; uncommon (Lapham and Geyer 1969, p. 38)

Chlorite group - green crystals associated with muscovite (Lapham and Geyer 1969, p. 38); var. pennine (Sims, 1968, p. 117)

Chrysotile - veins in serpentine; soft, fibrous tufts (fig. 421)

Clinochlore - greenish-gray foliae up to several centimeters across; often distorted, but exhibiting perfect cleavage (Smith, 1978, p. 81); sometimes zoned (fig. 420)

Chrysocolla - reported as coatings

Cobaltite - reported in microcrystals

Covellite - tentatively identified by Sims (1968, p. 117) at an interface between marcasite and pyrrhotite

Datolite - massive; light green, often exhibiting some crystal faces (fig. 422)

Diginite - identified with chalcopyrite in fracture fillings; uncommon (Sims, 1968, p. 117)

Diopside - scattered grains and nodular aggregates in limestone associated with serpentine (Sims, 1968, p. 116)

Dolomite (Tsusue, 1964, p. 5)

Epidote - small green crystals, associated with hematite and andradite; confirmed by SEM-EDS analysis on a specimen from the Sloto collection

Erythrite - purple coateings

Feldspar - commonly pink orthoclase from above the ore zone (Lapham and Geyer 1969, p. 38)

Fluorite - fracture fillings in diabase, limestone, and sedimentary rock (Sims, 1968, p. 117)

Forsterite - in limestone (Sims, 1968, p. 116)

Galena - rare; cleavages and small crystals (Lapham and Geyer 1969, p. 38; Smith, 1976, p. 232); also occurs as fracture fillings in pyrrhotite and pyrite (Sims, 1968, p. 117) cuboctahedra to 1.2 cm (fig. 423)

Figure 416. Apophyllite from the Grace mine, New Morgan Borough, 2.1 cm. Carnegie Museum of Natural History collection CM 31994 (Brookmeyer collection 3394).

Figure 417. Augite crystals from the Grace mine, New Morgan Borough. Sloto collection.

Figure 419. Calcite from the Grace mine, New Morgan Borough. The large crystal is 2.1 cm. Carnegie Museum of Natural History collection CM 30046 (Brookmeyer collection 1081).

MINERALS FROM THE GRACE MINE

Figure 418. Apophyllite crystals from the Grace mine, New Morgan Borough. Crystals are 1 cm. Steve Carter collection.

Figure 420. Zoned clinochlore crystals from the Grace mine, New Morgan Borough, magnified. Specimen provided by Larry Eisenberger.

Figure 421. Crysotile from the Grace mine, New Morgan Borough. Steve Carter collection.

Figure 422. Datolite from the Grace mine, New Morgan Borough, 3 cm. Ron Kendig collection.

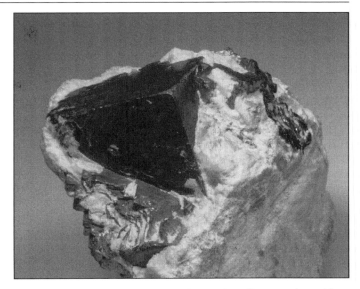

Figure 423. Galena crystal from the Grace mine, New Morgan Borough, 1.2 cm. Carnegie Museum of Natural History collection CM 30044 (Brookmeyer collection 1079).

Figure 424. Heulandite from the Grace mine, New Morgan Borough. Crystals to 7 mm. Joseph Varady collection.

Figure 425. Gypsum from the Grace mine, New Morgan Borough, 5.5 cm. Steve Carter collection.

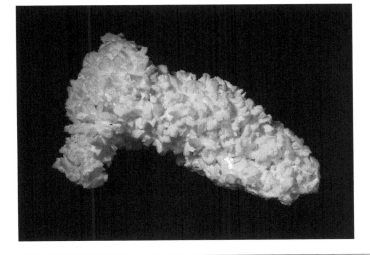

Figure 426. Laumontite from the Grace mine, New Morgan Borough, 6 cm. Carnegie Museum of Natural History collection ANSP 26251.

Figure 427. Magnetite from the Grace mine, New Morgan Borough, 7.2 cm with crystals to 1.5 cm. Carnegie Museum of Natural History collection CM 31998 (Brookmeyer collection 3398).

Garnet group - massive and crystals (Lapham and Geyer 1969, p. 38); andradite and grossular (?)

Goethite - secondary mineral formed from the alteration of pyrrhotite (Sims, 1968, p. 117)

Gypsum - selenite cleavages (Lapham and Geyer 1969, p. 38); joint filling in Triassic sedimentary rock (Sims, 1968, p. 116) (fig. 425)

Hematite - in crystalline plates, which may be replaced by magnetite; also var. specular (Lapham and Geyer 1969, p. 38); small rosettes

Heulandite - light tan microcrystals (fig. 424)

Hornblende - reported

Hypersthene - in diabase (Tsusue, 1964, p. 4)

Ilmenite - accessory mineral in diabase (Sims, 1968, p. 115)

Jarosite - reported

Laumontite - ofter powdery and fragile; snow white crystals to 5 mm (fig. 426)

Malachite - reported

Magnesite - reported

Magnesio-riebeckite - reported

Magnetite - massive and in fine crystals; dodecahedrons to 5 cm coated with white antigorite (Smith, 1978, p. 232); also octahedra (Sims, 1968, p. 116-117); crystals sometimes have stepped faces; single crystals to 3.3 cm (figs. 428-433)

Marcasite - found in fractures in pyrrhotite (Tsusue, 1964, p. 6); secondary mineral formed from the alteration of pyrrhotite (Sims, 1968, p. 117)

Melanterite - reported

Muscovite - var. sericite, in the chilled contact zone (Tsusue, 1964, p. 4)

Natrolite - prismatic needles associated with other zeolites (Lapham and Geyer 1969, p. 38) (fig. 434)

Olivine - in diabase (Tsusue, 1964, p. 4)

Orthoclase - reported

Palygorskite - reported

Plagioclase - a constituent of diabase (Tsusue, 1964, p. 4)

Phlogopite - in veins in diabase associated with serpentine and chlorite (Sims, 1968, p. 115); short prismatic crystals to 5 mm associated with fine-grained serpentine (Tsusue, 1964, p. 6)

Pigeonite - a constituent of diabase (Sims, 1968, p. 115)

Prehnite - massive, fan-like aggregates, light to dark green, crystals to 7 mm (fig. 436)

Pyrite - octahedra, cubes, and pyritohedra; some cubes show penetration twins; contains an average of 0.51 percent cobalt (Sims, 1968, p. 116-117); single crystals to 2.8 cm (figs. 435 and 437)

Pyroxene - a constituent of diabase (Sims, 1968, p. 115)

Pyrolucite - reported

Pyrrhotite - platy crystals with a hexagonal outline; shows polysynthetic twinning (Sims, 1968, p. 117); uncommon; excellent, well-formed, single crystals to 1.2 cm (figs. 438 and 439)

Quartz - generally massive (Lapham and Geyer 1969, p. 38); blue quartz, massive and small crystals

Rutile - reported in microcrystals

Figure 428. Magnetite crystals with striated faces from the Grace mine, New Morgan Borough. Steve Carter collection.

Figure 429. Magnetite crystals coated with tremolite from the Grace mine, New Morgan Borough, 3.3 cm. Carnegie Museum of Natural History collection CM 30063 (Brookmeyer collection 1099).

Figure 430. Magnetite crystals from the Grace mine, New Morgan Borough. Large crystal is 2 cm. Sloto collection 1243.

Figure 431. Magnetite crystal group from the Grace mine, New Morgan Borough. Crystals to 1.5 cm. Steve Carter collection.

Figure 432. Magnetite crystals from the Grace mine, New Morgan Borough. Crystals to 9 mm. Carnegie Museum of Natural History collection CM 30068 (Brookmeyer collection 1105).

Figure 433. Magnetite from the Grace mine, New Morgan Borough, 3.5 cm. Steve Carter collection.

Figure 434. Natrolite on prehnite from the Grace mine, New Morgan Borough, 10 cm. Carnegie Museum of Natural History collection CM 30049 (Brookmeyer collection 1084).

Figure 435. Cubic pyrite from the Grace mine, New Morgan Borough. Crystals to 6 mm. Carnegie Museum of Natural History collection CM 30033 (Brookmeyer collection 1068).

Figure 436. Prehnite from the Grace mine, New Morgan Borough. Joseph Varady collection.

Figure 437. Octahedral pyrite from the Grace mine, New Morgan Borough, 2.8 cm. Carnegie Museum of Natural History collection CM 30067 (Brookmeyer collection 1103).

Figure 438. Pyrrhotite from the Grace mine, New Morgan Borough. Crystals to 1.2 cm. Steve Carter collection.

Figure 439. Pyrrhotite from the Grace mine, New Morgan Borough. Crystals to 1.2 cm. Carnegie Museum of Natural History collection ANSP 5178.

Figure 440. Talc pseudomorphs after magnetite from the Grace mine, New Morgan Borough. Crystals to 8 mm. Steve Carter collection.

Figure 441. Talc from the Grace mine, New Morgan Borough, 8 cm. Sloto collection 130.

Figure 442. Talc pseudomorph after magnetite from the Grace mine, New Morgan Borough, 1.5 cm. Steve Carter collection.

Figure 443. Tochilinite in calcite from the Grace mine, New Morgan Borough. Carnegie Museum of Natural History collection CM 30081 (Brookmeyer collection 1120).

Figure 444. Tochilinite from the Grace mine, New Morgan Borough, 5 mm. Carnegie Museum of Natural History collection CM 30032 (Brookmeyer collection 1067).

Riebeckite (?)

Siderite (?)

Sphalerite - small crystals; very rare (Lapham and Geyer 1969, p. 38): small masses included in chalcopyrite (Tsusue, 1964, p. 6); occurs as fracture fillings in pyrrhotite and pyrite, uncommon (Sims, 1968, p. 117)

Serpentine - possibly lizardite and chrysotile (Tsusue, 1964, p. 5)

Stilbite - radial clusters (Lapham and Geyer 1969, p. 38)

Stilpnomelane - small, dark, brassy flakes

Talc - occurs in 2 to 3 cm wide calcite veins (Smith, 1976, p. 88 and 258); massive light green (fig. 441); octahedral, generally light green pseudomorphs after magnetite (?) (figs. 420 and 422); confirmed as talc by SEM-EDS analysis on a specimen from the Sloto collection

Titanite - a constituent of the pegmatite facies of diabase (Sims, 1968, p. 115); blue prismatic microcrystals; confirmed by SEM-EDS analysis on a specimen provided by Larry Eisenberger

Tochilinite - black to bronzy, paper thin, tabular crystals to 1 cm; massive aggregates; crystals are deeply striated parallel to the long axis and may radiate or be twisted (Smith, 1976, p. 256); first found by Bryron Brookmeyer; associated with clear and white calcite, some of which form euhedral crystals; black metallic masses of tochilinite to 5 by 7 mm in calcite. Some coats calcite crystals and can be scraped off in minute sheets resembling black foil paper. Some of the coatings have a distinctly bronze cast; also occurs as fibrous, partly radiating aggregates of elongate, paper-thin grains, some up to 2 mm long striated parallel to the elongation; confirmed by microprobe and X-ray diffraction analysis by Jambor (1976, p. 65-69); crystal sprays to 5 mm (figs 443 and 444)

Tourmaline group - in Triassic-age sedimentary rock (Sims, 1968, p. 116); probably schorl and/or dravite

Tremolite - prismatic; less common than actinolite (Lapham and Geyer 1969, p.38)

Vesuvianite - reported in microcrystals

Zircon - reported in microcrystals

A VERY BRIEF HISTORY OF THE BETHLEHEM STEEL CORPORATION

The Bethlehem Steel Corporation, based in Bethlehem, Pennsylvania, was once the second-largest steel producer in the United States and was second only to the Pittsburgh-based U.S. Steel Corporation. Bethlehem Steel also was one of the largest shipbuilding companies in the world.

The roots of the Bethlehem Steel Corporation extend to 1857, when the Saucona Iron Company was organized by Augustus Wolle. After encountering difficulty organizing and financing, the company moved to South Bethlehem and changed its name to the Bethlehem Rolling Mill and Iron Company. On June 14, 1860, the board of directors elected Alfred Hunt president. On May 1, 1861, the company's name was changed to the Bethlehem Iron Company. Construction of the first blast furnace began on July 1, 1861, and it went into operation on January 4, 1863. The first rolling mill was built between the spring of 1861 and the summer of 1863, with the first railroad rails rolled on September 26, 1863. A machine shop was completed in 1865, and another blast furnace was constructed in 1867.

Although the company continued to prosper during the early 1880s, its share of the rail market began to decline in the face of competition from growing Pittsburgh-based firms, such as the Carnegie Steel Company. The Nation's decision to rebuild the U.S. Navy with steam-driven, steel-hulled warships helped to reshaped the Bethlehem Iron Company. The company began to produce armor plating for the U.S. Navy, and, in 1887, it produced guns for the first American battleships.

In 1899, the company assumed the name Bethlehem Steel Company. In 1904, Charles M. Schwab, formerly with U.S. Steel, and Joseph Wharton formed the Bethlehem Steel Corporation, with Schwab becoming its first president and chairman of its board of directors. In addition to armor, weaponry, and rails, Bethlehem Steel also supplied machinery for other steel companies and forged parts for the Niagara Falls hydroelectric generating plant. The corporation installed the revolutionary grey rolling mill (named for Henry Grey) and produced the first wide-flange structural shapes to be made in America. These shapes were used to build skyscrapers and established Bethlehem Steel as the construction industry's leading supplier of steel. Notable landmarks constructed with Bethlehem steel include Rockerfeller Center, Madison Square

Figure 445. Stock certificate of the Bethlehem Steel Corporation, 1953.

Garden, the Golden Gate and George Washington Bridges, Alcatraz, and the Hoover Dam.

In the early 1900s, the corporation branched out from steel, with iron mines in Cuba and shipyards around the country. In 1913, it acquired the Fore River Shipbuilding Company of Quincy, Massachusetts, thereby becoming one of the world's major shipbuilders. In 1917, it incorporated its shipbuilding division as the Bethlehem Shipbuilding Corporation, Limited. In 1922, it purchased the Lackawanna Steel Company, which included the Delaware, Lackawanna, and Western Railroad, as well as extensive coal holdings.

During World War II, as much as 70 percent of airplane cylinder forgings, one-quarter of the armor plate for warships, and one-third of the big cannon forgings for the U.S armed forces were manufactured by Bethlehem Steel. In 1943, Eugene Grace promised President Roosevelt one ship per day, and exceeded the commitment by 15 ships. Bethlehem Shipbuilding Corporation's 15 shipyards produced a total of 1,121 ships, more than any other builder during the war and nearly one-fifth of the U.S. Navy's fleet.

The steel industry in the U.S. prospered during and after World War II, while the steel industries in Germany and Japan lay devastated by Allied bombardment. Bethlehem Steel's high point came in the 1950s, when the company produced 23 million tons of steel per year. The U.S. advantage lasted about two decades, during which the U.S. steel industry operated with little foreign competition. Eventually, the foreign firms were rebuilt with modern techniques, such as continuous casting, while profitable U.S. companies resisted modernization. Meanwhile, U.S. steelworkers were given rising benefits. By the 1970s, imported foreign steel was generally less expensive than domestically produced steel.

In 1982, Bethlehem reported a loss of $1.5 billion and shut down many of its operations. Profitability returned briefly in 1988, but restructuring and shutdowns continued through the 1990s. In the mid-1980s, demand for the plant's structural products began to diminish, and new competition entered the marketplace. Lighter construction styles did not require the heavy structural grades produced at the Bethlehem plant.

In 1991, Bethlehem Steel discontinued coal mining. At the end of 1995, after roughly 140 years of metal production at its Bethlehem plant, Bethlehem Steel ceased operations. Bethlehem Steel ended the railroad car business in 1993 and ceased shipbuilding in 1997 in an attempt to preserve its steel-making operations. In 2001, Bethlehem Steel filed for bankruptcy. In 2003, the company's remnants, including its six massive plants, were acquired by the International Steel Group. In 2005, the International Steel Group merged with Mittal Steel (now ArcelorMittal), ending U.S. ownership of the assets of Bethlehem Steel.

In 2007, the Bethlehem property was sold to Sands BethWorks, and plans to build a casino where the plant once stood were drafted. Construction began in the fall of 2007, and the casino was completed in 2009. Ironically, the casino had difficulty finding structural steel for construction because of a global steel shortage.

NORTH HEIDELBERG TOWNSHIP

Tulpehocken Stone Company Quarry

The Tulpehocken Stone Company quarry was west of the intersection of North Heidelberg Road and Overlook Lane at about 40° 22′ 49″ latitude and 76° 07′ 54″ longitude on the USGS Strausstown 7.5-minute topographic quadrangle map (Appendix 1, fig. 22, location NH-1). The quarry was in the Hamburg Sequence. The Tulpehocken Stone Company, Inc. of Robesonia operated a quarry on the property of J. Paul Hertzog of Wolmelsdorf. John C. Showalter was the company president. The company declared bankruptcy in 1978 (*Reading Eagle*, June 23, 1977, and April 27, 1978).

OLEY TOWNSHIP

The Oley Forge was built in 1740 along the Manatawny Creek at what is now Spangsville. The first iron furnace in Oley Township, the Shearwell Furnace, was built by Diedrich Welker on land granted to him in 1744. The Oley Furnace, located along Furnace Creek, was built in 1772. Both furnaces were operated as late as 1783.

LIMESTONE QUARRIES

Eastern Industries Quarry

The Eastern Industries quarry (fig. 447), also known as the Oley quarry, closed in 2015. The quarry was north of Bieber Mill Road at about 40° 21′ 18″ latitude and 75° 48′ 05″ longitude on the USGS Birdsboro 7.5-minute topographic quadrangle map (Appendix 1, fig. 30, location O-1). The quarry was in the Annville Formation. Several older farm quarries were located in the area quarried by Eastern Industries. The Eastern Industries quarry is sometime called the Oley Valley quarry; however, there was an Oley Valley quarry in a different location that predated the Eastern Industries quarry by several decades.

In 1941, the Eastern Lime Corporation was incorporated in Delaware. The company was primarily a producer of chemical grade limestone for cement companies, crushed stone for ready-mix concrete, highway construction aggregate, and agricultural limestone. Its principle operation was the Hinterleiter quarry in Maxatawny Township, and its office was located in Kutztown.

Figure 446. Skip Colflesh (right) and Scott Snavely (left) collecting minerals in the Eastern Industries Oley quarry, Oley Township.

Figure 447. Ariel view of the Eastern Industries Oley quarry, Oley Township, April 2016.

In 1951, the Eastern Lime Company leased the 172-acre Harry Renninger farm and began a core drilling program. The goal of core drilling was to determine if there was sufficient limestone that contained a minimum of 92 percent calcium carbonate necessary for cement manufacture (*Reading Eagle*, February 19, 1952), The core drilling program was successful, and in 1954, Eastern Lime acquired 464 acres of land. The corporation purchased the farms of Harry and Sallie Renninger (177 acres for $45,000), Henry Eyrich (184 acres for $32,000), and David Lutz (20 acres for $6,000 and 83 acres for $24,000) (*Reading Eagle*, April 22, 1954). The Eastern Lime Corporation quarry opened in 1957 (*Reading Eagle*, October 6, 1957). In 1965, the company's name was changed to Eastern Industries, Inc.

The quarry was developed along an anticline trending about N. 60° W. and plunging 11° NW. Sphalerite-fluorite-barite mineralization occured in interbedded limestone and dolomite that had been brecciated. Fluorite and other minerals were deposited in open cavities within the matrix of the breccia. The paragenesis is: golden sphalerite, black sphalerite, barite, pyrite, marcasite, fluorite, chalcopyrite, and calcite with considerable overlap after the barite (Smith, 1978, p. 124).

Minerals

From Smith (1977a, p. 112-114) and Smith (1978, p. 120-124) plus personal observations.

Anorthite - euhedral phenocrysts in a diabase dike at the west end of the quarry

Aurichalcite - blue-green tufts associated with sphalerite; confirmed by SEM-EDS analysis on a specimen from the Sloto collection (fig. 451)

Figure 448. Sphalerite from the Eastern Industries quarry, Oley Township. Crystals to 7 mm. Steve Carter collection.

Barite - milky-white tabular prisms to 2 mm with etched faces that taper to a point at both ends; clusters of small, dark yellow crystals (fig. 450)

Calcite - transparent, colorless scalenohedrons with rounded faces to 5 cm; some pyrite and fluorite are included on or in the outer zones of calcite crystals (figs. 4489, 452, 454, and 458)

Celestine - reported

Chalcopyrite - small crystals on calcite

Fluorite - moderately common; 0.5 to 3 mm inky-blue to purple cubes (fig. 455) with irregular pyramid-shaped capping on crystal faces. The color is irregular with the outer zones containing most of the purple coloration. Gamma-ray spectrometry by Smith (1978, p. 120-121) showed that randomly distributed, intensely colored, roughly spherical dots (fig. 456) in the outer zones were caused by minute, uranium-bearing nuclei.

Hemimorphite - fan-like aggregates of small tan crystals (fig. 457)

Malachite - weathering product associated with chalcopyrite

Marcasite - thin, twinned, tin-white, striated, metallic plates (1 mm) with serrated, cockscomb edges in calcite and rectangular prisms on calcite

Palygorskite, var. mountain leather - white mats on limestone (fig. 453)

Pyrite - tiny crystals on calcite (fig. 449)

Quartz - clusters of crystals; sometimes as an overgrowth on calcite with amethyst

Figure 449. Calcite and pyrite from the Eastern Industries quarry, Oley Township, 8 cm. Skip Colflesh collection.

MINERALS FROM THE EASTERN INDUSTRIES OLEY QUARY

Figure 450. Barite from the Eastern Industries quarry, Oley Township. Steve Carter collection.

Figure 451. Aurichalcite from the Eastern Industries quarry, Oley Township, magnified. Confirmed by SEM-EDS analysis. Sloto collection.

Figure 452. Calcite from the Eastern Industries quarry, Oley Township, 5 cm. Sloto collection 3310.

Figure 453. Palygorskite from the Eastern Industries quarry, Oley Township, 6 cm. Sloto collection 3633.

Figure 454. Calcite from the Eastern Industries quarry, Oley Township, 5 cm. Sloto collection 3225.

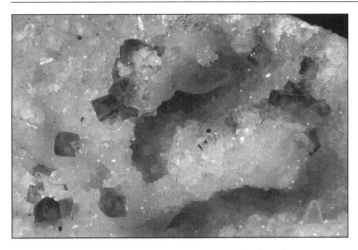

Figure 455. Fluorite from the Eastern Industries quarry, Oley Township. Crystals are about 1 mm. Bryn Mawr College collection Quickel 452.

Figure 456. Fluorite with uranium-bearing round spots from the Eastern Industries quarry, Oley Township. Sloto collection 3311.

Figure 457. Hemimorphite from the Eastern Industries quarry, Oley Township. Crystals are about 1 mm. Carnegie Museum of Natural History collection CM 30714 (Brookmeyer 1963).

Figure 458. Quartz, var. amethyst, on calcite from the Eastern Industries quarry, Oley Township, 9 cm. Skip Colflesh collection.

Figure 459. Sphalerite from the Eastern Industries quarry, Oley Township. Steve Carter collection.

Quartz, var. amethyst - light purple crystals and as an overgrowth on calcite crystals; some are doubly terminated (fig. 458)

Sphalerite - crude terahedra to 2 cm with triangular growth markings on some faces; thin (~0.2 mm), black overgrowths on golden-brown crystals (figs. 448 and 459)

Lehigh Cement Company Quarries

The three abandoned and flooded Lehigh Cement Company quarries (fig. 461) are on the USGS Birdsboro 7.5-minute topographic quadrangle map (Appendix 1, fig. 30, locations O-2A, O-2B, and O2-C). The 201-acre Oley West quarry (O-2-A) is at 40° 21′ 38″ latitude and 75° 46′ 21″ longitude on the west side of Pennsylvania State Route 662, north of Oley Turnpike Road. The 64-acre Oley No. 1 quarry (O-2C) is at 40° 20′ 55″ latitude and 75° 46′ 31″ longitude on the south side of Oley Turnpike Road, west of Pennsylvania State Route 662. The 136.7-acre Oley No. 2 quarry (O-2-B) (fig. 460) is at 40° 21′ 34″ latitude and 75° 45′ 41″ longitude on the east side of Pennsylvania State Route 662, north of Oley Turnpike Road. It is the site of the former L. DeTurk quarry (described below), which was located at 40° 21′ 38″ latitude and 75° 46′ 06″ longitude.

Figure 460. Lehigh Cement Company Oley No. 2 quarry, Oley Township, April 2015.

Exploration for a new source of stone began after the Allentown Portland Cement Company quarry at Kirbyville ran out of high calcium limestone. The company purchased 297 acres that included the farms of Harry and Sallie Renninger (131 acres for $78,000), C. Clay Brown (107 acres for $66,294), and John and Bertha Wetzel (59 acres for $59,000). An extensive core drilling program was carried out before construction of the crushing plant. The primary crusher was built by the Birdsboro Foundry and Machine Company (*Reading Eagle*, April 22, 1954).

In 1954, the Oley No. 1 quarry was about 75 feet deep. Stone was trucked from Oley Township to the Allentown Portland Cement Company plants in Evansville (see Maidencreek Township) and West Conshohocken. The quarry employed 23 people (*Reading Eagle*, April 22, 1954). In June 1960, the Allentown Portland Cement Company was acquired by the National Gypsum Company. About 2004, the quarries were acquired by the Lehigh Cement Company LLC (*Reading Eagle*, April 12, and May

Figure 461. Ariel view of the three Lehigh Cement Company quarries, Oley Township, April 2016. 1, Oley West quarry; 2, Oley No. 2 quarry; 3, Oley No. 1 quarry.

Figure 462. Stock certificate of the Lehigh Portland Cement Company, 1914.

3, 2004). Portland-Zementwerke Heidelberg A.G., a unit of the German building-materials company Heidelberger Zement A.G., purchased Lehigh Cement in 1977 for $85 million.

L. DeTurk's Quarries

L. DeTurk's quarries (fig. 466) were on the east side of Pennsylvania State Route 662, north of Oley Turnpike Road at about 40° 21′ 38″ latitude and 75° 46′ 06″ longitude on the USGS Birdsboro 7.5-minute topographic quadrangle map (Appendix 1, fig. 30, location O-2B). The property was acquired by the Allentown Portland Cement Company, and the quarries were incorporated into the Oley No. 2 quarry.

There were two quarries on L. DeTurk's farm. The northernmost quarry was situated on the left bank of small tributary to Manatawny Creek about two hundred yards north of DeTurk's house. The quarry exposed about 12 feet of brecciated limestone dipping S. 60 W. 30°. The southernmost quarry was located on a private road 325 feet from Pennsylvania State Route 662. D'Invilliers reported that, at the time of his visit, the quarry was "*considerably developed, though idle.*" There was about 20 feet of blue and white banded magnesian limestone exposed that dipped S. 18° W. 55° (D'Invilliers, 1883, p. 171).

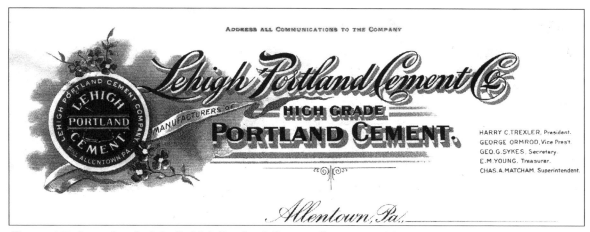

Figure 463. Letterhead of the Lehigh Portland Cement Company, 1901

Figure 464. Lime kiln on Bertolet Mill Road, Oley Township, April 2015

Houck Quarry

The S. Houck quarry (fig. 208) was on the west side of Pennsylvania State Route 662, south of the intersection of Pennsylvania State Route 662 and Mine Lane at about 40° 22′ 12″ latitude and 75° 46′ 34″ longitude on the USGS Birdsboro 7.5-minute topographic quadrangle map (Appendix 1, fig. 30, location O-3). The quarry, in the Beekmantown Group, exposed 10 feet of blue and white limestone, which dipped S. 60° W. 38°. An analysis of the limestone was published by D'Invilliers (1883, p. 171): 88.9 percent calcium carbonate, 3.8 percent magnesium carbonate, and 6.8 percent silica.

Peter Guldin Quarry

The Peter Guldin quarry (fig. 208) was east of Pennsylvania State Route 662, and southeast of the intersection of Pennsylvania State Route 662 and Mine Lane at about 40° 22′ 19″ latitude and 75° 46′ 24″ longitude on the USGS Birdsboro 7.5-minute topographic quadrangle map (Appendix 1, fig. 30, location O-4). The small quarry, in the Beekmantown Group, exposed 15 feet of limestone, which dipped S. 60° W. 40° (D'Invilliers, 1883, p. 172).

Levi Hartman Quarry

The Levi Hartman quarry was on the east side of Pennsylvania State Route 662, and southeast of the intersection of Pennsylvania State Route 662 and Oley Turnpike at 40° 21′ 08″ latitude and 75° 45′ 53″ longitude on the USGS Birdsboro 7.5-minute topographic quadrangle map (Appendix 1, fig. 30, location O-5). The quarry was in the Beekmantown Group. D'Invilliers (1883, p. 172) described the quarry: *"Half mile west of Griesermersville, and just south of the Oley pike, a new opening has been made by Levi Hartman, exposing about eighteen feet of good blue and white stone, dip N. 10 W. 64°."*

S.P. Guldin Quarry

The S.P. Guldin quarry was west of Pennsylvania State Route 662, northwest of the intersection of Pennsylvania State Route 662 and Covered Bridge Road at 40° 19′ 53″ latitude and 75° 45′ 30″ longitude on the USGS Birdsboro 7.5-minute topographic quadrangle map (Appendix 1, fig. 30, location O-6). The small quarry, in the Beekmantown Group on the S.P. Guldin estate, exposed 20 feet of *"first-rate"* blue quartzose limestone. An analysis was published by D'Invilliers (1883, p. 172): 90.4 percent calcium carbonate, 2.8 percent magnesium carbonate, and 6.6 percent silica.

The quarry was locality number 14 of Gray (1951, p. 39). The quarry exposed massive, gray dolomite with some interbedded limestone. The limestone was distinctly banded and contained chert lenses 1 to 2 inches thick and several feet long. Scattered sand grains were also present in the limestone. The

Figure 465. Lime kiln near the E. Schaeffer quarry, Oley Township, April 2015.

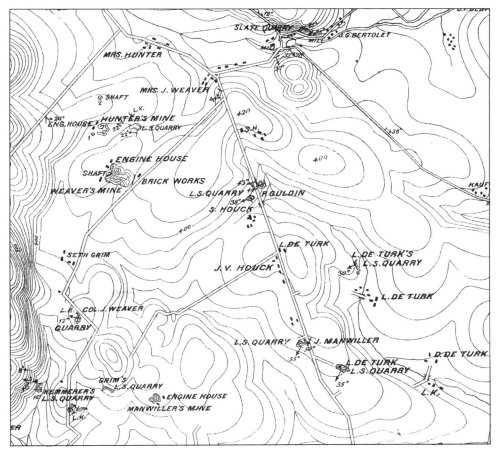

Figure 466. Quarries and iron mines in central Oley Township, 1882. From Pennsylvania Geological Survey (1883).

strike of the beds was N. 30° E., and the dip was 30° to 35° NW. Gray (1951, p. 39) presented analyses of stone from the quarry.

E. Schaeffer Quarry

The E. Schaeffer quarry was west of Covered Bridge Road, north of its intersection with Pennsylvania State Route 662 and Covered Bridge Road at 40° 20′ 02″ latitude and 75° 45′ 20″ longitude on the USGS Birdsboro 7.5-minute topographic quadrangle map (Appendix 1, fig. 30, location O-7). The quarry, in Beekmantown Group, exposed about 30 feet of blue limestone dipping N. 78-88° W. 35-37° capped with 3 feet of slate (D'Invilliers, 1883, p. 172-173).

Colonel J. Weaver's Quarry

Colonel J. Weaver's quarry (figs. 466 and 467) was on the west side of Mine Lane at about 40° 21′ 40″ latitude and 75° 47′ 13″ longitude on the USGS Birdsboro 7.5-minute topographic quadrangle map (Appendix 1, fig. 30, location O-8). The quarry, in the Beekmantown Group, exposed about 10 feet of massive blue dolomite dipping N. 65° W. 12°. D'Invilliers (1883, p. 173) published an analysis: 71.2 percent calcium carbonate, 23.3 percent magnesium carbonate, and 4.8 percent silica.

The quarry was locality number 24 of Gray (1951, p. 48). The quarry exposed thick-bedded limestone with dolomite interbeds. The limestone was dark blue, very fine grained and faintly laminated. The uppermost bed in the quarry was weathered to friable, shaley limestone. The strike of the beds was N. 72° W., and the dip was 10° S. Gray (1951, p. 48) presented analyses of the limestone and dolomite.

Seth Grim Quarry

The Seth Grim quarry (fig. 466) was on the east side of Mine Lane at 40° 21′ 31″ latitude and 75° 47′ 06″ longitude on the USGS Birdsboro 7.5-minute topographic quadrangle map (Appendix 1, fig. 30, location O-9). The quarry, in the Beekmantown Group, was 500 yards south of Colonel J. Weaver's quarry. The small quarry exposed 15 to 20 feet of blue and white limestone. D'Invilliers published an analysis: 96.7 percent calcium carbonate, 2.1 percent magnesium carbonate, and 1 percent silica. D'Invilliers (1883,

Figure 467. Colonel J. Weaver's quarry, Oley Township, April 2015.

township." The Reading Eagle (May 26 and June 16, 1896) reported that William K. Grim was operating an extensive stone quarry on his farm at Grim's Mill and was producing cellar stones.

The quarry was locality number 25 of Gray (1951, p. 49). The quarry exposed laminated blue limestone with interbedded gray dolomite. The strike of the beds was N. 10° E., and the dip was 5° NW. Gray (1951, p. 49) presented analyses of rock from the quarry.

Kemmerer Quarries

D. Kemmerer's quarries (fig. 466) were on the east side of Mine Lane. The western quarry (location O-10A) was at 40° 21′ 20″ latitude and 75° 47′ 20″ longitude, the northern quarry (location O-10C) was at 40° 21′ 21″ latitude and 75° 47′ 14″ longitude, and the eastern quarry (location O-10B) was at 40° 21′ 17″ latitude and 75° 47′ 14″ longitude on the USGS Birdsboro 7.5-minute topographic quadrangle map (Appendix 1, fig. 30). The quarries were in the Beekmantown Group.

The western quarry, on the west side of Mine Lane at a lime kiln close to Slate Hill, exposed about 16 feet of compact blue limestone that dipped into the hill S. 70° W. 10°. The limestone was conformably overlain by about 4 feet of slate. An analysis of the limestone was published by D'Invilliers (1883, p. 174): 95.5 percent calcium carbonate, 2.7 percent magnesium carbonate, and 1.9 percent silica.

D'Invilliers (1883, p. 174) reported: *"There is a unique wood-burning kiln here upon which Mr. Kemmerer spent considerable thought and money with an idea of improving draught. It is made of brick about 25' high, cylindrical in shape, banded on top with an iron ring. Spliced into the brick walls with iron L-shaped guides, there is a sheet-iron chimney 3' 6" high, also cylindrical, and with the same circumference as the kiln. The upper part of this chimney—also about 3' 6" high — is cone-shaped, its sides sloping on top to a width of about 3'. From this rises a 25' stack of sheet-iron, 3 feet in diameter, devised to increase the draught. The lower part of chimney is provided with 4 curved doors and which open out and rest on the L-shaped bars through which the charge is made. Neither kiln or quarry were being worked at time of visit, so that no practical results can be reported."*

The small northern quarry exposed the same strata as the western quarry. The limestone dipped into the hill S. 72° W. 8°. The larger quarry to the south was at the base of a hill near the creek where the stone was harder and more compact. There, the limestone dipped S. 48° W. 16° (D'Invilliers, 1883, p. 174). The eastern quarry exposed a light blue, rather silicious limestone that was 15 feet thick with 6 feet of top soil above it. There, the limestone dipped N. 10° W. 18°. At the time of D'Invilliers visit in 1882, the quarry was abandoned and overgrown (D'Invilliers, 1883, p. 174-175).

Figure 468. Lime kiln near the Levi Knabb Quarry, Oley Township, April 2015.

P. Sneider Quarry

The P. Sneider quarry (fig. 466) was on the east side of Bieber Mill Road at approximately 40° 21′ 04″ latitude and 75° 47′ 44″ longitude on the USGS Birdsboro 7.5-minute topographic quadrangle map (Appendix 1, fig. 30, location O-11). The quarry, in the Annville Formation, exposed 30 feet of hard blue limestone dipping N. 65° W. 30° at the west end of quarry and overlaid by slate. The east end of the quarry exposed a slaty limestone that dipped S. 45° W. 48° (D'Invilliers, 1883, p. 175). The quarry was assimilated into the Eastern Industries quarry.

Hine Quarry

The D. Hine quarry was northeast of the intersection of Oley Turnpike and Mine Lane at 40° 21′ 02″ latitude and 75° 47′ 20″ longitude on the USGS Birdsboro 7.5-minute topographic quadrangle map (Appendix 1, fig. 30, location O-12). The quarry was in the Beekmantown Group. The name of the quarry owner was spelled Hine by D'Invilliers (1883, p. 175), but was spelled Nine on the 1882 topographic map (fig. 466). The D. Hine quarry exposed a blue and white limestone that dipped S. 20° E. 33°. D'Invilliers (1883, p. 175) noted: *"This quarry is only worked for private use on the farm but the stone is massive and pure."*

Levi Knabb Quarry

The Levi Knabb quarry (fig. 466) was southeast of the intersection of Reifsnider and Hunter Roads at 40° 20′ 37″ latitude and 75° 47′ 21″ longitude on the USGS Birdsboro 7.5-minute topographic quadrangle map (Appendix 1, fig. 30, location O-13). The small quarry, in the Annville Formation, exposed soft, blue limestone dipping S. 32° W. 30-38° (D'Invilliers, 1883, p. 176).

Raudenbusch Quarry

The Raudenbusch quarry was was southeast of the intersection of Reifsnider and Hunter Roads at about 40° 20′ 33″ latitude and 75° 47′ 12″ longitude n the USGS Birdsboro 7.5-minute topographic quadrangle map (Appendix 1, fig. 30, location O-14). The quarry, in the Annville Formation, exposed 25 feet of a magnesian, generally white to dove color limestone dipping 25-30° S. D'Invilliers (1883, p. 176) noted that the limestone was *"greatly broken up and seamed with quartz, making it siliceous and of poor quality."* D'Invilliers published an analysis: 51.3 percent calcium carbonate, 32.7 percent magnesium carbonate, and 15.2 percent silica.

Ezra Griesermer Quarry

The Ezra Griesermer quarry was north of Hunter Road, on the Lehigh Cement Company property at 40° 20′ 51″ latitude and 75° 46′ 40″ longitude on the USGS Birdsboro 7.5-minute topographic quadrangle map (Appendix 1, fig. 30, location O-15). The quarry, in the Beekmantown Group, exposed 22 feet of an *"excellent"* blue and white compact limestone dipping S. 60° W. 12°. D'Invilliers (1883, p. 176) noted that the limestone broke into *"large slabs good for foundation walls."* D'Invilliers published an analysis of the limestone: 92.1 percent calcium carbonate, 1.8 percent magnesium carbonate, and 5.8 percent silica. The 1876 atlas listed Ezra Z. Griesemer, 112 acre farm with a fine limestone quarry. The Mine and Quarry News Bureau (1897) listed the Ezra Greisemer quarry as a lime producer in 1897.

J.G. Fischer Quarry

The John G. Fischer quarry was north of the intersection of Hunter and Blacksmith Roads at 40° 20′ 30″ latitude and 75° 46′ 16″ longitude n the USGS Birdsboro 7.5-minute topographic quadrangle map (Appendix 1, fig. 30, location O-16). The quarry was in the Beekmantown Group. D'Invilliers (1883, p. 176-177) stated: *"J. G. Fischer has opened three small quarries, all showing excellent soft blue stone, burnt on the ground at his own kiln. The dips are all towards the hill as follows: S. 63° W. 20°, fifteen feet thick, in east quarry; S. 75° W. 25°, thirty feet thick, in middle quarry with ten feet of slaty top covering; S. 65 N. 24°, twenty five feet thick in west quarry, six feet of slate top."* D'Invilliers (1883, p. 177) published an analysis of a composite taken from all three quarries, which were on the same outcrop: 95.6 percent calcium carbonate, 1.5 percent magnesium carbonate, and 3 percent silica.

Levi Herbein Quarry

The Levi Herbein quarry was west of the intersection of Hunter and Blacksmith Roads at 40° 20′ 19″ latitude and 75° 46′ 22″ longitude on the USGS Birdsboro 7.5-minute topographic quadrangle map (Appendix 1, fig. 30, location O-17). The quarry, in the Martinsburg Formation at the contact with the Beekmantown Group, was on the south side of Slate Hill. D'Invilliers (1883, p. 178) noted at the time of his visit: *"It was abandoned and partly closed up, but is said to have lead into a cave in its west end, where some fine stalactites were found. The cave was explored for 125 feet —all in good blue limestone. The general dip was S. 50 W. 33°.* "D'Invilliers published an analysis of the limestone: 90.2 percent calcium carbonate, 6.4 percent magnesium carbonate, and 3.5 percent silica.

L.J. Bertolet Quarry

The L.J. Bertolet quarry was north of Limekiln Road at 40° 20′ 00″ latitude and 75° 47′ 20″ longitude on the USGS Birdsboro 7.5-minute topographic quadrangle map (Appendix 1, fig. 30, location O-18). The small quarry, in the Beekmantown Group, exposed about 10 feet of compact limestone dipping S. 30° E. 40°. The quarry was formerly owned by Samuel Marquart. D'Invilliers (1883, p. 178) published an analysis of the limestone: 81.3 percent calcium carbonate, 10.4 percent magnesium carbonate, and 7.8 percent silica.

John Snyder Quarry

The John Snyder quarry (fig. 208) was northeast of the intersection of Limekiln and Reifsneider Roads at 40° 20′ 16″ latitude and 75° 48′ 03″ longitude on the USGS Birdsboro 7.5-minute topographic quadrangle map (Appendix 1, fig. 30, location O-20). The small quarry, in the Beekmantown Group, exposed 25 feet of a hard, blue, silicious limestone dipping S. 36° W. 55° (D'Invilliers, 1883, p. 178).

Isaac Brumbach Quarry

The Isaac Brumbach quarry was south of Bortz Road at 40° 24′ 14″ latitude and 75° 45′ 03″ longitude on the USGS Fleetwood 7.5-minute topographic quadrangle map (Appendix 1, fig. 14, location O-21). The quarry was in the Beekmantown Group. D'Invilliers (1883, p. 166) described the quarry as *"showing dips of S. 10° W. 60° near the road and S. 25° W. 64° further in the quarry. These dips however are only approximations, as the limestone is greatly jointed by S. E. cleavage planes. The quarry has been long abandoned and is grass grown. The limestone exposed measures about 20 feet, blue and gray, and apparently of good quality as far as could be judged. No analysis was taken at this quarry, nor at the small quarry or pit 200 yards further north-west along south side of road, owned by the same party."*

David Yoder Quarry

The David Yoder quarry was north of the intersection of Yoder and Oysterdale Roads at 40° 23′ 18″ latitude and 75° 43′ 52″ longitude on the USGS Manatawny 7.5-minute topographic quadrangle map (Appendix 1, fig. 17, location O-22). The 1876 atlas listed Franklin Yoder as a *"dealer in all kinds of building and lime stones, for furnaces and other purposes."* The quarry was in the Allentown Formation.

D'Invilliers (1883, p. 164) described the quarry: *"David Yoder's quarry about 500 yards north east of Pleasantville, which shows about 30 feet of excellent blue and white limestone, with some slightly silicious beds, but mostly a soft blue stone carrying calcite. The quarry has been considerably developed, and as it has little or no top soil it can be easily worked. The dips are all to the west as follows: In the north end S. 60° W. 70°; in the south end N. 67° W. 63°; west end N. 60° W. 40°, showing therefore some little irregularity of bedding."* D'Invilliers (1883, p. 165) published an analysis of a composite sample of stone taken from different parts of the quarry: 92 percent calcium carbonate, 2.3 percent magnesium carbonate, and 5.9 percent silica.

Reuben Shearer Quarry

The Reuben Shearer quarry was southeast of the intersection of Snyder Road and Pennsylvania State Route 73 at 40° 22′ 52″ latitude and 75° 44′ 41″ longitude on the USGS Manatawny 7.5-minute topographic quadrangle map (Appendix 1, fig. 17, location O-23). The quarry, in the Allentown Formation, exposed massive grey limestone dipping N. 50° W. 36°. In 1882, the small quarry was about 10 feet deep (D'Invilliers, 1883, p. 165). In 1897, the Reading Eagle (October 16, 1897) reported that William Fisher of Amity was crushing stone in Reuben Schearer's quarry for township supervisor William Hartman.

Wilman Quarry

The Wilman quarry was on the south side of Manatawny Creek, northwest of Covered Bridge Road at 40° 23′ 11″ latitude and 75° 44′ 26″ longitude on the USGS Manatawny 7.5-minute topographic quadrangle map (Appendix 1, fig. 17, location O-24). The quarry was in the Allentown Formation. D'Invilliers (1883, p. 165) described the quarry: *"Wilman's quarry midway between Connard's paper-mill and Pleasantville, which, though but slightly developed for private use, shows some of the best stone in the township. There is a little slate bedding in the east end but elsewhere the stone is a smooth fine-grained blue limestone, massive, and about 20 feet thick. It has little or no top covering and is readily quarried. The dip is S. 12° W. 54°."* D'Invilliers published an analysis of the limestone: 95.1 percent calcium carbonate, 1 percent magnesium carbonate, and 3.7 percent silica.

Schollenberger Quarry

The D.M. Schollenberger quarry was north of the intersection of Hoch and Cleaver Roads somewhere around 40° 23′ 35″ latitude and 75° 44′ 06″ longitude on the USGS Manatawny 7.5-minute topographic quadrangle map (Appendix 1, fig. 17, location O-25). The small quarry, in the Allentown Formation, exposed about 12 feet of blue, grey, and white banded magnesian limestone. D'Invilliers (1883, p. 165-166) published an analysis of the limestone: 59.9 percent calcium carbonate, 36 percent magnesium carbonate, and 2.8 percent silica

William Weidner Quarry

The William Weidner quarry was east of Manatawny Creek at 40° 23′ 53″ latitude and 75° 44′ 17″ longitude on the USGS Manatawny 7.5-minute topographic quadrangle map (Appendix 1, fig. 17, location O-26). The quarry, in the Allentown Formation, exposed 18 feet of interbedded limestone and *"limestone slate from 2 to 8 inches thick, conformable with the limestone"* (D'Invilliers, 1883, p. 166). The strata dipped S. 80° W. 35-40°. D'Invilliers published an analysis of the limestone: 66.5 percent calcium carbonate, 21.6 percent magnesium carbonate, and 9.6 percent silica.

Deisher Quarry

The Deisher quarry was north of the intersection of Mud Run Road and Jefferson Street at 40° 23′ 56″ latitude and 75° 45′ 40″ longitude on the USGS Fleetwood 7.5-minute topographic quadrangle map (Appendix 1, fig. 14, location O-27). The quarry, in the Beekmantown Group, exposed about 20 feet of massive blue and white limestone dipping N. 74° W. 32°, conformably overlaid by about 2 feet of slate. D'Invilliers (1883, p. 168) published an analysis of the limestone: 87.6 percent calcium carbonate, 9.8 percent magnesium carbonate, and 2.2 percent silica.

J.G. Bertolet Quarry

The J.G. Bertolet quarry was between Pennsylvania State Route 73 and Bertolet Mill Road at about 40° 23′ 04″ latitude and 75° 46′ 02″ longitude on the USGS Fleetwood 7.5-minute topographic quadrangle map (Appendix 1, fig. 14, location O-29). The area is developed, and the location of the quarry is not apparent. The quarry, in the Beekmantown Group was a double quarry that exposed about 12 feet of limestone, dipping in the east end of the quarry S. 78° W. 38°, in the center of the quarry S. 85° W. 35°, and in the west end of the quarry S. 60° W. 35°. D'Invilliers (1883, p. 169) published an analysis of the *"blue and white soft stone"*: 92.2 percent calcium carbonate, 5 percent magnesium carbonate, and 2.8 percent silica. The 1882 topographic map shows a lime kiln south of the quarry.

D.F. Bertolet Quarry

The D.F. Bertolet quarry was south of Manatawny Creek at 40° 22′ 49″ latitude and 75° 45′ 21″ longitude on the USGS Fleetwood 7.5-minute topographic quadrangle map (Appendix 1, fig. 14, location O-30). The quarry, in the Beekmantown Group, exposed about 40 feet of blue banded and massive quartzose limestone dipping S. 20° W. 60° (D'Invilliers, 1883, p. 170). The quarry was not being worked at time of D'Invilliers visit in 1882.

F.V. Kauffman Quarries

Two quarries were located on the F.V. Kauffman farm. The northern quarry (location O-31A) was north of Kauffman Road at about 40° 22′ 14″ latitude and 75° 45′ 24″ longitude, and the southern quarry (location O-31B) was south of Kauffman Road at 40° 22′ 05″ latitude and 75° 45′ 10″ longitude on the USGS Birdsboro 7.5-minute topographic quadrangle map (Appendix 1, fig. 30). The quarries and a lime kiln are shown on the 1882 topographic map. The quarries were in the Beekmantown Group.

The northern quarry exposed a slaty limestone dipping N. 84° W. 38°. D'Invilliers (1883, p. 170) published an analysis of the limestone: 63.1 percent calcium carbonate, 27.4 percent magnesium carbonate, and 8.2 percent silica. The southern quarry was 400 yards southeast of the northern quarry on the south side of small creek. The southern quarry exposed about 25 feet of hard blue magnesian limestone dipping N. 83° W. 82° (D'Invilliers, 1883, p. 170). In 1908, the Reading Eagle (January 25, 1908) reported that Irwin Shane, Oley roadmaster, with a force of men was quarrying limestone on F.V. Kaufman's land for public road purposes.

Thomas P. Lee Quarry and Farm Locality

The Thomas P. Lee quarry was on the Lee farm, southeast of the intersection of Pennsylvania State Route 73 and Snyder Road at 40° 23′ 11″ latitude and 75° 45′ 04″ longitude on the USGS Fleetwood 7.5-minute topographic quadrangle map (Appendix 1, fig. 14, location O-33). The small quarry exposed about 10 feet of a blue limestone dipping S. 85° W. 35°. The limestone had numerous quartz veins (D'Invilliers, 1883, p. 168).

The quarry and surrounding area is the Lee farm quartz crystal locality. Genth (1875, p. 56) reported quartz crystals up to 2 inches long with good terminations on Thomas Lee's farm. D'Invilliers (1883, p. 400) reported quartz in transparent crystals from 1 to 2 inches long at Lee's farm. The area has been developed and is no longer a farm.

The quarry was locality number 35 of Gray (1951, p. 56). The quarry exposed a blue, fine-grained, laminated, locally mottled limestone. The laminations were prominent to indistinct on weathered surfaces. The strike of the beds was N. 5-20° W., and the dip was about 10° SW. Gray (1951, p. 56) presented analyses of the limestone.

Reiffe Quarry

The Reiffe quarry was on the east side of Old State Road at 40° 22′ 50″ latitude and 75° 48′ 25″ longitude on the USGS Fleetwood 7.5-minute topographic quadrangle map (Appendix 1, fig. 14, location O-39). The quarry is shown in the 1876 atlas, and a lime kiln at the quarry is shown on the 1860 and 1862 maps.

Yale Quarry

The Yale quarry was north of Bertolet Road, northeast of its intersection with Main Street at about 40° 22′ 47″ latitude and 75° 46′ 20″ longitude on the USGS Fleetwood 7.5-minute topographic quadrangle map (Appendix 1, fig. 14, location O-19). The quarry was locality number 38 of Gray (1951, p. 58). The quarry exposed massive gray dolomitic limestone. The strike of the beds was N 10° E, and the dip was 25° NW. Gray (1951, p. 58) presented an analysis of rock from the quarry.

Slate Quarries

Israel Bertolet Quarry

The Israel Bertolet quarry was north of the intersection of Mill and Reider Roads at 40° 24′ 39″ latitude and 75° 47′ 17″ longitude on the USGS Fleetwood 7.5-minute topographic quadrangle map (Appendix 1, fig. 14, location O-32). The quarry exposed about 30 feet of dark greenish-grey slate overlain by a soft, brownish, talcose slate dipping S. 68-74° E. 35-45°. The slate sold for about $2.50 per cord and was locally used for lining lime kilns (D'Invilliers, 1883, p. 182-183).

Wellington B. Griesemer Quarry

About 1854, William Knabb noticed slate at the bottom of Monocacy Creek and opened a quarry for roofing slate. Enough slate was taken out to roof a house and barn in the vicinity. However, Knabb soon abandoned the quarry. The property was afterwards sold to Jessiah Manwiller who later sold it to Aaron Miller. Miller sold the property to Wellington B. Griesemer, who mined some roofing slate from a 5-foot deep opening (*Reading Eagle*, September 21, 1879). The quarry was abandoned until September 1893, when Griesemer's sons, Charles and Edward, began working the quarry. In 1893, the quarry was 8 feet wide, 16 feet long, and 18 feet deep. Water that accumulated in the quarry overnight was pumped out in 4 hours by a steam engine. The slate was hoisted from the quarry by horse and derrick. Some pieces were 3 feet thick and 5 to 6 feet long, but split easily (*Reading Eagle*, February 4, 1894).

IRON MINES

Hunter Mine

The flooded Hunter mine was between Moravian School Road and Mine Lane at 40° 22′ 22″ latitude and 75° 47′ 15″ longitude on the USGS Birdsboro 7.5-minute topographic quadrangle map (Appendix 1, fig. 30, location O-34). The mine is shown on the 1885 topographic map (fig. 466). The mine was in the Jacksonville Formation.

The Hunter mine was on the Daniel Hunter farm about 900 feet northwest of the Weaver mine. The Hunter mine consisted of an open cut and three shafts. It was abandoned before 1882 when it was considered to be exhausted. When the mine was visited by D'Invilliers in June 1882, it was filled with water to within 10 or 12 feet of the surface. The sides of the small open cut were composed of a buff-colored clay (D'Invilliers, 1883, p. 366).

The mine was originally leased by the Clymer Iron Company around 1865. The company sank a few trial pits but did no mining. Next, Mr. Bailey of the Pine Iron Works obtained a 10-year lease with a 50 cent per ton royalty. Bailey mined ore from a 50-foot deep open cut in clay. At the bottom of the cut, he sank the 50-foot deep shaft No. 1 in mixed clay and limestone. He also sank the 84-foot deep shaft No. 2 about 250 feet north of the engine house (fig. 466). The shaft penetrated 20 feet of limestone clay and soil, 2 feet of ore, and 62 feet of slate and bluish gravel. Bailey also sank the 125-foot deep shaft No. 3 at the engine house to obtain water. The shaft penetrated 20 feet of clay and 105 feet of limestone. Bailey surrendered the lease after 5 years of mining (D'Invilliers, 1883, p. 366-367).

The mine was then leased by Griffith Jones of the Pottstown Iron Company, who sank a shaft about 50 feet southwest of the open cut. This shaft, which was about 100 feet deep, penetrated about 90 feet of yellow clay, 1 foot of kaolin, and a 1 to 2 foot thick bed of limonite. A black clay containing siderite balls was encountered below the limonite. A thin layer of limonite mixed with black clay was encountered at the bottom of the shaft. When Jones failed to find the 7-foot thick ore bed reported by the property owner, he surrendered the lease (D'Invilliers, 1883, p. 366).

About 1870, the Pottstown Iron Company discovered a deposit of white clay while prospecting for iron ore on the Hunter farm. The clay was found in great abundance overlying the ore. Theodore W. Ludwig of Douglasville and D.B. Mauger made arrangements with the Pottstown Iron Company to mine the clay and iron ore together (*Reading Eagle*, September 13, 1870, and November 22, 1873).

Ludwig divided the clay into three grades according to quality: No. 1, white; No. 2, mixed white and yellow; and No. 3, a buff-colored argillaceous clay colored by iron. The mine was idle when visited by D'Invilliers in June 1882, but there were a few tons of No. 2 white and yellow clay in the sheds. The best quality No. 1 white clay, which was used in weighting paper, sold for $7 to $15 per ton. The white clay was used at Connard's paper mill in Pleasantville and at the Burgess paper mill in

Figure 469. Limonite from the Hunter mine, Oley Township, 5.6 cm. Reading Public Museum collection.

Spring City, Chester County. Burgess purchased the first 100 tons of No. 1 white clay produced at $9 per ton (D'Invilliers, 1883, p. 367). The Phoenix Pottery works in Phoenixville used the white clay to make pottery that competed with English ironstone china.

The three grades of clay in the deposit totaled about 800 tons. The No.1 clay was mined from a face 30 feet long and 20 feet wide under 6 feet of overburden; it was mined out by 1882. A considerable quantity of No. 3 buff-colored clay remained (D'Invilliers, 1883, p. 367). D'Invilliers (1883, p. 368) published an analysis of the clay.

Weaver Mine

The Weaver mine (fig. 466) was south of the Hunter mine at about 40° 22′ 12″ latitude and 75° 47′ 06″ longitude on the USGS Birdsboro 7.5-minute topographic quadrangle map (Appendix 1, fig. 30, location O-35). The mine was on the property of Mrs. C. Weaver. The mine, in the Jacksonville Formation, was opened about 1865 as an open cut.

The 1880 Census (Pumpelly, 1886, p. 960) listed the Clymer Iron Company as the operator of the Weaver mine. Production between July 1, 1879, and June 30, 1880, was 1,253 tons of limonite. The ore was shipped to Clymer's Oley and Temple Furnaces. In 1882, the mine consisted of two shafts and an open cut. D'Invilliers (1883, p. 363-364) noted: "*The ore—which generally throughout the mine is composed of 25 per cent. lump and 75 per cent. wash, though operations now are confined to surface digging which is mostly wash ore.*"

In 1878, shaft No. 1 was sunk to a depth of 66 feet. At 49 feet, a 2-foot thick bed of hard ore dipping about 30° NW. was encountered. At 56 feet, an 8-foot thick ore bed was encountered; the bed was almost flat, with a slight dip to the northwest. A drift was driven northwest on this ore bed, which

Figure 470. Weaver mine, Oley Township, April 2015.

pinched out. However, the drift encountered the 2-foot thick ore bed 30 feet from the shaft (D'Invilliers, 1883, p. 364).

Shaft No. 2, located at the west end of the engine house, was sunk in search of water. The shaft penetrated 116 feet of clay and encountered limestone at the bottom of the clay. At 104 feet, a drift was driven 32 feet to the east through clay. Work stopped when the drift encountered water and immediately collapsed. At 100 feet, a drift was driven 32 feet to the west and encountered iron ore. The ore bed was followed for 16 feet. D'Invilliers (1883, p. 365) noted: *"This is very likely the same ore struck in No. 1 shaft between 50 and 60 foot level, as the distance between shafts is great enough to allow of this ore dipping at 30° to come in at 100-foot level in No. 2. The ore body in the latter shaft was of good quality and dipped between 30° and 40° N. W."*

When the attempt to find water for washing the ore failed, it was decided to utilize the water found in the east drift by a *"circuitous and carefully constructed gangway."* A drift was started due north at 104 feet below land surface. It turned gradually to the east in a half-moon shape. At 71 feet from the shaft, the drift successfully struck a plentiful supply of water that was used for washing the ore (D'Invilliers, 1883, p. 365).

In 1882, ten men were employed at the mine, and production averaged 10 to 15 tons of ore per day. A 20-horse power steam engine, built by Wren & Brother of Pottsville, was used for pumping water and running the ore washer. The engine used both wood (1.5 cords per day at $1 per cord) and coal (1/2 ton anthracite coal per day at $3.60 per ton delivered at the mine) (D'Invilliers, 1883, p. 364).

The ore was limonite that occurred in a bed 2 to 8 feet thick underlain by clay or decomposed damourite slate. D'Invilliers noted: *"In the past many beautiful specimens of lump and shell ore have been gathered from here."* The ore dipped from 30° to 50° NW. away from the hill. In 1882, all of the ore was smelted at the Oley Furnace. The ore was hauled to the Oley furnace, 3.5 miles away, by four single teams owned by the Cly-mer Iron Company. There was a royalty of 35 cents per ton on the ore (D'Invilliers, 1883, p. 364). D'Invilliers (1883, p. 365) published an analysis of the ore: 36.7 percent metallic iron and 18.6 percent silica.

The damourite slate beds decomposed to an argillaceous, white- and buff-colored clay. The clay was used at the mine to manufacture an excellent quality building brick. Most of the brick was used in Oley (D'Invilliers, 1883, p. 364-365).

Smith (1977, p. 114) presented an analysis of limonite from the Weaver mine. The sample contained 0.19 percent zinc.

Manwiller Mine

The Manwiller mine (fig. 466) was east of Mine Lane at about 40° 21' 19" latitude and 75° 47' 06" longitude on the USGS Birdsboro 7.5-minute topographic quadrangle map (Appendix 1, fig. 30, location O-36). Limonite was mined from the Beekmantown Group. The open-cut mine was opened in 1873 and abandoned in 1878. Most of the mining was done by the Warwick Iron Company of Pottstown. Production was 20 tons per day (*Reading Eagle*, October 23, 1881), and total production was estimated at 2,000 tons. The royalty was 35 cents per ton. D'Invilliers (1883, p. 368-369) noted: *"There is a fair showing of lump ore in the cut, but the whole deposit is only a pocket, similar to those opened weekly, one might say, in the middle of the great valley, and the best of it is worked out."* MacLachlan (1979, p. 60) reported a very high zinc concentration (0.34 percent) in the limonite from the mine.

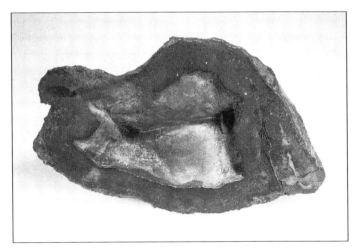

Figure 471. Limonite from the Weaver mine, Oley Township, 5.7 cm. Reading Public Museum collection 2000C-027-163.

Figure 472. Limonite from the Manweiller mine, Oley Township. Reading Public Museum collection 2000c-27-165.

Hertzel and Swoyer Mine

The Hertzel and Swoyer mine was northwest of the intersection of Reider Road and Water Street at about 40° 24' 36" latitude and 75° 46' 42" longitude on the USGS Fleetwood 7.5-minute topographic quadrangle map (Appendix 1, fig. 14, location O-37). The mine is labeled as the Swoyer mine on the 1882 topographic map. The mine was in hornblende gneiss.

There were three shafts at the Hertzel and Swoyer mine, which was located just south of Furnace Hill. D'Invilliers (1883, p. 274) noted: "*No work was being done at these openings — which are all shafts — at the time of visiting them* [summer of 1882]; *but considerable ore piles still remained at the shafts.*" The ore mined from the three shafts was similar and resembled the ore from the nearby the Clymer open cut. D'Invilliers (1883, p. 274) observed: "*It is rather more brown in color than the black powder of the Clymer mine, not so rich in iron, and shows a higher percentage of silicious matter.*"

The sinking of the two western shafts began in 1874. The mine was leased by Dr. Hertzel and Frank Swoyer in 1879. By 1879, the eastern shaft was 24 feet deep, and a drift was run from the bottom for 33 feet along the strike of the ore bed, which dipped 40° SE. The foot wall was a black hornblende gneiss. The third shaft was known as the Talley mine, which is described below (D'Invilliers, 1883, p. 274-275). The ore, which was filled with brownish mica scales, was shipped to Fleetwood. D'Invilliers (1883, p. 274) published an analysis of the ore: 25.2 percent metallic iron and 25.4 percent silica.

Talley Mine

The third shaft of the Hertzel and Swoyer mine was on the property of Benjamin Yoder and was known as the Talley mine. It was on a hill about 100 feet from Reider Road, and close to the Rockland Township border. The Talley mine was northwest of the intersection of Reider Road and Water Street at about 40° 24' 33" latitude and 75° 46' 17" longitude on the USGS Fleetwood 7.5-minute topographic quadrangle map (Appendix 1, fig. 14, location O-38). The mine was in hornblende gneiss.

The shaft was started in 1879 and was sunk to a depth of 100 feet. D'Invilliers (1883, p. 274-275) noted: "*The dump shows a good deal of fine black and brown ore, and near by a considerable amount of black hornblendic rock.*" D'Invilliers published an analysis of the ore: 29.6 percent metallic iron, 20.8 percent silica, 0.014 percent sulfur, and 0.044 percent phosphorus.

Other Localities

Oley (Friedensburg) Area

Oley was formerly known as Friedensburg. Minerals from the Oley area are listed below.

Minerals
Quartz - blue, amethystine (D'Invilliers, 1883, p. 400)
Quartz, var. chalcedony (Genth, 1876, p. 218)
Quartz, var. hornstone (Genth, 1876, p. 218)
Silicified (petrified) wood (Genth, 1876, p. 218) (fig. 474)

Oley Uranium Occurrence

The Oley uranium occurrence is located on both sides of Pennsylvania State Route 73, northwest of the intersection of Pennsylvania State Route 73 and Friedensville Road, centered

Figure 473. Magnetite from the Hertzel and Swoyer mine, Oley Township, 8.5 cm. Reading Public Museum collection 2004C-2-207. Collected in 1876.

around 40° 23′ 23″ latitude and 75° 48′ 05″ longitude on the USGS Fleetwood 7.5-minute topographic quadrangle map (Appendix 1, fig. 14, location O-40).

In 1978, the Pennsylvania Geological Survey discovered an area enriched in uranium and possibly thorium near Oley while examining outcrops of Precambrian gneiss along Pennsylvania State Route 73 with a Geiger counter (Pennsylvania Geological Survey, 1978). An area of salmon pink albite gneiss about 0.3 by 0.6 mile was identified. A sample of weathered rock analyzed contained 67 ppm U_3O_8. The presence of fluorescent green (under ultraviolet light) hyalite opal also was noted on two of the samples collected. Also observed were golden-brown radioactive grains with a resinous luster and a zircon-like structure and thin yellow films of an unidentified yellow secondary mineral along fractures.

Oley (Shearwell) Furnace

Oley Furnace was built in 1759-60 by a company composed of Benedict Swope; Dietrich Welcker, a blacksmith; and Peter Harpel, an innkeeper. Welcker was the ironmaster. The furnace was initially known as Shearwell Furnace or the Shearwell Oley Furnace. In 1768, Welcker borrowed £100 from John Lesher, who was the forge master at Oley Forge. The furnace owners were apparently under-capitalized from the beginning and ran into financial difficulty fairly quickly. Welcker also borrowed money from others, and, by 1772, was in receivership.

In 1772, Jacob Winey and Christian Lower obtained the furnace. Winey, a Philadelphia merchant, was involved in a number of iron works in Chester and Berks County and owned Moselem Forge. The new owners rebuilt the furnace, and, by 1774, Winey was selling Oley-made stoves at his store in Philadelphia. John Patton appears to have rented Oley Furnace in conjunction with his Berkshire Furnace during the early part of the Revolutionary War. Oley Furnace was described by Hermelin (1783) as in blast in 1783 and producing 400 tons of pig iron annually. Schoepf (17887) extensively discussed Oley Furnace (see next page).

By 1778, Winey's nephew, Daniel Udree, who was the clerk at his uncle's Moselem Forge, was hired to run Oley Furnace. He remained at Oley Furnace and eventually obtained ownership, which he kept until his death. Udree was elected to the Pennsylvania House of Representatives (1799-1805) and was elected to the U.S. Congress to fill vacancies in 1813, 1820, and 1822. In 1825, he returned to Oley Furnace. He died in Reading on July 15, 1828. His nephew, J. Udree Snyder, was the furnace owner in 1828. At that time, the furnace employed 153 workmen, had 75 horses, and used 10,500 cords of wood to produce 1,050 tons of pig iron and 360 tons of castings.

In the 1830s, the furnace was leased to Nicholas Hunter, who eventually acquired it. Hunter's son, Daniel Hunter, managed the furnace for the family. Lesley (1859, p. 38) noted that the furnace stood idle for about 12 years and was run for a short time in 1853. In 1857, Oley Furnace was owned by Murkells & Levan and managed by Samuel Murkells. In 30 weeks in 1857, the furnace made 757 tons of car wheel iron using magnetite from Deishler's mine mixed with magnetite from Zinner's mine in Rothrockville.

Figure 474. Polished section of petrified wood from Oley Valley, Oley Township, 6 cm. Ron Kendig collection.

In 1860, William H. Clymer, Sr. and his brother, Edward M. Clymer, purchased the Oley Furnace. In 1880, the brothers organized the Clymer Iron Company, a corporation which included the Oley Furnace, Mt. Laurel Furnace, limestone quarries, and iron ore mines. In 1884, the annual capacity of Oley Furnace was 2,000 tons, and the furnace's specialty product was No.1 dead gray iron. The furnace was dismantled in 1888 (American Iron and Steel Association, 1884, 1886, and 1892).

ONTELAUNEE TOWNSHIP

Ontelaunee Quarry

The Berks Products Company Ontelaunee quarry (fig. 475) is a large, active (2016) quarry at the confluence of Maiden Creek and the Schuylkill River. It is west of Pennsylvania State Route 61 at 40° 25′ 29″ latitude and 75° 56′ 37″ longitude on the USGS Temple 7.5-minute topographic quadrangle map (Appendix 1, fig. 9, location ON-1). The quarry is in the Allentown Formation.

The quarry was opened in March 1990 on a 500-acre tract, and, by 1991, the quarry occupied 25 acres and was 100 feet deep (*Reading Eagle*, June 16, 1991). The quarry suffers from a sinkhole problem; many sinkholes have opened up to allow river or creek water into the quarry. In 1999, during Hurricane Floyd, a sinkhole opened up in Maiden Creek, causing the quarry to flood to a depth of 50 feet. Mining stopped until the quarry could be pumped out (*Reading Eagle*, October 14, 1999, and June 6, 2000).

Oley Furnace

Johann David Schoepf (1752-1800) was a German botanist, zoologist, and physician. He traveled to New York in 1777 as the chief surgeon for a regiment of Hessian troops fighting for King George III. After the Revolutionary War ended,, he traveled for 2 years in the United States, East Florida, and the Bahamas. During his travels through Berks County in 1783, he recorded the following description of the Udree mine. From Schoepf (1877, p. 197-200).

"Mr. Daniel Udree's iron-works lie 10 miles from Reading, in a narrow valley among the Oley Hills. The mine which supplies the iron ore is five miles beyond, and has a depth of not more than 6-7 fathoms. Recently ore has been discovered still nearer, which in several respects is better than the first, and in future this will be used in mixture; hitherto they have not known how to apply the advantage to be gained from a mixture of several ores.

Nearly at the top of the hill and immediately behind the high-furnace, a mine was formerly worked which is rich in the best and most compact ore. The rock in this hill is a coarse-grained wacke, lying in thick beds running almost north and south. The ore is found at a depth of only 12-20 ft. below the surface mould, and in places along the hill even shallower. A gallery-stoll had been driven in the hill, some 12 ft. high, 15 ft. broad, and about 300 ft. long, and then a 60 ft. shaft was sunk, and a beautiful, compact, quartz-ore, shimmering green and blue was taken out, which was the richest and most easily fluxed of any ore in that whole region. But water broke in too strongly and drowned out the work. And besides, the ore having to be blasted, at the beginning of the [Revolutionary] war powder was too dear and work-people scarce, they were compelled to give over this mine, but will now take it up again.

A reddish fine-grained sand-stone which stands the fire excellently is brought to the high-furnace from beyond the Schuylkill, and is called merely Schuylkill stone. Formerly they tried at a loss the wacke found on the nearest hills; this split and burst in the fire. The cost of setting up the interior of the furnace, including the expense of breaking and hauling the stone, amounts always to about 100 [pounds Pennsylvania currency]; but the furnace often bears two smeltings.

Some 10,000 acres of forest are attached to this high-furnace. The oaks on these dry hills are small, to be sure, but there are among them many chestnuts which make the best coals. The furnace consumes 840 bushels of coals in 24 hours, for which 21-22 cords of wood are necessary. It is estimated that 400 bushels of coals are used in getting out one ton of bar-iron. A turn of coals, about 100 bushels, costs about 20 shillings [Pennsylvania currency]. (The guinea at 35 shillings.) Wages for wood-cutting are two shillings three pence the cord. A man chops two and a half to four cords a day, and so can earn 6-9 shillings.

At present only six men work at the mine; but they supply more than the furnace can consume. If the work was uninterrupted there could be turned out yearly between 2-300 tons of iron. A hundredweight of the ore worked at this time yields 75 pounds of cold iron. A miner receives 40 shillings a month and rations. The furnace men, founders and hammer men, are paid by the ton. For a ton of pig—5 shillings; for a ton of furnace iron or other ware—40 shillings. In this way, if much is worked, the first founder stands to receive several pounds in the week. Nowhere among the sundry mines and forges of America had wages become fixed as yet, the custom being to treat with each man conformably and according to his abilities. Miners by profession worked commonly by the fathom.

The price of a ton of pig-iron (which on account of the easier transport is cheaper in America) is 10 pounds current. A ton of furnace iron, kettles or other utensils, 20-25 pounds. Bar-iron, in the good times before the war, cost the iron-masters 22-23 pounds a ton; they sold it at 25 pounds cash money or 30 pounds at six months credit. But at present they cannot deliver a ton for less than 32-37 pounds.

If the furnace is not properly managed the slag is pale green and coarse, but otherwise a fine sky-blue. There lay at the furnace more than 200 tons of such slag, which Mr. Udree had turned over to a man who was to give him 15 tons of iron for the privilege of breaking it up, washing it, and getting it worked over at a bloomery; his estimate is that it will take him two years to clear out this slag.

Mahogany wood is used for mould-forms at furnaces, because it is the least subject to warping and splitting,

Formerly Mr, Udree dealt with his workmen as is customary in Germany; that is, he furnished them with all necessaries on account. They made use of the opportunity to run up their accounts, and not being trammeled with families got out of the way; and so he changed his method.

Figure 475. Ariel view of the Ontelaunee quarry, Ontelaunee Township, April 2016.

Maidencreek Station Quarry

The Maidencreek Station quarry was northwest of the intersection of U.S. Route 222 and Dries Road, on the north side of Willow creek at about 40° 26′ 18″ latitude and 75° 54′ 39″ longitude on the USGS Temple 7.5-minute topographic quadrangle map (Appendix 1, fig. 10, location ON-2). The quarry is shown on the USGS 1913 Reading 15-minute topographic map. The quarry is in the Allentown Formation. Miller (1934, p. 215) described the Maidencreek Station quarry as an old quarry a short distance north of Willow Creek where lime was once burned. The rock was gray to grayish blue and fairly high in magnesia. The beds were massive and dipped 25° N. Some layers contained the algal fossil *Cryptozoon proliferum*.

Leesport Iron Company Quarries

The Leesport Iron Company operated three quarries; two were southwest of the intersection of Pennsylvania State Route 61 and Orchard Lane, and one was northeast of the intersection. The largest quarry, which is flooded (location ON-3A) (fig. 476), was at 40° 26′ 29″ latitude and 75° 57′ 21″ longitude on the USGS Temple 7.5-minute topographic quadrangle map (Appendix 1, fig. 7). A small quarry was located to the south (location ON-3B) at 40° 26′ 23″ latitude and 75° 57′ 18″ longitude, and the smallest quarry was located to the northwest (location ON-3C) at 40° 26′ 43″ latitude and 75° 57′ 16″ longitude. The quarries are in the Epler and Ontelaunee Formations.

The quarries were opened around 1871 and operated by the Leesport Iron Company for furnace flux. In 1873, the Reading Eagle (December 13, 1873) reported that a branch railroad leading to the Leesport Iron Company quarry was finished. In 1884, the New York Times (December 16, 1884) reported an engine pulling rail cars in the Leesport Iron Company quarry derailed and fell upon the rocks below. The engine and a number of cars were destroyed, and the hands narrowly escaped by jumping. That same year, the New York Times (March 13, 1884) reported that a *"premature explosion in a stone quarry of the Leesport Iron Company hurled a man named Aaron Clay 20 feet into the air. His face was crushed to jelly and an arm and both legs were broken. He lived but a few hours."* Such graphic newspaper reports were common during that time period. The quarries likely were abandoned in 1895 when the Leesport Iron Company Furnace ceased operations.

The quarry is locality 12 of Gray (1951, p. 37). The quarry exposed interbedded limestone and dolomite showing a variety of lithologies. The limestone was blue to gray, dense with black laminations, and crystalline with only faint lamination. Some beds showed distinct fossil fucoid (seaweed) mottling. The limestone may grade into dolomite, with strong mottling in the transition zone. The dolomite was chiefly gray, finely crystalline, and massive. One bed showed an indistinct conglomeratic texture. The quarry followed the strike of the beds. The average strike was N. 45° E., and the dip was about 65° NE. An overturned synclinal drag fold occurred in the south wall near the center of the quarry. Gray (1951, p. 37) provided chemical analyses of the rock.

Figure 476. Southernmost quarry of the Leesport Iron Company, Ontelaunee Township, April 2015.

PERRY TOWNSHIP

Glen-Gery Quarries

The Glen-Gery Company quarried clay in several places. Clay quarries were located south of the intersection of Pennsylvania State Route 61 and Shoemakersville Road at 40° 28′ 53″ latitude and 75° 57′ 44″ longitude (location PY-1A) and northeast of the intersection of PA 61 and Shoemakersville Road at 40° 29′ 26″ latitude and 75° 57′ 17″ longitude (location PY-1B) on the USGS Temple 7.5-minute topographic quadrangle map (Appendix 1, fig. 7). A clay quarry, now flooded, was northeast of the intersection of Pennsylvania State Route 61 and Ridge Road at 40° 30′ 05″ latitude and 75° 57′ 33″ longitude (location PY-1C) on the USGS Hamburg 7.5-minute topographic quadrangle map (Appendix 1, fig. 7). The quarries were in the Hamburg Sequence.

In the late 1800s, the Glen-Gery property was owned by the Moll family. Franklin B. Moll manufactured pottery from clay dug on the property as late as 1885. In 1897, the Moll farm was purchased by the Schuylkill Valley Clay Manufacturing Company, a Philadelphia company established for the manufacture of glazed sewer pipe. Large buildings were erected on the farm and along Pennsylvania State Route 61. This business, also known as the Shoemakersville Clay Works, continued for a period of 10 years. During a period of difficult financial times, the property was purchased by the Glen-Gery Shale Brick Company (Noecker and others, 1965).

In 1890, Albert A. Gery, nicknamed A.A., decided to try his hand at making fire brick. With financing from his father-in-law, Mathan Harbster, Gery purchased 32 acres of land rich in clay near the Montello station of the Reading Railroad. Gery built two rectangular kilns and formed the Montello Clay and Brick Company. Soon Gery was contacted by a Philadelphia contractor who needed a million bricks to build the Wernersville State Hospital, 5 miles from Montello. In 1898, a new plant was built in Wyomissing that featured the largest brick kiln in the United States.

In 1908, Albert A. Gery and his brothers Frank and Will formed the Glen-Gery Brick Company and purchased the sewer-pipe plant at Shoemakersville. The Glen-Gery Shale

Figure 477. Stock certificate of the Glen-Gery Shale Brick Company, 1929. The certificate, signed by Albert A. Gery, was issued to Dr. Edward H. Brown just 8 days before the stock market crash of October 29, 1929.

Figure 478. Ariel photograph of the Glen-Gery quarries and brick plant, Shoemakersville, Perry Township, October 31, 1958. Courtesy of Bureau of Topographic and Geologic Survey, PennPilot Historical Aerial Photo Library.

Brick Company (fig. 477) was incorporated on March 15, 1912, by Albert A. Gery, Frank S. Gery, and William A. Gery. The Shoemakersville plant became known as the Glen-Gery Brick Company Shoemakersville Brick Division (Fox, 1925, p. 349).

The Glen-Gery Brick Company revamped the Shoemakersville plant to manufacture face and paving brick. The company employed as many as 110 people and produced 30 million bricks annually. Face brick was manufactured by the extruded method and was used extensively in the construction of multistory buildings. Paving bricks were first extruded, and then repressed to form beveled edges, Paving bricks were in great demand for street paving. While much of the production was shipped to New York and Philadelphia, great quantities of bricks also were used by nearby communities, such as Reading and Pottstown. Most of the streets in Pottstown still have their original Shoemakersville paving bricks beneath the macadam (Noecker, 1965). The clay to make the bricks was quarried locally (fig. 478).

As a result of the economic crash of October 1929, the plant ceased operations. In 1934, the company went into receivership and was operated by the receivers from 1934 to 1939 with Russel Eshenaur as general superintendent. In 1939, the company was reorganized as the Glen-Gery Shale Brick Corporation with 180 employees. When the company was reorganized, Eshenaur became president.

By 1949, there were 550 employees at 8 factories. The Shoemakersville brick plant, which was closed during the depression, was reconstructed and modernized. It began operating again in 1947 (*Reading Eagle*, January 28, 1968). The new facilities included seven down draft kilns, new brick machinery, and the introduction of modern materials handling equipment. In June 1958, all brick kilns were changed to gas fired kilns (Noecker, 1965).

With the 1986 purchase of Hanley Brick, Inc. in Summerville, Pennsylvania, the company expanded into the area of high-quality architectural brick. Many well-known buildings, such as World Wide Plaza, the Chrysler Building, Mt. Sinai Hospital in New York City, and Washington Harbor, utilized Hanley products. Glen-Gery expanded westward in 1988 with the purchase of Midland Brick Co. and its manufacturing facilities in Chillicothe and Ottumwa, Missouri, and the Redfield plant in Iowa. Also in 1988, Glen-Gery purchased the New Jersey Shale Brick Manufacturing Corporation in Somerville, the only brick plant in New Jersey. In 1999, the Glen-Gery Corporation was acquired by CRH PLC of Ireland and became part of its Architectural Products Group. In 2014, the Glen-Gery Corporation was acquired by Bain Capital. By then, Glen-Gery operated 10 manufacturing plants and 10 retail centers and was one of the largest brick manufacturers in the U.S. (Glen-Gery Corporation, 2014).

O'Neil and others (1965, p. 85) provided an analysis of clay from the Glen-Gery quarry. The clay, composed of quartz, mica, kaolinite, and feldspar, was mined from yellow-gray shale beds, which generally ranged from 2 to 6 inches thick. A few siltier beds were Interbedded with the shale. The shale weathered to various shades of yellow and brown. By 1965, the quarry workings were extensive, measuring more than 500 feet by 1,000 feet and averaging over 20 feet deep.

Jacob Leiby Flagstone Quarry

The Jacob Leiby Flagstone quarry was east of Pennsylvania State 143 somewhere around 40° 32' 05" latitude and 75° 52' 46" longitude on the USGS Hamburg 7.5-minute topographic quadrangle map (Appendix 1, fig. 2, location PY-2). It is shown in the 1876 atlas.

Jacob Leiby (1798-1884) purchased a a farm and settled in Perry Township in 1825. He built a stone house there in 1829. He was a blacksmith, stone mason, stone dresser, and farmer. He opened a flagstone quarry on the farm. Jacob's son, Issac U. Leiby (1830-1910), was born on the farm and spent his entire life there. He was a stone mason by trade and operated Leiby's Flagstone Quarry. He sold flagstone in Philadelphia and Schuylkill and Carbon Counties. Fagstone from the quarry was used for curb stones in local cities and towns (J.L. Floyd and Company, 1911, p. 692-93). The 1876 atlas listed Isaac W. Leiby as "Proprietor of Leiby's Celebrated Stone Quarry."

The quarry produced dark gray flagstone 2 feet wide by 3 feet long by 3 inches thick. Some were 10 feet long (Sanders, 1883, p. 131-132). The stone also was used for curbing and masonry work at bridges. The quarry operated as late as 1908 (*Reading Eagle*, January 22, 1908). In 1921, the 97-acre Isaac U. Leiby homestead and quarry was offered for sale to settle his estate (*Reading Eagle*, August 25, 1921).

Collier's Flagstone Quarry

W. Collier's flagstone quarry was 0.75 mile northeast of Shoemakersville. The quarry was 150 feet long with 10 feet of flagstone exposed. Dark gray flagstones from 2 to 4 feet wide by 5 to 8 feet long were produced. They were generally 2 inches thick with smooth faces. The joints were not regular, causing a loss of about one third when they were squared (Sanders, 1883, p. 132).

Weidman Farm Shale Prospect

Yellow ocherous shale was quarried about 1909 along the Pennsylvania Railroad tracks opposite the Perry station of the Lehigh Valley Railroad on the farm of J.K. Weidman of Shoemakersville. The shale was light yellow or buff in color and contained little grit. Where quarried, the shale dipped about 25° SE. Two car loads were dug at this locality and shipped to the J. Wilbur Company in Providence, Rhode Island. The shale was found to be of considerable value as a base for oilcloth and linoleum, but no more was quarried because the land owner and quarry operator failed to agree on the amount of royalty (Miller, 1911, p. 491).

PIKE TOWNSHIP

Rohrback Mine

The Lewis Rohrback mine was east of Mine Road at about 40° 25' 23" latitude and 75° 41' 47" longitude on the USGS Manatawny 7.5-minute topographic quadrangle map (Appendix 1, fig. 17, location P-4). The mine was at the contact between hornblende gneiss and felsic to mafic gneiss. There was a 100-foot deep shaft, and a 150-foot long adit was driven nto the hill on the Lewis Rohrbach property. The iron ore bed was reported to be only 2 feet thick (D'Invilliers, 1883, p. 303). The mine supplied ore to the District Furnace and later to the Bethlehem Steel Corporation (Orth, 2009). The ore was titaniferous magnetite (*Reading Eagle*, August 3, 1933). The Reading Eagle (May 25, 1916) reported that the Lewis Rohrbach quarry at Pikeville supplied stone for a new county bridge.

S. Yoder Mine

The S. Yoder mine was in Pine Waters, northeast of the intersection of Mine and Heiligs School Roads at about 40° 25' 09" latitude and 75° 42' 09" longitude on the USGS Manatawny 7.5-minute topographic quadrangle map (Appendix 1, fig. 17, location P-5). The mine was in hornblende gneiss. Several pits were sunk near Yoder's house in search of iron ore. A tunnel 200 feet long was driven into a hill. The mine was abandoned around 1875. The last mining was by the Clymer Iron Company, which reportedly mined "*considerable ore*" (D'Invilliers, 1883, p. 303-304).

Mines South of Lobachsville

LIDAR imagery shows two mines on the northwest side of a hill, about 0.5 mile south of Lobachsville, west of Lobachsville Road at about 40° 24' 13" latitude and 75° 43' 58" longitude on the USGS Manatawny 7.5-minute topographic quadrangle map (Appendix 1, fig. 17, location P-2). The mines were in the Hardyston Formation. Rogers (1841, p. 43) reported that limonite was mined along Pine Creek about 0.5 mile southeast of Lobach's mill. Rogers stated that "*the ore has the aspect of a talcose slate, charged with the oxide of iron; it has a laminated or rather a fibrous structure.*" The mine had been long abandoned by 1882 (D'Invilliers, 1883, p. 304). D'Invilliers (1883, p. 398) reported massive hematite from Lobachsville.

Figure 479. Ilmenite from Lobachsville, Pike Township. Skip Colflesh collection.

Hill Church Riebeckite Localities

Bliss (1913) described two riebeckite, var. crocidolite, localities in the vicinity of Hill Church. Locality 5 was 0.25 mile northeast of Hill Church, and locality 6 was 0.75 mile northeast of Hill Church. Locality 5 was a roadcut east of the intersection of Hill Church and Hess Roads at about 40° 23′ 07″ latitude and 75° 39′ 57″ longitude on the USGS Manatawny 7.5-minute topographic quadrangle map (Appendix 1, fig. 18, location P-6). Locality 6 was a roadcut south of Hill Church Road at about 40° 23′ 01″ latitude and 75° 39′ 27″ longitude on the USGS Manatawny 7.5-minute topographic quadrangle map (Appendix 1, fig. 18, location P-7). The localities were in felsic to mafic gneiss.

Blue riebeckite was disseminated throughout the rock as one of the chief constituents and as a thick coating on weathered surfaces, thereby giving the appearance of a weathering product. The coating, which varied from a deep blue or almost black to an ultramarine blue, was up to 5 mm thick. It was readily removed by scraping with a knife and crushed to an earthy powder. The luster was normally dull; occasionally a silky lustre was seen, which was caused by the presence of minute prismatic flakes of hornblende.

Bliss (1913) stated that riebeckite occurred for a distance of 0.25 mile in the two road cuts. The riebeckite formed a coating on the gneiss and also was the dominant constituent of a schistose rock that readily weathered. The riebeckite occured abundantly throughout the mass of the rock in small, dark blue grains with an irregular outline. The grains, which did not exceed 0.2 mm in diameter, had a dull lustre, a hardness from 5 to 6, and showed neither crystal outline nor cleavage faces. In some instances, they wee associated with prismatic crystals of an almost black hornblende. Other minerals occurring at these localities included albite (Eyerman, 1889, p. 26) and oligoclase (Eyerman, 1889, p. 22).

Mica Prospect Number 3

The mica prospect number 3 of Buckwalter (1954) was southwest of the intersection of Mountain Mary and Stone Roll Creek Roads at about 40° 23′ 42″ latitude and 75° 40′ 50″ longitude on the USGS Manatawny 7.5-minute topographic quadrangle map (Appendix 1, fig. 18, location P-1). Biotite occured in pegmatite float in a cultivated field in books to 3/4 inch in diameter and 1/4 inch thick. Mica books up to 4 inches have been reported (Buckwalter, 1954, p. 20).

John Keim Quarries

The John Keim quarries were northeast of Boyer Road, northeast of its intersection with Hoch Road at about 40° 24′ 39″ latitude and 75° 45′ 13″ longitude on the USGS Fleetwood 7.5-minute topographic quadrangle map (Appendix 1, fig. 14, location P-8). The quarries were in the Allentown Formation.

Limestone was quarried in two places, on Boyer Road and on the north-west side of a knoll. Neither of the quarries were being worked at time of D'Invilliers visit on June 19, 1882. The quarry on the road exposed about 40 feet of blue limestone about 10 feet thick. The limestone was slaty at the top. The quarry on the knoll exposed 25 feet of bluish-gray limestone interstratified with bands of slate. The limestone dipped S. 28° E. 40°. Both quarries were owned by J. Keim. D'Invilliers (1883, p. 167) published an analysis of limestone from quarry on the knoll: 79.1 percent calcium carbonate, 14.2 percent magnesium carbonate, and 5.7 percent silica.

Rolling Rock Building Stone Quarry

The Rolling Rock Building Stone, Inc. quarry is an active (2016) quarry southeast of the intersection of Oysterdale and Rolling Rock Roads at 40° 24′ 17″ latitude and 75° 40′ 48″ longitude on the USGS Manatawny 7.5-minute topographic quadrangle map (Appendix 1, fig. 18, location P-3). In 1955, Homer Weller and a hired hand began quarrying building stone by hand with two hammers and one truck as their only equipment. They quarried and sold 100 tons of building stone in the first year of operation. In December 1974, the company incorporated under the name Rolling Rock Building Stone, Inc. In 2010, the annual production was 60,000 tons (Rolling Rock Building Stone, Inc., 2010).

The Eckert Family

Issac Eckert

Isaac Eckert was born in 1800 in Womelsdorf. After attending the University of Pennsylvania, he and his older brother William took over their father's grocery business. In 1828, they moved their grocery business to Reading. In 1836, Isaac withdrew from the partnership to form an iron manufacturing company in partnership with his younger brother, Dr. George N. Eckert. From 1842 to 1844, they built the Henry Clay Furnace, named for Henry Clay, a champion of high tariffs on imported goods to protect American industry. At the time of its construction, the Henry Clay Furnace was one of the largest anthracite furnaces in the country. In 1855, a second stack was completed. After George died on June 28, 1865, Isaac became the sole proprietor. In 1852, Isaac became president of the Leesport Iron Company, of which he remained the chief executive until his death. In 1838, he was elected president of the Farmers Bank, where he served as president for 35 years. At his retirement in 1873, his sons, Henry S. and George B. Eckert took over the iron business. Upon his death on December 13, 1873, Isaac was succeeded by his son, Henry S. Eckert.

Henry S. Eckert

Henry S. Eckert, the son of Isaac, was born in Reading. He graduated from Franklin and Marshall College and joined his father in the family iron business. He took over as manager of the iron works on July 1, 1873. Shortly before taking over, he formed a partnership with his brother George under the firm name of Eckert & Brother. In 1857, Henry married Carrie Hunter, daughter of ironmaster Nicholas Hunter. Besides being in charge of Eckert & Brother, he served as president of the Topton Furnace Company. He succeeded his father as president of the Farmers Bank in 1873, and he continued to hold that position until his own death in 1893, when his son Isaac succeeded to the position. He also served as a director of the Wilmington & Northern Railroad Company. Henry represented the Eighth Ward in Reading in the Select Council from 1872 to 1875 and served as Council president from 1873 to 1875. He also represented the Eighth Ward on the school board and served as president of the school board from 1872 to 1888. In 1873, the public school on North Tenth Street near Washington Street was named after him.

George Brown Eckert

George Brown Eckert, the son of Isaac, was born in Reading on September 5, 1840. He attended an advanced preparatory school in Danbury, Connecticut. He married Mary Ann Trexler, daughter of ironmaster Horatio Trexler. He enlisted as a volunteer second lieutenant in the 3rd U.S. Infantry during the Civil War. Lieutenant Eckert was honorably discharged on November 10, 1864. Upon his return home, he joined the family iron business as a clerk. In 1873, he and his brother Henry formed a partnership under the name of Eckert & Brother and purchased the furnace property. They carried on the family iron business until his brother's death in 1894. Afterward, he continued its operation for himself and his brother's estate until May 1899, when the plant was sold to the Empire Iron and Steel Company of New York. Eckert became a director of the Farmers National Bank in 1874. Upon the death of his brother Henry (who had served as president of the bank from 1873 to 1894), he became the president, but resigned after 17 months because of poor health. He died on July 5, 1899.

Figure 480. Portraits of Issac Eckert (left), Henry Eckert (center), and George Eckert (right). From Montgomery (1886).

READING

Mount Penn Iron Mines

The Mount Penn iron mines were on or near the old Reading fair grounds, which were east of the intersection of Washington and 11th Streets on the USGS Reading 7.5-minute topographic quadrangle map (Appendix 1, fig. 27, location R-1).

The 1857 wall map of Reading (Geil and Shrope, 1857) (fig. 481) shows four iron mines in the vicinity of the old Reading Fairgrounds—one located inside the oval horse track and three southeast of the fairgrounds on the Miller, Oakley and Richards, and Issac Eckert properties.

The Scientific American (May 19, 1849, p. 274), referring to an article in the Reading Gazette, reported: *"On Penn's Mount, a mountain known to contain vast quantities of the [iron] ore, the most extensive and valuable veins of various kinds of iron have been discovered. For many years these rich deposits were abandoned and the openings had been entirely neglected, either for want of capital or the absence of a proper spirit of enterprise, until they attracted the attention of our enterprising fellow-townsman, George W. Oakley, Esq., who appreciated their value, and in the face of most discouraging barriers, sufficient to retard the progress of one less determined, he went to work with his men, and by personal efforts and skill, succeeded in drawing from the bowels of the mountain, ore as rich and as valuable as ever was found in the placers of Sacramento."*

There was a stationary steam engine in a field between Hill Road and Clymer Street, which pumped water and hoisted iron ore out of shafts sunk in the side of Mount Penn. Prior to installation of the engine, ore was hoisted in buckets by horse power with a rope wound around a large drum. There was a drift extending a considerable distance from Mineral Springs Road into the side of the hill. A car on iron rails was used to haul the ore through the drift to Mineral Springs Road, where it was loaded into wagons.

The first mining on Mount Penn was done by George W. Oakley, who was a well-known druggist and chemist. Mining on the Oakley property also was done by the Phoenix Iron Company, which shipped the ore to their furnaces in Phoenixville.

William Rowe sank the first shaft in search of iron ore on Mount Penn around 1843. Issac Eckert owned a considerable tract of land along the side of Mount Penn and paid Rowe a salary of $1,000 a year to prospect for iron ore, sink shafts, and act as superintendent of mining. The ore was hauled to Eckert's Henry Clay Furnace. Mining ceased about 1873 partly because of the Panic of 1873, but principally because of the large quantity of groundwater that infiltrated into the mines, which required almost constant pumping. John P. Miller was engaged in mining on Mount Penn and was one of the principle organizers of a mining and smelting company around 1850 (*Reading Eagle*, September 20, 1903). This likely was the Penn Mining and Smelting Company, which was chartered on March 8, 1850.

Rogers (1858, vol. 2, p. 716) described the mines that existed during the time of the First Pennsylvania Geological Survey: *"There is an important vein of igneous iron-ore, which has been wrought for some years. It is opened on Penn's Mount, about half a mile E. of Reading. The vein apparently is injected conformably to the bedding of the Primal white sandstone, and the ore is not accompanied by any bounding wall of igneous rock, but is in immediate contact with the rock itself. The latter rock disintegrates quickly on exposure to the atmosphere, and develops innumerable small grains of hornblende, which speckle the yellowish-grey sand. The ore vein ranges from the Reading Fair ground a little S. of E., dipping 45° S. Its thickness is seldom less than 18 inches, and has been as great as 28 feet. Under this enlargement it does not seem to suffer in qual-*

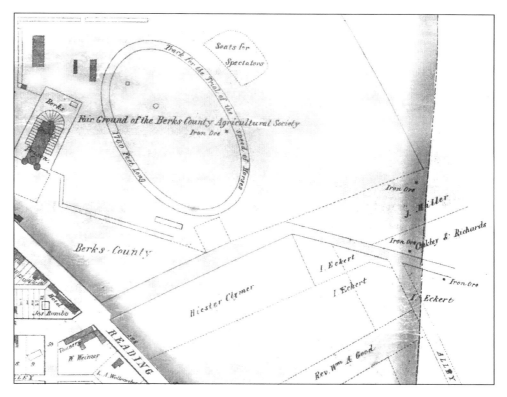

Figure 481. The Mount Penn iron mines, Reading, 1857. From Geil and Shrope (1857). From the collection of the Historical Society of Berks County Museum and Library, Reading, Pa.

MINERALS FROM MOUNT PENN AND READING

Figure 482. Actinolite from Mount Penn, Reading, 8.1 cm. Reading Public Museum collection. 2004C-10-247.

Figure 484. Limonite from Mount Penn, Reading, 3.2 cm. Reading Public Museum collection 2000C-027-099.

Figure 486. Limonite from near Reading, 8.7 cm. Bryn Mawr College collection Vaux 3221.

Figure 483. Goethite from Mount Penn, Reading, 4.9 cm. Reading Public Museum collection 2000C-027-093.

Figure 485. Fibrous goethite from Mount Penn, Reading, 7.8 cm. Reading Public Museum collection 2000C-027-803.

Figure 487. Scepter quartz from Reading, 5.5 cm. Ron Kendig collection.

ity. *The ore itself is of the granitoid variety, highly crystalline, containing quartz and feldspar, especially the latter, in great abundance; hornblende and apatite enter also into its composition. The vein has been wrought at its surface, outcropping in the Reading Fair ground and for one-third of a mile E., by Eckert & Brother; the Phoenix Iron Company, and others, on the lands of Mr. Oakley and B. Davis The principle mine is the vertical shaft of Eckert & Brother; this is sunk 142 feet to the level of a tunnel, which is cut N. 28 feet through rotten sandstone to the top of the vein. From this tunnel the vein is followed by a gangway 30 feet E. and 115 feet W. The ore is worked along the foot-wall rising towards the surface, the hanging wall or roof being supported by timbers. The length of the breast to the old surface workings is 72 feet. The ore from this old level was obtained to a depth of 82 feet. The Phoenixville Company are now obtaining their ore from a surface-level and whim-shaft E. of Eckert's Mine. In Eckert' s old level, 100 feet west of the whim-shaft, the vein split, but the north branch vein thinned away in 100 feet.*"

The mines were abandoned long before 1882. D'Invilliers (1883, p. 277) wrote: "*Within the city limits of Reading there are various old openings exposed on the south flank of Mount Penn, between the fair ground and the Mineral Spring hotel, and from all accounts the locality must have been an active one some years ago. Everything is abandoned and shut up now.*"

The Reading Eagle (October 17, 1954, and March 2, 1984) reported that one of the Mount Penn iron mines was beneath the Central Catholic High School. The 168-foot deep mine penetrated an iron ore vein 28 feet thick. Prior to construction of the high school, the Luden mansion stood on the site. The contractor building the Luden mansion attempted to fill the shaft, and he finally closed it by putting railroad rails across it and closing it with concrete. The shaft was under the main stairway entrance of the mansion.

Rogers (1858, vol. 2, p. 717) presented an analysis of Mount Penn iron ore, which he described as "*Crystalline magnetic ore, with hornblende, feldspar, quartz, and apatite.*" The ore was 37 percent metallic iron and 24.17 percent silica.

Minerals from Mount Penn and Reading

Actinolite (fig. 482)

Allanite - with zircon (Dana, 1868, p. 777)

Apatite (Rogers, 1858, p. 717)

Calcite - north Reading (Genth, 1876, p. 228)

Goethite (figs. 483 and 485)

Gold (?) - in ferruginous quartz, found at 8th and 9th Streets, Reading, and near the suburb of Hampden, at the western base of Mount Penn (Wetherill, 1854, p. 234; D'Invilliers, 1883, p. 397)

Kaolinite - on the eastern slope of Mount Penn (D'Invilliers, 1883, p. 398).

Lepidocrocite - on P.D. Wanner's farm at the head of Walnut street (D'Invilliers, 1883, p. 397)

Limonite - north Reading (D'Invilliers, 1883, p. 398) (figs. 484 and 486)

Limonite, var. turgite - on P.D. Wanner's farm (D'Invilliers, 1883, p. 402)

Orthoclase - on the eastern slope of Mount Penn (D'Invilliers, 1883, p. 396)

Quarttz (fig. 487)

Quartz, vars. agate-jasper and jasper (Dana, 1868, p. 777; Genth, 1876, p. 60)

Quartz, var. smoky - crystals (Dana, 1868, p. 777)

Silver (?) - traces reported by Wetherill (1854, p. 234)

Zircon (Dana, 1868, p. 777)

Mount Penn White Spot Quarry

LIDAR imagery shows two large quarries in the white spot on Mount Penn. The northern quarry (location R-2A) was at 40° 20′ 36″ latitude and 75° 54′ 23″ longitude. The southern quarry (location R-3) was the Whitman quarry described below. The 1882 topographic map shows a third sandstone quarry (location R-2B) around 40° 20′ 13″ latitude and 75° 54′ 33″ longitude. The quarries are on the USGS Reading 7.5-minute topographic quadrangle map (Appendix 1, fig. 27). The three quarries are in the Hardyston Formation.

As early as 1761, a road was established from the east end of Penn Street to the "white spot" to quarry stone for the Reading courthouse, a church, and other public buildings. The quarry was known as the town quarry. Julius Keiter and Frederick Goodhand quarried rough stone in 1761 from the town quarry and received payment of 3 shillings and 10 pence per perch according to an agreement dated October 20, 1761.

Whitman Quarry

The Whitman quarry is the southern quarry in the "white spot" west of Mount Penn Boulevard at 40° 20′ 32″ latitude and 75° 54′ 24″ longitude on the USGS Reading 7.5-minute topographic quadrangle map (Appendix 1, fig. 27, location R-3).

William Abbott Whitman opened a quarry on Mount Penn around 1886 to produce white sandstone. The stone was used to construct numerous churches and the fronts of over 200 buildings in Reading. The Whitman quarry supplied stone for St. Matthew's Lutheran Church on the northwest corner of 5th and Elm Streets in Reading and for the St. Thomas Reformed Church at 11th and Windsor Streets.

After the stone was quarried and broken, it was conveyed by an inclined railroad to the bottom of the hill. Waste rock was crushed into sand. Between 1886 and 1896, 35,000 tons of stone and 15,000 tons of sand were produced. Howard Ahrens and George Tobias, who leased the Whitman quarry, installed a $5,000 stone crusher with a capacity of 150 tons in a 10-hour day, a 50-horsepower steam engine, and a 75-horsepower boiler. Whitman received a royalty on all stone and sand sold (*Reading Eagle*, July 1896).

In 1896, Whitman and his brother Jonathan (also known as John) sued the City of Reading when the construction of Mount Penn Boulevard cut off access to the the quarry, driving them out of business. The Whitmans were awarded $3,500 in

The Pagoda

William A. Whitman Sr. (1860-1936) was born on October 19, 1860, the eldest of six children. After finishing high school, Whitman worked for the Scott Works as a machinist. In 1886, with his brother Jonathan, he formed A. Whitman & Brother, a coal yard business. He later formed a construction company and purchased a 10-acre parcel of land on Mount Penn. In 1886, Whitman began quarrying stone on Mount Penn, which resulted in a white scar that was visible from all over Reading (Heizmann, 1971).

In 1886, Whitman began his political life when be became a Reading City councilman from the Eighth Ward. During his first year in city government, Whitman served on the Highways and Paving Committee. After completing his second term as Eight Ward councilman, Whitman moved to the Thirteenth Ward. In 1892, he ran for the office of Thirteenth Ward councilman, but was narrowly defeated. Whitman appealed the election results, and 3 years later won the appeal and was appointed councilman.

Whitman served the city government as a councilman from 1894 to 1907, earning a reputation as a flamboyant and fiery fighter and a holder of important committee chairmanships. The city began planning to build a park on Mount Penn, which would include a new road. In 1896, Whitman presented a bill to stop construction of the new road, which would end his quarry operations. The bill requested that the city amicably settle the damages accrued from the closing of the quarry. The bill was not passed. Whitman sued the city and eventually took the case to the Pennsylvania Supreme Court, where he lost.

In 1906, Whitman was targeted by the Taxpayer's League, a group dedicated to eliminating corruption from government. After forcing three corrupt councilmen out of the Reading City government, they focused on Whitman. They accused Whitman of defrauding the city in a sand and crushed stone contract. Under the municipal ordinance, a councilman could not be party to a contract. In court, Whitman explained that he owned the quarry, and that he and his brother jointly owned the land around the quarry. Whitman sold the sand and crushed stone to his brother Jonathan, who then sold them to the city. Jonathan acted as an independent agent in his business with the city. Whitman was found guilty and had to relinquish his seat on City Council.

Whitman unsuccessfully ran for mayor in 1908, 1911, 1919, and 1923. Between his mayoral campaigns in 1908 and 1911, Whitman built the Pagoda on the site of his quarry, using stone from the quarry in its construction. Witman purchased his brother's interest in the property prior to the start of construction. Local lore holds that Witman suffered much public criticism as a result of his stone quarry, which had defaced Mount Penn's western slope. His announcement on August 10, 1906, that he intended to build the Pagoda on the quarry site was made in an effort to win favor with voters. Whitman intended to operate the Pagoda as a luxury resort hotel and tourist attraction to compete with several other mountain retreats on Mount Penn and Neversink Mountain.

The Pagoda was designed and constructed by father-son contractors James and Charles Matz. The Pagoda was supposed to be a replica of a Japanese Shogun's palace, but was actually based on a structure in a Japanese tea garden at Coney Island, New York. The Pagoda is 7 stories high, 28 feet wide, and 50 feet long. The five overhanging roofs with upswept corners are covered by 88 tons of red tiles from Saint Mary's, Georgia. A bronze Japanese gong cast in 1739 hangs from the ceiling.

The Pagoda was completed in 1908 at a cost of $33,000 (or $50,000 depending on the source). Whitman applied for a liquor license, but it was denied. Because he did not have a liquor license, Whitman never opened the Pagoda. In 1910, the Farmer's National Bank foreclosed on the property. Jonathan Mould, owner of the Bee Hive Department store in Reading and a director of the Farmer's National Bank, purchased the Pagoda and 10 surrounding acres of land to save the bank from a loss. In 1911, Mould and his wife presented the Pagoda as a gift to the City of Reading. The city accepted the gift as an addition to their park system on Mount Penn. It was added to the National Register of Historic Places in 1972.

Summarized from Cassidy (1958), Heizmann, (1971), and Lynch (1995).

Figure 488. The Pagoda on Mount Penn, Reading, June 2016.

Figure 489. Gring's quarry, Reading, April 2015. A soccer field occupies the quarry floor.

damages; however, they were not satisfied with the award and filed an appeal, which they lost (*Reading Eagle*, March 17 and November 14, 1898).

Mount Penn Quarries

The Mount Penn quarries were east of Oley Street at about 40° 20′ 46″ latitude and 75° 54′ 33″ longitude on the USGS Reading 7.5-minute topographic quadrangle map (Appendix 1, fig. 27, location R-4). The quarries were in the Hardyston Formation.

In the early 1920s, the City of Reading began to acquire land on Mount Penn for a park. The Reading City council condemned August F. Kostenbader's land and offered him $11,000 for 13 acres that contained his quarry on the western slope of Mount Penn between Green and Oley Streets. The City also offered $34,000 to the Reading Sand and Stone Company for their land. Kostenbader and the Reading Sand and Stone Company appealed to the Pennsylvania State Supreme Court (*Reading Eagle*, October 13, 1922, and October 25, 1923).

The 18-acre Kirschmann stone and sand quarry property, which adjoined the Kostenbader quarry property, was purchased by the City of Reading for $50,000. The city auctioned off the machinery and buildings at the quarry for $1,596.53 in July 1924 (*Reading Eagle*, March 20, 1922, and July 2 and 12, 1924). The Kirschmann quarry was owned and operated by Edward C. Kirschmann ca. 1916 (Pennsylvania Department of Labor and Industry, 1916), Harry E. Kirschman ca. 1922 (Pennsylvania Department of Labor and Industry, 1922), and Elizabeth Kirschmann (*Reading Eagle*, October 13, 1922). The quarry was known as the Mount Penn Sand and Stone Works (fig. 490). In 1909, stone from Kirschmann quarry was used to build the St. James Reformed church in West Reading (*Reading Eagle*, October 4, 1909).

In 1900, A.K. Stauffer sued the City of Reading, alleging that construction of Mount Penn Boulevard cut off access to his stone and sand quarry. Stauffer sought $15,000 in damage, but was paid only $3,200. The city cut off access to the quarries because it did not want heavy teams on Mount Penn Boulevard (*Reading Eagle*, February 2, 1900, and May 29, 1906).

Long's Quarry

Long's quarry was on the west bank of the Schuylkill River at about 40° 19′ 24″ latitude and 75° 55′ 42″ longitude on the USGS Reading 7.5-minute topographic quadrangle map (Appendix 1, fig. 27, location R-5). The location of the quarry was determined from the 1937 aerial photography. The quarry was destroyed when U.S. Route 422 was constructed. The quarry is shown on the 1882 topographic map. It was in Lower and Middle Cambrian rocks.

George Long operated two quarries in a gray, massive limestone. The quarries were opened prior to 1868 (*Reading Eagle*, August 7, 1868). The stone was used as furnace flux by the Monocacy Iron Company, which furnished the following analyses: blue stone, calcium carbonate, 52.7 percent; magnesium carbonate, 44.6 percent; and silica, 1.4 percent; limestone, calcium carbonate, 47.3 percent; magnesium carbonate; 40.5 percent; and silica, 9.9 percent.

The larger of the two quarries was about 100 yards north of the railroad bridge. The larger quarry exposed about 30 feet of limestone with 10 feet of overburden. The limestone, which was greatly altered, displayed irregular cleavage and dipped about S. 40° E. 50°. The smaller of the two quarries was located north of the larger one. The smaller quarry exposed massive limestone that dipped S. 40° E. 40°-84° (D'Invilliers, 1883, p. 195).

Gring's Quarry

Gring's quarry (fig. 489) was on the east bank of Tulpehocken Creek, south of Pennsylvania State Route 12 at 40° 21′ 03″ latitude and 75° 57′ 20″ longitude on the USGS Reading 7.5-minute topographic quadrangle map (Appendix 1, fig. 26, location R-6). The quarry was in the Allentown Formation. Quarry operations began in the 1800s. The abandoned quarry

Mt. Penn Sand and Stone Works
E. C. KIRSCHMANN, Prop.
BOTH PHONES AT OFFICE
Office: 216 North Tenth Street
Works: 14th and Buttonwood Streets

Figure 490. Advertisement for Kirschmann's Mount Penn Sand and Stone Works, 1914. From Boyd's Reading Directory.

is now the Stonecliffe Recreation Area, which is part of the Berks County Parks system.

Gold in Reading

In 1853, Dr. Charles Mayer Weatherill (1853, p. 351) wrote in the Transactions of the American Philosophical Society: *"During a stay at Reading, in the summer of 1851, I noticed a vein of decayed ferruginous quartz, very much resembling the auriferous quartz of North Carolina. It was uncovered in exploring the deposits of iron ore in Penn's Mount behind the city. I neglected at the time to secure specimens, and upon a second visit to the locality this spring, to obtain a quantity for analysis, I found it covered. I obtained, however, from the vicinity a quartz rock, quartz and feldspar mingled, and sand, which, on analysis, yielded an exceedingly minute quantity of a brownish powder after treating the silver button resulting from cupellation by nitric acid; but which were too minute from which to derive any definite conclusion as to the presence or absence of gold. A former pupil of mine in an examination of the pyrites of the same locality, thought to have detected traces of gold. I have no doubt, that a more careful examination of the rocks in the vicinity would yield affirmative results in an examination for this metal."*

In 1854, Weatherill (1854, p. 233-234) wrote in the Proceedings of the Academy of Natural Sciences of Philadelphia: *"In a paper upon the occurrence of gold in Pennsylvania, read before the American Philosophical Society, and published in vol. x. of their Transactions, I alluded to an auriferous quartz in the neighborhood of Reading, Pa., and the examination of which afforded me slight, through uncertain traces of gold. I stated at the close of the article, that 'I had no doubt that a more careful examination of the rocks in the vicinity, would yield affirmative results in an examination for this metal.' The views then expressed have proved to be correct. Last summer, Mr. Philipps, a mining geologist, in searching for iron ore on the farm of Mr. Entlich, a few miles eastward from Reading, and of Mr. Jonathan Deininger, about a mile from the same place on the western slope of Penn's Mount, detected gold by washing specimens of the ferruginous quartz. I called upon Mr. Deininger, who showed me the specimens in his possession, and gave me some of the quartz rock from his farm. Mr. Deininger showed me a specimen of gold, in weight I should judge between one and five centigrammes, which was broken by himself out of the rock.*

At the angle formed by the intersection of 8th and 9th streets, Reading, there is a heap of stones gathered from the adjoining fields, containing about two percent, of pieces of this quartz rock; I brought home specimens with me for examination. These specimens, together with those obtained from Mr. Deininger's field by myself, were pulverized and washed, but without, in any instance, detecting gold. They were then smelted with litharge and charcoal, and the button of lead cupelled. Of course, the litharge was examined for gold. The 30 gramme button of lead from about 100 of litharge gave a silver button of 0.00575 grms., and which contained no gold.

A. 8th street quartz - 65 grammes + 130 litharge + 10 black flux gave a lead button of 14 grms., and silver 0.0075, which contained gold beyond a doubt, as judged from its lustre and resistance to nitric acid.

B. Another portion of quartz from the same locality - 200 grms + 400 lithrage + 0.5 charcoal dust, gave lead 17 grms.; silver 0.00875 containing gold, though not as distinctly as the last.

C. Quartz from Mr. Deininger's fields - 185 grms. + 370 litharge + 0.5 charcoal gave 20 grms. of lead containing 0.00825 silver, in which no gold could be detected."

Quarries in the City of Reading

- Quarry at Elm and Rose Streets furnished stone for the Elm Street sewer (*Reading Eagle*, October 8, 1881)

- Judge Stitzel's quarry on Woodward Street employed six men in 1881

- Simon Kline operated a quarry at the corner of Center and 6th Streets (ca. 1877)

- Major Kestner operated a quarry at 14th and Oley Streets (*Reading Eagle*, July 27, 1919)

- In 1871, the Reading City council appropriated $20 to fill up the old stone quarry on Moss street between Windsor and Green Streets (*Reading Eagle*, August 1, 1871)

- Morris Keehn operated a quarry on Pear Street between Walnut and Washington Streets (*Reading Eagle*, August 24, 1895)

- A boy drowned in an abandoned quarry at Buttonwood and 14th Streets (*Reading Eagle*, September 4, 1931)

- Becker, Bauer & Company operated a quarry at Buttonwood and 6th Streets (ca. 1877)

- John F. Huber operated a quarry at the corner of Buttonwood and 2nd Streets that produced curbing and building stone (ca. 1877)

- The David Benson quarry was located on Front Street (ca. 1885)

- Thalheimer's sand hole was located at the head of Green Street

- In 1882, the City of Reading offered the use of the quarry at Washington and Pear Streets for a royalty of 6 cents per ton for all stone quarried (*Reading Eagle*, February 14, 1882)

- An accident occurred at the Melchoir Brown sand quarry at 14th and Douglas Streets (*Reading Eagle*, December 11, 1901)

- Quarry on 12th street between Spruce and Muhlenberg became the site of St. Mary's Church,

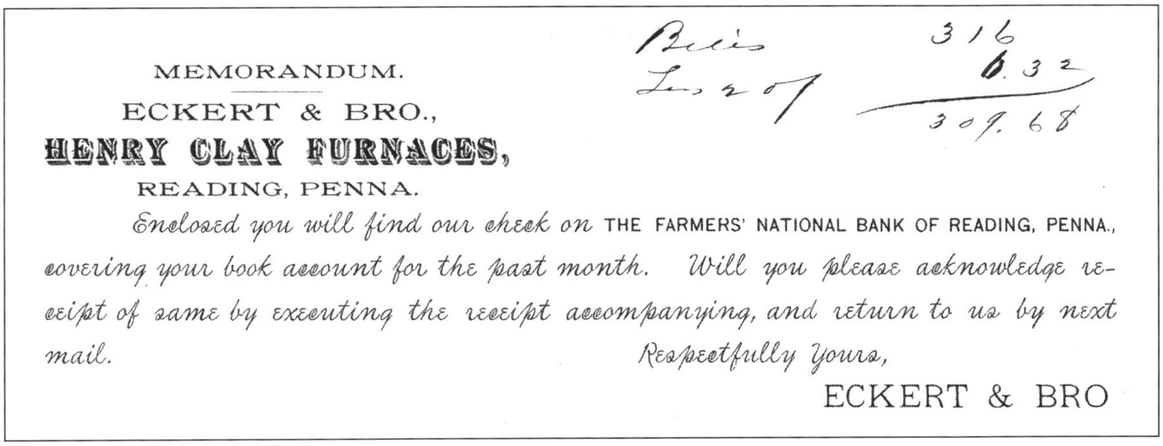

MEMORANDUM.

ECKERT & BRO.,

HENRY CLAY FURNACES,

READING, PENNA.

316
6.32
309.68

Enclosed you will find our check on THE FARMERS' NATIONAL BANK OF READING, PENNA., *covering your book account for the past month. Will you please acknowledge receipt of same by executing the receipt accompanying, and return to us by next mail.*

Respectfully Yours,

ECKERT & BRO

Figure 491. Eckert & Brother Henry Clay Furnace letterhead.

high school, and the Felician sisters convent (*Reading Eagle* May 12, 1937)

- Reuben Klapp quarry at North 2nd and Walnut streets (ca. 1877)
- H.E. Ahren's quarry was along the west bank of the Union Canal (*Reading Eagle*, May 24,1889)
- John H. Sternburg limestone quarry opened in 1840
- Charles Davis quarry on South 8th Street (ca. 1877)
- Christian Eben quarry at South 6th and Buttonwood Streets (ca. 1877)

Henry Clay Furnace

The Henry Clay Furnace in Reading was built by Isaac Eckert and his brother George Nicholas Eckert and operated under the firm name Eckert & Brother. It was located on the Pennsylvania Railroad on the south end of Reading. Furnace No. 1 was built in 1842-44 and was named for Henry Clay At the time of construction, Furnace No. 2 was built in 1855. The furnace used a combination of limonite and magnetite. In 1856, the furnace produced 4,729 tons of pig iron in 46 weeks (Lesley, 1859). After George's death in 1865, Isaac operated the furnace for another 7 years.

In 1869, the furnace employed 127 men and used 29 horses. It consumed 12,000 tons of anthracite coal, 6,000 tons of limestone, and 9,000 tons of iron ore to produce 3,000 tons of pig iron per year (*Reading Eagle*, January 30, 1869).

Shortly before Isaac's death in 1873, his sons Harry S. Eckert and George B. Eckert took over the furnace and continued its operation under the firm name of Eckert & Brother.

The Henry Clay Furnace smelted limonite and magnetite from Berks and Lebanon counties and produced foundry and gray forge pig iron sold under the brand name "Henry Clay." The total annual capacity was 36,000 tons (American Iron and Steel Association, 1884, 1892, and 1896). In 1899, the Empire Steel and Iron Company acquired the furnace. The furnace went out of blast in 1916.

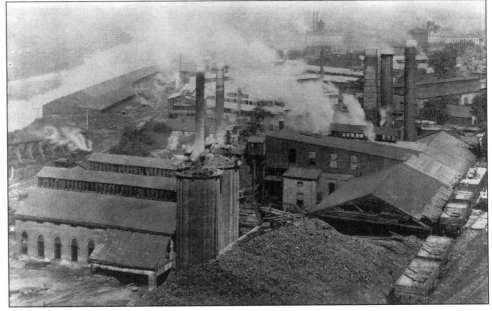

Figure 492. Henry Clay Furnace, Reading. American Iron and Steel Institute photograph collection 1968268_0141, Hagley Museum and Library, Wilmington.

Figure 493. Reading Iron Company Furnace, Reading, 1907.

Keystone Furnace

The Keystone Furnace in Reading was built by the Keystone Furnace Company. The first stack was built in 1869, and the second stack was built in 1872-73. The total annual capacity was 20,500 tons. Jacob Bushong was president of the Keystone Furnace Company, and H.M. Bushong, was secretary and treasurer (American Iron and Steel Association, 1876, 1880, 1884, 1886, and 1888). The Bushong Brothers operated the Keystone Furnace and were the owners of three large paper mills and several large farms, They were largely responsible for building the Berks County Railroad, later known as the Schuylkill & Lehigh Railroad. They also were instrumental in the construction of the Wilmington & Northern Railroad and were engaged in other business enterprises. In 1889, the furnace was acquired by the Reading Iron Company.

Reading Iron Company Furnace

The Reading anthracite furnace at the south end of Reading was built in 1853-54 by Seyfert, McManus & Company. In 39 weeks in 1856, the furnace produced 5,9721 tons of iron from a mixture of limonite and magnetite (Lesley, 1859). A second stack was constructed in 1873 and was blown in on October 5, 1874. The total annual capacity in 1876 was 20,000 tons (American Iron and Steel Association, 1876).

In 1878, the Reading Iron Works was incorporated and acquired the furnace. The furnace used mainly limonite from Lehigh and Lebanon Counties. The principal product was foundry and mill pig iron Two new stacks were constructed in 1872 and 1873, and the older stacks were remodeled in 1886. In 1888, the total annual capacity was 85,000 tons (American Iron and Steel Association, 1888).

The Reading Iron Works went bankrupt in 1889 and was put up for sale. Forty minutes after the bidding began, the property, which included furnaces, rolling mills, tube mills, pipe mills, a foundry, forges, and real estate, was sold to William P. Bard, a Reading lawyer, for $150,500, subject to a mortgage of $600,000. Bard made the purchase for the Philadelphia and Reading Coal and Iron Company. On August 12, 1889, the company was reorganized and incorporated under the name of the Reading Iron Company (fig. 494). The company acquired the Reading Keystone Furnace in 1889.

In 1890, the furnace used ore from the Lake Superior region, local limonite, and magnetite from New Jersey and New York. The furnace produced foundry and mill pig iron and had a total annual capacity of 85,000 tons (American Iron and Steel Association, 1890).

In 1938, the Philadelphia and Reading Coal and Iron Company ordered the immediate sale of the Reading Iron Company's stock of finished products and materials. On the last day of 1938, permits were obtained to raze 85 buildings at three plants of the Reading Iron Company, which eliminated, with one exception, all remaining operations of the company in Reading. The exception was the nail mill at the South 7th Street pipe mill, the only unit in regular operation in 1938.

Figure 494. Reading Iron Company letterhead, 1897.

RICHMOND TOWNSHIP

Dragon Cave

Dragon Cave, about 1 mile northeast of Virginville, received its name from an old legend. According to the legend, one day at dusk, native Americans living along the banks of Saucony Creek saw a dragon surrounded by lightning, descend swiftly out of thunderheads above Blue Mountain and descended with a roar into the cave (Barnsley, 1932, p. 23-24).

According to local lore, fishermen from Oley fishing in Maiden Creek discovered the cave. They did not explore the cave at the time of discovery; however, they returned at a later date. They entered the cave and failed to return at the specified time. A rescue party found the men lost in the cave. Their lights had gone out, and they were unable to find their way back to the entrance. The rescuers avoided the same fate by taking a long rope and fastening one end at the cave opening (Wagner and others, 1913, p. 148-149).

Rupp (1843, p. 240-241) provided an early description of Dragon's Cave: "*The entrance to the cave is on the brow of a hill, in the edge of a cultivated field. Passing into it, the adventurer descends about fifty yards through narrow passage, and then turns to the left, at an angle with the passage hitherto pursued. After proceeding about thirty yards farther, he enters the great chamber, fifty feet long, twenty wide, and fifteen to twenty feet high, a rock of limestone. Near the end of this chamber, opposite the entrance, is the altar, a large mass of Stalactite, which rings under the hammer, and is translucent.*

Figure 495. Calcite from Dragon Cave, Richmond Township, 6.4 cm. Sloto collection 3317.

Formation stalactites are found in other parts of the cave, though not as large as the mass just mentioned."

The mouth of the cave is five feet wide, almost 10 feet long, and nearly horizontal. From the perpendicular entrance a sloping passage leads about 150 feet to a small opening. To pass this, one must crawl a short distance. For some 300 feet there runs a passage in which there is sufficient room to walk erect. The way is winding and is connected to chambers of various dimensions. The largest room is about 25 to 30 feet wide,

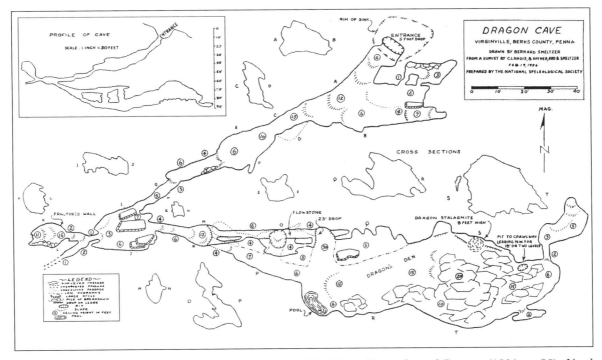

Figure 496. Map of Dragon Cave, Richmond Township. From Mostardi, and Durant (1991, p. 26). Used with permission from the Mid-Appalachian Region of the National Speleological Society.

Figure 497. Aragonite from Dragon Cave, Richmond Township, 4.5 cm. Sloto collection 3316.

approximately the same height, and about 75 to 90 feet in length. It was given the name "The Temple of the Dragon." At one end of the Temple, a flight of natural stone steps leads to a gallery above, from which point a person can look down into all parts of the Temple (Smeltzer, York Grotto Newsletter, vol. 13, no. 1. p. 10)).

C.K. Williams & Company Ocher Mines

The C.K. Williams & Company ocher mine was east of the intersection of Fleetwood-Lyons and Dryville Roads on the USGS Fleetwood 7.5-minute topographic quadrangle map (Appendix 1, fig. 12, location RH-5). The mines were located on the slope of South Mountain.

The C.K. Williams works at Fleetwood included three adjacent properties—the Schafeffer, Williams, and Schollenberger properties The Schaeffer property contained the best ocher and was the oldest of the properties (Meiser and Meiser, 1982, p. 43).

C.K. Williams & Company operated two ocher mines east of Fleetwood. When first visited by Benjamin Miller in May 1909, one of the mines was operated by the Keystone Ocher Company. It was acquired by C.K. Williams in February 1910. After that date, only one mine at a time was in operation, although both could be worked at the same time, if necessary. Miller (1911, p. 39) reported that after the consolidation of the companies, a new mill was constructed to replace the two old mills. The new mill contained a Sprout-Waldron grinder with a capacity of 10 tons per day. The planned output of the new mill was 2,000 to 2,500 tons per year. Steam drying was abandoned because of the danger of the ocher contacting the steam pipes, which caused it to turn red.

The mines were 3/8 mile from the Philadelphia and Reading Railroad. The finished ocher was hauled in wagons to the railroad and loaded into cars for shipment. At the time of Miller's last visit in August 1911, work had just ceased in one mine, and poor ventilation in the other mine rendered it unsafe to enter. Consequently, the descriptions given by Miller were based on his visit in May 1909. Other descriptions of the mines were published by Stoddard and Callen (1910, p. 427-433) and were repeated by Miller (1911, p. 36).

The following description of the Keystone Ocher Company mine was provided by Stoddard and Callen (1910, p. 431-433): *"The deposit is opened up through two shafts within 60 feet of each other, one being used as a hoisting and pump shaft and the other as an air and timber shaft. The former is 70 feet deep and extends down to the lower level, from which all the ocher is hoisted. The method of carrying on the underground*

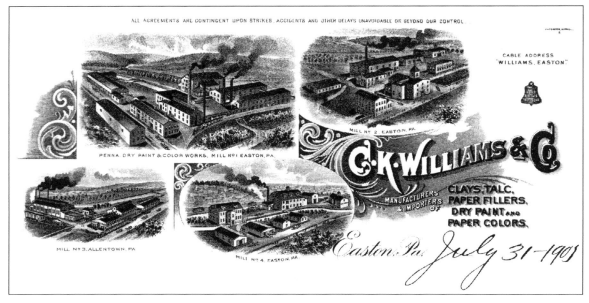

Figure 498. Letterhead of C.K. Williams & Company, 1901.

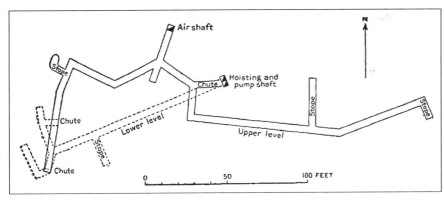

Figure 499. Plan of the Keystone Ocher Company underground workings, Richmond Township. From Stoddard and Callen (1910, p. 431).

work is to drift along and follow the pockets and stringers of ocher, mining them out in stopes or breasts, and then to drift indefinitely until other deposits are found.

The accompanying plan [fig. 499] shows the approximate layout of the underground workings. The drifts or gangways are 6 to 7 feet high and 5 feet wide, being provided with four-piece round timbering to resist the squeezing action of the clay. Lagging of sawed slabs is laid close on the tops and sides, and the bottom is plank floored for the passage of wheelbarrows. Chutes are provided, as shown [fig. 499], for dumping the ocher from the upper to the lower level, whence it is wheeled to the shaft and hoisted. The stopes are turned off where pockets are encountered, and, if their size demands it they are timbered up with square sets.

The ocher occurs either as small masses in pockets in the clay, or interstratified with the clay, as shown in the accompanying sketch [fig. 500]. It is separated by hand from the clay in the mine, and the clay is used to fill up the old workings. The impurities in the ocher are particles of quartzite, cherty limestone, flakes of shaly limestone, and fragments and nodules of limonite. The limonite is picked out on the surface and is saved until a sufficient quantity for shipment has accumulated. No bed rock has been encountered in the mine workings, but a well drilled down the hoisting shaft struck loose bowlders of sandstone at 257 feet, which prevented drilling deeper.

The method of treating the ocher for the market is essentially the same as the methods previously described for the Reading plant [in Muhlenburg Township], but the equipment is more complete.

The ocher is hoisted from the mine by an engine hoist and then dumped into a log washer, from which it passes to a series of 28 floating troughs. These troughs are 14 to 16 feet long and 13 inches square in cross section. The fine sand is separated out in the first 12 or 13 troughs, and the final separation is accomplished in the smaller set of 15,

after which the mixture is run through a long trough to the settling ponds. Here it is left to partly dry as a preliminary to its transfer to the drying sheds. After it has thoroughly dried in the sheds it is ground in French buhr mills as the final treatment for the market. The best sienna from this plant brings from $30 to $40 per ton, and the washed ocher brings $15 to $18 per ton.

The land is usually leased for a period of fifteen or twenty years, one year or six months being allowed for exploration before the lease is executed finally. A royalty is paid to the owner either at a nominal rate or according to the amount of ocher taken out at a fixed price per ton."

Beginning in 1883, the C.K. Williams & Company mine was worked intermittently for iron ore, which was found in the upper levels. Old drifts and shafts indicated that considerable work was done on the property in mining the limonite deposit. In 1910, the No. 2 mine hoisting shaft extended downward 91 feet to the bottom of the lower level and 126 feet to the bottom of the sump, which received the mine water and was pumped out at intervals. Fifty feet from the main hoisting shaft, there was an air shaft 46 feet deep connected with the upper level of the mine. The underground workings were similar to those of the Keystone Ocher Company, but were larger. The two levels were connected by chutes and by an old shaft, which was retimbered and repaired for the passage of the miners.

The mining method was described by Warren Boyer, Jr., a former employee of the C.K. Williams & Company mine. A team of three miners worked together. The miners were lowered down the shaft into the mine in buckets about 30 inches deep and 36 inches across. One miner used a pick to dislodge ore, one shoveled the ore into a wheeled bucket, and a third wheeled the bucket to the shaft, where it was hoisted to the surface. To lessen the chance of a cave in, drifts were cut in the shape of an inverted keystone—5 feet wide at the top, 6 feet wide at the bottom, and 6.5 feet high. Timbers were set every 3

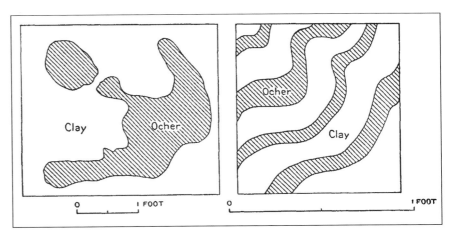

Figure 500. Occurrence of ocher and clay in the Keystone Ocher Company mine, Richmond Township. From Stoddard and Callen (1910, p. 432).

Figure 501. Settling troughs and drying sheds, C.K. Williams ocher mine, Richmond Township, 1911. From Miller (1911, plate 6).

Figure 502. Settling pond, drying shed, and mill, C.K. Williams ocher mine, Richmond Township, 1911. From Miller (1911, plate 9).

Figure 503. Settling boxes, C.K. Williams ocher mine, Richmond Township, 1911. From Miller (1911, plate 8).

Figure 504. Drying sheds filled with ocher, C.K. Williams ocher mine, Richmond Township, 1911. From Miller (1911, plate 11).

Figure 505. Shaft house and settling troughs, C.K. Williams ocher mine, Richmond Township, 1911. From Miller (1911, plate 7).

Figure 506. Engine house and tracks, C.K. Williams ocher mine, Richmond Township, 1911. From Miller (1911, plate 10).

RICHMOND TOWNSHIP

Figure 507. Removing partially dried ocher from the settling ponds, C.K. Williams ocher mine, Richmond Township, 1911. From Miller (1911, plate 11)

Figure 508. Drying sheds and settling pond, C.K. Williams ocher mine, Richmond Township. Undated photograph. Courtesy of Tom Pracher.

Figure 509. Drying shed, C.K. Williams ocher mine, Richmond Township. Undated photograph. Courtesy of Tom Pracher.

Figure 510. Drying sheds, C.K. Williams ocher mine, Richmond Township, 1940s. Courtesy of Tom Pracher.

Figure 511. Ocher pits, C.K. Williams ocher mine, Richmond Township, April 2015.

Figure 512. Tom Pracher and the rock house, C.K. Williams ocher mine, Richmond Township, April 2015. The building is faced with colorful red, orange, brown, and yellow ocher rock.

feet, and lagging (slab wood) was place behind the tops and sides of the timbers (Meiser and Meiser, 1982, p. 43).

The plant was described by Stoddard and Callen (1910, p. 433): *"The washing and drying plant consists of a log washer, 26 floating troughs 16 feet long, four mud dams, and four drying sheds. The ocher and iron ore occur in pockets, the ore predominating in the upper levels and the ocher in the lower, with clay between. The deposit seems to be in the form of a horseshoe extending along the hill, with its greatest dimension parallel to the hill. The bands of iron ore, clay, and ocher appear to run horizontally. The underlying rock is quartzite, which outcrops along the ridge with a dip of 75° toward the bottom of the hill.*

The ocher found at this mine is of three grades, as follows:
1. Gold dust, called No. 1, the purest variety.
2. Gravel ocher, which is good ocher but contains particles of limonite that have to be washed out.
3. Clay and ocher, which is the poorest variety and contains pieces of chert up to 2 feet in diameter. The clay is red, yellow, white, and purplish and is of no value.

The time taken to treat the ocher varies, but in general it takes a month to completely fill the mud dams and three weeks more for the material to dry sufficiently to permit being shoveled. When it has dried to the consistency of a stiff mush it is put in the drying sheds, where it has to be left one month more before it is in condition to grind. The material is finished at the company's mills at Easton, Pa."

Stoddard and Callen (1910, p. 427) described the ocher: *"The ocher occurs in irregular masses in the clay. At the Keystone Ocher Company's mine at Fleetwood there was evidence of stratification in the clay and ocher. Some of the masses are large and can easily be worked for high-grade ocher, but many of them are simply small pockets, which can be used only for second or third grade on account of the large amount of clay which must be mined with the ocher. Most of this clay is so fine that it can not be separated by washing and settling and so lowers the quality of the finished product. As a rule there is a considerable thickness of clay above the rock. A well boring at Fleetwood gave over 250 feet of clay, below which a bed reported to be unconsolidated gravel was struck, which could not be penetrated by the drill."*

Fleetwood Area

Minerals reported from the Fleetwood area iron mines are listed below.

Epidote (D'Invilliers, 1883, p. 396) (fig. 518)
Kaolinite
Hematite, var. turgite - dark hematite with botryoidal surfaces that exhibit strong iridescence from yellow to blue to pink
Limonite (figs. 514 and 517)
Limonite, var. ocher
Limonite, var. xanthosiderite (Eyerman, 1889, p. 11)

Quartz - crystals to 4.2 cm; doubly terminated crystals (fig. 516); drusy (fig. 515)
Quartz, vars. chalcedony, jasper, flint (D'Invilliers, 1883, p. 400); pale blue chalcedony balls coating a cavities in hematite

Kirbyville Quarries

The two flooded Kirbyville quarries were east of the intersection of U.S. Route 222 and Richmaiden Road. The eastern quarry (location RH-3A) was at 40° 28′ 20″ latitude and 75° 51′ 23″ longitude, and the western quarry (location RH-3B) was at 40° 28′ 17″ latitude and 75° 51′ 33″ longitude on the USGS Fleetwood 7.5-minute topographic quadrangle map (Appendix 1, fig. 11). The quarries were in the Stonehenge Formation.

The Kirbyville quarries were worked primarily for rock used in blending for the manufacture of cement. The last operator was the Allentown Portland Cement Company, which abandoned the quarry around 1954 when the quarry ran out of high calcium limestone (*Reading Eagle*, April 22, 1954).

Moselem Mines

The Moselem mines covered a large area north of the intersection of Mine and Richmond Roads. A large, water-filled pit is at 40° 29′ 26″ latitude and 75° 51′ 48″ longitude on the USGS Fleetwood 7.5-minute topographic quadrangle map (Appendix 1, fig. 11, location RH-1). The mines are in the Ontelaunee Formation.

The Leesport Iron Company mine (fig. 519) is considered one of the Moselem mines. LIDAR imagery shows that the Leesport Iron Company mine was a large open-pit mine south of the main Moselem mine pit, located west of the intersection of Mine and Richmond Roads at 40° 29′ 13″ latitude and 75° 51′ 01″ longitude on the USGS Fleetwood 7.5-minute topo-

Figure 513. Kirbyville quarry, Richmond Township, April 2015.

MINERALS FROM THE FLEETWOOD AREA

Figure 514. Limonite from the Fleetwood area, Richmond Township, 6.7 cm. Reading Public Museum collection 2000c-27-091.

Figure 517. Limonite "bog ore" from the Fleetwood area, Richmond Township, 19 cm. Reading Public Museum collection 2005c-3-716.

Figure 515. Druzy quartz from the Fleetwood area, Richmond Township. Reading Public Museum collection 2005C-2-388.

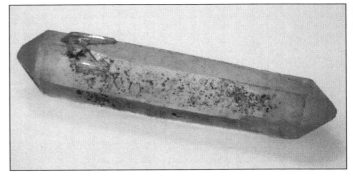

Figure 516. Quartz crystal from the Fleetwood area, Richmond Township, 3.3 cm. Reading Public Museum collection 2005c-2-187.

Figure 518. Epidote from the Fleetwood area, Richmond Township. Bryn Mawr College collection Vaux 5931.

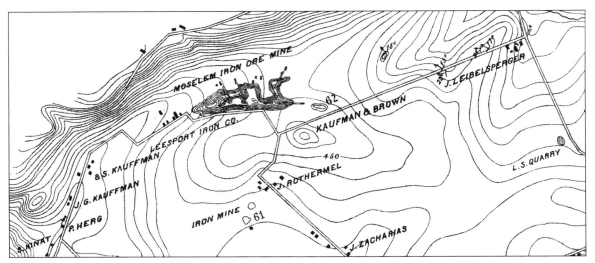

Figure 519. Map of the Moselem mines, Richmond Township, 1882. From Pennsylvania Geological Survey (1883).

graphic quadrangle map (Appendix 1, fig. 11, location RH-4). It is shown on the 1882 topographic map as mine number 61. The Rothermel iron mine is shown in the 1876 atlas (fig. 519). In 1859, S.H. Rothermel and his brother leased part of a general store building on the northwest corner of Franklin and Main Streets in Fleetwood to operate their iron ore business. They hired miners to mine the ore, hired teams to haul the ore, and sold and rented all kinds of mining equipment (Stitzer, 2013). In July 1880, the Rothermel Iron Company and the Rothermel Mining Company leased property from Reuben Dunkel (Berks County Deed book, vol. 24, p. 572).

Moselem forge was built prior to 1766 by Frederick Delaplank and probably Jacob Shaffer. In 1786, Christian Lower sold the forge and two tracts of land totaling 243 acres to Valentine Eckert. Eckert built a charcoal iron furnace, known as the Moselem Furnace, on this property. On January 1, 1799, Eckert sold the furnace and forge to his nephew Solomon Eckert, who operated the furnace until 1817. By that time Solomon was the owner of several thousand acres of land, and these acquisitions may have resulted in indebtedness. In November 1817, he assigned the furnace, forge, and 759 acres of land in Richmond Township to his creditors, Joseph Old, John Wily, George Gernant, and Samuel Baird.

On March 12, 1820, the Moselem Furnace property was sold at sheriff's sale to Jacob V.R. Hunter for $10,725. The purchase price included Moselem Furnace, a forge, a grist mill, a saw mill, and 323 acres of land. On May 27, 1820, the property was conveyed to Jacob's father, Nicholas Hunter. Nicholas Hunter (the elder) then appointed his son, Nicholas V.R. Hunter (the second) as the ironmaster at Moselem Furnace. On March 20, 1828, Nicholas Hunter conveyed the Moselem furnace and 143 acres to his son Nicholas V.R. Hunter. In 1830, the Moselem Furnace employed 18 workmen, had 15 horses, used 4,500 cords of wood, and produced 643 tons of pig iron.

In 1842, Nicholas V.R. Hunter purchased 3 acres of land known as the "Moselem mine holes" for $1,800 at sheriff's sale. It was formerly the property of the heirs of Daniel Udree. In 1846, Hunter purchased additional property in Richmond Township containing iron ore from the Mengle estate.

The furnace was remodeled when it was operated by Frederick S., Nicholas S., and Daniel S. Hunter under the company name of F.S. Hunter & Brothers. After several years, Nicholas S. Hunter became sole proprietor of the Moselem Furnace. In 1847, Frederick S. Hunter moved to Leesport where he became president and general manager of the Leesport Iron Company, a position he held until 1863.

In 1847, Nicholas S. Hunter completely rebuilt the furnace, increasing production to 100 tons of iron per week. In 1852, Hunter purchased a quarry in Ontelaunee Township between the Centre Turnpike and the Schuylkill River from William H. Clymer, who had purchased it 1 month earlier at the sheriff's sale of the property of ironmasters Darrah and Jones for $5,000 (*Reading Eagle*, March 10, 1918). In 1872, Hunter rebuilt the furnace as a one stack, hot blast, anthracite furnace with a capacity of 4,000 tons. The furnace went out of blast in 1883.

The Moselem iron ore deposit supposedly was discovered about 1750. Rogers (1858, vol. 1, p. 266) described the Moselem mine: "*The ore is obtained by sinking shafts through the soil, and is commonly reached at a depth of from 20 to 40 feet. The diggings are very extensive, covering an area of several acres, and have been wrought for very many years, yielding a large amount of ore. This is said to occur in nests and irregular layers, varying in thickness from 1 to 8 feet; it is of good quality. Some of it has a bluish tint, and contains a little manganese.*"

Rogers (1858, vol. 2, p. 726) further described the mine: "*At this locality there are two pits, an Eastern and a Western, 200 yards asunder, situated in a narrow valley between a conspicuous slate-ridge and a lower limestone hill, in both of which the dip is N. W., the limestone passing under the slate. The E. mine is wrought chiefly for the supply of the furnaces of Seyfert, McManus, & Co., at Reading. The surface excavation is not *large, and the ore is found in confused and irregular bunches of a few inches thickness up to 20 feet; it is intermixed with clay and earth. The ore is now obtained, in great measure, from a series of gangways diverging from a shaft at a depth of 120 feet. The ore is a brown compact haematite of the average quality, in lumps large and small. The annual yield of the mine is 10,000 tons of ore, raised at a cost of $1.50 per ton. The W. mine is exclusively an open pit, excavated 80 feet deep over half an acre. The ore is delivered to the Leesport furnaces, 9 miles above Reading. As at the former mine, the ore is found in irregular nests, and requires to be excavated and washed before it is fit for the furnaces. Yield, 12,000 tons per annum; cost of mining, $1.30 per ton under ordinary circumstances. In both of these mines, but especially the latter, many fragments of slate and cherty limestone are found only partially converted into ore."*

Figure 520. The Moselem mine, Richmond Township, April 2015.

In 1878, the mine was surveyed by A.P. Berlin. It was 2,000 feet long and 100 feet deep with five inclined planes (Lesley, 1892, vol. 1, p. 350-351). In 1885, the holdings of the Moselem Iron Company included furnace buildings, a grist mill, a saw mill, a stone hotel, 42 tenant homes, a fine mansion, 700 acres of land, three farms, and thousands of dollars of mining equipment. The 1880 Census (Pumpelly, 1886, p. 960) listed Librandt and McDowell as the operator of the Moselem mine. Production between July 1, 1879, and June 30, 1880, was 5,183 tons of limonite. The 1880 Census (Pumpelly, 1886, p. 960) listed D.B. Fisher as an operator of the Leesport Iron Company mine at Moselem. During the same time period, he mined 1,530 tons of limonite, which was shipped to the Leesport Iron Company.

The ore was hauled across the hill to the Moselem Furnace by horse and mule teams until the Moselem Ore Mine Railroad was built. After the Moselem Furnace went out of blast in 1883, ore was hauled to the wharf at Evansville for shipment. In later years, ocher also was mined and hauled to Fleetwood for used in paint manufacturing. Moselem Furnace was abandoned in 1892.

In the 1880s, the Moselem mine was operated by John J. Kauffman & Company. The mine supervisor was John K. Hawkins (*Reading Eagle* May 17, 1888). On November 18, 1896, the 147-acre farm including the "*celebrated Moselem iron ore banks of 30 acres*" was offered for public sale by the John Kaufman estate. The advertisement in the *Reading Eagle* (November 7, 1896) also stated, "*Machinery for operating the ore banks is on the premises and can be purchased at a later sale.*" Also offered for public sale at the same time was a 117-acre farm with a limestone quarry and kiln.

The Moselem mines were idle from 1891 to 1903. In 1902, the Moselem Mining Company was organized to resume mining. The incorporators were J.K. Spang of Reading, A.K. Hartzel of Allentown, Abraham C. Godshall and Harvey H. Godshall of Lansdale, and James R. Weimer of Leesport (*Reading Eagle*, October 21, 1902, and April 2, 1903). Mining resumed in 1903 and continued until 1906. In 1904, the *Reading Eagle* (February 6, 1904) reported that the Moselem Mining Company was shipping 10 car loads of ore per week.

An analysis of the ore was presented by McCreath (1879, p. 211): 40.8-87.7 percent metallic iron, 0.004-0.013 percent sulfur, 0.18-0.4 percent phosphorus, and 0.8-1.0 percent metallic manganese. D'Invilliers (1883, p. 397-401) reported oolitic chalcedony and goethite from the Moselem mine.

Figure 521. Moselem Furnace, Richmond Township, ca. 1930s. American Iron and Steel Institute photograph collection, 1968268_0196, Hagley Museum and Library, Wilmington.

Richmond Township

Figure 522. Goethite from the Moselem mine, Richmond Township, 3.6 cm. Reading Public Museum collection 2000c-27-072.

Virginville Area

D'Invilliers (883, p. 399) reported pyrite in globular and radiating forms from the Virginville area.

Kirbyville-Moselem Springs Quartz Crystal Locality

The Kirbyville-Moselem Springs quartz crystal locality is the area along U.S. Route 222 between Kirbyville and Moselem Springs and perhaps beyond in both directions. The area is on the USGS Fleetwood 7.5-minute topographic quadrangle map (Appendix 1, fig. 11, location RH-2). The quartz crystal area probably also extends into Maiden Creek Township to the west. Quartz crystals were found in farm fields and roadside gutters (Gordon, 1922, p. 157). The area is known for colorless, doubly terminated quartz crystals to 6.4 cm long (fig. 523).

Noll's Mine

Noll's mine was located near Moselem Springs. In 1879, Jerry Noll leased his property to the Moselem Iron Company (*Reading Eagle*, May 6, 1879). The mine also was known as Peter and Noll's mine. The Reading Eagle (January 6, 1881) reported that Peter and Noll's mine was an open cut, and William Bower was the mine engineer. Mine cars were hoisted up and down an inclined plane on a wire rope. D'Invilliers (1883, p. 399-400) reported doubly terminated quartz crystals at Noll's Mine in Fleetwood (fig. 524).

Charles Heffner's mine

LIDAR imagery shows the location of two mines east of Heffner and Deka Roads. The eastern mine (location RH-6A), labeled as mine number 56 on the 1882 topographic map, was at about 40° 28' 10" latitude and 75° 46' 01" longitude, and the western mine, labeled as mine number 57 on the 1882 topographic map, was at about 40° 28' 09" latitude and 75° 46' 04" longitude on the USGS Fleetwood 7.5-minute topographic quadrangle map (Appendix 1, fig. 12).

The 1880 Census (Pumpelly, 1886, p. 960) listed the Clymer Iron Company as the operator of Charles Heffner's mine. Production between July 1, 1879, and June 30, 1880, was 1,800 tons of limonite, which was shipped to the Clymer Iron Company furnace in Temple. Charles Heffner's mine was leased by the Temple Iron Company ca. 1886. In 1899, the mine was purchased by Issac D. Fegley (*Reading Eagle*, April 24, 1899).

Charles Heffner's mine produced a compact limonite ore that was sandy and full of ocher seams. Some of the limonite was cellular with clay-filled cells. The ore was various shades of light and dark brown. An analysis was presented by McCreath (1879, p. 212): 36.15 percent metallic iron, 0.026 percent sulfur, and 0.361 percent phosphorus.

Figure 523. Quartz crystals from Kirbyville, Richmond Township. Top, 6 cm. Bottom 3.4 cm. Collected in 1908 and 1910. Carnegie Museum of Natural History collection ANSP 24764.

Figure 524. Quartz crystals from Noll's mine, Richmond Township. Top, 4.1 cm. Bottom 4.2 cm. Bryn Mawr College collection Rand 4770.

One of Charles Heffner's mines was operated by Frank Brownback and was known as Brownback's mine. The 1880 Census (Pumpelly, 1886, p. 960) listed Frank Brownback as the operator of the Heffner mine. Production between July 1, 1879, and June 30, 1880, was 3,406 tons of limonite, which was shipped to the Lyons Furnace. An analysis was presented by Putnam (1886, p. 186): 41.25 percent metallic iron, 0.032 percent sulfur, and 0.741 percent phosphorus.

Merkel Quarries

LIDAR imagery shows several old quarries along the east side of Moselem Creek, north of the intersection of Pennsylvania State Route 662 and Mine Road, on the Moselem Springs golf course. The largest quarry was at 40° 29′ 53″ latitude and 75° 50′ 37″ longitude on the USGS Fleetwood 7.5-minute topographic quadrangle map (Appendix 1, fig. 11, location RH-7). The quarries were in the Epler Formation. The quarries were on the farm of A. Merkel, who produced stone and lime (Mine and Quarry News Bureau, 1897).

Richmond No. 5 Quarry

The Lehigh Cement Company Richmond No. 5 quarry is southwest of the intersection of Eagle and Kempsville Roads at 40° 30′ 23″ latitude and 75° 49′ 06″ longitude on the USGS Kutztown 7.5-minute topographic quadrangle map (Appendix 1, fig. 3, location RH-8). In 2007, the Lehigh Cement Company proposed to expand the Richmond No. 5 quarry to 206 acres (*Reading Eagle*, April 19, 2007). The company opened the quarry in 1962, but only mined the required 500 tons annually to keep the state mining permit. The company wanted to mine the Jacksonburg Formation there because its Evansville quarry in Maiden Creek Township was expected to be depleted in a few years (*Reading Eagle*, April 20, 2007). In 2010, Richmond Township voted to allow the Lehigh Cement Company to expand the Richmond No. 5 quarry from 8 acres to 206 acres (*Berksmont News*, May 14, 2010).

ROBESON TOWNSHIP

The John T. Dyer Company operated three diabase quarries in the Birdsboro area—the Gibraltar and Trap Rock quarries in Robeson Township and the Monocacy quarry in Union Township. The Monocacy quarry was the largest of the three plants, the Traprock quarry was the second largest, and the Gibraltar quarry was the smallest.

Dyer Gibraltar Quarry

The Dyer Gibraltar quarry has been known by many names: Birdsboro, Clingan, Gickerville, and Robeson quarry. It is called the Dyer Gibraltar quarry in this book because of its close proximity to the village of Gibraltar and to avoid confusion with the H&K Group Birdsboro Materials (Trap Rock) quarry in Birdsboro. The large, active (2016) Dyer Gibraltar quarry is southwest of the intersection of Rock Hollow and Quarry Roads at 40° 16′ 08″ latitude and 75° 51′ 34″ longitude on the USGS Birdsboro 7.5-minute topographic quadrangle map (Appendix 1, fig. 31, location RB-1). The quarry is in diabase.

The location of the quarry often is given as Birdsboro; the quarry is 2.7 miles northwest of the center of Birdsboro, which is the largest population center in the area. The village of Gickerville was at the intersection of Pennsylvania State Route 724 and Cedar Hill Road; it is no longer shown on maps. Clingan is shown southeast of Gickerville on the USGS 1915, 1944, and 1956 Reading 15-minute topographic quadrangle map; it is not

Figure 525. Richmond No. 5 quarry, Richmond Township, March 2016.

John T. Dyer

John T. Dyer, a prominent business man of Norristown, was born in Lehigh County on April 19, 1848. After completing school, Dyer was employed as a clerk in one of the slate quarries in Slatington, Lehigh County. He became interested in railroad construction and worked overseeing the construction of new rail lines. In 1880, he started his own railroad contracting company. His first large contract was the New York, Ontario and Western Railroad. He also did much of the construction work on the Schuylkill Valley Railroad, which led to more important contracts. Soon afterwards, he located permanently in Norristown. Other railroad lines that contracted with Dyer included the Bay Ridge and Annapolis Railroad, the Ohio River Railroad, and the Milwaukee and St. Paul Railroad. Dyer also did much construction work on trolley lines in Norristown and other locations. He built the terminal at Waterbury, Connecticut, and several sections of the Trenton Cutoff Branch of the Pennsylvania Railroad. Dyer also operated quarries near Norristown in Montgomery County, at Howellville in Chester County (see Sloto, 2009, p. 178-179), and near Birdsboro in Berks County. From Roberts (1904, p. 35-36)

Figure 526. Portrait of John T. Dyer.

shown on modern 7.5-minute topographic quadrangle maps. Gibraltar is a small village about 1.25 miles northwest of the quarry. It also was called the Robeson quarry because of its proximity to the Robeson railroad station. Robeson, no longer shown on maps, was between Gickerville and Gibraltar.

In 1911, the John T. Dyer Company purchased the Gibraltar quarry property to increase production and secure a shipping location on a railroad other than the Wilmington & Northern Railroad. The Gibraltar property was near the Pennsylvania Railroad. The quarry, known as the John T. Dyer Clingan plant, was in operation and producing crushed stone by 1913.

In the late 1920s and early 1930s, blasting at the quarry was a big event and was newsworthy. There was one large annual blast to loosen enough stone to last a year, and the public was invited to watch the spectacle. In 1926, the Community Hiking Club took the trolley to Gibraltar and hiked along the road to the northern slope of Sheep Hill where members observed the blast (*Reading Eagle*, October 24, 1926).

The blast of 1928 was the largest blast in the quarry's history. Twenty-three holes, 6 inches in diameter, and 200 feet deep, were loaded with 25 tons of dynamite. The Dyer company cleared a spot in a safe place for spectators to watch. The DuPont company, which furnished the dynamite, provided the explosives engineers who set off the blast. More than 500 people witnessed the blast, which occurred on Saturday, April 7 at 3:25 in the afternoon. The blast dislodged an estimated 241,000 tons of rock (*Reading Eagle*, April 5 and 8, 1928). The blast of 1929, which occurred on June 26, 1929, used 15 tons of dynamite to loosen 150,000 tons of rock (*Reading Eagle*, June 29, 1929).

In 1932, annual production was 200,000 tons of crushed stone. Levi Carson was the quarry superintendent. Drilling was done using two model B Armstrong drills. A model 28 Marion shovel was used for stripping overburden. Model 31 and model 37 Marion shovels were used to dig and load rock. Stone was transported from the quarry to the crushing plant by a fleet of eight 3-ton Autocar trucks. The primary crusher was a 42-inch by 26-inch Farrel jaw crusher. Secondary crushing was done by two 36-inch by 18-inch Farrel jaw crushers and a 4-foot Symons cone crusher. The crushed stone was shipped by rail (Pit and Quarry, 1932).

In 1932, Dyer laid off 48 workmen because of lack of business caused by the depression. To mitigate their misfortune, the company allowed them to cut and sell wood from its woodland tract near Douglassville. The company supplied a sawmill and a truck for wood delivery. Thomas C. McPoyle, the company's construction and civil engineer, provided management and marketing at no cost to the men (*Reading Eagle*, September 20, 1932). In 1935, the John T. Dyer Company filed a petition for reorganization in Federal court. The judge directed Dyer to remain in possession of the property and operate it (*Reading Eagle*, May 22, 1935).

In 1968, the Warner Company purchased the Dyer quarry and operated it until Warner was purchased by Waste Management, Inc. in 1990. In 1998, the quarry was purchased by

Figure 527. Dyer Gibraltar quarry, Robeson Township, September 2010.

James J. Anderson, and its name was officially changed to Dyer Quarry, Inc. The Dyer quarry produces about 2,000,000 tons of aggregate annually. The quarry property is 525 acres and is serviced by the Norfolk Southern Railroad. Dyer has its own locomotive, which can handle up to 200 cars at one time, loading at a rate of 1,000 tons per hour (Anderson Companies, 2012).

Minerals from the Dyer Gibraltar Quarry

The Dyer quarry is the only known locality for babingtonite in Pennsylvania. Its discovery and occurrence was described by Smith (1978, p. 51-54).

Many older mineral specimens are labeled only as "Dyer quarry" or "Birdsboro," and it may be difficult to tell from which of the three Dyer quarries they originated. Nearly all specimens collected since 1955 are from the Dyer Gibraltar quarry.

Actinolite - dark green blades; often associated with golden calcite (Geyer and others, 1976, p. 39) (fig. 529)

Actinolite, var. byssolite - acicular crystals; often associated with prehnite (Geyer and others, 1976, p. 39) (fig. 530)

Apophyllite - colorless or white tabular crystals (Gordon, 1922, p. 158); associated with byssolite (Geyer and others, 1976, p. 39); colorless psuedocubic crystals (Smith, 1978, p. 53) (fig. 528)

Babingtonite - tiny (less than 3 mm), splendant, striated, black, blade-like, asymmetric prisms; found in 5-cm wide coarse-grained felsic diabase veins altered by late-stage hydrothermal fluids; on the south wall of the lower level; found with crystallized epidote and zeolite minerals in small vugs (Smith, 1978, p. 51-54) (fig. 531)

Bornite - iridescent black coating; confirmed by SEM-EDS analysis on a specimen from the Sloto collection

Calcite - brilliant, pale to golden yellow rhombohedra (Gordon, 1922, p. 158; Geyer and others, 1976, p. 39); doubly terminated scalenohedrons, fluoresces green under ultraviolet light (Thomas, 1946, p. 142) (fig. 534)

Celestine - white, fibrous in veins in diabase; confirmed by SEM-EDS analysis on a specimen provided by Marge Matula

Chabazite-Ca - small colorless crystals (Gordon, 1922, p. 158); beautiful colorless, cream, and flesh-colored rhombohedral crystals up to 1/2 inch with stilbite (Geyer and others, 1976, p. 39); white cubic crystals to 1.1 cm (fig. 532). Confirmed as chabazite-Ca by SEM-EDS analysis on a specimen from the Sloto collection

Chalcopyrite - small blebs in veins in the south and east face of quarry; dispenoid crystals reported by Geyer and others (1976, p. 39)

Chlorite group (Smith and others, 1988, p. 327)

Chrysocolla (?) - (Gordon, 1920, p. 167); blue-green coatings with bornite (Geyer and others, 1976, p. 39)

Cobaltite - rare; in prehnite veins; associated with erythrite and chalcopyrite; probably the material listed as asenopyrite by Gordon (1922, p. 158) (Geyer and others, 1976, p. 39); determined by X-ray diffraction (Grant, 1974, p. 6); silvery-white metallic grains to 3 mm across; in veins associated with prehnite, babingtonite, quartz, calcite, chlorite, epidote, and apophyllite (Smith, 1978, p. 92 and 95)

Erythrite - coatings (Gordon, 1922, p. 157); rare; peach-blossom colored coatings with cobaltite (Geyer and others, 1976, p. 39); confirmed by SEM-EDS analysis on two specimens from the Sloto collection

Figure 528. Apophyllite from the Dyer Gibraltar quarry, Robeson Township. Steve Carter collection.

MINERALS FROM THE DYER GIBRALTAR QUARRY

Figure 529. Actinolite from the Dyer Gibraltar quarry, Robeson Township, 6.4 cm. Steve Carter collection.

Figure 530. Actinolite, var. byssolite, from the Dyer Gibraltar quarry, Robeson Township. Sloto collection 3380.

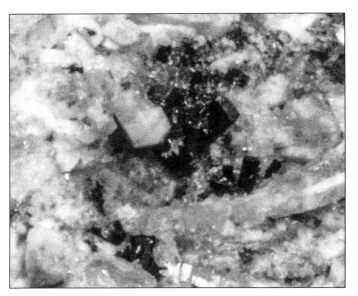

Figure 531. Babingtonite from the Dyer Gibraltar quarry, Robeson Township. Skip Colflesh collection.

Figure 532. Chabazite from the Dyer Gibraltar quarry, Robeson Township, 7.5 cm. Steve Carter collection.

Figure 533. Natrolite from the Dyer Gibraltar quarry, Robeson Township. Sloto collection 3191.

Figure 534. Calcite from the Dyer Gibraltar quarry, Robeson Township. Carnegie Museum of Natural History collection CM 32471.

Figure 535. Opal from the Dyer Gibraltar quarry, Robeson Township, 7.5 cm. Sloto collection 3596.

Figure 536. Quartz, var. agate, from the Dyer Gibraltar quarry, Robeson Township, 4.8 cm. Sloto collection 3193.

Figure 537. Prehnite from the Dyer Gibraltar quarry, Robeson Township. Steve Carter collection.

Figure 538. Stellerite from the Dyer Gibraltar quarry, Robeson Township. Carnegie Museum of Natural History collection CM 32751. Photograph by Debra L. Wilson.

Figure 539. Stilbite from the Dyer Gibraltar quarry, Robeson Township. Sloto collection 3601.

Epidote - pistachio-green microcrystals

Hematite - small, specular scales (Geyer and others, 1976, p. 39)

Heulandite - small yellow crystals (Gordon, 1922, p. 158); colorless to golden-yellow, coffin-shaped crystals (Geyer and others, 1976, p. 39)

Laumontite - chalk white, prismatic, radiating crystals (Gordon, 1922, p. 158)

Loellingite (?) - also called safflorite; silvery-white grains and masses (Smith, 1978, p. 53)

Magnetite - micro-octahedrons with byssolite (Geyer and others, 1976, p. 40)

Natrolite - white needle-like crystals in radiating sprays (Gordon, 1922, p. 158); colorless to white terminated needles (Geyer and others, 1976, p. 40) (fig. 533)

Opal - colorless to light blue to dark blue coatings with calcite; also massive (Geyer and others, 1976, p. 40) (fig. 535)

Okenite - very rare; white, cotton-like "fuzz" sometimes associated with chabazite in filled veins and lenses in diabase. Collected in January 2010 by Skip Colflesh and Scott Snavely and confirmed by Dr. Lance Kearnes of James Madison University (Friends of Mineralogy, 2011). This is the only known occurrence of okenite in Pennsylvania.

Prehnite - green, mammillary crystal aggregates (Gordon, 1922, p. 158); colorless to green crystals; mammillary clusters (Geyer and others, 1976, p. 40) (fig. 537)

Pyrite - massive and cubic microcrystals (Geyer and others, 1976, p. 40); SEM-EDS analysis indicates that pyrite contains small quantities of arsenic and cobalt

Quartz - colorless, milky, and amethystine microcrystals (Geyer and others, 1976, p. 40)

Quartz, var. agate - blue and white in veins in diabase (fig. 536)

Riebecite, var. crocidolite - reported as pale-blue fibrous coatings on altered diabase (Geyer and others, 1976, p. 40)

Skutterudite - small, bright silver, generally massive; confirmed by SEM-EDS analysis on a specimen showing a partial crystal face collected by Marge Matula in 1980

Sphalerite - black, very rare (Smith, 1978, p. 53)

Stilbite-Ca - colorless, white, and yellow; prismatic crystals, usually in sheaf-like aggregates, radiating (Gordon, 1922, p. 158); the epidesmine of Gordon (1920, p. 167); crystals to 1.5 cm with bow ties and clusters to 2.7 cm (fig. 539). Confirmed as stilbite-Ca by SEM-EDS analysis on a specimen from the Sloto collection

Stilpnomelane - reported as bronze platy crystals with calcite (Geyer and others, 1976, p. 40)

Stellerite - older specimens may be labeled as stilbite; confirmed by Dr. Lance Kearnes of James Madison University by EDS analysis on a specimen collected by Skip Colflesh (written communication, 2015)(fig. 538)

Tenorite (?) - the melaconite of Gordon (1922, p. 158); massive, black; coated with chrysocolla

Thaumasite - white, compact mass of acicular crystals (fig. 540)

Thompsonite (?) - colorless to white, slender prisms (Geyer and others, 1976, p. 40)

Titanite - microcrystals; golden flat wedges with byssolite and in altered diabase (Geyer and others, 1976, p. 40)

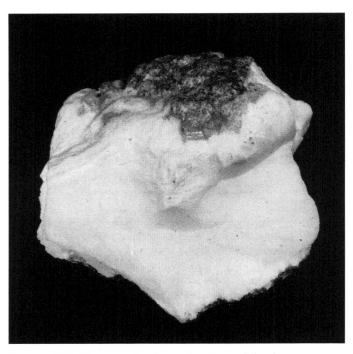

Figure 540. Thaumasite from the Dyer Gibraltar quarry, Robeson Township, 2 cm. Sloto collection 3558.

Dyer Trap Rock (Birdsboro Materials) Quarry

The Dyer Trap Rock quarry (fig. 541) is west of Pennsylvania State Route 82, 1.4 miles southwest of the center of Birdsboro at 40° 15′ 02″ latitude and 75° 49′ 27″ longitude on the USGS Birdsboro 7.5-minute topographic quadrangle map (Appendix 1, figs. 31 and 36, location RB-2). The active quarry (2016) is in diabase. The quarry was known as the Trap Rock, Pupek, and H&K quarry, and is currently (2016) the H&K Group Birdsboro Materials quarry.

In 1848, the Brooke family built the Hampton Furnace on Hay Creek in the area later known as Trap Rock. The cold blast charcoal furnace produced an average of 20 tons of iron each week. The furnace was dismantled in the 1880s but the land remained in the Brooke family. In 1880, George Brooke reorganized the Brooke family holdings into: (1) the Birdsboro Iron Foundry, which remained the manufacturing unit; (2) the E & G Brooke Iron Company, which was formed to operate the iron works and mines; and (3) the E & G Brooke Land Company, which was formed to oversee the family real estate holdings (Hoffman, 1976). The E & G Brooke Land Company and its successors leased property for quarrying on a royalty basis.

Figure 541. Birdsboro Materials (Trap Rock) quarry, Robeson Township, April 2015.

The Wilmington & Northern Railroad (fig. 542) extended north from Wilmington, Delaware, through Chester County to Birdsboro, where it joined the Reading Railroad's main line. The original rail line, the Berks & Chester Railroad, was incorporated in 1864. The rail line, which was constructed through the Hay Creek Valley, was opened in 1870. In early 1875, the

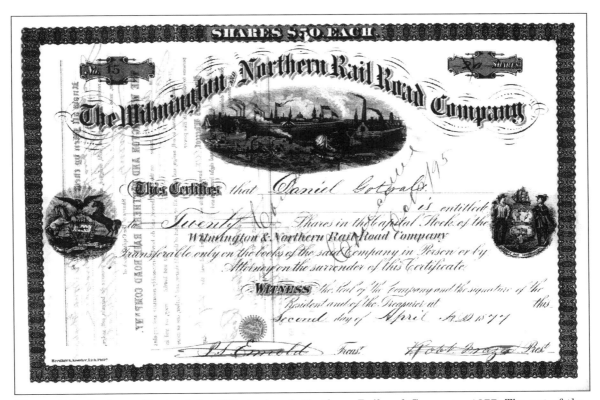

Figure 542. Stock certificate of the Wilmington & Northern Railroad Company, 1877. The cut of the railroad through the diabase south of Birdsboro led to the opening of the Dyer Trap Rock quarry.

The Mines and Minerals of Berks County

Figure 543. Number 3 and 4 crushers at the Dyer Trap Rock quarry, Robeson Township, early 1900s. Crusher number 3 is in the foreground, and crusher number 4 is in the background. Courtesy of the Amity Heritage Society.

Figure 544. Crusher Number 2 at the Dyer Trap Rock quarry, Robeson Township, early 1900s. Courtesy of the Amity Heritage Society.

Figure 545. Dyer Trap Rock quarry and crusher Number 2, Robeson Township, early 1900s. Courtesy of the Amity Heritage Society.

operation of the railroad was placed in the hands of three trustees. In 1876, the trustees sold the railroad to a bondholders' committee, which reorganized it under the name Wilmington & Northern Railroad. George and Edward Brooke were among the original investors in the company. The Wilmington & Northern Railroad reached Birdsboro in 1870 and the Reading Railroad main line in 1874. In 1880, a branch line was built from Elverson to the E & G Brooke Iron Company's French Creek mine in Saint Peters (see Sloto, 2009, p. 214). The Wilmington & Northern Railroad was acquired by the Reading Railroad Company in 1898.

In 1890, the John T. Dyer Company operated a limestone quarry in Howellville, Chester County (see Sloto, 2009, p. 178-179). The railroads demanded a rock harder than limestone for ballast, so Dyer assigned several men to search for a source of harder rock. One of those men, William Johnson, the superintendent of the Howellville quarry, was formerly a resident of Berks County. Johnson recalled that when the Wilmington & Northern Railroad was being constructed, several contractors encountered difficulty excavating a deep cut through hard rock about 1 mile south of Birdsboro. Johnson investigated this rock, which was diabase, and found it to be of excellent quality for railroad ballast. The railroads were satisfied with the hardness of the rock, and Dyer approved the location. In 1893, after securing a lease on the property from the E & G Brooke Land Company, Dyer opened the Trap Rock quarry (Conway, 1931). The stone was used for railroad ballast and producing macadam. It was trademarked as "Birdsboro Trappe Rock."

There was a small station on the Wilmington & Northern Railroad named Hampton station after the Hampton Furnace. It was less than a mile from the Dyer quarry. Shortly after the Dyer quarry opened, the station was moved closer to the quarry and was renamed the Trap Rock station. Trap rock is an old term for diabase. A settlement of Italian immigrant laborers,

Figure 546. Diabase exposed in the Birdsboro Materials (Trap Rock) quarry, Robeson Township, April 2015.

who worked in the nearby quarries, sprang up around the station, and the settlement also was called Trap Rock (*The Trade-Mark Reporter*, 1911, p. 68). The Dyer Company leased a tenant house from the E & G Brooke Land Company to house the workmen.

In 1894, a siding was constructed from the Wilmington & Northern Railroad to the quarry. The capacity of the plant was 375 tons per day including 13 car loads of prepared macadamizing material daily. In August, a blast dislodged about 3,000 tons of rock, which partly covered the railroad tracks and delayed the northbound passenger train by 30 minutes (*Reading Eagle*, May 3, August 4, and November 10, 1894).

In 1895, a second crusher, built by contractor L.H. Focht at a cost of $1,000, increased the capacity to 200,000 tons per year. A bridge was constructed across Hay Creek to link the

Figure 547. Workers at the Dyer Trap Rock quarry, Robeson Township. The photograph was taken in the 1920s or 1930s. Courtesy of the H&K Group.

crushers. Seven horses and carts were added so that both plants could run at full capacity (*Reading Eagle*, January 12 and March 22, 1895).

Dyer operated three quarries close to each other; two quarries were on the right bank of Hay Creek, and one was on the left bank. In 1898, the lower quarries exposed a breast of diabase about 100 feet high. Rand (1898) described the quarries: *"A striking feature was the cleanliness of these quarries. The blasting is done usually twice a day, but after each blast some of the quarrymen are detailed to fork up the fragments, so that the horses and carts travel over a smooth and level surface. The horses were of a quality rarely seen in quarry work, and evidently were well cared for and well treated, while the men work industriously, without the vigorous language too often heard in quarrying operations."* Rand reported heulandite and laumonite from the upper quarries.

In 1900, production was about 2,000 tons per week. Stone was shipped to locations in Pennsylvania, New Jersey, and Delaware. The quarry superintendent was Harry Swartz (*Reading Eagle*, September 27, 1900). By 1908, production had increased to 2,000 tons per day. Crushed stone was produced in seven sizes that ranged from screenings to 3 inches. The rock was used for macadam, ballast, and concrete.

Blasting occurred three times a day—in the morning, at noon, and in the evening. The blasting technique was called mud capping. A large pile of dynamite sticks, usually 500 to 2,000 pounds, were connected with fuses, covered with mud, and detonated. About 160 men were employed year round. There were four crushing plants with 10 crushers and a large stable on the premises with stalls for 35 horses, which were well fed and well cared for (*Reading Eagle*, October 3, 1908).

In 1908, the Dyer Company increased the wages of skilled workmen by 5 percent and increased the wages of laborers from $1.25 to $1.30 per day (*Reading Eagle*, July 6, 1908). In 1909, the wages of laborers were increased to $1.45 per day. However, the workmen demanded an increase to $1.60 per day. The Dyer Company refused to increase the wages. In response, the men went on strike. They were paid and informed that their connection with the company ceased at that moment. Many of the older employees expressed a willingness to return to work and were reinstated at their former wage (*Reading Eagle*, July 17, 1909). In 1912, the Dyer Company constructed a two story frame house as a dwelling for its workmen (*Reading Eagle*, July 19, 1912). In 1916, the Dyer Company employed 80 men at the quarry, and the daily output was about 5,600 tons of crushed rock (*Reading Eagle*, February 1, 1916).

In 1924, the 30-year lease

Figure 548. Advertisement for the Autocar Company featuring the Dyer Trap Rock quarry, 1920. Autocar trucks are being loaded by a steam shovel at the quarry. From the Literary Digest, March 29, 1920.

with the E & G Brooke Land Company expired. When the Dyer Company attempted to renew the lease, it learned that the Birdsboro Stone Company, which operated the nearby Monocacy quarry, had secretly negotiated and signed a lease with the Brooke Estate for the Trap Rock quarry. This situation forced Dyer to hunt for a new quarry site (*Reading Eagle*, February 21, 1925).

The Dyer Company decided to open a new quarry on Monocacy Hill, across the Schuylkill River from Birdsboro in Amity Township. Monocacy Hill is underlain by diabase. Dyer employed James Kline to purchase land, including land for a railroad siding. Dyer purchased 12 farms totalling several hundred acres on and surrounding Monocacy Hill. Dyer began moving its equipment from the Trap Rock quarry to the new quarry site on the east side of Monocacy Hill (*Reading Eagle*, October 24, 1924). It was expected that the new plant would be in operation by the spring of 1926.

Several roads were constructed in preparation for the new quarry. A bridge spanning Monocacy Creek was built, and a temporary crusher was installed for concrete work. A railroad track was laid across Monocacy Creek Road as far as U.S. Route 422. The railroad bed was graded to the crusher site, except for a small area on the north side of U.S. Route 422 where the property owners refused to sell and temporarily blocked the railroad siding. A bin storage building was constructed, and concrete foundations were laid for crushing and conveying machinery. When Dyer's purchase of Monocacy Hill became pubic, there were protests by many who though the company would remove the local landmark by quarrying it away. Dyer suspended work on the new quarry in response to the protests (*Reading Eagle*, February 20 and June 25, 1925).

During this time, the Dyer Company was engaged in a bitter court battle with the Birdsboro Stone Company over the Trap Rock quarry. The Birdsboro Stone Company filed suit against Dyer and the E & G Brooke Land Company. The E & G Brooke Land Company admitted that Dyer unlawfully quarried and removed stone after the November 1, 1924, lease expiration date and failed to remove its machinery within the 4 month time frame allowed by the lease. The E & G Brooke Land Company claimed that equipment had become part of the realty. There was an agreement for the Birdsboro Stone Company to buy Dyer's plant and equipment, but the agreement was never consummated (*Reading Eagle*, May 25 and July 2, 1925). The litigation ended when Dyer bought the Birdsboro Stone Company, which included the Monocacy quarry. After the purchase, all work stopped at the Monocacy Hill site, and Dyer removed its equipment. In 1968, 420 acres on Monocacy Hill were purchased from Dyer for open space.

In 1932, the annual production was 200,000 tons. Drilling was done using two model B Armstrong drills. Stripping was done with a model B Marion shovel. Stone was transported from the quarry to the plant using 12 three-ton Autocar trucks. Primary crushing was done by two 36-inch by 18-inch Farrel jaw crushers. Secondary crushing was done by two 20-inch by 13-inch Farrel jaw crushers. The crushed stone was shipped by rail (Pit and Quarry, 1932).

In 1933, Harry Schwartz, age 72, the superintendent of the quarry since it opened in 1893, was injured by a truck. Also that year, the No. 4 crusher, which had not been used for some time, burned down (*Reading Eagle*, February 13, 1933).

The A. Louise C. Brooke estate, successor to the E & G Brooke Land Company, owned the 364 acre property where the Trap Rock quarry was located. On September 2, 1936, the Girard Trust Company, trustees of the estate, leased the quarry to the Dyer Company for 20 years for a royalty of 5 cents per ton of stone quarried and shipped (*Reading Eagle*, September 27, 1948).

Bascom and Stose (1938, p. 107) described the quarries at Trap Rock. The quarry on the west side of the valley was about 2,400 feet long, 450 feet wide, and 75 feet deep. Two adjoining quarries were on the east side above and east of the railroad with a large open cut through the spur to the north and a face more than 50 feet high. One of the quarries was abandoned about 1933.

In 1952, the annual capacity was 250,000 tons. Jasper Wamshery was the superintendent (Pit and Quarry, 1952). In 1956, the A. Louise C. Brooke estate sued the Dyer Company to stop them from removing the railroad tracks it no longer used. Dyer wanted to move the tracks to its Gibraltar quarry plant. The Brooke Estate claimed it owned the railroad line from Six Penny Falls to the Pennsylvania Railroad main line and cited an August 2, 1906, agreement for construction of a Pennsylvania Railroad Company track giving the landowners free use of the railroad and the right to connect tracks with the line. A restraining order was issued to prevent Dyer from removing the tracks (*Reading Eagle*, July 3, July 12, and October 2, 1956).

Figure 549. Mineralogical Society of Pennsylvania mineral collecting field trip at the Dyer Trap Rock quarry, August 17, 1952.

Figure 550. Apophyllite from the Dyer Trap Rock (Birdsboro Materials) quarry, Robeson Township, 7 cm. The apophylite crystal is 2.5 cm. Reading Public Museum collection 2004C-7-640.

The court ruled that the rails should not be removed. In 1957, at the expiration of it's lease, Dyer abandoned the Trap Rock quarry and moved out of Union Township. The move necessitated an increase in the property tax (*Reading Eagle*, April 3 and March 8, 1957).

The abandoned quarry property was acquired by Benjamin A. and Margaret A. Pupek, operating under the name of Chestnut Hill B&M, Inc. of Phoenixville. Diamond Contractors, Inc. of King of Prussia proposed to use the quarry site for a landfill. The Robeson Township supervisors denied a request for a zoning change to allow a landfill on the site, and the Pennsylvania Department of Environmental Protection rejected the landfill proposal in July 1988 (*Reading Eagle*, May 22, 1986, and August 10, 1989). In 1988, the 425-acre quarry property was purchased from the Pupeks by Haines and Kibblehouse, Inc. of Skippack for $1.7 million. Haines and Kibblehouse, Inc. planned to reopen the quarry.

Reopening the quarry was fought by Robeson Township for 12 years. The fight went all the way to the Pennsylvania State Supreme Court, which refused to hear the case. A mining permit was issued by Pennsylvania Department of Environmental Protection in March 1999, and the quarry reopened (*Reading Eagle*, March 19, 2002).

Another fight ensued over the right of Haines and Kibblehouse to use the old Wilmington & Northern railroad right-of-way for shipment of stone by rail. In 1867, the E & G Brooke Land Company granted a right of way to the Wilmington & Reading Railroad who later granted it to Conrail, which abandoned the line in 1981. On June 15, 1999, a Pennsylvania State Superior Court judge granted Haines and Kibblehouse the right of way over the abandoned rail line (*Reading Eagle*, December 7, 1999). In 2000, Birdsboro appealed the ruling to the Pennsylvania State Supreme Court and lost (*Reading Eagle*, August 23, 2000).

Minerals from the Dyer Trap Rock quarry

Smith (1910) listed a number of minerals from "*a railroad cut east of Reading.*" Under the listing for "Trap quarries at Trap Rock on Hay Creek," Gordon (1922, p. 158, footnote 103), stated: "*This is undoubtedly the locality given by Smith as 'Railroad cut east of Reading,' although lying 9 miles southeast of Reading.*" Minerals were given to Smith 10 years prior (ca. 1900) by D.B. Brunner and Dr. Schoenfeld. "*which they gathered from a railroad cut east of Reading. Considerable blasting had been done, and a great deal of rock material had been removed.*" It likely was the cut of the Wilmington & Northern Railroad through the diabase south of Birdsboro. It is the only railroad cut through diabase in the vicinity of Reading, and it is the right time frame for construction of the spur from the railroad to the quarries. The list of minerals occurring in the cut is similar to those occurring in the nearby Trap Rock quarries. Smith provided an analysis of each mineral. They are included with the list of minerals from the Trap Rock quarries.

Apophyllite - colorless, or white crystals (Smith, 1910, p. 538, analysis); tabular crystals to 2.5 cm; colorless to white crystals with a vitreous luster and basal cleavage; sometimes associated with prehnite (figs. 550 and 551)

Calcite - clear cubic and acicular scalenohedrons (Groth, 1904, p. 160); small, brilliant yellow, rhombohedral and scalenohedral crystals (Gordon, 1918); doubly terminated crystals to 2.5 cm (fig. 552)

Calcite, var. ferrocalcite - honey brown crystals; SEM-EDS analysis indicates that it contains 0.4 weight percent iron

Chabazite - colorless or white rhombohedra (Smith, 1910, p. 538, analysis)

Garnet group - black to green, brown to gray

Figure 551. Apophyllite and prehnite from the Dyer Trap Rock (Birdsboro Materials) quarry, Robeson Township. Apophyllite crystals to 1.5 cm. Scott Snavely collection.

MINERALS FROM THE DYER TRAP ROCK (BIRDSBORO MATERIALS) QUARRY

Figure 552. Calcite from the Dyer Trap Rock (Birdsboro Materials) quarry, Robeson Township, 7 cm. Sloto collection 3328.

Figure 553. Prehnite from the Dyer Trap Rock (Birdsboro Materials) quarry, Robeson Township, 6.5 cm. Scott Snavely collection.

Figure 554. Natrolite from the Dyer Trap Rock (Birdsboro Materials) quarry, Robeson Township. Bryn Mawr College collection Heyl 2407.

Figure 555. Sphalerite from the Dyer Trap Rock (Birdsboro Materials) quarry, Robeson Township, 7 mm. Steve Carter collection.

Figure 556. Quartz from the Dyer Trap Rock (Birdsboro Materials) quarry, Robeson Township. Reading Public Museum collection 2005C-2-207.

Figure 557. Humphrey sandstone quarry adjacent to Hay Creek, Robeson Township, April 2015.

Heulandite (Groth, 1904, p. 160)

Hornblende (Groth, 1904, p. 159)

Laumontite - white, prismatic, radiating (Smith, 1910, p. 538, analysis); chalky white with a vitreous luster

Natrolite - crystal sprays to 3.2 cm (fig. 554)

Prehnite - deep green, associated with calcite or quartz crystals (Groth, 1904, p. 160); green, mammillary (Smith, 1910, p. 538); rarely forms spherical aggregates (figs. 551 and 553)

Pyroxene (Groth, 1904, p. 159)

Pyroxene - light green crystals with a violet luster

Quartz - small, colorless crystals on quartzite, also associated with prehnite (Groth, 1904, p. 160) (fig. 556)

Scolecite - masses of silky radiating needles, with minute calcite crystals on them (Smith, 1910, p. 538, analysis); masses of radiating needles with a silky luster mixed with small but distinct calcite crystals particularly at the base of the tufts of the needles

Stilbite - well-formed dark brown, yellow, and light green crystals, radiating crystals to 2 inches in diameter (Groth, 1904, p. 160); thin, radiating crystals (Smith, 1910, p. 538, analysis); groups of small, clear crystals (Gordon, 1918); associated with calcite (Thomas, 1946, p. 205)

Sphalerite - small, well-formed, dark green to black crystals (fig. 555)

Thaumasite - white, compact, acicular crystals

Thomsonite - minute crystals on natrolite

Hay Creek Prehnite Locality

The Hay Creek Prehnite locality described by Beard (2008) is along Hay Creek near the Trap Rock quarry. The locality is on the east side of Pennsylvania State Route 82 (Appendix 1, fig. 36) on the USGS Elverson 7.5-minute topographic quadrangle map. Beard (2008) described prehnite occurring in an outcrop along Hay Creek south of Birdsboro at about 40° 14′ 59″ latitude and 75° 49′ 11″ longitude (location RB-4A) and minor prehnite and stilbite in creek boulders at 40° 14′ 51″ latitude and 75° 49′ 15″ longitude (location RB-4B). The outcrop is diabase with light green prehnite veins in small radiating crystals. Diabase boulders in the creek contain prehnite veins and stilbite.

Humphrey Sandstone Quarries

Two Humphrey sandstone quarries are shown in the 1876 atlas as the Brusstar quarries (fig. 558). One quarry was northwest of Hay Creek at 40° 14′ 30″ latitude and 75° 50′ 15″ longitude on the USGS Elverson 7.5-minute topographic quadrangle map (Appendix 1, fig. 36, location RB-3A) and a quarry with a railroad siding was adjacent to Pennsylvania State Route 82 on the east side of Hay Creek at 40° 14′ 26″ latitude and 75° 49′ 42″ longitude (location RB-3B) (fig. 557). The quarries were operated by P. Brusstar and Brother in the 1870s. The quarry also was known as the Maiden quarry (Mine and Quarry News Bureau, 1897). The eastern quarry was in the Hammer Creek Formation, and the western quarry was in the Hammer Creek Conglomerate.

The quarries were operated by James Humphrey from the late 1880s to around 1897. Humphrey owned the Humphrey Stone Company, which operated the quarry at Green's siding on the Wilmington & Northern Railroad (location RB-3B). Humphrey employed 40 men, including stonecutters and laborers. In 1889, Humphrey was shipping 6 to 7 car loads of stone to Phoenixville weekly for the Pennsylvania Railroad. He also shipped several car loads of stone weekly to the E & G Brooke Iron Company.

Humphrey received an order from Philadelphia for 500,000 "black granite" (diabase) paving blocks. The contract called for shipping 75,000 paving blocks a month. He decided to abandon his sandstone quarries and move the machinery to his diabase quarry on Round Hill. When he took down the large derrick at his sandstone quarry, it fell and injured two workmen. The 1880 Census listed James Humphrey as a granite producer with a quarry located 2 miles south of Birdsboro.

Humphrey quarried red sandstone and conglomerate. The rock dipped north at about the same angle as the slope of the hill. The lower quarry was opened at the base of the hill and up the slope to a height of about 100 feet (fig. 557); it was 20 to

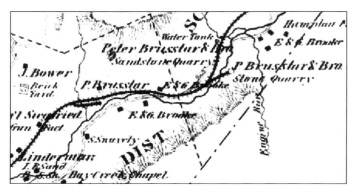

Figure 558. Locations of the two Brusstar quarries, Robeson Township, 1876. From Davis and Kochersperger (1876).

> **SANDSTONE, &c.**
>
> Grubb Benjamin.—Dealer in Birdsboro' and Lancaster Sandstone, takes out and makes to order all kinds of Platform Flagging, Caps, Sills, Lintels, Furnace Bottoms, Glass, Stone, &c. Furnace street, Birdsboro.
>
> Brusstar P. & Bro.—Dealers in the best Forest Sand and Red and White Sand Stone for furnace and building purposes. We are prepared to furnish any quality of sand or stone, such as Bridge Caps, Capping Ashlers, Belt Platforms, Sills, Steps and Building Stones, &c. Wilmington & Reading Railroad, Birdsboro'.

Figure 559. Sandstone quarries in Robeson Township advertising in the 1876 atlas. From Davis and Kochersperger (1876)

25 feet deep. The upper quarry was a little smaller. The stone had a uniform dark brown color and contained many pebbles, some up to 6 inches in diameter. The pebbles were arranged in irregular layers a few inches thick scattered through the strata. In some places, sandstone beds several feet thick were free from pebbles (Hopkins, 1897, p. 71-72).

Joanna Furnace

Joanna Furnace was in operation from 1792 to 1898. An excellent and very detailed history of the furnace was written by Jacob (1999). Rutter, May & Company began building Joanna Furnace in 1791 on Hay Creek. The company was comprised of Thomas Rutter, Thomas May, Samuel Potts, and Thomas Bull. Joanna furnace obtained ore from the Jones and Warwick (Chester County) mines. The furnace went into blast sometime between 1791 and 1794 (Jacob, 1999, p. 41). The furnace was managed until 1796 by David Potts, the son of Samuel Potts and nephew of Rutter and May. In 1796, Potts moved to the Warwick Furnace, and from 1796 to 1798, Thomas Rutter's son, Samuel Rutter, managed Joanna Furnace.

The death of Thomas Rutter and the bankruptcy of both Samuel Rutter and Samuel Potts' son, Thomas, who owned their families' shares in the furnace, caused the property to go up for sale. In 1797, one undivided fifth part of Joanna Furnace with about 1,400 or 1,500 acres of land together with the right of taking ore from the Warwick mine, was put up for sale. In 1798, Thomas May's executors also put his share of Joanna Furnace and about 1,352 acres of land up for sale. Thomas May's executor, Robert May, and Thomas Bull sold a one-third share of the furnace to Bull's son-in-law, John Smith, and Thomas Bull & Company was organized to run the furnace. Joanna pig iron was sent throughout eastern Pennsylvania, to Maryland, and as far away as South America.

Robert May died in 1812, and John Smith died in 1815. In 1820, Thomas Bull & Company disbanded, and Smith's son and Bull's grandson, Thomas Bull Smith, leased the furnace and managed it until 1825. At that time, Thomas Bull and William Darling rented the furnace for a short time to Buckley & Brooke of Hopewell Furnace. Levi Bull Smith, brother of Thomas Bull Smith, and grandson of Thomas Bull, joined the firm in 1830, and the firm name became Darling & Smith. The Darling & Smith years were the high point of Joanna Furnace's operation. During this period, they became known for their ten-plate stoves and pig iron used by the railroad. The furnace was rebuilt in 1847. In fifty weeks in 1855, the furnace produced 1,162 tons of iron (Lesley, 1859). Darling left the company in 1858, and the firm became Levi B. Smith & Company. In the 1860s, the sons of Levi Bull Smith joined the operation. The middle son, Levi Heber Smith, took over the operation of Joanna Furnace upon his father's death in 1876. In 1898 Smith died, and iron production stopped. The furnace complex is being restored by the Hay Creek Valley Historical Association.

ROBESONIA BOROUGH

Albert Wenrich Quarry

The Albert Wenrich quarry is in a stand of trees in an agricultural field, at the northern boundary of Robesonia Borough, east of Church Street, 0.35 mile northeast of the center of Robesonia at about 40° 21′ 22″ latitude and 76° 07′ 58″ longitude on the USGS Womelsdorf 7.5-minute topographic quadrangle map (Appendix 1, fig. 22, location RO-1). It was a small quarry in the Ontelaunee Formation worked for private farm use. The rocks dipped S. 20° W. 35° (D'Invilliers, 1886, p. 1555). Limestone and lime producers listed by Hice (1911) included Albert D. Wenrich.

Robesonia Furnace

The Robesonia Furnace, originally called the Reading Furnace, was built in 1794 by George Ege on a tract of land formerly owned by Conrad Weiser. Ore for the furnace came from the Cornwall Mine, 25 miles away, in Lebanon County. Ege

Figure 560. Joanna Furnace, Robeson Township, April 2015. The furnace and support buildings are being restored by the Hay Creek Valley Historical Association.

Figure 561. Robesonia Furnace, ca. 1910.

erected several log and stone tenant houses near the furnace and built an ironmaster's mansion around 1807. After Ege's death in 1829, a trustee ran the furnace. At that time, the property included 6,000 to 7,000 acres of land, an ore right to the Cornwall Mine, the iron masters mansion, a gristmill, a saw mill, and several other buildings. In 1832, it was the largest furnace in Berks County; it employed 228 workmen and produced 3,568 tons of pig iron and 95 tons of castings. The furnace, including the right to mine iron ore from the Cornwall mine, was sold at public sale in 1836. Between 1836 and 1844, the furnace changed hands several times.

In 1841, Clement Brooke of Hopewell Furnace obtained the furnace and land. In 1845, Brooke acquired a partner, Henry P. Robeson, an ironmaster from Manada Gap, Dauphin County, and the furnace was run under the firm name of Robeson & Brooke until it was abandoned in 1850. In 1845, Robeson & Brooke built an anthracite furnace capable of producing 50 tons of iron per week on the property; this furnace became known as the Robesonia Furnace. A second stack was constructed in 1857-58, which increased the capacity to 240 tons of iron per week. The ore for the furnace was obtained from the Cornwall mine.

Lesley (1859, p. 11) called the enterprise "*The Robesonia Iron Works, Furnace No.1 and 2, once called Reading Furnaces.*" He indicated that the furnace was owned by Robeson, White & Company since 1857. The Lebanon Railroad was completed through the area in 1858, which expedited and reduced the cost of shipping both ore from the Cornwall mine and furnace products. In 1856, furnace No. 1 produced 2,141 tons of pig iron in 48 weeks using magnetite from the Cornwall mine.

Robeson died in 1860, and the furnace came under the management of his son in-law, William White, who ran it under the firm name White, Ferguson & Company, which, by 1875, had become Ferguson, White & Company. Furnace No. 1 was blown out for the last time in 1874 and dismantled in 1884. Furnace No. 2 was enlarged in 1874 and rebuilt in 1884. The furnace produced red-short pig iron for Bessemer steel and bar iron sold under the brand name "Robesonia." The annual capacity was 25,000 tons in 1884 and 50,000 tons in 1886 (American Iron and Steel Association, 1884, and 1889).

In 1876, George Taylor became the superintendent of the furnace. Taylor was interested in further increasing production. In 1884, one stack was abandoned and the second stack was enlarged to increase its capacity to 1,000 tons of iron per week. At that time, the firm decided to manufacture only one kind of pig iron, "Robesonia," made exclusively of Cornwall ore.

In 1885, the furnace was sold for $850,000 to White & Company of Philadelphia, which became a stock company, the Robesonia Iron Company, Limited. W.C. Freeman of Lebanon became president of the company. George Taylor remained superintendent until his death in 1903. Also that year, a new furnace was built, which remained in service until 1914. In 1923, the Robesonia Iron Company was incorporated.

A large pile of slag grew around the furnace (fig. 562). In 1912, the company used some of the slag to make cinder block. They built a number of company buildings with the block, including a fire house, several tenant houses, and a chemical laboratory. Further improvements were made in 1921, when automatic fill was instituted, and in 1924, when an automatic pig-casting machine and a large electric gantry crane were installed.

The increased productivity of the furnace eventually led to its demise. When the furnace was established, it acquired a right to free ore from the Cornwall mine, which it used for about 130 years. Initially, the furnace used approximately 2,000 tons of ore per year. With increased iron production, the furnace required 100,000 tons of ore per year. Although the origi-

Figure 562. Large slag piles at the Robesonia Furnace, ca. 1920s. J. Victor Dallin Aerial Survey collection, 70.200.00273, Hagley Museum and Library, Wilmington.

nal agreement only allowed them ore for one furnace, by maintaining two furnaces, one was always in blast when the other was out of blast. Legal battles with the Cornwall mine proprietors resulted in the court establishing their right to ore for one furnace, but would not allow them to utilize a second furnace when the primary furnace was not in service. When the Bethlehem Steel Corporation acquired the Cornwall mine, it decided to end Robesonia furnace's right to free ore. This was accomplished by acquiring and dismantling the Robesonia furnace. Dismantling the furnace began on April 18, 1927. Subsequently, the homes and other buildings were sold individually (Friends of the Robesonia Furnace, Inc., 2015).

The Robesonia furnace slag pile accumulated during the 133 years of furnace operation. The slag pile towered over the homes along South Freeman Street. The company began making cinder blocks from the slag, but stopped for unknown reasons. After the furnace was dismantled, the site was purchased by the Wernersville Stone and Lime Company, which built a crusher on site and began selling the slag. The slag reportedly was used in building U.S. Route 422 from Wernersville to Womelsdorf. The crusher was in operation until the late 1960s. The slag pile was completely removed (Gartner, 2015).

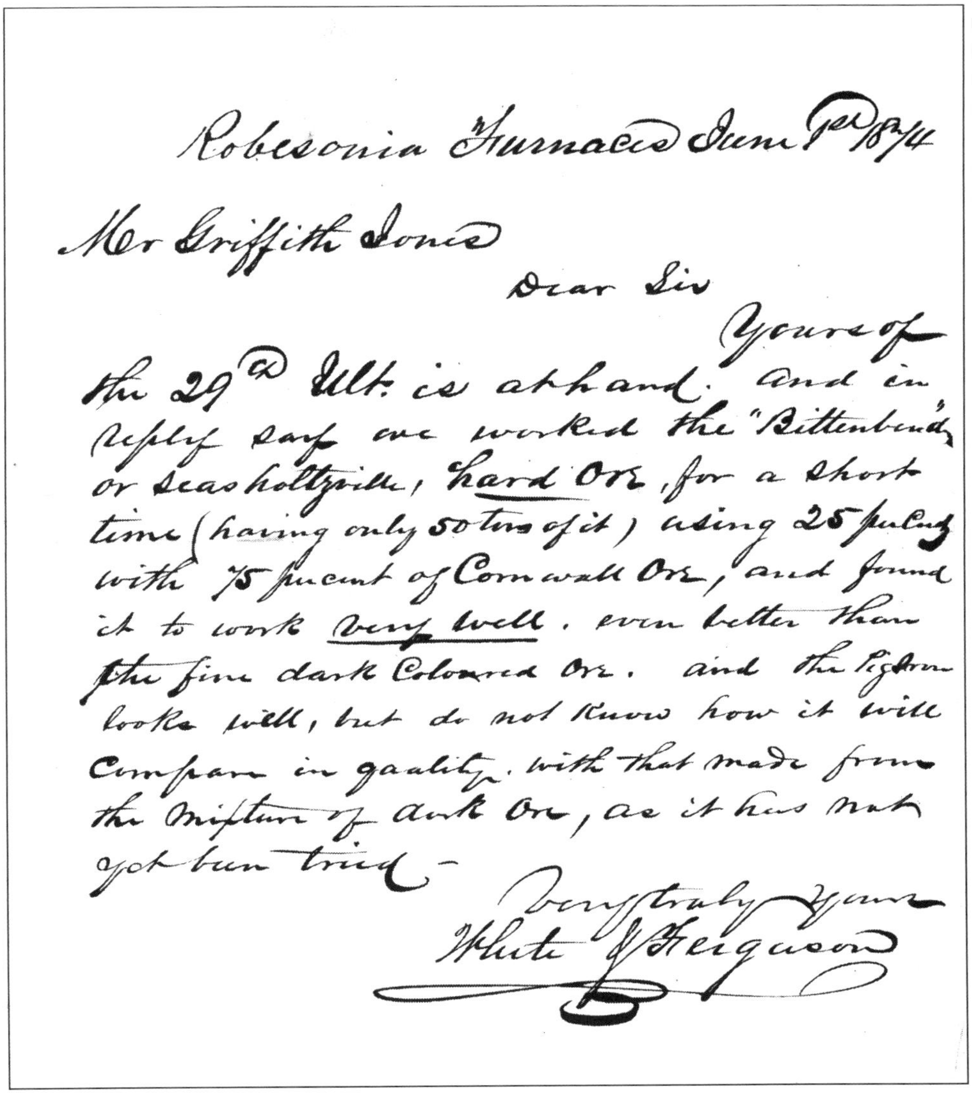

Robesonia Furnaces June 1st 1874

Mr Griffith Jones

Dear Sir

Yours of the 29th Ult. is at hand. and in reply say we worked the "Bittenbender" or Seasholtzville, hard Ore, for a short time (having only 50 tons of it) using 25 percent with 75 percent of Cornwall Ore, and found it to work very well. even better than the fine dark Coloured Ore. and the Pig Iron looks well, but do not know how it will compare in quality. with that made from the mixture of dark Ore, as it has not yet been tried—

Very truly yours
White & Ferguson

Figure 563. Letter from White Ferguson of the Robesonia Furnace to Griffith Jones of the Steel Ore Company concerning the use of ore from the Bittenbender mine, 1874. From the Griffith Jones Papers Concerning Iron Ore Furnaces (3615), Historical Collections and Labor Archives, Special Collections Library, Pennsylvania State University.

ROCKLAND TOWNSHIP

Beitler Mine

The Beitler iron mine (fig. 564) was east of Five Points Road, and south of its intersection with Fredericksville Road at about 40° 26′ 43″ latitude and 75° 41′ 55″ longitude on the USGS Manatawny 7.5-minute topographic quadrangle map (Appendix 1, fig. 15, location RK-1). The mine was in felsic to mafic gneiss.

The mine was owned by Absalom Beitler in 1882. Mining began in the 1700s, and the ore at one time was hauled to the Lyon Furnace. Thomas Weaver & Company of Allentown leased the mine in March 1880. Eighteen men were employed, and John Pascoe was the superintendent. When D'Invilliers visited the mine on June 6, 1882, it was operating but filled with water to the 60-foot level. When D'Invilliers visited the mine again in the fall of 1882, it was abandoned (D'Invilliers, 1883, p. 266-267).

There were two slopes on the property. The No. 1 slope followed ore from the outcrop to a depth of 240 feet. For the first 60 feet, the ore and rock dipped S. 15° E. 67°. The ore bed

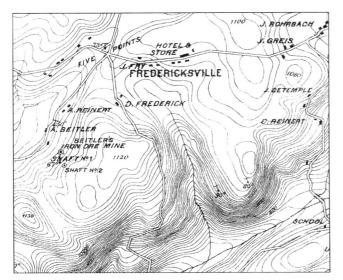

Figure 564. Locations of the Beitler mine, Rockland Township, 1882. From Pennsylvania Geological Survey (1883).

dipped about 30° from 60 feet to the 240-foot level on slope, which was about 150 feet vertically below land surface. A sump was constructed in the No. 1 slope at about 60 feet to catch water that infiltrated the mine. A No. 6 Knowles pump was used to pump water from the sump to the surface; however, its full capacity of 30 gallons per minute was never used. The water pumped from the mine was just about enough to meet the requirements for washing the ore. At the 240-foot level, which was reached on June 6, 1882, a water-bearing fracture was encountered that completely drained the No. 2 slope, which was higher up the hill. This sudden inflow of water flooded the No. 1 slope to the 60-foot level prior to D'Invilliers visit in June 1882. A 25-horsepower steam engine was used at the No. 1 slope; it used 1 ton of pea coal per day, which cost $3.25 a ton (D'Invilliers, 1883, p. 267-268).

The No. 2 slope was about 300 feet southeast of the No. 1 slope and was located vertically about 30 feet higher. The No. 2 slope was worked with a horse and windlass hoist. A third shaft was sunk in rock about 60 feet southwest of the No. 1 slope, but it was abandoned at a depth of 40 feet without striking ore (D'Invilliers, 1883, p. 268).

The two ore beds were about 60 feet apart horizontally. The ore bed mined by the No. 1 slope was about 10 to 12 feet thick at the surface. The foot wall was decomposed quartzose gneiss, and the hanging wall was harder hornblende gneiss. Frequent attempts were made to follow leaders of ore east and west, but the leaders pinched out 10 to 30 feet from the slope. The No. 2 ore bed was about 3 feet thick and had the same characteristics as the ore bed in the No. 1 slope, except the ore contained more gangue. At a depth of 50 feet, the dip deceased to half of what it was above 50 feet, similar to that in the No. 1 slope (D'Invilliers, 1883, p. 267-268).

The ore was a black, coarse-grained, very hard magnetite with a metallic luster. It required little washing. The ore was mixed with a considerable amount of pyrite and gangue rock, mostly feldspar and quartz, which meant that it had to be carefully picked and sorted before it was shipped. The ore sold for $3.50 per ton. The average production in 1881 was about 10 tons a day, but the output was limited by the configuration of the mine, which prevented more than four miners working at the same time. The ore was shipped from the Bowers station on the East Penn Railroad 4 miles away. The principal customers in 1882 were the Saucon Iron Company in Lehigh County, the Leesport Iron Company, and the Sheridan and Kutztown Furnaces (D'Invilliers, 1883, p. 266-268).

An analysis of the ore was reported by D'Invilliers (1883, p. 268-269): metallic iron, 41.3-67.4 percent; manganese, 0.936 percent; sulfur, 0.224-1.9 percent; phosphorus, 0.04 percent; and silica, 2.1-20.3 percent. The ore was titaniferous, which probably gave the ore its lustrous appearance.

Percival Brumbach's Mine

Percival Brumbach's mine (fig. 565) was west of the intersection of Forgedale and Bick Roads at about 40° 25′ 41″ latitude and 75° 46′ 21″ longitude on the USGS Fleetwood 7.5-minute topographic quadrangle map (Appendix 1, fig. 14, location RK-2). The mine is shown as an iron mine on the 1860 and 1862 maps and 1876 atlas. LIDAR imagery suggests extensive workings. The mine was in the Hardyston Formation near the contact with felsic to mafic gneiss.

Percival Brumbach's mine was on a hill close to the site of the old Rockland forges. The mines were abandoned before D'Invilliers visit in 1882. D'Invilliers (1883, p. 360) noted: *"Specular ore of a blood-red color and earthy texture, showing no crystallization whatever occurs in P. Brumbach's ore holes, in close proximity to limonite. The ore is of a good quality, but*

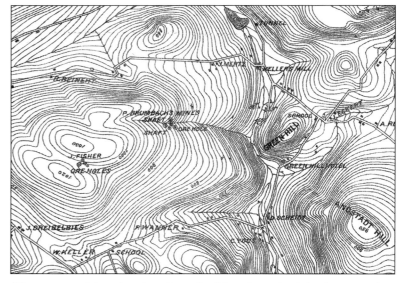

Figure 565. Map showing Brumbach's mine, J. Fisher's ore holes, and Green Hill, Rockland Township, 1882. From Pennsylvania Geological Survey (1883).

Figure 566. Goethite from Braumbach's mine, Rockland Township, 4 cm. Bryn Mawr College collection Rand 5594.

though the hill is riddled with small shafts and holes, no great quantity of ore seems to have been produced. As far as I could judge, it occurred more as surface pocket of ore than in any regular bed, though owing to the utter abandonment of the mines, no examination could be made." The ore was a mixture of red hematite and limonite.

Green Hill Area

The 1882 topographic map (fig. 565) shows Green Hill as the hill north of the intersection of Forgedale and Bick Roads on the USGS Fleetwood 7.5-minute topographic quadrangle map (Appendix 1, fig. 14, location RK-8). Gordon (1922, p. 157) indicated that Green Hill was the site of Rockland Forge.

Minerals

Muscovite, var. damourite - pale grayish-green to light brown, massive (Genth, Jr., 1882, p. 47-48)

"Psilomelane" (D'Invilliers, 1883, p. 399)

Quartz - lilac-colored (D'Invilliers, 1883, p. 400)

Old Millert Mine

The old Millert mine was south of the intersection of Pricetown and Forgedale Roads at about 40° 26' 13" latitude and 75° 46' 13" longitude on the USGS Fleetwood 7.5-minute topographic quadrangle map (Appendix 1, fig. 14, location RK-9). When the old Millert mine was visited by D'Invilliers on September 22, 1882, he found "*an abandoned cut of considerable size, about whose production nothing was learned*" (D'Invilliers, 1883, p. 372).

New Jerusalem Mines

The New Jerusalem mines were on both sides of Pricetown Road, 0.5 mile southwest of the intersection of Pricetown and Lobachsville Roads at about 40° 26' 35" latitude and 75° 45' 36" longitude on the USGS Fleetwood 7.5-minute topographic quadrangle map (Appendix 1, fig. 12, location RK-5). The mines were in felsic to mafic gneiss.

D'Invilliers (1883, p. 371) stated that some very good cabinet specimens of bomb shell ore (limonite) were found at the small mines about 0.5 mile southwest of New Jerusalem and about 300 feet northwest of Israel Heist's house. The mines were 30 to 40 feet deep and appeared to be mostly in clay. D'Invilliers also noted that further down the road, about 300 feet from the New Jerusalem mines, there were several trial pits and shafts, none of which seem to have produced ore.

Bieber Mine

The Bieber mine was 0.2 mile southeast of Boyers Junction, southeast of the intersection of Pricetown and Forgedale Roads at about 40° 26' 18" latitude and 75° 46' 08" longitude on the USGS Fleetwood 7.5-minute topographic quadrangle map (Appendix 1, fig. 12, location RK-3). The mine was in felsic to mafic gneiss. It is shown on the 1882 topographic map as H. Bieber's mine. The Bieber mine produced limonite. D'Invilliers (1883, p. 371) noted: "*Some ore has been mined here by Henry Bieber, but nothing could be learned about it, owing to everything being closed up, and Mr. Bieber's whereabouts not being known.*" According to the 1880 Census (Pumpelly, 1886, p. 960), the Bieber mine produced 335 tons of limonite between July 1, 1879, and June 30, 1880, which was shipped to the Topton and Kutztown Furnaces.

Daniel Fisher Mine

The Daniel Fisher mine was in the vicinity of the intersection of Highland Drive and Crown Lane, in the Forest Ridge development, about 1.2 miles southwest of Boyers Junction at

Figure 567. Limonite from the old Millert mine, Rockland Township, 4.4 cm. Reading Public Museum collection 2000C-027-814.

Figure 568. "Psilomelane" from near New Jerusalem, Rockland Township, 5.5 cm. Carnegie Museum of Natural History collection ANSP 7087.

about 40° 25′ 33″ latitude and 75° 46′ 48″ longitude on the USGS Fleetwood 7.5-minute topographic quadrangle map (Appendix 1, fig. 14, location RK-4). The mine was in the Hardyston Formation. The area is now a residential development. The Daniel Fisher mine likely was a small prospect. D'Invilliers (1883, p. 371) stated that further west of Percival Brumbach's mine and near summit of a hill, "*some small trial pits have been put down by Daniel Fisher, but there was not a hatfull of ore to be seen when visited.*" The mine is shown as J. Fisher's ore pits on the 1882 topographic map (fig. 565).

Schrading Mines

The Schrading mines were southeast of Bowers. The northern Schrading mine (location RK-6A) was north of the intersection of Mine and Cedar Mill Roads at 40° 29′ 03″ latitude and 75° 43′ 46″ longitude, and the southern Schrading mine (location RK-6B) was south of the intersection of Mine and Cedar Mill Roads at 40° 28′ 52″ latitude and 75° 43′ 49″ longitude, on the USGS Manatawny 7.5-minute topographic quadrangle map (Appendix 1, fig. 15). The mines were in the Leithsville Formation. The mines are shown as depressions on the topographic map. LIDAR imagery suggests these mines were large open pits. The property owner is shown as William Shradin on the 1860 and 1862 maps. The northern mine is shown as the W. Schrading mine in the 1876 atlas. The 1882 topographic map shows it as mine number 55.

Flint Hill Locality

Flint Hill (fig. 569) is about 1.2 miles southeast of Bowers on the USGS Manatawny 7.5-minute topographic quadrangle map (Appendix 1, fig. 15, location RK-7). Flint Hill is underlain by the Hardyston Formation.

An outcrop of jasper is located on the side of Flint Hill. The area was a Native American tool workshop. In 1930, the Reading Eagle (March 2, 1930) reported: "*The outcrop covers several acres and the ground is thickly strewn and mixed with flakes and rejected jasper. A small circular patch of woodland shows depressions made by Indian diggers. Hammerstones were abundant there years ago, but have since been carried away by collectors.*" Evidence of Native American encampments around Fleetwood are fairly numerous. Besides coming there to hunt and fish, Native Americans frequently followed a trail along the top of the mountain from Reading to Flint Hill to obtain flint for tools and arrow heads.

Minerals

From D'Invilliers (1883, p. 399-400) unless noted

Limonite - ocher

Limonite, var. xanthosiderite (Eyerman, 1889, p. 11)

Molybdenite (?)

Quartz - drusy, milky, blue, pink, ferruginous; smoky (Eyerman, 1889, p. 14)

Quartz, vars. chalcedonic jasper, jasper, agate, flint

Sally Ann (Hunter's, Rockland) Furnace

The Sally Ann Furnace was located on Sacony Creek, a branch of Maiden Creek, in Rockland Township. Montgomery (1884) indicated the furnace was built in 1791 by ironmaster Valentine Eckert; however, its early history is obscure. Eckert's involvement with the furnace is unclear, and, if Eckert was involved, it may have had a different name.

Nicholas Hunter is given credit in several secondary sources as having built the furnace in 1803, at which date he was said to be operating it. The furnace, if not built by him, was at least expanded under his ownership. Hunter is said to have renamed the furnace after his wife, Sarah (Sally) Ann Fisher, whom he married in 1814. Subsequently, Hunter transferred the furnace and lands to his son, Jacob V.R. Hunter, who by 1825 had obtained full ownership. In 1828-30, the furnace employed 150 workmen and used 10,800 cords of wood to produce 1,300 tons of pig iron and 252 tons of castings.

Lesley (1859, p. 39) called the furnace "Sally Ann or Rockland Cold-blast Charcoal Furnace," which was owned by Jacob V.R. Hunter of Reading and managed by J.N. Hunter of Dryville. In 1856, it made about 600 tons of iron using a mixture of limonite from Trexlertown or the Moselem mine and "flat ore."

Active operations were discontinued about 1869. The American Iron and Steel Association (1876) listed the Sally Ann Furnace as abandoned. Montgomery (1884, p. 71) noted that it was leased from the Hunter estate, rebuilt, converted to steam power, and reopened in 1879, but operated for only a short time. The American Iron and Steel Association (1880) reported that the "Rockland Furnace, formerly called Sally Ann Furnace," was operated by the Rockland Furnace Company, Limited, of Douglassville. D'Invilliers (1883, p. 234) indicated

FLINT HILL

Figure 569. Flint Hill, Rockland Township, April 2015. View is looking north toward Flint Hill.

Figure 570. Druzy quartz from Flint Hill, Rockland Township, 10 cm. Carnegie Museum of Natural History collection CM 1685 (Jefferis 9302).

Figure 571. Quartz, var. chalcedony, from the Flint Hill area, Rockland Township, 8.5 cm. Carnegie Museum of Natural History collection ANSP 6984.

Figure 572. Quartz, var. yellow jasper, from the Flint Hill area, Rockland Township, 7.4 cm. Sloto collection 3230.

Figure 573. Native American arrowheads made from yellow jasper. These arrowheads were found in Berks County. Sloto collection.

that the furnace was out of blast on December 31, 1880, and the American Iron and Steel Association (1886) listed the Rockland Furnace as "burned" in 1881.

RUSCOMBMANOR TOWNSHIP

Udree Mine

The Udree mine was on the north flank of Furnace Hill, northeast of the intersection of Fry and Walnuttown Roads, 1.05 miles east of Breezy Corner at 40° 24′ 55″ latitude and 75° 49′ 42″ longitude on the USGS Fleetwood 7.5-minute topographic quadrangle map (Appendix 1, fig. 13, location RM-1). The Udree mine, also known as the Udree ore bank, was the largest producer of limonite ore in the Oley Valley. The mine was in felsic to mafic gneiss.

Daniel Udree was born in Philadelphia on August 5, 1751. He came to the Oley Valley to work for his uncle, Jacob Winey, as a clerk at the Oley furnace. During the Revolutionary War, Udree served in the army and rose to the rank of major. In 1778, he became a partner with Jacob Winey, Jacob Lever Seyler, and others in the Oley Furnace. In 1801, he became the sole owner of the furnace (Croll, 1926, p. 105). During the war of 1812, Udree again served our Nation and rose to the rank of major general (Schultz, 1936).

Udree attempted to sink a shaft. However, the shaft was sunk only to a depth of 60 feet and was abandoned because of the abundant inflow of groundwater and the high cost of explosives.

In 1871, the mine was acquired by the Clymer Iron Company. The ore was very siliceous and produced brittle iron; this prevented its use in Clymer's nearby Oley charcoal furnace. Therefore, the ore was hauled 6 miles at a cost of 60 cents per ton to the company's Mt. Laurel (Temple) Furnace in Temple (D'Invilliers, 1883, p. 363). According to the 1880 Census (Pumpelly, 1886, p. 960), the mine produced 8,002 tons of limonite between July 1, 1879, and June 30, 1880, which was shipped to the Clymer Temple Furnace.

When D'Invilliers (1883, p. 362-363) visited the mine on June 19, 1882, Daniel Rauenzahn was the mine superintendent, and 12 men were employed at the mine. Iron ore was mined from an open cut 300 feet long and 70 feet deep. The ore was hoisted on a double track incline by a 35-horse power steam engine built by Archambault of Philadelphia. The hoisting engine had a double geared 15-foot diameter drive gear and a 39-foot diameter flywheel. The ore was dumped into a single 26-foot long washer with a capacity of 50 tons. Water to wash the ore was pumped from a well near the engine house. A small pump was used to keep the mine dewatered.

The ore body dipped about S. 20° E. 70° and averaged 20 feet thick. The ore bed contained horses of clay. Some of the limonite was bomb-shell ore, which was hollow, filled with clay and water, and coated inside with an incrustation of manganese oxide. The mine produced 18 to 20 tons of ore per day (D'Invilliers, 1883, p. 362-363). D'Invilliers published an analysis of the ore: metallic iron, 40.1 percent; metallic manganese, 3.3 percent; sulfur, 0.003 percent; phosphorus, 0.522 percent; and siliceous matter, 22.4 percent.

Minerals

From D'Invilliers (1883, p. 362 and 402) unless noted.

Goethite

Goethite, var. turgite - often iridescent

Limonite - red and yellow ocher; D'Invilliers reported *"handsome specimens of concretionary and stalactitic ore"* and bomb-shell ore

Lepidocrocite

Ocher - red and yellow

"Psilomelane" (Gordon, 1918)

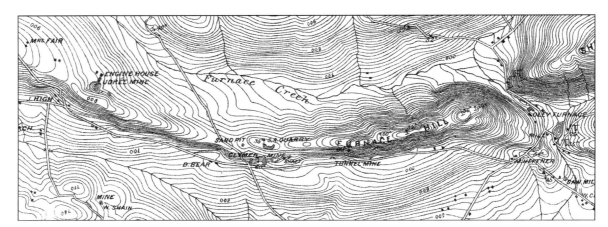

Figure 574. Locations of the Udree, Clymer, and old Tunnel mines, Ruscombmanor Township, 1882. From Pennsylvania Geological Survey (1883).

Clymer Mines

The Clymer mines were north of Fry Road, northeast of its intersection with Pennsylvania State Route 662, about 1.2 miles west-southwest of Oley Furnace at about 40° 24' 27" latitude and 75° 48' 46" longitude on the USGS Fleetwood 7.5-minute topographic quadrangle map (Appendix 1, fig. 13, location RM-2). LIDAR imagery shows a number of extensive open cuts on the south side of Furnace Hill. The W. Clymer ore mine was shown in the 1876 atlas. The mines were in the Hardyston Formation and hornblende gneiss.

The Clymer mine was an open cut on the south flank of Furnace Hill. When the mine was first visited by D'Invilliers in the summer of 1880, it was being actively worked by the Clymer Iron Company under superintendent Benjamin Wyle. The cut was about 250 feet long, 40 feet wide, and about 40 feet deep. The ore was removed from the open cut in carts and hauled to Clymer's Mt. Laurel Furnace. A small part of the ore was shipped to the Oley Furnace at the eastern end of Furnace Hill. The mine was abandoned when visited again by D'Invilliers on July 19, 1882, *"on the plea that no regular or persistent bed could be found there."* The workings had been extended along the hill towards the old Tunnel mine, and several shafts were sunk to test the dip of the gneiss. One of the shafts was about 40 feet deep and was equipped with a windlass (D'Invilliers, 1883, p. 269-271).

The ore bed was 1 to 4 feet thick, often dividing and forking into several branches. The ore was underlain by gneiss, dipping with the ore at S. 30° E. 60°. Above the ore was a decomposed greenish black slate, which was mixed with the ore. The ore near the top of the bed was harder and more highly crystallized than the ore near the bottom and frequently contained balls of porous hard ore with distinct octahedral crystallization. The magnetite ore was a soft, fine, dull black, earthy powder with numerous scales of black and brown mica. D'Invilliers (1883, p. 270-271) noted: *"The ore where found is of excellent quality and in its finely divided state and mixture with easily fluxed materials, it was readily used in the furnaces. Its high percentage of manganese makes it valuable for Bessemer iron."*

Figure 575. Goethite (left), 5.3 cm, and limonite pipe ore (right), 6.4 cm, from the Udree mine, Ruscombmanor Township. Reading Public Museum collection 2000C-027-851.

D'Invilliers published an analysis of the ore: metallic iron, 59.1 percent; sulfur, none; phosphorus, 0.068 percent; and siliceous matter, 0.86 percent. An analysis was presented by McCreath (1879, p. 212-213): metallic iron, 40.8 percent; metallic manganese, 1.7 percent; sulfur, 0.026 percent; phosphorus, 0.453 percent; and siliceous matter, 23.59 percent. McCreath described the ore as dark-brown, generally very compact and fine-grained lump and wash ore with a conchoidal fracture.

Minerals

Chlorite (D'Invilliers, 1883, p. 395)
Goethite, var. turgite
Muscovite (D'Invilliers, 1883, p. 271 and 399)
"Wad" (D'Invilliers, 1883, p. 402).

Old Tunnel Mine

The old Tunnel mine was east of the Clymer mines. It was north of Fry Road, northeast of its intersection with Pennsylvania State Route 662, about 0.9 mile west-southwest of Oley Furnace at about 40° 24' 28" latitude and 75° 48' 23" longitude on the USGS Fleetwood 7.5-minute topographic quadrangle map (Appendix 1, fig. 14, location RM-3). It is shown as the Tunnel mine on the 1882 topographic map (fig. 574). The mine was in the Hardyston Formation.

Figure 576. Clymer mine, Ruscombmanor Township, April 2015.

Figure 577. Limonite pipe ore from the Clymer mine, Ruscombmanor Township, 4.6 cm. Reading Public Museum collection 2000C-027-842.

Figure 578. Limonite from the Clymer mine, Ruscombmanor Township, 6.5 cm. Reading Public Museum collection 2000C-027-842.

The old Tunnel mine was on the property of the Clymer Iron Company on the south flank of Furnace Hill on the direct line of strike of the Clymer open cut and about 200 feet east of it. Mining began in the 1700s and ended prior to 1882. When the mine was visited by D'Invilliers (1883, p. 271-272) in 1882, it was "*entirely abandoned and fallen in so that no examination could be made of it.*" The mine was abandoned because the ore bed thinned, making it unprofitable to mine, and there was a large inflow of groundwater. Several attempts were made by the Clymer Iron Company to clean out the workings in the hopes of finding a large body of good ore, but ore was never found. The mine reportedly produced "*both hard and soft ore of excellent quality.*"

D'Invilliers (1883, p. 272) believed that this is the mine referred to by Rogers (1841, p. 43 and 1858, vol. 2, p. 716), who reported: "*At this spot (2 miles N. W. of Friedensburg [Oley]) is the old iron mine belonging to the Oley furnace. The ore was dug from immediately under an outcrop of Primal sandstone, the digging running parallel with it for more than 100 yards, and being 18 or 20 feet deep and 8 or 10 feet wide. This mine, now abandoned, furnished us some specimens from the sidewall of the excavation. These are argillaceous and laminated, and of a purplish-red color. A shaft unites the main excavation with another nearly under the first, having about the same direction, but descending more perpendicularly. The latter mine is from 3 to 5 feet wide; the wall is of metamorphic rock, chiefly feldspathic and hornblendic gneiss, but sometimes entirely micaceous, and it contains in certain places magnetic and micaceous iron ore.*" In 1841, Rogers (1841, p. 43) added: "*The proprietors of Oley furnace propose reopening this old mine, having nearly completed a tunnel now six hundred feet long, intended to reach the lower excavation. The rocks passed through in the tunnel, are gneiss, syenite, hornblende, and micaceous slates.*" Eyerman (1889, p. 22) reported muscovite from the Oley Tunnel mine.

Schittler (Hain) Mine

The Schittler mine was southwest of the intersection of Pricetown and Orchard Roads, about 1.5 miles southeast of Pricetown on the USGS Fleetwood 7.5-minute topographic quadrangle map (Appendix 1, fig. 14, location RM-4). It likely is the "*old iron mine situated on a hill 3/4 mile east-northeast of Pricetown*" described by Gordon (1922, p. 158), and the "Reading" locality of Smith (1855, p. 188). It also is known as the Hain or Haines mine. It is shown as "old mine" on the 1882 topographic map (fig. 583).

A mine, known as Schittler's Ore Bank, was located on land originally owned by Milton Schittler about 1.25 miles east of Pricetown. It was later owned by Mr. Haines, and in 1882, it was owned by Eckert & Brother of Reading (D'Invilliers, 1883, p. 273). This mine may be the "Road's vein" of Rogers (1858, vol. 2, p. 716). Rogers stated: "*At Road's [Rhode's], near 2 miles*

Figure 579. Magnetite from the Old Tunnel mine, Ruscombmanor Township, 7.5 cm. Collected in 1883. Reading Public Museum collection 2004C-2-207.

E. of Pricetown, there is a magnetite of superior quality, said to be between 5 and 6 feet thick. Its dip is perpendicular."

In 1882, D'Invilliers (1883, p. 273) found an abandoned shaft close to the creek. He found a considerable quantity of magnetite ore in a nearby field and on an old dump. The magnetite was highly crystallized, tough, and black, often showing bright metallic faces. Embedded in the ore were *"beautiful chocolate brown, lustrous and opaque crystals of Zircon, sometimes an inch in length, but rarely perfectly terminated."*

The zircon was analyzed by Wetherill (1853, p. 346-350), who described the crystals: *"The Zircon occurs in large crystals firmly imbedded in the ore, which are in some instances well terminated, but brittle, and detached with great difficulty from their matrix, to which they adhere with such tenacity, that the impression left in the matrix after detaching them is polished, of vitreous lustre and of the colour of the crystals, as if they had been melted in the ore after their formation. Their planes and angles are rounded off in places, as if they had been subjected to an incipient fusion. The largest crystals which I obtained after carefully breaking several pounds of the ore, measured one and a half inches, by one-quarter inch, by three-eighth inch (nearly). It was distinctly terminated at one end, and showed traces of termination at the other. The usual crystal form was a right prism terminated by corresponding pyramids, the angles of which were frequently modified. The colour, chocolate brown; opaque; lustre, adamantine; planes, as before stated, uneven. One specimen which was too much broken to form certain conclusions, appeared to be part of one of the terminal pyramids. It was highly modified, possessed perfectly sharp edges and glass smooth planes, of adamantine lustre; was on the edges transparent, like hyacinth, and of deeper colour than the other crystals. The cleavage of the mineral was indistinct and fracture very uneven, but apparently in planes perpendicular to the vertical axis. Hardness, between quartz and topaz, or 7-8 of Mohs' scale. Its powder was brownish yellow, very light. 0.965 grammes weighed in distilled water of temperature 26° C., 0.755 corresponding to a density of 4.595."*

Figure 580. Magnetite and zircon from the Hain (Schittler) mine, Ruscombmanor Township, 9 cm. Carnegie Museum of Natural History collection ANSP 1628.

Schweiter and Kutz Mine

The Schweiter and Kutz mines were south of Fleetwood, between Pennsylvania State Route 662 and Lake Clemmens Road, just south of the Fleetwood Borough boundary with Ruscombmanor Township on the USGS Fleetwood 7.5-minute topographic quadrangle map (Appendix 1, figs. 11 and 12). The mines are mines number 59 and 60 on the 1882 topographic map. Mine number 59 (location RM-7A) is a flooded pit at 40° 26′ 47″ latitude and 75° 48′ 43″ longitude. Mine number 60 (location RM-7B), which is west of mine number 59, is at 40° 26′ 47″ latitude and 75° 48′ 52″ longitude. The mines were in the Allentown and Leithsville Formations.

The Schweiter and Kutz mine was an open cut on the north flank of South Mountain. When visited by D'Invilliers on September 22, 1882, the mine had only been in operation

Figure 581. Schweitzer and Kutz mine open cut (left) and flooded open pit (right), Ruscombmanor Township, April 2015.

about 5 months and lacked a pump to dewater the cut. On that day, the bottom of the mine was flooded from recent rain. The engine house was equipped with one 10-horsepower pumping engine. D'Invilliers (1883, p. 371-372) noted: *"A considerable pit had been dug down against the face of the hill. The bed is said to be 12' thick, though no such thickness was exposed at time of visit."* The ore dipped about 45° S.E. into the hill. The strata below the ore bed was sandstone, and the strata above the ore bed was a decomposed muscovite slate and clay. The average production was 10 tons per day, which was shipped from Fleetwood to the Temple Iron Company. The ore cost 60 cents per ton to haul 2 miles to Fleetwood.

Pricetown Area

The Pricetown area (fig. 583) includes the farms along Pricetown Road, east of Pricetown on the USGS Fleetwood 7.5-minute topographic quadrangle map (Appendix 1, fig. 13, location RM-5). The area includes Schittler's mine described above. Genth (1875, p. 76) reported that zircons were found on the farms of W. Haines, Schroeder, and Mrs. D. Rhodes.

Minerals

Allanite - pitch-black, massive, with zircon and magnetite; often coated with a brownish decomposition product (Smith, 1855, p. 188)

Biotite - 1 mile northeast of Pricetown (Eyerman, 1889, p. 21)

Epidote - massive at Hain's (Schittler's) mine, near Pricetown (Genth, 1875, p. 220)

Garnet group (D'Invilliers, 1883, p. 397)

Limonite, var. xanthosiderite (Eyerman, 1889, p. 11)

Magnetite - titaniferous; sometimes exhibiting polarity (Genth, 1875, p. 21; D'Invilliers, 1883, p. 401)

Quartz, vars. chalcedony and jasper (Genth, 1875, p. 21; D'Invilliers, 1883, p. 400); drusy quartz, milky, occasionally tinted pale blue, pink. or ferruginous (Genth, 1875, p. 218)

Zircon - in magnetite; chocolate-brown, opaque, prismatic crystals to 3.8 cm, often with rounded terminations

Figure 582. Limonite from the Schweitzer and Kutz mine, Ruscombmanor Township, 6 cm. Reading Public Museum collection 2000c-027-845.

(Wetherill, 1853, p. 346-350; Genth, 1875, p. 76; D'Invilliers, 1883, p. 402) (figs. 584 and 585)

Other Localities

Sand Hill Sand Quarries

There were several sand quarries on Sand Hill on both sides of Watnuttown Road, near its intersection with Hartz Road, about 1 mile southwest of Pricetown. LIDAR imagery shows the largest quarry at about 40° 25' 19" latitude and 75° 50' 30" longitude on the USGS Fleetwood 7.5-minute topographic quadrangle map (Appendix 1, fig. 13, location RM-6).

The 1876 atlas shows a sand mine at this location. The area is underlain by the Hardyston Formation.

The Reading Eagle (December 4, 1892) reported that there were four or five sand quarries on Sand Hill. The sand was hauled 4 miles to Blandon where it was shipped by rail. O'Neil (1965, p. 37) reported that the Refractory Sand Company, Inc. of Ringtown, Schuylkill County, produced sand at its Pricetown plant. The Refractory Sand Company was incorporated in Pennsylvania on July 25, 1951.

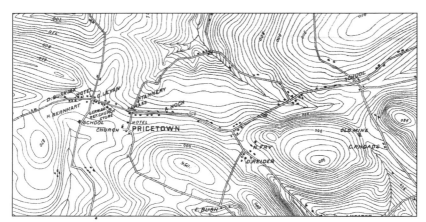

Figure 583. Location of the Pricetown area, Ruscombmanor Township, 1882. From Pennsylvania Geological Survey (1883).

Blandon Area Wavellite Locality

Gordon (1922, p. 159) reported wavellite on limonite from an old iron prospect on a hill 1.5 miles southeast of Blandon. Gordon obtained information on this locality from Edgar Wherry,

Jacob Fox Farm Locality

D'Invilliers (1883, p. 399-400) reported ocher, blue (amethystine) quartz, cellular quartz, and quartz, var. Lydian stone (flinty black jasper, also called basanite), red jasper, nodular flint, and chalcedony on the Jacob Fox farm. Gordon (1922, p. 159) stated that the Jacob Fox farm was underlain by quartzite and was located 2.5 miles south of Blandon.

Figure 584. Zircon from the Pricetown area, Ruscombmanor Township, 1.3 cm. Steve Carter collection.

Blue Quartzite Quarry

The blue quartzite quarry (fig. 586) was on the north side of South Mountain, east of the intersection of Delcamp Road and Beech lane at 40° 27' 13" latitude and 75° 47' 57" longitude on the USGS Fleetwood 7.5-minute topographic quadrangle map (Appendix 1, fig. 12, location RM-8). Blue quartzite was quarried from a hillside quarry in the Hardyston Formation for use as building stone in the Fleetwood area.

Bomegratz's Farm Quartz Locality

D'Invilliers (1883, p. 400) reported ferruginous quartz, drusy quartz, and chalcedony in globules on Bomegratz's farm in Ruscombmanor, 2 miles west of Pricetown.

Figure 585. Zircon from the Pricetown area, Ruscombmanor Township, 1.2 cm. Ron Kendig collection.

Figure 586. Blue quartzite quarry, Ruscombmanor Township, April 2015.

SINKING SPRING BOROUGH

Limestone quarries in Sinking Spring Borough were described by D'Invilliers (1886, p. 1560-1561). Quarry locations are shown in Pennsylvania Geological Survey (1891, reference map 11) (fig. 587).

Evans Quarry

The James Evans quarry was northwest of the intersection of Columbia Avenue and Commerce Street at about 40° 19′ 19″ latitude and 76° 01′ 53″ longitude on the USGS Sinking Spring 7.5-minute topographic quadrangle map (Appendix 1, fig. 24, location SS-1). It is quarry number 96 on reference map 11 in Pennsylvania Geological Survey (1891) (fig. 587). The Evans quarry was a large, old quarry in the Ontelaunee Formation on the north side of the Lebanon Valley Railroad. D'Invilliers (1886, p. 1560) noted: "*The quarry has not been worked for many years, and the product went to various points on the railroad for furnace use before it was necessary to have limestone of such purity as is now required for fluxing purposes.*" The rocks dipped toward the south at about 40°.

George Peipher Quarry

The George Peipher quarry was just south of U.S. Route 422, east of its intersection with Columbia Avenue at about 40° 19′ 30″ latitude and 76° 01′ 06″ longitude on the USGS Sinking Spring 7.5-minute topographic quadrangle map (Appendix 1, fig. 24, location SS-2). It is quarry number 98 on reference map 11 in (Pennsylvania Geological Survey (1891) (fig. 587). The small Peipher quarry was in the Ontelaunee Formation just inside Sinking Springs Borough. D'Invilliers (1886, p. 1561) noted: "*This quarry, like many others in the vicinity, has now been abandoned for a long time, mainly due to the leanness of the limestone, unfitting it for either furnace use or for making a first-class quality of burned lime.*"

Deckert Quarry

The Adam Deckert quarry was south of U.S. Route 422 and north of the railroad tracks, southwest of its intersection with Pennsylvania State Route 724 at about 40° 19′ 25″ latitude and 76° 01′ 06″ longitude on the USGS Sinking Spring 7.5-minute topographic quadrangle map (Appendix 1, fig. 24, location SS-3). It is quarry number 99 on reference map 11 in Pennsylvania Geological Survey (1891) (fig. 587). The 1876 atlas (fig. 588) shows a lime kiln and a railroad siding extending from the Lebanon Valley Railroad to the quarry. The quarry was opened in the Ontelaunee Formation by Danner S. Hamlin on

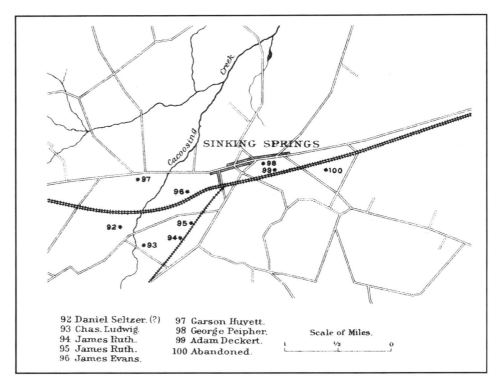

Figure 587. Limestone quarries in the vicinity of Sinking Spring Borough, 1891. From Pennsylvania Geological Survey (1891).

the Adam Deckert farm in 1873. At that time, the quarry employed 10 to 12 men (*Reading Eagle*, October 17, 1873). D'Invilliers (1886, p. 1561) stated: *"The limestone here is poor and the quarry was abandoned years ago."* The dip of the rocks was about S. 25° E. 36° to 50°.

Shillington Road Quarry

The Shillington Road quarry was east of the intersection of U.S. Route 422 and PA 724 at about 40° 19′ 25″ latitude and 76° 00′ 44″ longitude on the USGS Sinking Spring 7.5-minute topographic quadrangle map (Appendix 1, fig. 30, location SS-4). It is quarry number 100 on reference map 11 in Pennsylvania Geological Survey (1891) (fig. 587). The quarry, in the Richland Formation, was on the south side of the railroad tracks. D'Invilliers (1886, p. 1561) noted: *"The rock here is lean and thinly bedded and is hardly fit for any purpose."*

Thierwechler Mine

The Thierwechler mine was east of the intersection of Cacoosing Avenue and Reedy Road at about 40° 19′ 40″ latitude and 76° 01′ 21″ longitude on the USGS Sinking Spring 7.5-minute topographic quadrangle map (Appendix 1, fig. 24, location SS-5). The location is now the Village Green Golf Course; no trace of the former mining operation remains. The mine was in the Ontelaunee Formation.

The 1876 atlas shows the iron mine on the property of Mrs. A.E. Thierwechler with a pumping engine and a hoisting engine (fig. 588). This likely is the mine operated by the Bushong Iron Company of Reading in 1873 (*Reading Eagle*, October 17, 1873) and L.M. Kaufman of the Leesport Iron Company in 1880. Teams hauled ore from the mine to a warf on the Lebanon Valley Railroad for shipment (*Reading Eagle*, February 4, 1880). It also is the likely location for the "velvet iron ore" goethite reported by Genth (1875, p. 48) from Sinking Spring.

Figure 588. Thierwechler mine and Dechert quarry, Sinking Spring, 1876. From Davis and Kochersperger (1876).

Sinking Spring Area

Brucite was reported from an unidentified quarry near Sinking Spring. Genth (1886, p. 40) described the occurrence: *"The brucite from near Sinking Spring, as Dr. Smith states, occurs in thin colorless laminae in thin seams in the limestone, but also in silky fibrous masses or even pulverulent, with but a faint silky lustre. The brucite is associated with deweylite, coarse grained calcite and aragonite, in dolomite. I have analyzed a perfectly pure piece of the silky fibrous brucite.*

Deweylite, Aragonite, Calcite.—In the magnesian limestone occur these three minerals, more or less mixed together and associated with brucite. The deweylite is white, yellowish-white or brownish, amorphous, sometimes in rounded grains or in stalactites or botryoidal forms, in thin plate-like masses or slabs occasionally over one inch in thickness, or in irregular coatings. These slabs are often arranged in layers of white or

Figure 589. Actinolite from Sinking Spring. Reading Public Museum collection 2004C-10-236.

brownish deweylite of greater or less purity, often intimately mixed with aragonite, which sometimes separates in the form of radiating columnar masses, some of the individuals being over 50 mm in length. The layers often separate very easily and the surfaces of such planes of separation are covered with small brilliant crystals of aragonite. Calcite is also present, both in small and insignificant crystals and in coarse crystalline masses.

Pseudomorphs of deweylite after aragonite.—The needle-shaped crystals of aragonite and the radiating masses undergo a change and are gradually altered into brownish-yellow deweylite. It begins with a very thin coating of colorless and brownish-yellow deweylite upon the aragonite, which gradually becomes thicker and finally changes the entire aragonite into pure deweylite."

Specimens of aragonite (fig. 590) and brucite (fig. 591), labeled Fritztown, which is 3.1 miles southeast of Sinking Spring, are likely from the *"near Sinking Spring"* locality. These specimens were in the collections of father William A. Drown (1836-1890) and son Edward D. Drown (1861-1919), which were acquired by the Academy of Natural Sciences of Philadelphia and are now part of the Carnegie Museum of Natural Science collection.

South Heidelberg Township

Limestone quarries in South Heidelberg Township were described by D'Invilliers (1886, p. 1556-1559). Quarry locations are shown in Pennsylvania Geological Survey (1891, reference map 11) (fig. 592).

Levi Goul Quarries

The Levi Goul quarries were west of Wernersville, and northwest of the intersection of U.S. Route 422 and Hospital Road on the USGS Sinking Spring 7.5-minute topographic quadrangle map (Appendix 1, fig. 23). The Levi Goul No. 1 quarry (location SH-1) was between Old West Penn Avenue and the railroad tracks at 40° 20′ 24″ latitude and 76° 06′ 50″ longitude. The Levi Goul No. 2 quarry (location SH-2) was at 40° 20′ 16″ latitude and 76° 06′ 50″ longitude. The Levi Goul No. 3 quarry (location SH-3) was between the railroad tracks and a farm lane south of the railroad tracks at 40° 20′ 08″ latitude and 76° 06′ 25″ longitude. They are quarry numbers 84, 84a, and 84b, respectively, on reference map 11 in Pennsylvania Geological Survey (1891) (fig. 592). The quarries were in the Richland Formation.

The Levi Goul No. 1 quarry was a short distance north of the Lebanon Valley Railroad. D'Invilliers (1886, p. 1555-1556) noted: *"There is quite a good opening at this point showing a rather heavily bedded rock of fair quality. The stripping is light, and the stone could be economically quarried."* In 1886, the quarry face was about 20 feet high. On the north side, the rocks dipped S. 40° W. 55°. There were two kilns at the quarry indicating that the stone was quarried for lime production.

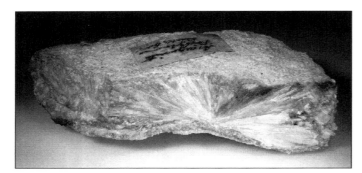

Figure 590. Aragonite from Fritztown, 12 cm. Carnegie Museum of Natural History collection ANSP 6935.

Figure 591. Brucite from Fritztown, 11 cm. Carnegie Museum of Natural History collection ANSP 6961.

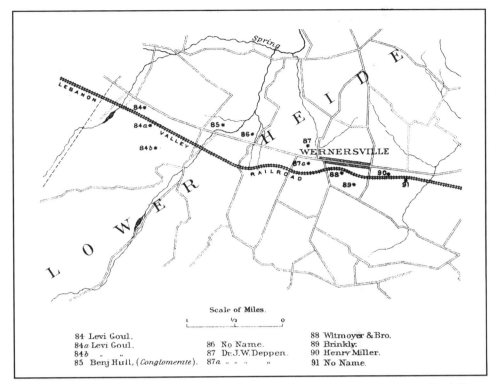

Figure 592. Limestone quarries in the vicinity of Wernersville, South Heidelberg Township, 1891. From Pennsylvania Geological Survey (1891).

The Levi Goul No. 2 quarry was on the south side of the Lebanon Valley Railroad immediately opposite the Levi Goul No. 1 quarry. D'Invilliers (1886, p. 1556) stated: *"There appears to have been a large quantity of stone excavated from this opening; where it went to and for what purpose it was used was not learned. A large cave-like opening was observed in the limestones. The whole operation is now abandoned and has been so for some years. The quality of the stone is only fair, and is reported to have been quite magnesian."* The rocks dipped S. 40° W. 30°.

The Levi Goul No. 3 quarry was a short distance southeast of the Levi Goul No. 2 quarry. It was a small quarry with one kiln to burn lime for local use (D'Invilliers, 1886, p. 1556).

The Reading Eagle (January 17, 1906) reported that a blast in Cyrus Gaul's quarry loosened enough stone to produce 5,000 bushels of lime. Limestone and lime producers listed by Hice (1911) included Cyrus W. Gaul and James M. Gaul in Wernersville.

Miller (1934, p. 219) reported that the A.V. Gaul quarry, located along the highway halfway between Wernersville and Robesonia, produced crushed stone. The rock was mainly thin bedded, dark colored, and low in magnesia. Loose blocks of white to pinkish gray stone were observed in the quarry. The quarry face was about 65 feet high, the strike was N. 23° W., and the dip was 55° SW.

Daniel Seltzer Quarry

The Daniel Seltzer quarry was northeast of Lincoln Road and Krick Lane, east of the South Heidelberg industrial Park at about 40° 19' 08" latitude and 76° 02' 51" longitude on the USGS Sinking Spring 7.5-minute topographic quadrangle map (Appendix 1, fig. 24, location SH-4). It is quarry number 92 on reference map 11 in Pennsylvania Geological Survey (1891) (fig. 587). The quarry was an old quarry in the Richland Formation. D'Invilliers (1886, p. 1559) observed: *"The quarry exhibits stone of a fairly good quality, but quite broken and weathered, dipping irregularly southward with an average inclination of 35°."* In 1905, the 76-acre Daniel Seltzer estate, which included "a stone quarry of limestone and flag stone" was offered for public sale (*Reading Eagle*, October 10, 1905).

Huyett Quarry

The Garson Huyett quarry was between U.S. Route 422 and Maywood Avenue, just west of Sinking Spring at 40° 19' 34" latitude and 76° 02' 26" longitude on the USGS Sinking Spring 7.5-minute topographic quadrangle map (Appendix 1, fig. 24, location SH-5). It is quarry number 97 on reference map 11 in Pennsylvania Geological Survey (1891) (fig. 587). The large quarry in the Richland Formation was abandoned in 1853. D'Invilliers (1886, p. 1560) described the limestone: *"There is a large opening at this point, showing a dark blue-gray limestone, rather lean and hard, and with strata occurring alternately in thick and thin beds, the latter being rather slaty, and all inclining southward about 30°."* The rock was used to make lime and also was shipped to Reading for furnace flux.

Limestone Quarry Number 2

Unnamed limestone quarry number 2 of D'Invilliers (1886, p. 1559) was between U.S. Route 422 and the railroad tracks, just east of Wernersville at about 40° 19' 39" latitude and 76° 04' 17" longitude on the USGS Sinking Spring 7.5-minute topographic quadrangle map (Appendix 1, fig. 23, location SH-6). It is quarry number 91 on reference map 11 in Pennsylvania Geological Survey (1891) (fig. 592). The area has been developed. The unnamed quarry, in the Hershey and Myerstown Formations, undivided, was on the north side of the railroad tracks. D'Invilliers (1886, p. 1559) noted: *"The rock exposed here is exceedingly hard, with an occasional slaty,* impure band, and the strata are very irregular. This opening has not been worked for many years." The limestone dipped about S. 20° E. 55°.

Fritztown Area

Fritztown is at the intersection of Fritztown, Galen Hall, and Mail Route Roads on the USGS Sinking Spring 7.5-minute topographic quadrangle map (Appendix 1, fig. 25, location SH-7). LIDAR imagery shows several excavations on the northwest side of Fritztown Road in South Heidelberg Township.

Minerals

From D'Invilliers (1883, p. 396-401), unless noted.

Calcite

Fluorite - pale yellow and topaz-colored cubes to 7/8 inch; associated with calcite (Eyerman, 1889, p. 7)

Limonite

Magnetite - massive

Muscovite, var. damourite - massive, lamellar, translucent (Genth, 1882, 47; analysis)

Pyrrhotite - nickeliferous (Eyerman, 1889, p. 5)

Quartz - blue amethystine

Quartz, vars. chalcedony, jasper, and flint

Cushion Mountain Locality

Cushion Mountain is northwest of Fritztown, and northwest of the intersection of Galen Hall and Preston Roads on the USGS Sinking Spring 7.5-minute topographic quadrangle map (Appendix 1, fig. 25, location SH-8). Cushion Mountain also is known as South Mountain.

Quartz, vars. chalcedonic jasper and jasper (D'Invilliers, 1883, p. 400)

Silver - Eyerman (1889, p. 3) stated: "According to Dr. D. B. Brunner, traces of silver occur in Berks county ... on Cushion Mt."

Spodumene - massive, white to pale brown (figs. 593 and 594)

Hain Mine

The Hain mine was the three flooded pits between U.S. Route 422 and the railroad tracks, west of Furnace Road at about 40° 19′ 50″ latitude and 76° 05′ 40″ longitude on the USGS Sinking Spring 7.5-minute topographic quadrangle map (Appendix 1, fig. 23, location SH-11). The mines were in the Epler Formation. The 1860 and 1862 maps show iron mines (labeled ore banks) at this location. The 1876 atlas map of Lower and South Heidelberg Townships shows the A. Hain iron mine (ore banks) next to the railroad west of Wernersville.

Cushion Mountain Iron Mines

The Cushion Mountain iron mines were northwest of the intersection of Galen Hall and Cushion Peak Roads at about 40° 17′ 48″ latitude and 76° 05′ 22″ longitude on the USGS Sinking Spring 7.5-minute topographic quadrangle map (Appendix 1, fig. 25, location SH-10). The mines were in the Hardyston Formation. The 1862 map and the 1876 atlas map of Lower and South Heidelberg Townships show two iron mines south of Cushion Hill. The mines were opened around 1760 (*Reading Eagle*, September 2, 2002).

SPRING TOWNSHIP

Some of the limestone quarries in Spring Township were described by D'Invilliers (1886, p. 1559-1560). Quarry locations are shown in Pennsylvania Geological Survey (1891, reference map 11) (fig. 587).

Limestone Quarries

Ludwig Quarry

The Charles Ludwig quarry was northeast of Montello Road at about 40° 18′ 55″ latitude and 76° 02′ 28″ longitude on the USGS Sinking Spring 7.5-minute topographic quadrangle map (Appendix 1, fig. 24, location S-1). It is quarry number 93 on reference map 11 in Pennsylvania Geological Survey (1891) (fig. 587). The quarry was in the Richland Formation. D'Invilliers considered the Ludwig quarry a "medium-sized" quarry. There were three kilns at the quarry, each of which produced 500 bushels of lime per week. The lime was sold to local farmers for 7 cents per bushel. D'Invilliers (1886, p. 1559) stated: "*The stone nearest the public road is slaty, much broken and has a curly structure, rendering it expensive and irregular to quarry. This structure, however, is only local, for the general dip in the vicinity seems quite regular to the south.*" Hice (1911) listed John R. Ludwig of Montello as a producer of limestone and/or lime.

Ruth Quarries

The James Ruth quarries were northeast of Montello Road. The James Ruth No. 1 quarry (location S-2) was at about 40° 18′ 58″ latitude and 76° 02′ 25″ longitude. The James Ruth No. 2 quarry (location S-3) was at 40° 19′ 01″ latitude and 76° 02′ 11″ longitude on the USGS Sinking Spring 7.5-minute topographic quadrangle map (Appendix 1, fig. 24). They are quarries number 94 and 95, respectively, on reference map 11 in Pennsylvania Geological Survey (1891) (fig. 587). The site of the James Ruth No. 2 quarry is now a fuel storage tank farm. The quarries also were known as the Ruth and Foreman quarries (*Reading Eagle*, August 16, 1881). The quarries were in the Richland Formation.

Figure 593. Spodumene from Cushion (South) Mountain (front and back), Spring Township, 4 cm. Reading Public Museum collection 2004C-10-185.

Spodumene from Cushion Mountain

Sometimes, there is an interesting story behind a label that takes a bit of detective work to unravel. One might assume that Dr. Smith's name on a Berks County mineral label refers to mineralogist Dr. E.F. Smith, who coauthored an article with Dr. D.B. Brunner on Berks County minerals (Brunner and Smith, 1883). However, the label does not refer to Dr. E.F. Smith. Further, what is water cure, and where is this specimen from?

The first person attracted to South Mountain to establish a health center was Dr. Charles E. Leisenring, a native of Frankfurt, Germany. He came to southeastern Pennsylvania looking for an ideal location to develop his water cure therapy and establish his own sanitarium. Around 1847, he found a location one-half mile from Cushion Peak on South Mountain and purchased a 52-acre property with ample fresh-water springs. Dr. Leisering's facility grew and became a self-contained health community known by various names, the most common of which was "Cushion Hill Water Cure." Leisenring's death in June 1857 cut short his plans.

Dr. Leisering's widow attempted to carry on the enterprise, but sold it to Dr. and Mrs. Aaron Smith in 1865. Both Dr. Smith and his wife graduated from a school of hydropathy. They renamed the facility "Hygiean Home" in order to distinguish its purposes from the "water cure" of Dr. Leisenring. The Smiths introduced many new practices such as massages, electrical treatments, and the like. They expanded the buildings, added new ones, and operated successfully from 1865 to 1873. It became known as Dr. Aaron Smith's Living Springs Water Cure. In 1879, Smith declared bankruptcy, and the facility was purchased by Dr. Reuben D. Wenrich and Dr. James W. Deppen, who named the facility "The Grandview House" (later known as Grand View Sanatorium and then Grand View).

After learning the historical background, we can determine that the spodumene specimen was collected on the grounds of the Grand View Sanatorium during the time it was owned by Dr. Aaron Smith between 1865 and 1873.

Figure 594. Grand View Sanatorium on South Mountain, Spring Township, 1908.

SPRING TOWNSHIP

Figure 595. Gring's sand quarry, Spring Township. From MacLachlan and others (1975). Courtesy of the Pennsylvania Geologic Survey.

The Ruth No. 1 quarry was on the west side of the Columbia Railroad. The quarry was last worked by Eckenrode and Schlott and was abandoned about 1882. The stone was shipped to the Reading Iron Company in Reading for furnace flux. The light blue, thin-bedded limestone dipped S. 15° E. 40° to 45° (D'Invilliers, 1886, p. 1559-1560).

The Ruth No. 2 quarry was a large quarry a short distance northeast of the No. 1 quarry on the west side of the Columbia Railroad. The quarry was last worked by Potteiger and Weitzel about 1881. The stone was shipped on the Philadelphia and Reading Railroad to the E & G Brooke Iron Company in Birdsboro for furnace flux. The rocks dipped toward the south at about 35°. D'Invilliers (1886, p. 1560) noted: *"the character of the stone exposed not any too good."*

John A. Ruth produced and sold lime made from the stone he quarried (*Reading Eagle*, March 17, 1897; Mine and Quarry News Bureau, 1897). The Reading Eagle (August 17, 1905) noted that John R. Ruth was a dealer in lime and building stone. Hice (1911) listed J.A. Ruth of Wernersville as a producer of lime and limestone products,

Breneman Quarry

The Breneman quarry is the flooded pit west of the intersection of State Hill Road and Yerger Boulevard at 40° 20′ 39″ latitude and 76° 00′ 37″ longitude on the USGS Sinking Spring 7.5-minute topographic quadrangle map (Appendix 1, fig. 24, location S-7). The quarry was in the Epler Formation.

Miller (1934, p. 218-219) referred to this quarry as "Old Quarry #3." In the 1930s, the quarry was worked for foundation and building stone. At that time, the quarry face was 125 feet long and 30 feet high. Massively-bedded, bluish-gray stone dipped gently to the southwest. Miller (1934, p. 219) presented an analysis of the stone.

The quarry was operated by Earle J. Breneman from 1941 to 1971 and became known as the Breneman quarry, In 1952, it had a capacity of 100 tons of crushed stone per hour. Eugene E. Uhler was the plant superintendent (Pit and Quarry, 1952).

When Breneman decided to retire in 1971, five employees—Joseph Schmidt, John McGowen, Raymond Beissel, George Williams, and Randal Bright—purchased the company for an undisclosed sum (*Reading Eagle*, March 21, 1971). The quarry ceased operations in November 1992.

The 81-acre quarry property was purchased by Spring Township for a park. The flooded quarry was named David Shoener Lake after the former supervisor who was instrumental in purchasing the property (*Reading Eagle*, January 1, 2008).

Gring's Sand Quarry

Gring's sand quarry (fig. 595) is a flooded pit northwest of the intersection of Grings Mill and Gelsinger Roads at 40° 18′ 04″ latitude and 76° 01′ 11″ longitude on the USGS Sinking Spring 7.5-minute topographic quadrangle map (Appendix 1, fig. 25, location S-11). Gring's quarry produced construction sand and gravel from weathered surficial material of the Hardyston Formation. In 1934, the Reading Street Department awarded the J.M. and J.A. Gring Company a contract for stone (*Reading Eagle*, March 28, 1934). In 1963, the Refractory Sand Company produced furnace sand from Gring's quarry (Kerr, 1963). The quarry was abandoned in 1964 and allowed to flood. In 1991, when CEEM, Inc. proposed to fill the quarry, the water was 15 feet deep. MacLachlan and others (1975, p. 206) noted that several large quarries on the north slope of Grings Hill were formerly worked for ganister (hard, fine-grained orthoquartzite) and sand for manufacturing silica brick typically used to line furnaces. LIDAR imagery shows two large quarries northwest of Gring's quarry.

Iron Mines

Wheatfield Mine

The Wheatfield mine was south of Wheatfield Road at about 40° 17′ 45″ latitude and 76° 02′ 02″ longitude on the USGS Sinking Spring 7.5-minute topographic quadrangle map (Appendix 1, fig. 25, location S-4). A landfill is now located on the site. The mines were in the Milbach Formation.

The mine was named after the discovery of iron ore in William Fisher's wheat field about 1851. The farm was purchased shortly afterwards by ironmasters Eckert, Clymer, Reed, Schwartz, and McManus, who installed machinery for mining the ore (D'Invilliers, 1883, p. 344). The ore body was mined by open pits and inclined slopes.

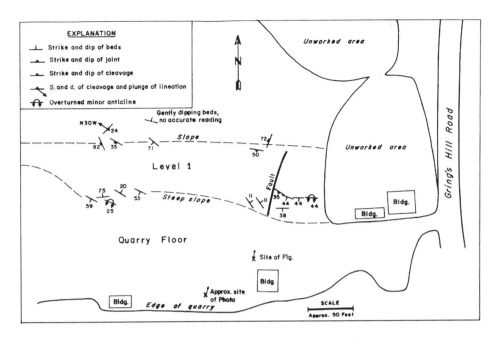

Figure 596. Sketch of Gring's quarry, Spring Township. From MacLachlan and others (1975). Courtesy of the Pennsylvania Geologic Survey.

(Rogers, 1858, vol. 2, p. 717) described the mine: *"The Wheatfield Mining Company have opened a series of veins about 5 miles W. of [Reading]. At this locality there are already proved about ten veins of igneous ore, ranging N. and S., and all included within a transverse distance of 150 feet. The maximum thickness attained by any one of these veins is 8 feet. They are opened from the surface over an irregular area to a depth of about 40 feet, and have been followed along the outcrop from 50 to 110 feet. They occupy loose unstratified ground, including igneous rocks, to the depth at which they have been mined; and below that, as they are included in the beds of the Mesozoic conglomerate, they become pyritous, and are not wrought. The ore is similar in general aspect to that of the [Mount Penn] mines. It frequently encloses fine crystals of calcspar. A narrow valley separates these veins as far as they have been traced N. from an E. and W. vein, which is worked by a slope. This vein dips 30° S. It is underlaid by a foot-wall of trap, and overlaid by white crystalline marble. The thickness ranges between 2 and 12 feet. There is a strong admixture of lime in the ore. The slope is sunk 78 feet, and a gangway is driven 110 feet E."*

The 1880 Census (Pumpelly, 1886, p. 961) listed Eckert & Brother as the operator of the Wheatfield mine. Production between July 1, 1879, and June 30, 1880, was 12,672 tons of magnetite.

When D'Invilliers (1883, p. 344) first visited the mine on July 26, 1882, the mine was being "cleaned up" and little ore was being mined. The mine employed 35 men, only 8 of whom were miners, under superintendents Robert Pickings and John Johnson. Two steam engines, one 15 horsepower and one 12 horsepower, were used for hoisting ore. Three steam pumps built by Allen of Tamaqua were used for dewatering the mines.

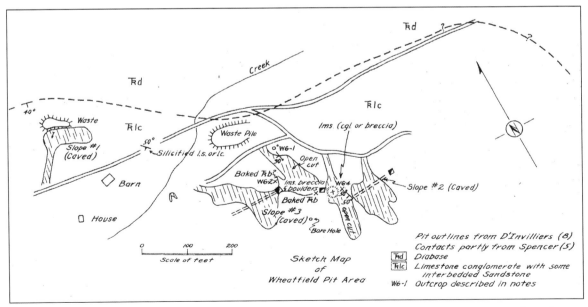

Figure 597. Plan of the Wheatfield mine, Spring Township, 1954. From Bever and Liddicoat (1954).

In July 1882, about 25 tons of ore per day was mined and divided among the lessees: Eckert, Brooke, Clymer, and Trexler. Most of the ore was shipped to the Eckert's Henry Clay Furnace (D'Invilliers, 1883, p. 344). Although the Wheatfield mine was abandoned about 1888, small quantities of surface ore were occasionally mined until 1906. In 1905, some ore was mined from a slope located a short distance east of the old workings (Spencer, 1908, p. 31).

The Bethlehem Steel Company held an option on the property and drilled some exploratory core holes (Bever and Liddicoat, 1954). Bever and Liddicoat (1954) indicated that the open pits were filled with debris in 1953 when they examined the property.

Iron ore was mined from about a dozen open pits at various times in addition to three slopes and several shafts. Weathered or soft ore was mined to a depth of 30 or 40 feet by open-pit methods. Below that depth, the unweathered or hard ore was mined by underground methods (D'Invilliers, 1883, p. 346).

The westernmost inclined shaft was known as slope No. 1 (fig. 598). The No. 1 slope followed an ore bed dipping 40° to the southeast for 100 feet. Gangways were driven to the east and west. The easternmost workings were marked by slope No. 2, which was driven on the outcrop 140 to 150 feet at a 45° angle, nearly due west. Gangways were driven north and south. These workings were abandoned by 1882. Slope No. 3 was 270 feet deep on a 35° dip. The ore body ranged from 2 to 12 feet thick (D'Invilliers, 1883, p. 346-347).

The ore body on north side of road (slope No. 1) was considered to be exhausted and was abandoned *"for a considerable time prior to 1879."* According to D'Invilliers (1883, p. 348), in the "boom" year of 1879, *"a large force of men was put on there and a good deal of ore raised. The mine was soon condemned, however, and the machinery removed."*

The surface rock was a dense, gray-white, brecciated limestone composed of angular fragments of limestone ranging in size from a pea to a man's fist. Occasionally, chunks of dark-green serpentine in the limestone gave the rock a checkered appearance. This rock, along with red shale, formed the hanging wall of the ore body. The gangue rock was limestone. Diabase occurred as numerous small dikes dipping 35° to 50° SW.

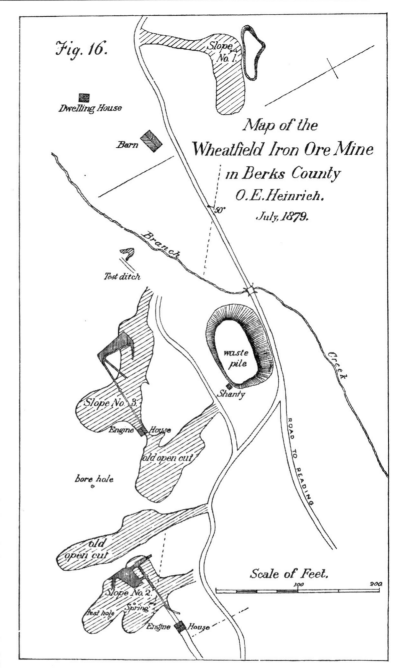

Figure 598. Map of the Wheatfield mine, Spring Township, 1879. From D'Invilliers (1883).

No ore was found north of the diabase (D'Invilliers, 1883, p. 344-346).

The ore occurred as irregular masses in layers that pinched and swelled. The ore was interbedded with limestone; however, the ore bodies were numerous rather than large, and a lack of persistency was a marked characteristic. The ore bodies ranged from 3 to 20 feet thick. The strike was about N. 10-15° E., and

APPROXIMATE SECTION OF PART OF THE WHEATFIELD MINE.
1 TRAP, 2 MAGNETIC ORE AND LIMESTONE, 3 MESOZOIC RED SANDSTONE.

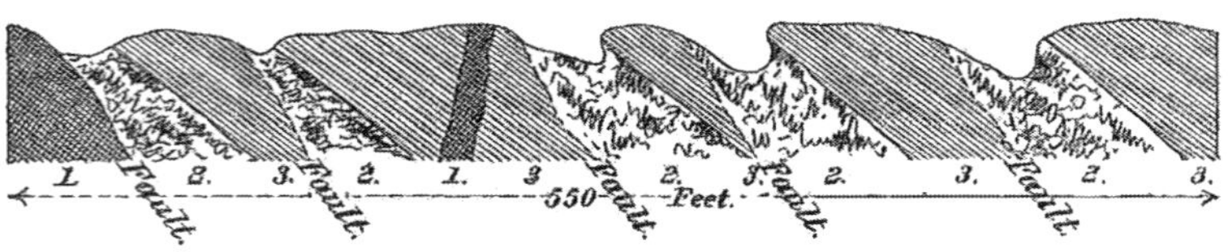

Figure 599. Cross section of part of the Wheatfield mine, Spring Township. From Willis (1880, p. 83).

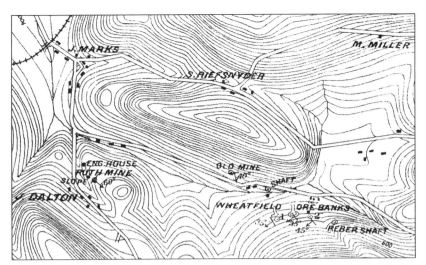

Figure 600. Location of the Wheatfield and Ruth mines and the Reber shaft, Spring Township, 1882. From Pennsylvania Geological Survey (1883).

dips were 30° or 35° NW. The ore was cut off by diabase to the northeast (D'Invilliers, 1883, p. 346; Spencer, 1908, p. 29-30).

Over a distance of 900 feet, there were seven outcrops of the Mesozoic red sandstone with the same number of marble (metamorphosed limestone) and ore beds, apparently stratified (fig. 599). The sandstone was metamorphosed to a dense quartzite, which at times resembled diabase. The ore occurred in irregular deposits between the marble and the overlying and underlying sandstone and within the marble (Willis, 1886, p. 228).

The ore to a depth of 30 or 40 feet was reported to be soft or earthy ("soft ore") and black in color. Considerable pyrite in the ore caused it to readily decompose near the surface. Ironmasters desired the soft ore for mixtures. The deeper, hard ore had a distinct blue cast and required blasting to mine. The hard ore was undesirable because it was low in iron content and so high in sulfur that it required roasting before it could be smelted (D'Invilliers, 1883, p. 346; Spencer, 1908, p. 31).

In 1882, D'Invilliers (1883, p. 347) observed: *"There is about 1,500 tons of ore lying on dump here yet, which shows a good deal of lime and sulphur."* The ore was hauled by private teams to the company's wharf on the Reading and Columbia Railroad at a cost of 40 cents per ton. D'Invilliers estimated production to 1883 at 300,000 tons of ore (D'Invilliers, 1883, p. 348).

D'Invilliers (1883, p. 348-349) published analyses of the ore: metallic iron, 37-41 percent; metallic copper, 0.005-0.12 percent; sulfur, 0.85-1.94 percent; phosphorus 0.02-0.08 percent; silica 17.96-20.2 percent; pyrite, 1.6-3.6 percent; and chalcopyrite, 0.15 percent.

Minerals

D'Invilliers (1883, p. 348) stated, *"The Wheatfield mines have also been quite a favorite locality for minerals, though as most of the debris and gangue rock is now left in the mine to fill up old chambers, very few specimens can be obtained at present."*

Aragonite (D'Invilliers, 1883, p. 394)

Brucite - pearly white cleavages (fig. 602)

Calcite - crystals (Brunner and Smith, 1883, p. 279); small crystals and cleavages; dark green cleavages (Lapham and Geyer 1969, p. 35)

Chlorite group - clinochlore, and chromian chlinochlore var. kämmererite (?) (D'Invilliers, 1883, p. 348 and 398-399); small crystal flakes (Lapham and Geyer 1969, p. 35); green, foliated hexagonal crystals, often with lines parallel to the sides

Copper - native, reported by D'Invilliers (1883, p. 348)

"Deweylite" - brownish, resinous, coatings (Gordon, 1922, p. 159)

Fluorite - amber-colored crystals; rare (D'Invilliers, 1883, p. 397)

Limonite (D'Invilliers, 1883, p. 398) (fig. 604)

Garnet group - grossular; possibly melanite and andradite (Lapham and Geyer, 1959, p. 21) (fig. 601)

Hematite - crystal plates; rare (Lapham and Geyer 1969, p. 36); micaceous (fig. 603)

Magnetite - massive; octahedral crystals (Gordon, 1922, p. 159); octahedrons with rare dodecahedral faces (Lapham and Geyer 1969, p. 36)

Malachite (Eyerman, 1889, p. 45)

Muscovite, var. damourite

Pyrite - abundant (D'Invilliers, 1883, p. 399); small crystals (Lapham and Geyer 1959, p. 36)

Pyrrhotite (D'Invilliers, 1883, p. 400)

Figure 601. Garnet from the Wheatfield mine, Spring Township. Sloto collection 3319.

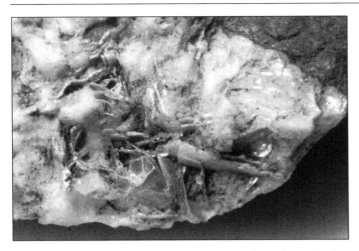

Figure 602. Brucite from the Wheatfield mine, Spring Township. Sloto collection 3318.

Figure 604. Limonite from the Wheatfield mine, Spring Township, 10.3 cm. Reading Public Museum collection 2000c-27-052.

MINERALS FROM THE WHEATFIELD MINE

Figure 603. Hematite from the Wheatfield mine, Spring Township, 6 cm. Reading Public Museum collection 2005c-1-693.

Figure 605. Quartz from the Wheatfield mine, Spring Township, 3 cm with crystals to 1 cm. Reading Public Museum collection 2005c-2-217.

Figure 606. Wavellite from the Wheatfield mine, Spring Township. Reading Public Museum collection 2005c-1-362.

Quartz - amethystine (D'Invilliers, 1883, p. 348); clear crystals to 1 cm (fig. 605)

Quartz, var. amethyst

Serpentine group - yellow, green, and black; var. retinalite (D'Invilliers, 1883, p. 348 and p. 401; Genth, 1885, p. 42; analysis)

Stilbite - pearly white, radiating, fibrous masses on diabase (Brunner and Smith, 1883, p. 279; analysis p. 280); rare

Vermiculite - crystal flakes interleaved with chlorite; brownish green to light green (Lapham and Geyer 1969, p. 36)

Wavellite - (D'Invilliers, 1883, p. 402) crusty, white to brown spheroids (fig. 606)

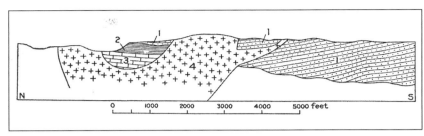

Figure 607. North-south structure section 100 feet east of the Ruth mine, Spring Township. 1, Mesozoic beds; 2, Paleozoic slate; 3, Paleozoic limestone; 4, diabase. From Spencer (1908, p. 31).

Ruth Mine

The Ruth mine was southeast of the intersection of Wheatfield and Chapel Hill Roads at about 40° 17′ 47″ latitude and 75° 03′ 39″ longitude on the USGS Sinking Spring 7.5-minute topographic quadrangle map (Appendix 1, fig. 25, location S-5). The Ruth Mine was in the Hammer Creek Conglomerate on the property of Henry Ruth.

Iron ore was discovered about 1847, which was 4 years prior to the discovery of the Wheatfield mine ore body. Exploration was accomplished by sinking a number of 20-foot deep shafts. When the ore body was located, an engine house was erected, and machinery was installed to pump water and raise ore. Ore was mined from an open cut and from a 35° inclined slope 190 feet deep. The mine was operated with few interruptions until 1863. Daniel Ruth reported that about 10,000 tons of ore were mined prior to 1864 (D'Invilliers, 1883, p. 350-351).

Rogers (1858, vol. 2, p. 717) believed that the ore bed at the Ruth mine was same ore bed mined in the Wheatfield mine. Rogers stated: *"In the range of the E. and W. vein of the Wheatfield Company, but a fourth of a mile farther W., is situated the Henry Ruth Mine. The vein, which is, in all probability, the same, having similar walls both above and below, dips 25° S., and has exceeded 15 feet in thickness. The slope upon it is 120 feet long, and gangways are driven 45 feet W. and 60 feet E. It has been also wrought at the surface outcrop."*

Further exploration done in 1878 was confined to the surface; the old workings were not pumped out. A two-thirds interest in the property was acquired by J.S. Livingood of Reading; the remaining one-third interest was retained by Daniel and Aaron Ruth. When D'Invilliers visited the mine in July 1882, the property was leased to the Monocacy Iron Company and Daniel Fisher of Harrisburg, who began mining on June 19, 1882. A new engine house was built and equipped with a 15-horsepower steam engine, but no ore was produced because of the large quantity of water that had to be pumped from the mine. D'Invilliers (1883, p. 350-351) observed: *"Strangely enough, the lessees went to the great and rather unwarrantable*

Figure 608. Magnetite from the Ruth mine, Spring Township, 11 cm with octahedral crystals to 3 mm. Reading Public Museum collection 2004C-2-218.

Figure 609. Limonite from the Ruth mine, Spring Township, 8 cm. Reading Public Museum collection 2000c-27-047.

Figure 610. "Deweylite" and aragonite from the Ruth mine, Spring Township. Steve Carter collection. Frederick A. Genth label on back of specimen (right).

expense of sinking a shaft here south of the old slope head, and at such a distance that, all said and done, they could only hope to strike the bottom of the existing slope. Since time of visit some 60 tons of ore have been raised, which appears to be about all that remained in the mine, and which could have been just as well won from the old slope workings as from the expensive shaft that was put down. Several leaders of ore were followed out on each side of the old slope, but all pinched into rock in short distances. The mine at present is abandoned and filled with water.... Ore has been extensively mined here, and, after a rather checkered history, it has been recently condemned as played out and abandoned."

The gangue rock and hanging wall were limestone or limestone breccia. The footwall was diabase. The ore bed was reported to dip 30° S., indicating that it was not conformable to the bedding of the enclosing rocks. The ore bed varied from 3 to 12 feet thick (D'Invilliers, 1883, p. 351). D'Invilliers published an analysis of the ore: metallic iron, 42.6 percent and silica, 21.9 percent.

Minerals

Apatite (?) (Eyerman, 1889, p. 40)

Aragonite (D'Invilliers, 1883, p. 394) (fig. 610)

Brucite - silky, fibrous masses, or thin, colorless laminae (Schoenfeld and Smith, 1884, p. 281; analysis; Genth, 1885, p. 40; analysis)

Calcite (Genth, 1885, p. 40)

Chlorite group - curved hexagonal crystals (D'Invilliers, 1883, p. 399)

Chabazite - reported

Chrysotile - in thin layers

Copper - native, reported by D'Invilliers (1883, p. 348)

"Deweylite" - white and brownish resinous coatings on aragonite (Smith, 1883, p. 280; analyses) (fig. 610)

Garnet group - reported

Limonite (fig, 609)

Magnetite - massive; octahedral crystals to 4 mm (fig. 608)

Pyrite

Serpentine, var. retinalite (D'Invilliers, 1883, p. 401; Genth, 1885, p. 42; analysis)

Raub Mine

The Raub mine was south of Wheatfield Road, and east of the Wheatfield mine at about 40° 17' 39" latitude and 76° 01'

Figure 611. Limonite from Lincoln Park, Spring Township, 3 cm. Lincoln Park was the location of the Eureka mine. Ron Kendig collection.

46″ longitude on the USGS Sinking Spring 7.5-minute topographic quadrangle map (Appendix 1, fig. 25, location S-6). The mine was in the Millbach Formation at the contact with diabase. The Raub mine, also known as the Reber shaft (fig. 600), was uphill and about 400 feet east-southeast of the Wheatfield mine slope No. 2. D'Invilliers (1883, p. 349) noted during his visit: *"Everything was idle there on July 26, 1882, and but little iron ore on dump."*

Grill Mine

The Grill mine was on the property of Joseph Grill. It was south of Wheatfield Road, and east of the Raub mine at about 40° 17′ 38″ latitude and 76° 01′ 24″ longitude on the USGS Sinking Spring 7.5-minute topographic quadrangle map (Appendix 1, fig. 25, location S-9). The mine was in diabase. Joseph Grill's iron ore mines are shown in the 1876 atlas. In 1887, the Reading Eagle (May 15, 1887) reported that Kepple and Reber opened a mine on the Grill farm and were stockpiling ore.

Figure 612. Albert A. Gery, 1890. Courtesy of Glen-Gery.

Eureka Mine

The Eureka mine was east of the intersection of U.S. Route 222 and Pennsylvania State Route 724 at about 40° 18′ 54″ latitude and 75° 59′ 38″ longitude on the USGS Reading 7.5-minute topographic quadrangle map (Appendix 1, fig. 26, location S-10). The mine was in the Hardyston Formation. It is the westernmost of the three open-pit mines shown on the 1882 topographic map, on which the Eureka mine is labeled as a hematite mine. The Eureka mine is shown in the 1876 atlas labeled with Wilson Sweitser, superintendent. The mine was worked by the Eureka Ore Company.

The Eureka mine was an open cut on the Jager (also spelled Yager) farm 3.5 miles west of Reading. When D'Invilliers (1883, p. 373) visited the mine in 1882, he observed: *"The ore seen, however, was cellular and cleaner and showed little or no pyrite. The cut is about 40 feet deep, though abandoned now. Some manganese oxide coatings are seen here also."* D'Invilliers (1883, p. 374) published an analysis of the ore: metallic iron, 48.4 percent; sulfur, 0.018 percent; phosphorus, 0.132 percent; and siliceous matter, 14.4 percent.

Montello Brick Works

The Montello Brick Works was located about 1 mile south of Sinking Spring. A man named Yaeger discovered a clay deposit on a 32-acre property near Fritztown below the Montello station on the Reading Railroad. He had the clay tested, and it was found suitable for the manufacture of fire brick. In 1890, Albert A. ("A.A.") Gery (fig. 612), his father-in-law Matthan Harbster, and a lawyer named Stevans purchased the 32 acres and Yaeger's interest in the property (*The Glen-Gery Review*, 1952, vol. 2, no. 1).

The Montello Clay and Brick Company was started in 1891 by A.A. Gery and Howard L. Boas. The company officers were: Matthan Harbster, president; Howard L. Boas, secretary, and A.A. Gery, superintendent. The company was incorporated with a capital of $150,000 to produce building, fire, and paving brick. In 1902, additional land was purchased for $115 an acre from the Charles S. Ludwig estate (Omler and others, 1976).

Figure 613. Montello Brick Works and quarry, Spring Township, 1890. Courtesy of Glen-Gery.

SPRING TOWNSHIP

Figure 614. Montello Clay and Brick Company quarry, Spring Township, early 1900s. Courtesy of Glen-Gery.

The company began to develop the property, and, by 1892, two rectangular kilns were constructed, and a road was built from the quarry to the brick plant (fig. 613). During 1892, brother William Gery joined the firm, and the company switched from making fire brick to making common red brick. This change was brought about by a Philadelphia contractor who needed 500,000 bricks to build the nearby Wernersville State Hospital (*The Glen-Gery Review*, 1952, vol. 2, no. 1).

About 1892, a white clay deposit was discovered on a hillside about 0.5 mile south of the Montello Brick Company's plant. Clay from this deposit was used to make fire brick; however, the manufacture of red brick was more profitable, so fire brick production ceased. When visited by Hopkins about 1899, the clay pit was about 100 feet in diameter and 20 to 25 feet deep. Borings made to 40 feet showed the clay was consistent. In order to drain water and expedite mining the clay, the company started an open cut on the hillside 20 feet below the old quarry. In 1899, the company reopened the white clay pit for use in the manufacture of white paving brick to meet the demand in Philadelphia (Hopkins, 1900, p. 22 and 37).

In 1893, brother Frank Gery joined the firm to construct and operate a continuous brick kiln. In 1898, the firm expanded and built a new plant immediately adjacent to the Reading Shale Brick Company in Wyomissing (fig. 615). A continuous kiln capable of making 120,000 bricks per day was designed and put into operation by Frank Gery. This plant was known as the Wyomissing plant of the Montello Brick Company. At that time, the Montello Brick Company was reorganized to form a holding company called the United States Brick Company. A new kiln erected at the Wyomissing plant cost $700,000 and was 576 feet long (*The Glen-Gery Review*, 1952, vol. 2, no. 1).

In addition to six downdraft kilns, the company operated a continuous kiln that was the largest of its kind in the U.S., capable of producing 45,000 bricks a day. In 1898, there were 110 employees, and production reached 3,000,000 bricks annually, which were sold chiefly in Reading, Philadelphia, and Wilmington. In April 1897, the Philadelphia Public Works Department tested the bricks, along with many others, for use in public works, and the Montello bricks were ranked number one (Omlor, and others, 1976).

The company initially manufactured bricks from residual limestone clay on a soft-mud machine. Beginning in 1899, the bricks were made from red Triassic shale, which was mined about 3 miles southwest of the plant in a railroad cut on the Lancaster branch of the Philadelphia and Reading railroad (fig. 614). The shale occurred with beds of sandstone and conglomerate, which were quarried along with the shale and used for railroad ballast and building stone (Hopkins, 1900, p. 37). Daily output was 50,000 bricks, about half of which were building brick and half were paving brick.

By 1907, the Montello Brick Company operated eight plants and produced 140 million bricks per year. When the financial panic of 1907 struck, approximately 75 percent of building stopped. The company went into receivership, and the Gery brothers nearly lost everything. In 1908, they formed the Glen-Gery Brick Company and purchased a sewer pipe plant in Shoemakersville (see Perry Township) (*The Glen-Gery Review*, 1952, vol. 2, no. 1).

In 1913, the Montello Brick Works was sold at sheriffs sale to the Glen-Gery Brick Company. In April 1920, the Glen-Gery Brick Company sold the property to Ellen Gery, wife of

Figure 615. Wyomissing plant of the Montello Brick Company, 1898. Courtesy of Glen-Gery.

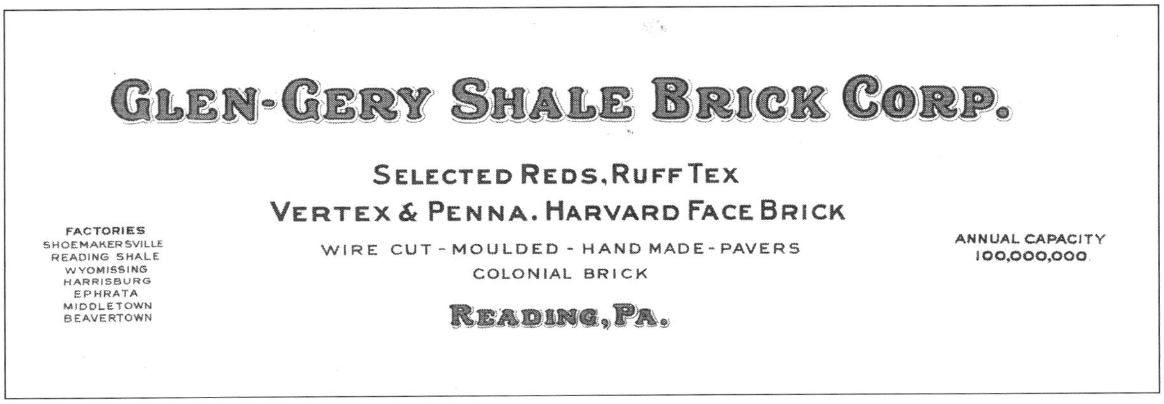

GLEN-GERY SHALE BRICK CORP.

SELECTED REDS, RUFF TEX
VERTEX & PENNA. HARVARD FACE BRICK
WIRE CUT - MOULDED - HAND MADE - PAVERS
COLONIAL BRICK
READING, PA.

FACTORIES
SHOEMAKERSVILLE
READING SHALE
WYOMISSING
HARRISBURG
EPHRATA
MIDDLETOWN
BEAVERTOWN

ANNUAL CAPACITY
100,000,000

Figure 616. Letterhead of the Glen-Gery Shale Brick Corporation, 1947. Courtesy of Glen-Gery.

A.A. Gery. In July 1931, she sold the property to the Atlantic Refining Company.

Fritztown Area

The Fritztown area is in South Heidelberg and Spring Townships; Fritztown is at the border in South Heidelberg Township. The Fritztown area is described under South Heidelberg Township.

ST. LAWRENCE BOROUGH

W.M. Stauffer Quarry

The W.M. Stauffer quarry was opened in 1881. In 1898, William Stauffer leased the quarry formerly belonging to his father to produce crushed stone for the Oley Turnpike Company. The quarry was owned by Schiller Building Associates (*Reading Eagle*, June 29, 1898). The quarry exposed a thin- to thick-bedded, blue limestone (Hawes, 1884, p. 70-71).

Esterly Area

St. Lawrence Borough was formerly known as Esterly. The borough was incorporated in 1927. D'Invilliers (1883, p. 400) reported smoky quartz on the Boyertown Road (present day Pennsylvania State Route 562) 4 miles east of Reading.

TILDEN TOWNSHIP

The Schuylkill Valley Sand Company of Berne operated a sand pit for molding sand on the east bank of the Schuylkill River opposite the Berne railroad station. The pit was next to the Pennsylvania Railroad tracks in the flood plain a few feet from the river. The alluvial sand was 5 feet thick and was excavated from a straight cut parallel to the river (*Stone*, 1928, p. 47-48).

Figure 617. Pyrite nodule coated with limonite from Tilden Township, 4 cm. Reading Public Museum collection 2000C-027-167.

Figure 618. Topton Furnace, 1906.

TOPTON BOROUGH

Topton Furnace

The Topton Furnace (fig. 618) was located on Railroad Street in Topton. Isaac Eckert built the one stack, hot blast, anthracite furnace in 1873 and managed it in conjunction with his Henry Clay Furnace in Reading. Shortly before Isaac's death in 1873, his two sons Harry S. Eckert and George B. Eckert, took over the furnace and administered it under the firm of Eckert & Brother. It was later operated by Harry and his son Isaac under the firm of H.S. Eckert & Son (ca. 1893). The Topton Furnace was remodeled in 1888, rebuilt in 1892, and repaired in 1902. In 1907, the annual capacity was 36,000 tons. It smelted local ore and Lake Superior region hematite and produced foundry and forge pig iron.

The Empire Steel and Iron Company was incorporated in New Jersey on March 14, 1899, for the purpose of merging a number of smaller iron furnaces that were in danger of failing because of the formation of large, integrated steel companies. The Topton Furnace and the Henry Clay Furnace in Reading were incorporated into the Empire Steel and Iron Company. On July 1, 1922, the Empire Steel and Iron Company was acquired by the Replogle Steel Company. The Topton Furnace was dismantled in 1927.

TULPEHOCKEN TOWNSHIP

Frystown Fetid Barite Nodule Localities

The Frystown fetid barite nodule localities are in a 6-square-mile area in Bethel and Tulpehocken Townships southeast of Frystown. The localities were described in detail by Berkheiser (1984). Only the localities in Tulpehocken Township are discussed in this section. This area also is the location of the Mt. Etna area fetid barite locality (fig. 620) of D'Invilliers (1883, p. 395).

Barite occurs as nearly pure, fist-size nodules or lensoidsal float, which is medium gray to medium dark gray. The nodules are composed of dense, compact, 2-mm size, subrounded

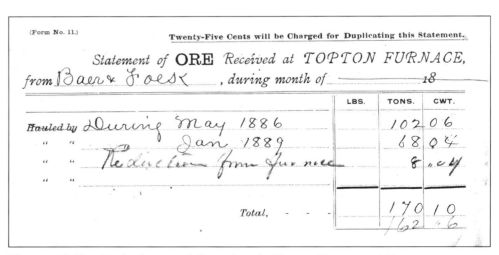

Figure 619. Receipt for iron ore delivered to the Topton Furnace, 1889.

Figure 620. Fetid barite from Mt. Aetna, Tulpehocken Township. Attached is the label of Daniel E. Brunner, son of David B. Brunner. West Chester University collection.

rence of Gyer and others (1976, p. 42) includes cultivated fields on the Roy McLain farm (owners in 1976), formerly the R. and M. Frantz farm. The main barite occurrence is about 1,000 feet west of the McClain house, and there is a lesser concentration of barite about 500 feet north of the house.

Burkholder Farm Fetid Barite Occurrence

The Burkholder farm fetid barite occurrence is west of Wintersville Road, on the east side of Little Swatara Creek at about 40° 26′ 55″ latitude and 76° 16′ 59″ longitude on the USGS Bethel 7.5-minute topographic quadrangle map (Appendix 1, fig. 6, location T-2). Float of medium- to fine-grained barite occurs in a 200-foot by 100-foot area elongated along strike in black shale in a cultivated field on the Burkholder farm (owner in 1984). The barite is characterized by crystalline, compact aggregates of barite crystals that appear to be indistinctly laminated. The barite is associated with minor pyrite (Berkheiser, 1984, p. 21-22).

Moore Farm Fetid Barite Occurrence

The Moore farm fetid barite occurrence (Berkheiser, 1984) is east of Pennsylvania State Route 645, and south of Little Swatara Creek at about 40° 26′ 24″ latitude and 76° 18′ 57″ longitude on the USGS Bethel 7.5-minute topographic quadrangle map (Appendix 1, fig. 6, location T-3).

to subangular barite crystals, which are generally surrounded by an envelope of radially-bladed barite crystals (fig. 621). Some core fragments appear to have a faint lamination with a pale yellowish orange to gray to grayish black argillaceous material occurring at crystal boundaries and as occasional interstitial fillings (Berkheiser, 1984, p. 24)

The host rock is mapped as the Martinsburg Formation, but it may include older, transported carbonate rocks. The nodules were formed in deep, stagnant, sulfate-bearing marine sediments. They were apparently permeated by barium chloride-bearing hydrothermal solutions of unknown origin. The barite is more resistant to weathering than the shale, so it is found loose in the soil. Some barite occurs in calcite veins cutting limestone. The barite emits a foul odor (Gyer and others, 1976, p. 42).

McLain Farm Fetid Barite Occurrence

The McLain farm fetid barite occurrence is northeast of the intersection of Pennsylvania State Route 645 and Mill Road at about 40° 26′ 06″ latitude and 76° 19′ 07″ longitude on the USGS Bethel 7.5-minute topographic quadrangle map (Appendix 1, fig. 6, location T-1). The Frystown barite nodule occur-

Figure 621. Radially bladed barite nodule showing zoned tabular crystals from the Bohn farm, Tulpehocken Township, 7 cm. From Berkheiser (1984, p. 27). Courtesy of the Pennsylvania Geologic Survey.

Gibble Farm Fetid Barite Occurrence

The Gibble farm fetid barite occurrence (Berkheiser, 1984) is east of Pennsylvania State Route 645 at about 40° 26′ 11″ latitude and 76° 19′ 33″ longitude on the USGS Bethel 7.5-minute topographic quadrangle map (Appendix 1, fig. 6, location T-4).

Elvin Kurtz Farm Fetid Barite Occurrence

The Elvin Kurtz farm fetid barite occurrence is north of Mill Road and south of Little Swatara Creek at about 40° 26′ 06″ latitude and 76° 18′ 56″ longitude on the USGS Bethel 7.5-minute topographic quadrangle map (Appendix 1, fig. 6, location T-5). The barite occurs in a 300-foot by 350-foot area (Berkheiser, 1984, p. 21). This Kurtz farm is included in the Frystown fetid barite occurrence described by Geyer sand others (1976, p. 42).

Kurtz-Landis Farm Fetid Barite Occurrence

The Kurtz-Landis farm fetid barite occurrence (Berkheiser, 1984) is north of Mill Road, and south of Little Swatara Creek at about 40° 26′ 04″ latitude and 76° 18′ 04″ longitude on the USGS Bethel 7.5-minute topographic quadrangle map (Appendix 1, fig. 6, location T-6).

Bohn Farm Fetid Barite Occurrence

The Bohn farm fetid barite occurrence (Berkheiser, 1984) is located along an unnamed tributary to Little Swatara Creek, north of Deck Road at about 40° 25′ 49″ latitude and 76° 17′ 13″ longitude on the USGS Bethel 7.5-minute topographic quadrangle map (Appendix 1, fig. 6, location T-7).

UNION TOWNSHIP

Dyer Monocacy Quarry

The Dyer Monocacy quarry is west of Pennsylvania State Route 345, east-southeast of the intersection of Pennsylvania State Route 345 and Geigertown Road at about 40° 14′ 12″ latitude and 75° 47′ 13″ longitude on the USGS Elverson 7.5-minute topographic quadrangle map (Appendix 1, fig. 36, location U-1). The diabase quarry also was known as the Six Penny quarry because of its location near Six Penny Creek.

The Schuylkill Stone Company, also known as the Schuylkill Valley Stone Company, was organized by investors from Birdsboro, Norristown, and Philadelphia to quarry stone in Union Township 1 mile south of Monocacy. In 1905, Mrs. Edward Brooke leased 3,000 acres to the Schuylkill Stone Company (*Reading Eagle*, December 30, 1905).

In 1907, the Schuylkill Stone Company began construction of a crushing plant. Italian immigrants made up the greater part of the construction workforce (*Reading Eagle*, November 4, 1907). Contractors Harker A. Long and Paschal Wamser began construction of a railroad spur from the Pennsylvania Railroad to the quarry. However, the project was abandoned because of a lack of laborers. The contract was then given to a Philadelphia company, which completed the rail line (*Reading Eagle*, June 21 and August 9, 1907).

In 1908, the company's office building was damaged by fire. The building was a two-story wood building measuring 40 feet by 80 feet with roof and sides of galvanized corrugated iron. The interior was divided into partitions with an office, drawing room, sitting room, dining room, and kitchen. All heating and cooking was done by wood stoves. Some of the pipes went through the first floor ceiling to heat the rooms on the second floor. It was believed that one of the pipes overheated and set fire to the joists. The employees formed a bucket brigade to bring water from Six Penny Creek and confined the fire to the interior of the second floor and part of the roof. Damage was estimated at $600 (*Reading Eagle*, January 26, 1908).

In 1908, the Schuylkill Stone Company opened the Monocacy quarry, and the crushing plant began operation (*Reading Eagle*, July 4, 1908). The company obtained a $60,000 mortgage on its property (*Stone*, August 1908, vol. 29, no. 3, p. 123). Irvin Griesemer replaced Mr. Egan as the quarry superintendent (*Reading Eagle*, March 28, 1908). The demand for crushed stone was high, and the company could not keep up with the orders. In 1909, the plant was wired for electricity so that a night shift could be added (*Reading Eagle*, July 11, 1909).

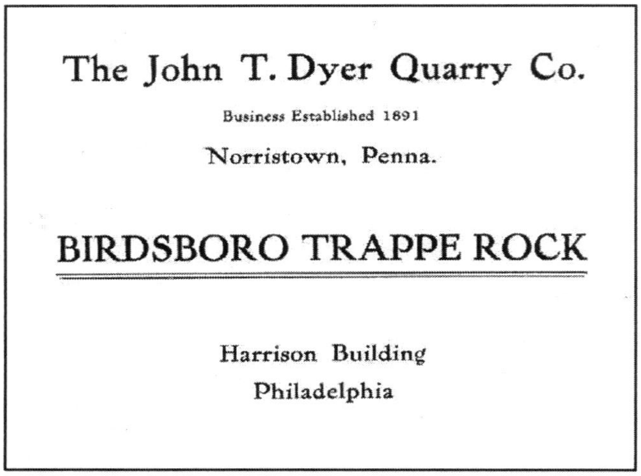

The John T. Dyer Quarry Co.

Business Established 1891

Norristown, Penna.

BIRDSBORO TRAPPE ROCK

Harrison Building

Philadelphia

Figure 622. Advertisement for Dyer quarry Birdsboro Trappe Rock, 1912.

Figure 623. Birdsboro Stone Company Monocacy quarry, Union Township, 1911. The photograph shows the quarry locomotive, stone cars, and two steam shovels.

Figure 624. Crusher and stone storage sheds at the Birdsboro Stone Company Monocacy quarry, Union Township, 1911.

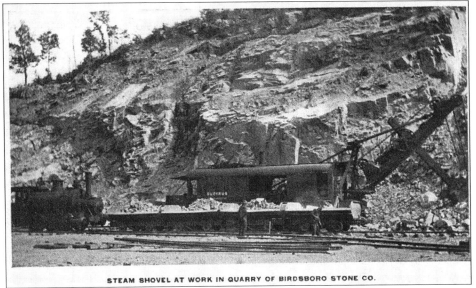

Figure 625. Steam shovel, locomotive, and stone cars at the Birdsboro Stone Company Monocacy quarry, Union Township, 1911.

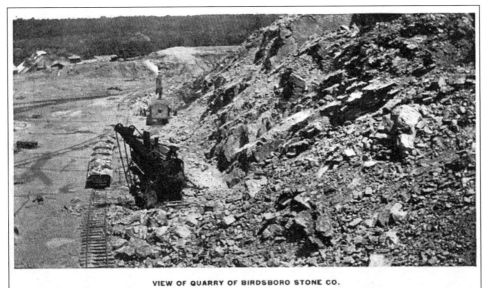

Figure 626. Steam shovel loading stone cars at the Birdsboro Stone Company Monocacy quarry, Union Township, 1911.

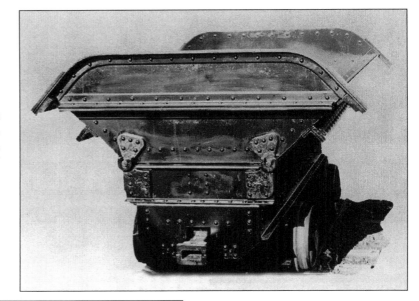

Figure 627. Stone car used at the Birdsboro Stone Company Monocacy quarry, Union Township, 1921. Hagley digital image collection, HF_J15_1_020, Hagley Museum and Library, Wilmington. (BE-319)

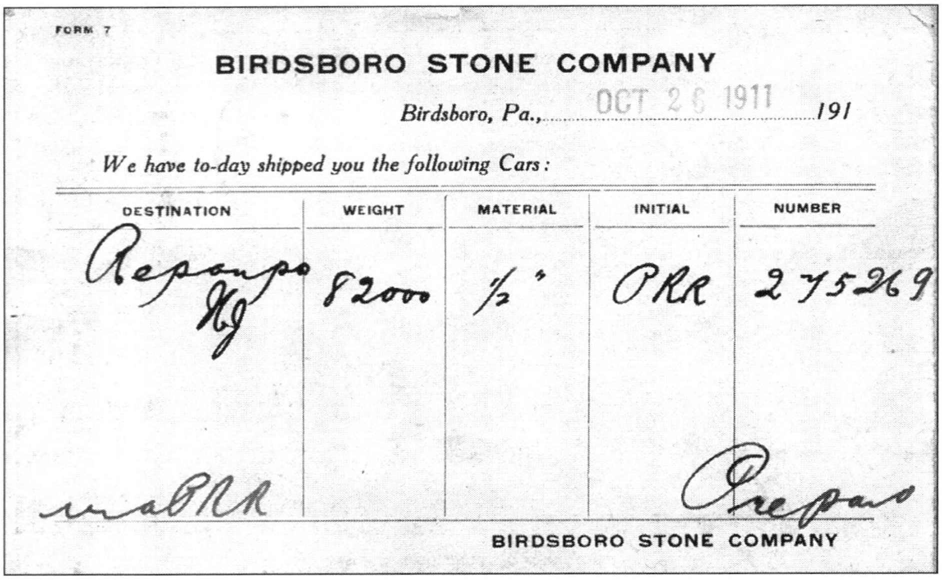

FORM 7

BIRDSBORO STONE COMPANY

Birdsboro, Pa., OCT 26 1911 *191*

We have to-day shipped you the following Cars:

DESTINATION	WEIGHT	MATERIAL	INITIAL	NUMBER
Repanpo NJ	82000	1/2"	PRR	275269

via PRR

BIRDSBORO STONE COMPANY

Figure 628. Receipt for stone shipped from the Birdsboro Stone Company Monocacy quarry, Union Township, to Repampo, New Jersey, 1911.

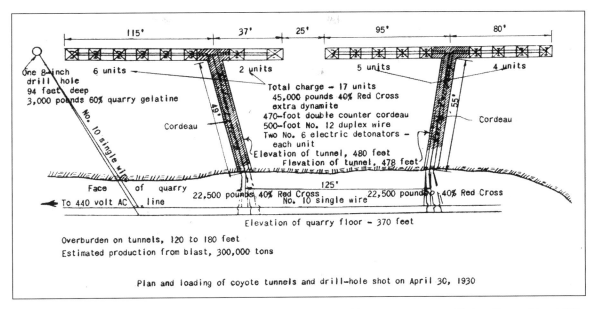

Figure 629. Diagram showing blasting tunnels at the Dyer Monocacy quarry, Union Township, 1930. From Conway (1931).

The Birdsboro Stone Company was organized and incorporated in Pennsylvania under a charter dated July 29, 1909. Charles Bergdoll, who later legally changed his name to Charles Braun, was the company president. Bergdoll was a member of a prominent Philadelphia family that owned the Louis Bergdoll and Sons Brewing Company. Around September 1, 1909, the Birdsboro Stone Company acquired the Schuylkill Stone Company. The property consisted of the quarry, power plant, rock crushing and screening equipment; storage and loading building; carpenter, blacksmith, and machine shops; office and boarding house; a single track standard gage railroad about 1.5 miles long; ground storage plant; five well drills; five air drills; two steam shovels; five locomotives; one steam locomotive crane; 20 dump cars; one automobile truck; two horses; a water-supply system; compressed air system; and various small tools (Barnett, 1922, p. 135). In 1910, the Birdsboro Stone Company employed 150 men and produced about 3,000 tons of crushed stone per day for concrete work and railroad ballast. Charles Snyder was the plant superintendent (*Reading Eagle*, December 29, 1910).

The Schuylkill and Birdsboro Stone Companies sold stone under the name "Birdsboro Trap Rock." In 1911, they were sued by the John T. Dyer Company for infringement of the Dyer trademarked "Birdsboro Trappe Rock" and unfair competition. Dyer sued for cessation of the use of "Birdsboro Trap Rock" and requested $10,000 in damages. Dyer quarried stone from the Trap Rock quarry and marketed it as "Birdsboro Trappe Rock" (fig. 622). Labels with this designation were affixed to railroad cars until 1907 when the railroad companies objected, and the labels were removed. The Dyer Company registered "Birdsboro Trappe Rock" as a trade mark for quarried stone on September 10, 1907. In 1908, the Schuylkill Stone Company began selling crushed stone as "Birdsboro Trap Rock." The U.S. Circuit Court ruled against the John T. Dyer Company on the basis that a geographical term cannot be the subject of an exclusive appropriation as a trademark and that "Birdsboro Trappe Rock" is the combination of a geographic with a descriptive term and, as such, is not a valid trademark (*Trade-Mark Reporter*, 1911, p. 63-89; *Federal Reporter*, 1911, p. 558-576).

In 1914, a blast loosened 150,000 tons of rock when 6,600 pounds of dynamite were detonated in 11 holes drilled from 60 to 93 feet deep. The holes were drilled with a new Keystone blast hole drill No. 3-1/2 purchased for $1,500 and loaded by Thomas Ryan, the quarry foreman. The plant had six large crushers, the largest of which weighed 225 tons and was capable of crushing a rock 5 by 7 feet in size. The crusher had a capacity of 3,500 tons per day, and, according to the Reading Eagle, was the largest in the world. The rolling stock consisted of six locomotives, 30 railroad cars, 12 yard cars, and three steam shovels. The shovels were 90, 95, and 125 ton capacity. The superintendent was C.E. Goodrich, and 165 men were employed (*Reading Eagle*, May 10, 1914).

In 1914, the Birdsboro Stone Company sued the Philadelphia and Reading Railroad Company. The shipping rate for the J.T. Dyer Trap Rock quarry from Trap Rock to Birdsboro where the railroad connected with the Pennsylvania Railroad was 15 cents per ton or $7.50 per car plus a per diem of 45 cents for use of the car for stone shipped out of the area and a rate of 30 cents per ton plus a per diem of 45 cents per car for stone for local use. However, the rate for the same stone for railroad ballast was only $1.50 per car with no per diem charge. The rates charged by the Reading Railroad for the Monocacy quarry were 30 cents per ton for stone shipped both for local delivery and for points beyond Birdsboro. The Birdsboro Stone Company felt that this gave an unfair competitive advantage to the Dyer Company because they sold large quantities of railroad ballast and, therefore, could produce stone for commercial purposes at a lower cost. The Birdsboro Stone Company sued for the same rate for all stone. The Public Utilities Commission ordered that

Figure 630. Dyer Monocacy quarry, Union Township, prior to blasting, ca. 1950. DuPont Company Product Information photographs, 1972341_0730, Hagley Museum and Library, Wilmington.

beginning on January 15, 1915, the Wilmington & Northern and Philadelphia & Reading Railroads adjust their charges so that the rate from Trap Rock to Birdsboro was the same for railroad ballast as for crushed stone purchased for other uses (*Pennsylvania Corporation Reporter*, 1915, p. 264-271) In 1916, production was 3,500 tons per day. The largest quantity of stone was purchased by the Pennsylvania Railroad Company for ballast. Stone was shipped as far as Pittsburgh (*Reading Eagle*, February 1, 1916),

Beginning about 1915, the quarry set off one huge annual blast to dislodge enough stone to last a year. In 1919, after preparing for a blast for a year, the Birdsboro Stone Company set off a blast that dislodged 200,000 tons of stone that was expected to keep 125 employees busy for a year. Sixteen holes, 6 inches in diameter, were drilled across the 280-foot quarry face to an average depth of 165 feet and loaded with 28,750 pounds of dynamite. William Kelly was the plant superintendent (*Reading Eagle*, March 24, 1919).

In 1922, 18 holes, 6 inches in diameter and 250 feet deep, were loaded with about 20 tons of dynamite. The explosion dislodged 350,000 tons of stone (*Reading Eagle*, July 3, 1922). In 1924, the Birdsboro Stone Company worked day and night drilling blast holes with steam and compressed air drills. The arc lamps used by the drillers working at night were seen shining on the hill. Local residents, unaware of the drilling, thought it was Ku Klux Klan activity (*Reading Eagle*, February 22, 1924).

In 1924, when the John T. Dyer Company attempted to renew the lease for the Trap Rock quarry, it found that the Birdsboro Stone Company had secretly negotiated and signed a lease for the quarry with the Brooke Estate. A bitter court battle ensued. The litigation ended in 1925 when Dyer bought the Birdsboro Stone Company and gained possession of the Monocacy quarry.

Blasting at the Monocacy quarry made big news in the late 1920s and early 1930s. The Memorial Day blast on May 30, 1927, loosened 1,000,000 tons of rock and was witnessed by hundreds of spectators. Twenty-seven drill holes were shot with 169,000 pounds of dynamite (Conway, 1931).

A new method of blasting at the Monocacy quarry was developed by William Kelly, the plant superintendent. On April 30, 1929, two top level tunnels were shot. The tunnels were driven in at 120 feet above the quarry floor. A steam shovel outfitted with a boom was placed at the top of the quarry to lower dynamite to the tunnel entrances. There, the boxes were opened, and the 5-inch by 16-inch cartridges were stacked in wheelbarrows and wheeled over plank flooring into the tunnels. Each wheelbarrow load moved 200 to 350 pounds of dynamite, which was piled like cord wood in the tunnels. Afterwards the tunnels were backfilled with dirt. The blast loosened 250,000 tons of rock (Conway, 1931).

Another big blast at the quarry occurred on April 30, 1930; however, it was not as big as previous ones. The blast dislodged only 300,000 tons of rock. Two tunnels, 4 feet wide and 5 feet high, were driven 50 feet into the quarry face. At their terminus, tunnels were driven perpendicularly for 200 feet,

Figure 631. Blast at the Dyer Monocacy quarry, Union Township, ca. 1950. DuPont Company Product Information photographs, 1972341_0730, Hagley Museum and Library, Wilmington.

Figure 632. Dyer Monocacy quarry, Union Township,, August 14, 1955. The photograph was taken 1 year before the quarry was abandoned.

making a T shape (fig. 629). Twenty-five tons of dynamite were placed in the tunnels, which were filled with dirt and packed shut. The blast was not well publicized, so there were only about 150 spectators. A large rock thrown by the blast crushed a truck. After the blast, the company served refreshments in the large boarding house at the plant (*Reading Eagle*, April 28, 1930).

In 1930, annual production was 500,000 tons. The daily capacity of the crushers was 4,000 tons, but production could be increased to 6,000 tons per day when necessary. The quarry employed 75 to 100 men. Work was suspended for a while during the summer of 1930 because of a lack of water in Six Penny Creek. The reservoir behind the storage dam became depleted (*Reading Eagle*, April 28 and July 16, 1930). In 1931, William Kelly, plant superintendent, died and was replaced by H.W. Craig of Clarksburg, West Virginia (*Reading Eagle*, February 20, 1931).

The Dyer Company operated a boarding house for its employees. The boarding house provided a hot dinner for employees for 35 cents. In addition, the company owned 27 company houses in the vicinity that it rented to employees. After 6 months working for the company, employees were eligible for life and health insurance. Life insurance premiums were paid by the company, and each employee paid $1 per month for health insurance. Employees, except drillers, were paid an hourly wage, which averaged 54 3/4 cents per hour in 1931 (Conway, 1931).

In 1931, a detailed description of the mining and crushing methods of the Dyer Monocacy quarry was published by the U.S. Bureau of Mines and U.S. Department of Commerce in an information circular (Conway, 1931). Stone was loaded into quarry cars by two steam shovels: a Marion Osgood model 120 and a Bucyrus model 950. Each shovel was operated by 3 men—an engineer, a craneman, and a fireman. Each shovel could load 8 cars of 9 tons capacity in 5 minutes and loaded about 2,000 tons of rock per day. The cars were moved from the quarry floor to the crusher, a distance of about 1,000 feet, by four H.K. Porter 18-ton, coal-burning steam locomotives, each hauling 3 to 5 steel cars. The cars were designed and built at the quarry. The plant produced 450,000 to 500,000 tons of crushed stone per year in five sizes (Conway, 1931).

In 1932, stripping was done with a 3/4 ton Osgood crawler shovel. Drilling for blasting was done by two well drilling rigs and jackhammers. A model 120 Osgood shovel and a model 100 Bucyrus shovel were used for digging and loading stone. Six standard gage locomotives with cars delivered stone from the quarry to the plant. The primary crushers were a 60-inch by 84-inch jaw crusher and a 30-inch gyratory crusher. The secondary crushers were a 30-inch McCully crusher, four 10-inch McCully crushers, two Weston crushers, and a 4-foot Symons crusher. H.W. Craig was the superintendent. Annual production was 400,000 tons (*Pit and Quarry*, 1932).

In 1936, a big blast dislodged 100,000 tons of rock using 11.5 tons of a new safety explosive placed in a T-shaped tunnel driven into the face of the quarry during slack time over the previous 6 years. The tunnel went 50 feet straight into the quarry wall and then perpendicular for 60 feet. Newsreel men from New York and Philadelphia were on hand to film the blast. They were positioned in the open and wore small metal boxes on their heads to protect against flying rocks. All other spectators were kept at a safe distance (*Reading Eagle*, August 2,

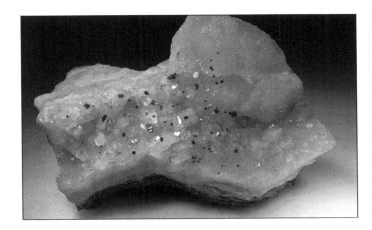

Figure 633. Apophyllite and pyrite on prehnite from the Dyer Monocacy quarry, Union Township, 5.5 cm with apophyllite crystals to 3 mm. Ron Kendig collection.

Hopewell Furnace

William Bird was born in 1703. Bird initially worked for Thomas Rutter, a pioneer iron master, at nearby Pine Forge. When Bird later went into business for himself, he acquired extensive lands near Hay Creek where he built the New Pine Forge in 1744. By 1756, he had acquired 12 tracts of land totaling about 3,000 acres. Mark Bird, the son of William Bird, took charge and expanded the family business upon his father's death in 1761. In 1770-71, Bird constructed Hopewell Furnace on French Creek in Union Township. The location provided ready access to the raw materials necessary for maintaining an iron furnace—iron ore, limestone, and hardwood forests for making charcoal.

The opening of Hopewell Furnace coincided with the beginning of the Revolutionary War. Bird produced cannon shells and other supplies for the war, including casting 115 cannons for the Colonial Navy. In 1775, Bird served as lieutenant colonel of the Second Battalion, Berks County Militia. In August 1776, Colonel Bird outfitted 300 men of the battalion with uniforms, tents, and provisions at his own expense. This force marched under his command to Washington's relief after the Battle of the Brandywine in late 1777. Bird was a member of the Provincial Conference of 1776 and was elected to the Provincial Assembly. Many of Bird's ironworks, gristmills, and sawmills supplied the Continental Congress with the materials for the Revolutionary War (Kurjack, 1954). Bird's patriotic endeavor nearly ruined him when the new government was unable to repay its debts following the Treaty of Paris. Compounding this, he made several unsuccessful investments, including an iron works on the Delaware River in Bucks County. (Graham, 2016)

In 1783, Hopewell Furnace produced 749.5 tons of pig iron and finished castings. Pig iron was its principal product. Finished castings included pots, kettles, stoves, hammers, anvils, and forge castings. While 1783 was a good year for iron production, the following years were not. In April 1786, Bird unsuccessfully tried to sell Hopewell Furnace. In April 1788, Bird assigned Hopewell Furnace and his 5,000 acres of property to John Nixon, a Philadelphia merchant who had lent him money, and moved to North Carolina where he died in poverty (Walker, 1966).

The furnace property changed ownership at least five times before 1800. In that year, Daniel Buckley and his brothers-in-law Thomas and Matthew Brooke purchased the property at sheriff's sale. The property included 5,000 acres, two mines, and the furnace. It was managed by the company of Buckley & Brooke. Hopewell Furnace reached its greatest prosperity under the Brooke family. They updated the furnace technology, improved and extended the boundaries of the property, and rebuilt the waterwheel. From 1816 to 1848, Clement Brooke, son of Thomas Brooke, operated the furnace, first as resident manager and later as ironmaster. Clement Brooke was able to turn Hopewell into a major supplier of iron products. By the 1820s, the furnace was operating at its peak, generally in excess of 300 days per year. Castings were the most profitable product, especially the popular Hopewell Stove. Hopewell produced as many as 23 types and sizes of cooking and heating stoves, which found a ready market in Philadelphia. Over 80,000 stoves were cast at Hopewell. In addition, Hopewell also cast pots, pans, kettles, bake plates, mortars, and waffle irons for the household; moldboards, corn-shelling machines, and windmill irons for the farmer; and machinery castings for industry.

The most productive years for Hopewell Furnace were from 1830 to 1837. Hopewell's products were so much in demand that the furnace was often forced to turn down orders from new customers. The depression of 1837, coupled with the successful introduction of the hot-blast method and the substitution of coke for charcoal in iron smelting, signaled the end of the charcoal-burning iron furnace era. By 1845, finished castings on a commercial scale were discontinued at Hopewell because of competition from the newer furnaces. Thereafter, the furnace concentrated on pig iron for which there was still a market. Clement Brooke retired in 1848. His successors found, despite a short reprieve during the Civil War, they could not compete against the new iron and Bessemer steel industries. In 1849, they erected an experimental anthracite furnace, which was to replace the old charcoal furnace. However, a poor design resulted in a catastrophic collapse of the furnace. In addition, the expense of hauling anthracite coal to Hopewell put an end to the experiment.

Figure 634. Stoves cast at Hopewell Furnace, Union Township. The stoves are on display in the visitor center at the Hopewell Furnace National Historic Site.

Figure 635. Cast house at Hopewell Furnace National Historic Site, Union Township. The top of the furnace is visible to the left of the cupola on the cast house.

The demand for iron during the Civil War and the simultaneous expansion of the railroad during the 1850s and 1860s temporarily brought the charcoal furnace back to life. When the iron and steel industries consolidated in urban manufacturing centers like Pittsburgh, Bethlehem, and Chicago, small independent rural enterprises like Hopewell could no longer compete, and the furnace ceased operating in 1883 (Kurjack, 1954).

In August 1935, the Federal government purchased over 4,000 acres that included the ruins of the iron furnace and community. The property was purchased for the French Creek Recreation Demonstration Area, one of five areas in Pennsylvania in the Recreation Demonstration Area Program, which was one of the many New Deal initiatives to help the nation recover from the Great Depression.

The Civilian Conservation Corps began to convert the land into a public recreation area and engaged the National Park Service to evaluate the furnace ruins found on the property. National Park Service historians recognized the value of the buildings in preserving the story of iron making in America and convinced the Department of the Interior that the furnace should be preserved and reconstructed. In 1938, the Acting Secretary of the Interior designated part of the land acquired for the French Creek Recreation Demonstration Area as Hopewell Village National Historic Site under the authority of the Historic Sites Act. The name was changed to Hopewell Furnace National Historic Site in 1985. The land south of the furnace, which included the Hopewell mines, was given to the Commonwealth of Pennsylvania and became State Game Lands Number 43. Other land given to the Commonwealth of Pennsylvania became French Creek State Park (Glaser, 2005).

The Iron-Making Process at Hopewell Furnace

The furnace was a truncated pyramid of thick stone 32 feet high and 22 feet square at the base built near the side of a small hill. Iron ore, charcoal, and limestone were carried across a wooden bridge that led from the hill to the opening of the furnace stack, into which they were dumped in alternating layers (fig. 636). Charcoal was the fuel used to smelt the iron ore. Blast for the furnace was produced with a water wheel. The blast was turned in, burning the charcoal at white heat (2,600 to 3,000° F) and melting the iron, which then dropped down to the hearth below. The slag, formed by the chemical fusion of the limestone with the impurities in the ore, floated on top and was drawn off from time to time and dumped outside the furnace. About twice a day, sometimes more often, the molten iron was run into a casting bed of sand. It required about 2 tons of ore, 1 to 2 tons of charcoal, and a few shovelfuls of limestone to make 1 ton of pig iron (Kurjack, 1954).

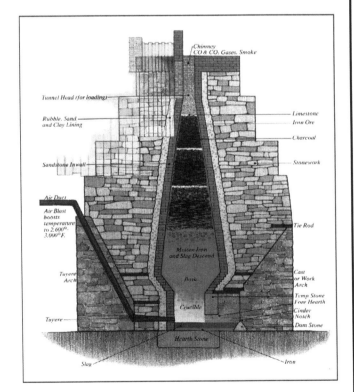

Figure 636. Cross section of a charcoal iron furnace. Courtesy of the National Park Service.

UNION TOWNSHIP

Figure 637. Stilbite from the Dyer Monocacy quarry, Union Township, 8.5 cm with crystal clusters to 1 cm. Steve Carter collection.

1936). In 1937, H.A. Rowan became the superintendent (Pit and Quarry, 1937).

The quarry stopped working on August 2, 1956. In 1974, French Creek State Park acquired 90 percent of a 1,200 acre tract adjoining the park that included the Monocacy quarry to protect the land from commercial development. Pennsylvania State Police divers began recovering cars from the quarry. The State Police estimated that there were many as 45 cars at the bottom of the quarry. Most of the first 20 cars pulled out were stolen. To keep cars away from the quarry property, the State eliminated all access to the quarry except for one road with a locked gate. There were eight drowning in the quarry between 1964 and 1974 (*Reading Eagle,* April 17 and December 10 and 23, 1974).

Minerals

From Thomas (1946, p. 278-279; 1948, p. 914-915; 1950, p. 356-360; and 1961a, p. 8) and personal observations

Apophyllite - cubic crystals to 6 cm, sometimes associated with prehnite; tabular crystals with a pearly luster; fluorescent under long-wave ultraviolet light (fig. 633)

Calcite - white to brown rhombehedrons to greater than 1 inch; fluoresces cream or light green under ultraviolet light; elongated diamond-shaped crystals associated with prehnite; scalenohedral crystals

Chlorite group

Datolite

Laumontite

Natrolite - transparent and translucent needles; associated with chlorite

Prehnite - light green aggregates of crystals (fig. 633)

Pyrite - associated with apophyllite (fig. 633)

Quartz - clear crystals associated with prehnite

Stilbite - colorless to white crystals in diabase cavities (fig. 637 and 638)

Titanite

Birdsboro Sandstone Quarries

Hopkins (1897, p. 71) noted four classes of building stone quarried in the vicinity of Birdsboro: (1) hard red shale, (2) coarse red sandstone, (3) light pink sandstone, and (4) diabase. Building stone was quarried in Birdsboro and for a mile or more south of the town. There were many small quarries but no large ones. The stone was used for more than 30 buildings, mostly dwellings, in Birdsboro and for farmhouses in the surrounding area. The shale contained considerable sand and was very durable.

Sandstone quarries south of Birdsboro included the following.

- The George Brook quarry supplied stone for the Birdsboro Episcopal Church (dedicated in 1853).

- David Lykens operated a quarry near south Birdsboro (1895). The quarry was sold to Grubb and Geigly, who supplied stone for curbs and streets in Birdsboro.

- Benjamin Grubb was listed as a sandstone producer in the 1877 Berks County Business Directory (Phillips, 1877) (fig. 559).

- The Grubb Brothers quarry supplied stone for the gutters on the north side of Flint Street between Furnace and Mill Streets in Birdsboro (1894).

- John Geigley quarry (1911)

- Wells and Hoffman quarry in Geigertown (1879)

- D. Harvey Whitman building stone quarry (1907)

- The William K. Lux and Michael Relser quarry was sold at sheriff's sale to John W. Slipp (1911).

Figure 638. Stilbite from the Dyer Monocacy quarry, Union Township. Joseph Varady collection.

Hampton Furnace

Hampton Furnace on Hay Creek, 2 miles south of Birdsboro, was constructed as a cold blast charcoal furnace in 1846 and was rebuilt as a hot-blast anthracite furnace in 1872. The furnace was owned by the E & G Brooke Iron Company. In 1854, the furnace used ore from the Jones and Warwick (Chester County) mines to produce 1,760 tons of iron (Lesley, 1859). In 1882, the furnace produced 254 tons of car-wheel iron (D'Invilliers 1883, p. 235; American Iron and Steel Association, 1884). The furnace was listed as abandoned in 1890 by the American Iron and Steel Association (1890).

WASHINGTON TOWNSHIP

The earliest mines in Washington Township supplied iron ore to the Mount Pleasant Furnace, which was built in 1737-38 by Thomas Potts, Jr. and Company and was active as late as 1790. The Mount Pleasant Furnace was located at the site of the Barto Stone and Cement Block Company, which was west of the intersection of Forgedale Road and old Route 100. Part of the old furnace walls were used in construction of the block plant (*Reading Eagle*, May 21, 1916). The first blast was made on October, 12, 1738. Six blasts made from then to July, 20, 1741, produced 690 tons of iron (Montgomery, 1884, p. 59-60), Despite being located next to iron ore deposits, much of the ore smelted by the Mount Pleasant Furnace came from the Boyertown mines, of which Thomas Potts, Jr. was part owner.

Mount Pleasant (Barto) Iron Mines

Barto was originally named Mount Pleasant. When the Colebrookdale Railroad was extended to Mount Pleasant, a post office was established. The name Mount Pleasant was already in use by a town in Westmoreland County, so the post office was named in honor of the Barto family (*Reading Eagle*, May 21, 1916). The Mount Pleasant mines were a group of mines in Barto that included the Landis, Barto, Berthou, and Desher mines. At Mount Pleasant, Rogers (1839, p. 14) stated: *"the ore is extensively excavated. It would appear to lie in beds, or rather in regular veins, and to be in truth, the magnetite ore of igneous origin in a rotten or decomposed condition."* There is much confusion in the scientific and historical literature, in newspaper reports, and on maps as to the names of the mines and who operated them.

Landis Mine

The Landis mine was south the intersection of Old Route 100 and Forgedale Road at about 40° 23′ 35″ latitude and 75° 36′ 37″ longitude on the USGS East Greenville 7.5-minute

Figure 639. Iron mines in the vicinity of Barto, Washington Township, 1876. From Davis and Kochersperger (1876).

topographic quadrangle map (Appendix 1, fig. 20, location W-1). The mine also was known as the Grove Brothers mine. The mine was in the Leithsville Formation near the contact with felsic to mafic gneiss.

Rogers (1841, p. 41) described the Landis mine: *"It includes two large excavations, one of them twenty feet wide and sixty or seventy feet long, and sixty feet deep, pursuing apparently a regular vein or bed parallel with the strata. The ore removed at various times, amounts to about three thousand tons. The second excavation, bears a little N. of E. from the above."*

The Landis Mine was on the property of John Landis about 1,500 feet west of the Barto railroad station. In the 1870s and 1880s, the Landis mine was operated by William Rowe, Sr., of Reading for the Grove Brothers of Danville, Pa., who leased about 60 acres of the property. Elias and Jacob Grove owned blast furnaces at Danville. From 1871 to 1883, the mine was operated by Rowe under a sub-lease from the Grove Brothers.

The 1880 Census (Pumpelly, 1886, p. 960) listed William G. Rowe as the operator of the Mount Pleasant mine. Production between July 1, 1879, and June 30, 1880, was 4,508 tons of magnetite, which was shipped to Danville. After being idle for many years, the mine was leased by a Philadelphia company in 1899 (*Reading Eagle*, November 4, 1899).

Two ore beds were mined—the smaller or "back bed" and the larger or "front bed." The back bed dipped more steeply than the front bed. The thickness of the front bed ranged from 6 to 30 feet. There were two shafts and an open cut north of the shafts. Each shaft had a 30-horsepower steam engine pumping water 24 hours per day. The newer No. 2 shaft was about 35 feet northwest of the shallower No. 1 shaft. The No. 2 shaft was

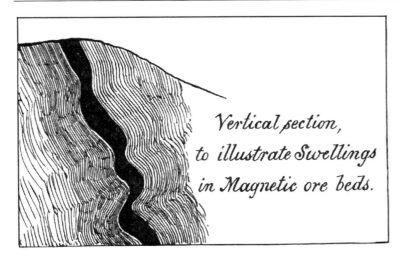

Figure 640. Diagram showing the pinching and swelling of iron ore beds. From D'Invilliers (1883, p. 191).

sunk 60 to 65 feet through rock—the divide between the two beds—and 80 feet through ore (D'Invilliers, 1883, p. 295-297).

When D'Invilliers visited the mine for the first time in May 1880, H.G. Taylor was the superintendent, and the No. 1 shaft had been sunk 116 feet through the hanging wall, which was a light grey gneiss. A slope was then extended from the 116-foot level to follow the ore bed for 100 feet. At the bottom of the slope, a drift was driven west for 130 feet along the foot wall. A gangway was driven east from the drift on the front bed, which was penetrated by the No. 2 shaft at 140 feet below land surface. The thickness of the back bed at this depth ranged from 5 to 15 feet; however, it often pinched out.

When D'Invilliers visited the mine again on October 19, 1882, Robert Laidy, the mine superintendent, reported that the No. 1 shaft had been sunk to 300 feet, where the ore bed dipped 30° SE. No active mining was taking place on that level, which was used for collecting the ore that came down through the chutes from the stopings. From the middle level, a 35-foot drift was driven northwest and struck the back bed, which was 6 feet thick and dipped about 50° SE. (D'Invilliers, 1883, p. 295-297). The No. 1 shaft reached a depth of 325 feet by 1884 (Montgomery, 1884, p. 1002).

At the surface, the ore dipped S. 15° E. 70°. The dip decreased at 116 feet to about 45°. With depth, the ore became exceedingly hard and contained an appreciable quantity of sulfur because of the pyrite crystals disseminated throughout the ore. The horse of rock dividing the two beds formed the foot wall of the front bed and the hanging wall of the back bed. The horse was a coarse gneiss with large crystals of pink feldspar.

The ore body frequently bulged into lenticular-shaped masses (fig. 640). Immediately northwest of the shaft, the foot wall took the shape of an inverted W, with ore occurring in the loops; this increased the thickness of the ore to nearly 40 feet. Timbering could not be used because of the swelling of the ore beds. Large pillars of ore 15 feet thick were left to support the mine. One of the pillars contained a large quantity of excellent ore immediately behind the No. 1 shaft, but could not be robbed for fear of endangering the safety of the miners (D'Invilliers, 1883, p. 295-297).

In 1882, the mine employed 24 men and produced 15 tons of ore per day. The ore sold for $4.75 per ton. It was loaded onto cars at the wharf at the Barto railroad station. The ore was shipped to the Montgomery Furnace at Port Kennedy (D'Invilliers, 1883, p. 295-298). D'Invilliers published analyses of the ore: metallic iron, 46.1-53.4 percent; sulfur 1-1.2 percent; phosphorus, 0.083-0.087 percent; and silica, 10.3-19.7 percent.

Barto Mine

The Barto mine was northwest of the intersection of Old Route 100 and Barto Road at about 40° 23′ 34″ latitude and 75° 36′ 46″ longitude on the USGS East Greenville 7.5-minute topographic quadrangle map (Appendix 1, fig. 20, location W-2). The Barto mine was east of the Landis mine and situated on the property of Abraham Barto 900 feet northwest of the Barto railroad station (D'Invilliers, 1883, p. 298). The mine was in felsic to mafic gneiss.

The Pottstown Iron Company was the first lessee of the Barto mine and held the lease until 1873. A shaft was sunk on the outcrop. The company purchased a 4-acre property in Barto in 1866 and a 50-acre property in 1867 (Washington Township Anniversary Fest Committee, 2004). In 1869, the Reading Eagle (May 13, 1869) reported that the Pottstown Iron Company had "*a mountain of iron ore lying at their mine on Barto's place awaiting transportation.*" Joseph L. High was the chief mine engineer.

In 1869, construction was completed on a 13-mile railroad line from Pottstown to Barto to reach the Pottstown Iron Company property. The railroad line serviced the Mount Pleasant iron mines and hauled ore from Barto to Pottstown. Shortly after construction was completed, the line was leased to the Philadelphia and Reading Railroad as the Colebrookdale Branch. When the iron mines closed in the early 1900s the Reading Company abandoned the 4.4 miles of rail line from Barto to Boyertown.

From 1873 to 1876, the Barto mine was leased by the E & G Brooke Iron Company. James Gayley of E & G Brooke reported that several thousand tons of ore were mined from a 250-foot deep shaft. A winze sunk on the ore bed 75 to 80 feet from the bottom level encountered ore, but not enough to mine profitably. The ore body became much smaller with depth and more mixed with rock, which required careful sorting. D'Invilliers (1883, p. 298-299) described a 150-foot deep shaft that was sunk by the E & G Brooke Iron Company with a drift on the 100-foot level. From bottom of the shaft, a cross cut to the northwest intersected two ore beds, one at 45 feet that was 5 feet thick and one at 60 feet that was about 6 feet thick. A gangway was driven northeast and southwest on the front bed. The southwest gangway was about 50 feet long, and it was stoped up to the 100-foot level.

Washington Township

Figure 641. The Dysher mine in Barto, Washington Township, March 2015.

When D'Invilliers visited the mine on October 19, 1882, the lease was held by Young and Dowdy. They had extended the northeast gangway 30 feet on the front bed. They did no mining, but kept the mine pumped out, probably with the intention of selling the lease (D'Invilliers, 1883, p. 298-299).

The ore occurred in gneiss, dipping about 65° SE. The dip decreased in the lower level similar to the Landis mine. D'Invilliers (1883, p. 299) published an "approximate analysis" of the ore: metallic iron 38-40 percent; phosphorus 0.6-0.7 percent; and silica 10-12 percent.

Berthou Mine

The Isaac Berthou Mine was southeast of the intersection of Old Route 100 and Forgedale Road at about 40° 23′ 37″ latitude and 75° 36′ 30″ longitude on the USGS East Greenville 7.5-minute topographic quadrangle map (Appendix 1, fig. 20, location W-3). The mine was in the Leithsville Formation. The Berthou mine was a few hundred feet northeast of the Landis mine. It likely was a small and short-lived operation.

Rogers (1841, p. 40-41) noted: "*Magnetic iron ore occurs on the border of Colebrookdale and Hereford townships, in the Mount Pleasant iron mines. In the north-eastern excavation, belonging to Isaac Berthou, the ore occurs between syenitic rocks, and is itself a mixture of rotten syenite and magnetic oxide. It is worked open to the air, in a drift ten or twelve feet wide, ranging E. of N. The dip here is 65° and a little S. of E. This mine, in the course of seven weeks of active operations, has yielded seven hundred tons of rich ore. The quality of the ore is however variable.*"

Dysher Mine

The Peter Dysher mine was south of the intersection of Old Route 100 and Barto Road at about 40° 23′ 30″ latitude and 75° 36′ 45″ longitude on the USGS East Greenville 7.5-minute topographic quadrangle map (Appendix 1, fig. 20, location W-4). The name of the mine also is spelled Disher. The mine was in felsic to mafic gneiss.

Rogers (1841, p. 41) noted: "*Two other excavations, on property belonging to Peter Disher, occur about two hundred yards W. of S. from [the Landis mine]. Here the bed has a nearly east and west direction, and may probably be the same which contains the [Landis] mines just previously spoken of. This ore, more compact than that of the other mines, is stated to have made, a rather redshort iron. The excavations are on a less scale than the largest one on Landis' place.*" When D'Invilliers (1883, p. 298) visited the Deysher mine in 1882, he reported an abandoned shaft without any indications of ore having been found.

The Deysher mine was leased by the Reading Iron Company. The mine was abandoned around 1886 because of too much groundwater inflow. In 1896, William G. Rowe prospected for iron ore on the Deysher estate. The Reading Eagle (September 2 and December 3, 1896) reported that Rowe struck a "*rich vein of hematite*" and put a force of men to work driving a 1,000-foot tunnel, which started alongside the Colebrookdale Railroad. The tunnel was expected to reach a large body of ore that could not be mined because of a large quantity of groundwater inflow; Rowe expected the tunnel to drain the water from the mine. In 1899, 20 men were mining ore from a 200-foot depth. The ore was hauled to Eshbach, where it was loaded into railroad cars and shipped to the Reading Iron Company in Reading (*Reading Eagle*, July 1, 1899).

A fire occurred at the Deysher mine in October 1899. The engine and hoisting house, which measured about 30 feet by 120 feet, caught fire. The building had been constructed by Daniel Boyer in 1896. Fortunately, the powder house was some distance away. The fire started about 3 AM. Mahlon Mutter was the engineer on duty. The miners working underground were notified of the fire. They were not able to be hoisted up, but they managed to get out of the mine safely. The shaft burned out and had out to be retimbered. The loss of the building and damage to the machinery was estimated at $3,000. Rowe replaced the burned building with a new building over 100 feet long and 30 feet wide and retimbered the shaft (*Reading Eagle*, October 6 and 28, 1899).

Other Iron Mines

L. Gilbert (Edison) Mine

The L. Gilbert mine, also known as Gilbert's ore pit, was east of Heydts Schoolhouse Road at about 40° 23′ 41″ latitude

Figure 642. The L. Gilbert Mine, which was operated by Thomas Edison, Washington Township, June 2016.

and 75° 38′ 11″ longitude on the USGS Manatawny 7.5-minute topographic quadrangle map (Appendix 1, fig. 18, location W-6). The mine is shown as L. Gilbert's mine on the 1882 topographic map. The mine was in felsic to mafic gneiss.

In 1882, the L. Gilbert mine consisted of two or three small open cuts and one shaft about 50 feet deep. The mine was idle when visited by D'Invilliers in the summer of 1882. Mining resumed around December 1, 1882, by its lessee, E. Lewis of Eshbach (D'Invilliers, 1883, p. 301).

The ore occurred between walls of hard gneiss dipping S. 35° E. and 38°. D'Invilliers (1883, p. 301) noted: *"The ore is a tough titaniferous magnetite with a bright luster, but fully one half of that taken out of the mine is composed of nodules of a white soda feldspar, rendering the ore very refractory. It was tried in the Bechtelsville furnace, where it was said to make a good, tough iron; but the slag formed from such a large quantity of silicious matter was very hard to work off, and could be spun out like so much glass."* Mining was expensive because the ore had to be very carefully sorted. D'Invilliers (1883, p. 301) published analyses of the ore: metallic iron, 24.8-49.6 percent and silica, 19.5-43 percent.

The L. Gilberg mine was later operated by Thomas Edison as part of an experiment to test his magnetic concentrating process. On January 3, 1889, Thomas Edison incorporated the New Jersey and Pennsylvania Concentrating Works in New Jersey (fig. 644). This marked the beginning of Edison's interest in mining and concentrating iron ore. On July 24, 1889, Edison purchased five acres of land from Abraham Benfield for $1,500, and on July 1, 1889, Edison purchased two properties of 9.5 acres each from George G. Greiss. Edison also secured leases from both men to select and purchase additional land. The lease from Benfield granted Edison the exclusive right *"to explore, excavate, and carry away the minerals and to acquire by deed a fee simple title to any portion of the premises, except that occupied by the buildings, upon payment of sixty dollars for each acre."* Edison constructed an ore concentrating plant on the Benfield property (Leidy and Shelton, 1958). A later deed transferred the five acres purchased from Benfield to the New Jersey and Pennsylvania Concentrating Works, which was Edison's operating company for iron mining. Ore for the concentrating plant was mined at the adjacent L. Gilberg mine.

In Edison's magnetic concentrating process, iron ore was ground to a powder and passed down a chute beside a permanent magnet (or a series of them). Edison's major contributions to the process were the conveyor belts, the giant roller-grinders, and a huge crane to lift the ore into the first roller-crusher. Magnetic separation reduced the cost of transporting ore to the furnace and facilitated smelting. The roller-crushers provided additional economic benefit by reducing the use of dynamite and labor in the mine. Edison minimized expenses by developing crushers that could pulverize large boulders. The concentrating plant used a huge 9-ton permanent magnet that was 2 feet thick, 9 feet long, and 5 feet wide. Eight horses were required to pull the wagon that moved the magnet from the railroad station to the plant (Leidy and Shelton, 1958).

The rock loosened by blasting was raised to the surface by a horse whim. The ore was loaded on a horse-drawn wooden sled and pulled downhill to the plant. The plant building, which

Figure 643. Iron ore from the L. Gilbert (Edison) Mine, Washington Township, 7 cm. Sloto collection.

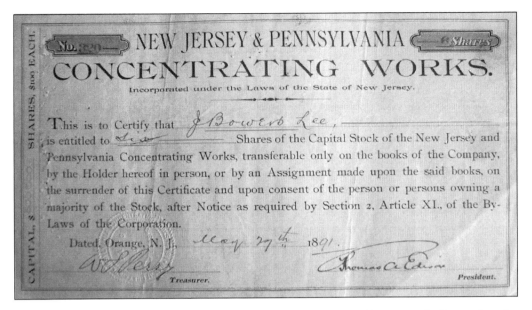

Figure 644. Stock Certificate of Thomas Edison's New Jersey and Pennsylvania Concentrating Works, 1891. The stock certificate was signed by Edison.

was constructed of lumber, was an impressive structure for its day. Huge steam-powered stone crushers were installed inside the plant. The ore was passed through a grinder and then through a series of rollers, which reduced it almost to a powder. The ground rock was conveyed on a belt to a chute through which it dropped beside the giant magnet. The waste rock fell on a conveyor belt beneath the magnet. The finely ground iron particles were attracted toward, but not to, the magnet and fell onto another conveyor belt, which carried it to containers for the finished product. Most of the iron concentrates were shipped away, but some were smelted at the nearby Bechtelsville Furnace. At the time of the plant's operation, a sand pile 75 feet high had accumulated and extended almost to the road (Leidy and Shelton, 1958). No trace of the buildings or the sand pile remains.

Edison's concentrating plant operated for two years. According to the Reading Eagle (March 21, 1930), Edison's explanation for the closing was that there was insufficient iron in the ore for the plant to be profitable. Local residents felt that a fatal injury at the plant was the reason it closed. Philip Dougherty was killed in an accident at the plant while repairing one of the large crushers. Leidy and Shelton (1958) speculated that Edison solved the major problems with his process and was ready to implement it at the Ogden mine in Sussex County, New Jersey. Edison did not sell his Berks County properties until 21 years after the plant closed.

Gilbert Mine

The Gilbert mine, also known as the Gilbert shaft, was east of Washington Road at about 40° 21' 40" latitude and 75° 37' 10" longitude on the USGS Sassamansville 7.5-minute topographic quadrangle map (Appendix 1, fig. 34, location W-10). The mine is labeled as the Gilbert mine on the 1882 topographic map; there is an unnamed mine to the south. Spencer (1908, plate 9) shows the Gilbert mine, which is not labeled, and the unnamed mine to the south in approximately the same locations as the 1882 topographic map. The mine was at the contact between diabase and the Brunswick Formation.

D'Invilliers (1883, p. 206), in describing the diabase at the Washington-Colebrookdale Township boundary, stated: *"Several shafts have been put down in this hill for iron ore and some very excellent material obtained from the Gilbert mine located here, though no regularity is reported in the deposit."* Spencer (1908, p. 63) was not able to find the shafts. Hawkes

Figure 645. Limonite from the Gilbert Mine, Washington Township, 5 cm. Reading Public Museum collection 2000c-27-155.

and others (1953, plate 17) found a magnetic anomaly at the 1882 map location of the Gilbert mine during an aeromagnetic survey of the Boyertown area.

Stauffer Mine

The Stauffer mine was southeast of the intersection of Hill Church and Moyer Roads, at the Washington Township-Bechtelsville border at about 40° 22′ 37″ latitude and 75° 37′ 58″ longitude on the USGS Manatawny 7.5-minute topographic quadrangle map (Appendix 1, fig. 18, location W-5). The Stauffer mine was on the Benfield property north of Bechtelsville. The mine, shown in the 1876 atlas, was in hornblende gneiss.

The mine was owned by John C. Stauffer. In an undated advertisement for a 40-year lease on his mine, Stauffer stated that a 22-foot thick vein of fine iron ore was struck at 26 feet in the shaft. A 2-foot thick vein of hard ore was beneath the fine ore. The vein cropped out 3 feet thick at the bottom of a creek. The mine was conveniently located about 0.25 mile from the Colebrookdale Railroad.

The shaft was in a valley at the base of small hill. The mine was idle during the many visits D'Invilliers made to the region between 1880 and 1882. In March 1883, the Phoenix Iron Company began work at the mine. The shaft was originally sunk to a depth of about 50 feet on the vein. Not finding the more desirable hard ore, the miners drifted south along the strike of the bed; however, they still did not encounter hard ore. Next, two drifts were driven east and west without finding hard ore. A winze was then sunk from the main gangway where the side drifts had been driven. There, the hard ore was found and followed down dip about 25 feet, where it turned west for 8 to 10 feet and encountered the hanging wall dipping in the same direction (to the west). The mine utilized a small steam pump to pump about 50 gallons of water per minute (D'Invilliers 1883, p. 300-301).

The ore vein was about 18 to 20 feet wide and contained a great deal of quartz, feldspar, and hornblende. D'Invilliers (1883, p. 300-301) published analyses of the ore. The hard ore contained: 38.9 percent metallic iron, 1.03 percent sulfur, 0.106 percent phosphorus, and 23.1 percent silica. The soft ore contained: 33.5 percent metallic iron, 0.039 percent sulfur, 0.141 percent phosphorus, and 31.9 percent silica. The ore was titaniferous.

Gilberg Mine

The Jacob Gilberg mine was east of the intersection of Lenape and Deer Run Roads at about 40° 24′ 32″ latitude and 75° 37′ 41″ longitude on the USGS Manatawny 7.5-minute topographic quadrangle map (Appendix 1, fig. 18, location W-7). The location of the mine is shown on the 1882 topographic map. The mine was near the contact of hornblende gneiss and felsic to mafic gneiss.

The Gilberg mine was on the property of E. Nestor. Mining began about 1865 and continued intermittently. In 1876, the Colerain Iron Company of Redington, Northampton County, sunk an 18-foot deep shaft to the ore vein and was sinking a second shaft near the first. Gilberg asserted that he had leased the land first (*Reading Eagle*, July 26, 1876). A preliminary injunction was awarded on behalf of Gilberg to restrain workmen of the Colerain Iron Company from "*prospecting, shafting and digging for iron ore.*" In 1880, the mine was acquired by Gilberg. In 1882, the property was jointly owned by Jacob Gilberg and D. Benfield.

Like many of the Berks County iron mines, it was idle when visited by D'Invilliers in 1882, although there was a considerable quantity of ore on the dump. The shaft was about 80 feet deep, and the mine was equipped with a 25-horsepower steam engine (D'Invilliers, 1883, p. 302).

The ore occurred in hard gneiss that dipped about S. 30° E. 40°. The ore was a titaniferous magnetite filled with feldspar nodules, similar to that at the L. Gilbert mine. However, the ore was a better quality requiring less sorting to free it from the gangue. The ore cost 40 cents per ton to haul 2 miles from the mine to Bechtelsville, where it was shipped to the Pottstown Iron Company Furnace in Pottstown. D'Invilliers (1883, p. 302) published an analysis of the ore: 49.5 percent metallic iron, 9.7 percent titanic acid, 0.007 percent sulfur, 0.021 percent phosphorus, and 16.9 percent silica.

Sparr Mine

The Sparr mine was southwest of the intersection of Crow Hill and Hillcrest Roads at about 40° 24′ 36″ latitude and 75° 36′ 18″ longitude on the USGS East Greenville 7.5-minute topographic quadrangle map (Appendix 1, fig. 20, location W-8). The mine was in felsic to mafic gneiss. The Sparr mine was on the property of J. Sparr of Churchville. The mine consisted of two shallow windlass shafts. Mining began in 1874 by Robert Hobart of Pottstown and Harry Nauman of Reading. They shipped a few car loads of ore to New Providence, Lancaster County. Gregory Greis of Alburtis was the last operator. The mine was idle when visited by D'Invilliers in 1882. The ore was magnetite mixed with hornblende (D'Invilliers, 1883, p. 303). D'Invilliers (1883, p. 399) reported the occurrence of pyroxene at the mine.

Eline Mine

The Eline mine was southwest of the intersection of Crow Hill and Hillcrest Roads at about 40° 24′ 38″ latitude and 75° 36′ 15″ longitude on the USGS East Greenville 7.5-minute topographic quadrangle map (Appendix 1, fig. 20, location W-9). The mine was in felsic to mafic gneiss. The Eline mine was opened in 1877 by Thomas Gay of Pottstown. D'Invilliers (1883, p. 303) found an old shaft sunk close to creek in the southwest corner of the township and stated: "*Nothing could be learned of this project, but the dump showed several tons of a highly-crystallized hornblendic rock called Black Jack by the* miners. *The rock dips east about 46° and shows but little indications of ore."*

Fegley mine

The Fegley mine was east of Washington Road at about 40° 21′ 49″ latitude and 75° 37′ 06″ longitude on the USGS Sassamansville 7.5-minute topographic quadrangle map (Appendix 1, fig. 34, location W-11). The mine location is shown in Spencer (1908, plate 9) and Hawkes and others (1953, plate 17). The mine was in diabase.

'Spencer (1908, p. 64-65) described the Fegly mine: *"Iron ore is said to have been extracted from workings known as the Fegley mine, situated well within the diabase area on the north side of the little brook about one-fourth mile north of the Gilbert shaft. A large fragment of diabase was found, on one side of which there was a coating of ore 1 inch thick composed of crystalline magnetite intergrown with a minor amount of feldspar. Appearance suggests that the magnetite was segregated into a crevice traversing the diabase."*

Quarries

Schall Quarry

The Schall quarry was about 0.5 mile north of Dale, west of the intersection of Dale Road and School Lane at about 40° 25′ 20″ latitude and 75° 37′ 00″ longitude on the USGS East Greenville 7.5-minute topographic quadrangle map (Appendix 1, fig. 20, location W-12). The quarry was in the Leithsville Formation.

Rogers (1841, p. 37) briefly mentioned the Schall quarry: *"Limestone ... is again exposed in two fine quarries, one on each side of the creek, at David Schall's forge (Thompson's on the map). In the western of these quarries, the limestone considered the best, is of a dark blue color... The dip in both quarries is to the N. W."* In 1874, the quarry was operated by Judge Shall (*Reading Eagle*, May 23, 1874). D. Horace and William Schall were listed as a lime producers in the 1877 Berks County Business Directory (Phillips, 1877).

D'Invilliers (1883, p. 146) noted: *"The next opening is ... on the Schall estate. This quarry was not being worked but is said to have yielded an excellent furnace limestone. It is white in color, hard and compact, showing 20 feet of limestone overlaid conformably by 4 to 5 feet of worthless grey cherty rock, both dipping S. 27° E. and 69°. Over the latter there is a layer of 3 feet of bastard limestone largely mixed with slate, and above this 6 feet of black and grey slates, greatly crushed."* A 7-foot thick bed of limonite was exposed in the upper Schall quarry.

Schall's largest quarry was located below a mill dam. The quarry produced a large quantity of blue, highly magnesian dolomitic limestone. The quarry exposed about 30 feet of massive limestone, dipping N. 35° W. 45°, and conformably capped with 4 to 6 feet of black slate. The stone was burned in a double kiln, and the lime was used by local farmers.

Oberholtzer Quarry

The Oberholtzer quarry was southwest of Eshbach, east of Old Route 100 at about 40° 22′ 45″ latitude and 75° 37′ 27″ longitude on the USGS East Greenville 7.5-minute topographic quadrangle map (Appendix 1, fig. 20, location W-13). The quarry is shown on the 1882 topographic map. It was in the Leithsville Formation.

The quarry was owned and worked by Jacob Oberholtzer. The stone varied from a grey through dove color, cherty limestone to a hard, massive, blue stone. Rogers (1841, p. 37) briefly mentioned the quarry. In 1873, Samuel Herb and John Reitenauer quarried stone for the lime kilns of Jacob and Amos Oberholtzer (*Reading Eagle*, August 29, 1873). Oberholtzer & Brother was listed as a lime dealer in the 1877 Berks County Business Directory (Phillips, 1877). In 1890, the quarry was operated by Menno Oberholtzer (*Reading Eagle*, August 25, 1890). In 1903, the quarry was owned by Daniel D. Heins (*Reading Eagle* September 1, 1903). Hice (1911) listed H.D. Heins of Bechtelsville as a limestone and lime producer. In 1927, Mrs. Olivia Heins and Clarence Haas, executors of the Horace D. Heins estate, offered the 75-acre farm, quarry, and two lime kilns for sale by (*Reading Eagle*, October 4, 1927).

D'Invilliers (1883, p. 150) noted: *"The limestone is a true dolomite. The beds are conformable, lying very flat, with a dip of S. 40° E. 12°-15°. The greatest exposure is about 30 feet thick. Good limestone is exposed in the south and east sides of the quarry, and at the time of visit, June 7, 1882, the north side was being uncovered. There is about 12 feet of a yellow clay over the limestone."*

From 500 to 1,500 bushels of stone per week were quarried and burned in two lime kilns with a capacity of 1,300 to 1,400 bushels per week. The lime was used for agriculture and sold for 12 cents per bushel at the kiln. D'Invilliers (1883, p. 151) published an analysis of the rock: 51.8 percent calcium carbonate, 40.5 percent magnesium carbonate, and 7.2 percent silica.

Dielh Quarry

The Dielh quarry was southeast of the intersection of Pennsylvania State Route 100 and Kutztown Road at 40° 24′ 49″ latitude and 75° 34′ 21″ longitude on the USGS East Greenville 7.5-minute topographic quadrangle map (Appendix 1, fig. 20, location W-14). The Dielh quarry is shown as an unnamed quarry on the 1882 topographic map. The quarry was in limestone fanglomerate.

Michael Dielh's quarry, on the property of Henry Dielh, exposed a limestone of mixed quality—a good dolomite in one location and a very siliceous limestone in another. The best stone was in the western end. The dips were all southeast, varying in different parts of the quarry. The quarry was shallow and exposed about 25 feet of white and bluish-gray limestone. Two kilns on the property produced between 1,500 and 1,600 bush-

WASHINGTON TOWNSHIP

Figure 646. Barr quarry, Barto, Washington Township.

els of lime per week, which sold for 11 cents per bushel (D'Invilliers, 1883, p. 149).

D'Invilliers (1883, p. 149) published an analysis of a composite sample taken from all beds in the quarry, which may account for the great amount of silica shown in the analysis: 26.1 percent calcium carbonate, 19.1 percent magnesium carbonate, and 50.6 percent silica.

Barr Quarry

The Barr quarry (fig. 646) was west of the intersection of Old Route 100 and Forgedale Road at about 40° 23′ 41″ latitude and 75° 36′ 38″ longitude on the USGS East Greenville 7.5-minute topographic quadrangle map (Appendix 1, fig. 20, location W-15). The quarry was in hornblende gneiss.

William Hallman, who owned 54 acres, purchased an 8-acre property from the Pottstown Iron Company in 1906. The existence of an old iron mine has been reported on this property. In 1914, he sold his holdings to the Barto Stone and Cement Block Company, which manufactured cement blocks from the crushed stone quarry it operated across the road. The Barto Stone and Cement Block Company was incorporated in 1914 in Pottstown with a capital of $30,000. Directors of the company were William Hallman, B.W. Luckenbill, and Livingston Saylor of Pottstown; John W. Fisher of Douglassville; and Joseph Barr of Allentown. Barr was president of the company, and Hallman was treasurer.

In 1922, the Barto Stone and Cement Block Company defaulted on it's mortgage, and the property was sold at sheriff's sale to R. Etta Barr of Allentown.

The quarry was operated by Barr until it closed in 1930 (*Reading Eagle*, November 4, 2002). The railroad siding was removed, and the quarry and remaining land were bought by Ivan J. Snyder, Sr. (Washington Township Historical Committee, 2014).

The quarry produced a hard blue rock and a softer brown rock. Stone was blasted, and the large rocks were split into sizes to be loaded into the cars for the crusher. All work was done by hand. The small cars were loaded on the quarry floor and then were pulled along tracks to a crusher on the top of the hill. Pieces too large to fit in the crusher were broken down by the laborers. From the crusher (fig. 647), stone was loaded into railroad cars on a private siding of the Reading Railroad. The engine was backed from the Barto station up the private siding to the crusher (Blackwell, 2004, p. 66).

Rush Quarries

The Rush quarries were southwest of the intersection of Forgedale and Lenape Roads at about 40° 25′ 08″ latitude and 75° 34′ 21″ longitude on the USGS East Greenville 7.5-minute topographic quadrangle map (Appendix 1, fig. 20, location W-17). The quarries were in the Leithsville Formation. The quarries are shown on the 1882 topographic map. There were several small quarries on the J. Rush property. The largest one exposed 25 feet of a dark blue, massive limestone, similar to that in the Schall quarry. The limestone dipped S. 6° E. 60°, and was overlaid by 6 feet of decomposed brownish red muscovite slate dipping S. 24° E. 16° (D'Invilliers, 1883, p. 147).

Figure 647. Crusher at the Barr quarry, Barto, Washington Township.

Gabel Quarry

The Gabel quarry is west of the intersection of Old Route 100 and Robin Hill Road at 40° 21′ 37″ latitude and 75° 37′ 46″ longitude on the USGS Boyertown 7.5-minute topographic quadrangle map (Appendix 1, fig. 18, location W-18). The quarry is underlain by felsic to mafic gneiss, graphitic felsic gneiss, and the Hardyston Formation. The quarry is operated intermittently by Martin Stone Quarries, Inc. for crushed stone.

Eshbach Area

Eshbach is at the intersection of Old Route 100 and Lenape Road on the USGS East Greenville 7.5-minute topographic quadrangle map (Appendix 1, fig. 20, location W-16). D'Invilliers (1883, p. 397-399) reported garnet and pyroxene from the hill just southwest of Eshbach. Spencer (1908, p. 64) reported that loose fragments of limestone conglomerate containing spangles of hematite were abundant on the hill slope about 1 mile southeast of Eshbach and also were found at several places along the ridge east of Middle Creek. Bliss (1913, p. 519) reported riebeckite, var. crocidolite, in outcrops 0.5 mile west of Eshbach.

WERNERSVILLE BOROUGH

Limestone quarries in Wernersville Borough were described by D'Invilliers (1886, p. 1557-1559). Quarry locations are shown in Pennsylvania Geological Survey (1891, reference map 11) (fig. 592).

Deppen Quarries

Dr. J.W. Deppen operated two quarries in Wernersville. Deppen's quarry No.1 is the water-filled pit northwest of the intersection of Elm and Fairview Streets at 40° 19′ 54″ latitude and 76° 05′ 03″ longitude (location WE-1). It is quarry number 87 on reference map 11 in Pennsylvania Geological Survey (1891) (fig. 592). Deppen's quarry No. 2 was between the railroad tracks and U.S. Route 422, southeast of the intersection of U.S. Route 422 and Furnace Road at about 40° 19′ 50″ latitude and 76° 05′ 12″ longitude (location WE-2). It is quarry number 87a on reference map 11 in Pennsylvania Geological Survey (1891) (fig. 592). Both quarries are located on the USGS Sinking Spring 7.5-minute topographic quadrangle map (Appendix 1, fig. 23). One of the quarries was known as the Keystone quarry. The quarries were in the Epler Formation.

Dr. Deppen operated a thriving lime business in the 1870s directly north of the Lebanon Valley Hotel, which he built and owned. Pay for the quarry workers was set at $1.05 a day for a 10-hour day. In 1876, Deppen constructed a reservoir beside his quarry (*Reading Eagle*, March 2, 1875, and January 21, 1876).

D'Invilliers (1886, p. 1557) described the Deppen No. 1 quarry: *"The quarry, with three kilns, has been in operation from time to time for many years, and an enormous quantity of stone has been taken from it. The product has been largely burned for lime, but a considerable amount has been used in the past for flux at different furnaces, although the general appearance of the stone would not suggest its being sufficiently pure for that purpose now."* In 1886, the quarry was about 550 feet wide and 900 feet long, but only about 150 feet of the length was actively quarried. The quarry exposed about 25 feet of silicious, pale blue to gray limestone. The rocks dipped about S. 20° W. 20° along the south side of the quarry.

After it was abandoned, the water-filled quarry No. 1 was known as "Devil's Lake." In 1904, it was owned by William L. Bower of Shillington, proprietor of the Lebanon Valley Hotel. Bower threw dynamite into the lake to kill carp, which were donated to the nearby South Mountain Insane Asylum (*Reading Eagle*, September 28, 1904). Bower owned the quarry and hotel until about 1928.

The Deppen No. 2 quarry was idle in 1886. D'Invilliers (1886, p. 1557) stated: *"There are eight or nine good kilns at this quarry and apparently every facility for a large output of lime or fluxing stone."*

Witmoyer and Brother Quarry

The Witmoyer and Brother quarry was south of the railroad tracks, between Beckley and Werner Streets at about 40° 19′ 40″ latitude and 76° 04′ 48″ longitude on the USGS Sinking Spring 7.5-minute topographic quadrangle map (Appendix 1, fig. 23, location WE-3). It is quarry number 88 on reference map 11 in Pennsylvania Geological Survey (1891) (fig. 592). The quarry was in the Epler Formation. The Witmoyer and Brother quarry was just south of and opposite the Wernersville railroad station, on the south side of the Lebanon Valley Railroad. In 1886, the quarry was operated by J. Knorr. The large quarry exposed a face of limestone 15 to 20 feet high. The entire output of the quarry was shipped to Reading and Pottstown for use as furnace flux. In 1886, production was about 28 tons per day. The rock dipped S. 35° W. 20° to 40°. D'Invilliers (1886, p. 1557-1558) noted: *"the character of the stone, both in color, quality and thickness of beds, is quite similar to the Deppen quarry lying immediately west and opened apparently in the same strata of rock."*

Henry Miller Quarry

The Henry Miller quarry was southeast of the intersection of U.S. Route 422 and Church Road at about 40° 19′ 40″ latitude and 76° 04′ 22″ longitude on the USGS Sinking Spring 7.5-minute topographic quadrangle map (Appendix 1, fig. 23, location WE-5). It is quarry number 90 on reference map 11 in Pennsylvania Geological Survey (1891) (fig. 592). The quarry was in the Hershey and Myerstown Formations, undivided. In 1881, Miller's quarry was reopened by Hassler and Lerch.

Stone was shipped to the Keystone Furnace in Reading for flux (*Reading Eagle*, January 6, 1881). The quarry was abandoned about 1883. The thin bedded limestone dipped S. 20° W. 25° (D'Invilliers, 1886, p. 1558-1559).

Wernersville Lime and Stone Company Quarry

The Wernersville Lime and Stone Company quarry, formerly the Brinkly and Zinn quarry, was north of the intersection of Lincoln Drive and Church Road at about 40° 19′ 35″ latitude and 76° 04′ 32″ longitude on the USGS Sinking Spring 7.5-minute topographic quadrangle map (Appendix 1, fig. 23, location WE-4). It is quarry number 89 on reference map 11 in Pennsylvania Geological Survey (1891) (fig. 592). The quarry was in the Hershey and Myerstown Formations, undivided.

The Brinkly and Zinn quarry was a short distance southeast of the Wernersville railroad station and a short siding connected it to the main line of the Lebanon Valley Railroad. The stone was used to make agricultural lime and also was shipped by rail to Reading and Pottstown for furnace flux. The quarry was about 125 feet long and 100 feet wide. The working face on the west and north sides was 20 to 25 feet high. The strata of the limestone was greatly curved, forming a compressed anticlinal roll leaning toward the north, with both legs dipping about S. 20° W. 45°. D'Invilliers (1886, p. 1558) noted: "*This structure would seem to render the stone somewhat expensive to quarry and renders it liable to break up in irregular pieces generally of small size. The entire thickness of limestone exposed is not much over ten feet, owing to the doubling of the beds.*"

The Brinkly and Zinn quarry was acquired by the Wernersville Lime and Stone Company, which was incorporated in 1921 with a capital of $25,000 to manufacture lime and related products. Carroll D. Winters and Albert E. Herbine were the owners of the company. Winters was the president, and Herbine was the general manager. David W. Brubright of Reading was the treasurer (*Pit and Quarry*, 1921, vol. 6, no. 3, p. 81). In 1930, the company produced crushed limestone and hydrated lime. The lime plant utilized one hydrator and three 10-foot by 35-foot shaft kilns and was capable of producing 5,000 tons of lime per year. (Shaw, 1930, p. 176). In 1928, 6,000 tons of stone were loosened by a blast of 1,200 pounds of dynamite placed in three holes at the quarry (*Reading Eagle*, June 4, 1928).

In 1932, the crushing plant had a capacity of 40,000 tons per year. Stone was excavated with a gasoline powered shovel. Digging and loading was done by hand. An Orr and Sembower incline and hoist transported the stone to the crusher. The primary crusher was a No. 6 Champion crusher, and the secondary crusher was an Acme 8-1/2 A crusher (Pit and Quarry, 1932).

Miller (1934, p. 219) noted that the Wernersville Lime and Stone Company quarry was working the quarry for crushed stone; lime was no longer being produced. The stone was mainly dark gray or blue, fairly high in magnesia, and hard. There were a few beds of light gray to pinkish, fine-grained

Figure 648. Promotional ink blotter from the Wernersville Lime and Stone Company, Wernersville, ca. 1940s.

marble. Veins of calcite were abundant. The quarry face was about 65 feet high.

The Wernersville Lime and Stone Company also sold slag from the nearby Robesonia Furnace under the name of the Robesonia Slag Company. Robert L. Kintzer was the plant manager in 1932 (Pit and Quarry, 1932). The furnace produced about 7,000,000 tons of slag, which was piled onsite. Slag was sold as late as 1959.

When Herbine retired in 1959, Carroll Winters and his wife Ruth took over the quarry. The Winters also owned the Keystone Quarry Company, which operated the Keystone quarry in Maxatawny Township (Reading Eagle, December 11, 1966).

WEST READING BOROUGH

Limestone quarries in West Reading Borough were described by D'Invilliers (1886, p. 1561-1562). Quarry locations are shown in Pennsylvania Geological Survey (1891, reference map 11) (fig. 649).

McQuade Brothers (Frill) Quarry

The McQuade Brothers quarry, also known as the Frill quarry, Leinbach's Hill quarry, and the Leinbach's Hill locality, was on Leinbach's Hill on the west side of the Schuylkill River, where the present day U.S. Route 422 bypass is located between Penn Street and the Lebanon Valley Railroad bridge on the USGS Reading 7.5-minute topographic quadrangle map. Leinbach's Hill was formerly known as Yeager's Hill. The George R. Frill quarry No. 1 was immediately opposite the George Drexel quarry about 1,000 feet south of the railroad bridge. It is quarry number 103 on reference map 11 in Pennsylvania Geological Survey (1891) (fig. 649). The George R. Frill quarry No. 2 was at the west end of the Lebanon Valley Railroad bridge close to the Schuylkill River. It is quarry number 104 on reference map 11 in Pennsylvania Geological Survey (1891) (fig. 649).

Quarrying on Leinbach's Hill began about 1870. As early as 1871, Frederick R. Frill operated a 25-horsepower stone crusher with a capacity of 50 tons per day. In 1871, rock from a blast at the Frill quarry penetrated the quarry building and fell behind the clerk on duty. Another rock struck the quarry wagon used for hauling stone (Reading Eagle, August 31, 1871). In 1873, the Reading Eagle (December 22, 1873) reported that lime kilns were in operation at the Frill quarry.

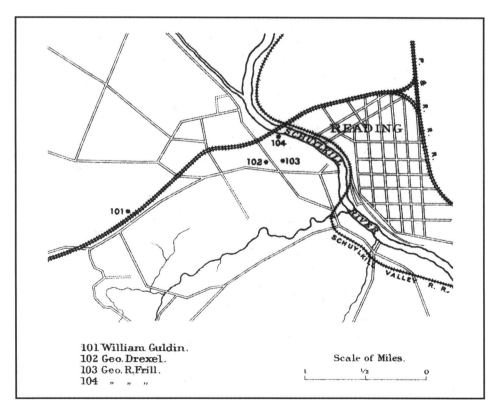

Figure 649. Limestone quarries in the vicinity of West Reading, 1891. From Pennsylvania Geological Survey (1891, reference map 11).

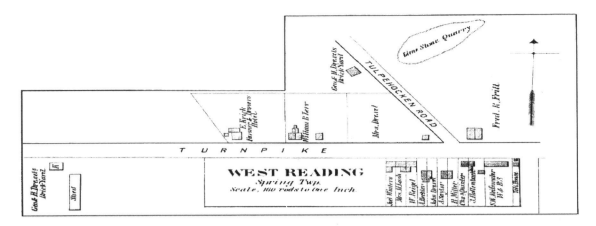

Figure 650 Location of the Frill quarry, West Reading, 1876. From Davis and Kochersperger (1876).

Frill owned all the land from Penn Street to the Lebanon Valley Railroad bridge. Quarries were located at both ends of the tract. The stone in West Reading was unsuitable for building, but was quarried for crushed stone. Thousands of tons were crushed and used on the streets of Reading (*Reading Eagle*, September 6, 1925).

D'Invilliers (1886, p. 1561-1562) described the Frill quarry No. 1: "*The opening shows a rich looking limestone with an even, dark-blue color, particularly well adapted for building purposes.*" The rock dipped S. 25° E. 40°. Around 1886, the quarry was operated by Henry Reifsnyder. Building stone was sold for 40 to 60 cents a perch depending on the quality and thickness. One perch is 16.5 feet long and is equal to 272.3 square feet. Average annual production was about 600,000 square feet. Most of the stone was used for buildings in Reading. The quarry provided stone for the Eckert and Posey mansions in Reading.

The Frill quarry No.2 was operated by William Adams around 1886. The color and quality of the rock was similar to that exposed in the Frill quarry No. 1. D'Invilliers (1886, p. 1562) noted: "*The rock quarried here is excellent for building, paving or curbing, and can be obtained in slabs of any desirable thickness. Possibly 50 or 60 feet in thickness is suitable for quarrying purposes, containing beds of great regularity, which can be economically quarried and handled.*" The rock dipped S. 30° E. 40°.

Frill supplied stone for the Reading City reservoir at the head of Penn Street. Some of the stone delivered there weighed 9 to 10 tons before being dressed. Much of the limestone curbing in Reading was produced at the Frill quarry. The curbing was up to 24 feet long. The curbing disappeared when the streets were paved (*Reading Eagle*, September 6, 1925). The City of Reading used crushed stone from Frill's quarry to surface roads and patch holes.

When Frill died in 1896, he willed the quarry to the Reverend Father Bornemann, pastor at St. Paul's Catholic Church. Bornemann valued the land at $3,000 an acre, but could not find a buyer at that price. The Frill quarry No. 2 was abandoned and never worked again (*Reading Eagle*, September 6, 1925).

William M. Fulton operated the Frill quarry No.1 from 1896 to about 1906, and it was known as Fulton's quarry (*Reading Eagle*, July 26, 1896). The Pennsylvania Railroad Company filed suit against Fulton alleging that blasting at the quarry threw stones across the river onto the railroad's property causing damage and endangering lives. The suit alleged that a stone weighing several pounds was thrown half a mile across the river and landed on the Pennsylvania Railroad Company platform (*Reading Eagle*, July 31, 1897). In 1897, an injunction was granted against Fulton restraining quarry operations so that stones would not be thrown upon the Pennsylvania Railroad Company station. In 1898, a petition was filed alleging that blasting threw stones that fell at the feet of Superintendent Myers and smashed train car windows. The judge ruled that if blasting could not be done without danger to life and property, it would have to cease (*Reading Eagle*, November 14, 1898). Quarry operations continued. In 1904, 37 laborers at the quarry went on strike demanding their wages be increased from $1.25 to $1.35 per day (*Reading Eagle*, May 15, 1904).

The Frill quarry No. 1 was acquired by Michael M. and James P. McQuade, trading under the name McQuade Brothers. The McQuade Brothers leased the quarry in 1906 and operated it as lessees for 8 years. In 1915, they purchased the 13-acre quarry property from the Northern Realty Company for $30,000. In 1915, production was 275 tons of crushed stone per day (*Rock Products and Building Materials*, December 22, 1915, p. 38).

Many newspaper reports concerned damage caused by blasting at the quarry. For example, a blast in 1912 broke window panes (*Reading Eagle*, February 1, 1912). A large blast took

WYOMISSING BOROUGH

Figure 651. Fluorite from Leinbach's Hill, West Reading. Reading Public Museum collection 2004c-007.

place in the quarry in 1916; 12,000 tons of limestone were dislodged by 1,200 pounds of dynamite packed into four 5-inch diameter holes 55 feet deep. The rock was loaded on quarry cars by a steam shovel with a 800 ton per day capacity (*Reading Eagle*, April 18, 1916). The McQuade Brothers quarry probably closed during the consolidation of quarries by the Berks Products Corporation prior to 1934. The site was purchased by the Boron Products Company of New York in 1951 (*Reading Eagle*, January 3, 1951).

Minerals

Gordon (1922, p. 152) reported the location of Leinbach's Hill as five miles northwest of Reading. It is north of the Penn Street bridge in West Reading.

Fluorite - deep blue to purple cubes (D'Invilliers, 1883, p. 397); massive (fig. 651)

Calcite - pink, in bluish quartz (D'Invilliers, 1883, p. 395)

Quartz - lilac colored (D'Invilliers, 1883, p. 400)

Drexel Quarry

The George Drexel quarry was north of Penn Street and west of the Schuylkill River around 40° 20′ 17″ latitude and 75° 56′ 57″ longitude on the USGS Reading 7.5-minute topographic quadrangle map (Appendix 1, fig. 26, location WR-2). It is quarry number 102 on reference map 11 in Pennsylvania Geological Survey (1891) (fig. 649). The quarry was in the Richland Formation. The Drexel quarry was a short distance southwest of the Lebanon Valley Railroad bridge over the Schuylkill River near Reading. The quarry was operated intermittently. D'Invilliers (1886, p. 1561) stated that the quarry "*seems to contain a fair quality of building stone... The entire exposure shows an evenly bedded limestone which can be quite readily quarried and handled.*" The limestone dipped S. 20° E. about 40° with bedding 1 to 4 feet thick.

WYOMISSING BOROUGH

Three open-pit iron (limonite) mines—Muhlenberg, Seitzinger, and Eureka—were located west of Reading and just north of the Spring-Cumru Township boundary. The Muhlenberg and Seitzinger mines were in Wyomissing, and the Eureka mine was in Spring Township. They are shown as hematite mines on the 1882 topographic map.

Seitzinger Mine

The Seitzinger mine was northwest of the intersection of Seventh Street and Parkside Drive at 40° 19′ 55″ latitude and 75° 57′ 12″ longitude on the USGS Reading 7.5-minute topographic quadrangle map (Appendix 1, fig. 26, location WY-1). The Seitzinger mine also was known as the West Reading mine hole and Weiser Lake. It was named Weiser Lake by the Wyomissing Development Company. The mine was in the Richland Formation.

The Seitzinger mine was an open-pit mine operated intermittently by the Wetherhold, Reichert, and Seitzinger families. The Seitzinger mine operated from the late 1860s to 1876. Daniel K. Sell was the mine superintendent. William K. Kauffman & Company had the exclusive rights to mine ore for a $1 per ton royalty. The exclusive contract to haul ore was held by Aaron Adams, Sr. of West Reading. Adams was the tenant farmer on the 307-acre Seitzinger farm. Adams hauled the ore, estimated to total between 10,000 and 15,000 tons in drays pulled by six-mule teams. Production was as much as 50 tons of ore per day (*Reading Eagle*. December 19, 1979).

The Seitzinger mine employed between 25 and 35 men. The 50-foot deep open-pit mine was worked by pick, shovel, and wheel barrow. Ore was removed in carts and pulled up ramps by mule-powered cables. The ore-laden carts were transported by six-mule teams to the warf located along the Lebanon Valley Railroad near present day Park Road in Wyomissing. The ore was shipped to the Kauffman & Company Sheridan Furnace in Lebanon County (Koch, 1949, p. 43-44).

The workmen arrived at the mine the morning of September 28, 1876, to find groundwater flowing into one part of the 3-acre mine pit. While 35 men were at work, water burst through the bottom of the pit filling the mine with 35 feet of water. Despite the quick flooding, men and mules were able to scramble out of the mine with no loss of life. However, the tools and carts were left in the pit (*Reading Eagle*, July 16, 1967; Koch, 1949). Two 1,700 gallon per minute pumps were run continuously for 65 hours with little effect on the water level. An advertisement for bids to pump out the mine hole resulted in the lowest bid of $35,000 (*Reading Eagle*. December 19, 1979). The mine was abandoned.

Sell's version of the mine flooding was more spectacular. He claimed there was a 4-foot square spot in yellow clay at a depth of 52 feet that had been wet for some time. In Sell's version of the story, there was a loud noise like an explosion, and a stream of water as thick as a man's body shot 20 feet in the air

Figure 652. Aerial view of the Seitzinger mine (Weiser Lake), Wyomissing, 1925. J. Victor Dallin Aerial Survey collection, HF_H72_001, Hagley Museum and Library, Wilmington.

making a sound like a geyser (*Reading Eagle*, December 19, 1979).

When D'Invilliers (1883, p. 373) visited the mine in 1882, he noted: "*It is likewise in limestone, which outcrops in its east and south sides. The rest of the bank has all fallen in, and the sides show loose limestone soil and reddish slate. No ore was seen here except a little on the dump, which shows a good deal of iron pyrites highly decomposed.*"

Ice was harvested from the flooded mine pit by nearby farmers and their hired hands, as well as employees from nearby brickyards, until artificial refrigeration phased out ice making. At one time, a frame ice house 200 feet long, 75 feet wide, and 30 feet high stood at the northeast corner of the mine. Much of the ice was sold to Reading breweries (*Reading Eagle*, July 16, 1967).

The Wyomissing Land Development Company owned the mine tract from about 1918 to 1930. In 1918, the company started a major tree planting and landscaping operation to turn the flooded mine pit into a park area (*Reading Eagle*, April 8, 1918). The landscaped mine pit was named Weiser Lake by the company (fig. 652). As the trees and shrubs grew up around the lake, it became a summer fishing and swimming place and a popular winter ice skating rink. In 1957, the mine was included in the 20 acres of land acquired from Wyomissing Industries by the Masonic Center Foundation to erect a $800,000 Masonic Temple. The temple was dedicated in 1964 (*Reading Eagle*, July 16, 1967). In 1966, the property owners decided to fill the lake to reduce the liability risk. The lake was filled, and only a few old trees remained to mark its location (*Reading Eagle*, December 19, 1979).

D'Invilliers published an analysis of the ore: 43.8 percent metallic iron, 1.017 percent sulfur, 0.104 percent phosphorus, and 21.1 percent silica.

Muhlenberg (Beidler) Mine

The Muhlenberg or Beidler mine was west of the intersection of Reading Boulevard and Evans Avenue at about 40° 19′ 31″ latitude and 75° 58′ 19″ longitude on the USGS Reading 7.5-minute topographic quadrangle map (Appendix 1, fig. 26, location WY-2). The area is completely developed. The mine is shown as the R.K. Heister (Neister?) mine in the 1876 atlas and as a hematite mine on the 1882 topographic map. The mine was in the Millbach Formation. In 1882, the Muhlenberg (Beidler) mine was the only iron mine in operation in West Reading. The mine was an open cut about 30 feet deep. The ore was in limestone and was siliceous hematite mixed with slate and clay (D'Invilliers, 1883, p. 373). D'Invilliers published an analysis of the ore: 41.5 percent metallic iron, 0.019 percent sulfur, 0.205 percent phosphorus, and 25.3 percent silica.

William Gudlin Quarry

The William Gudlin quarry was in the vicinity of the present day U.S. Route 422 interchange with State Hill Road on the USGS Reading 7.5-minute topographic quadrangle map. It is quarry number 101 on reference map 11 in Pennsylvania Geological Survey (1891) (fig. 649). The quarry was on the north side of the railroad, about half way between Sinking Springs and Reading. D'Invilliers (1886, p. 1561) noted: "*It is an insignificant exposure here showing limestone but a few feet thick, dipping steeply toward the south-east.*" The quarry was abandoned before 1886.

Figure 653. Pipe ore limonite from the Sitzinger mine, Wyomissing, 12.4 cm. Reading Public Museum collection 2000c-27-009.

BIBLIOGRAPHY

Alexander Drafting Company, 2006, Berks County Pennsylvania: 120 p.

Agarwal, R.K., Eben, F.C., and Taylor, C.E., 1973, Rock mechanics program at Grace mine: Mining Engineering, vol. 25, no. 8, p. 33.

American Iron and Steel Association, 1876, The iron works of the United States: Philadelphia, 136 p.

American Iron and Steel Association, 1878, Directory to the iron and steel works of the United States: Philadelphia, 159 p.

American Iron and Steel Association, 1880, Directory to the iron and steel works of the United States: Philadelphia, 183 p.

American Iron and Steel Association, 1882, Directory to the iron and steel works of the United States: Philadelphia, 192 p.

American Iron and Steel Association, 1884, Directory to the iron and steel works of the United States: Philadelphia, 202 p.

American Iron and Steel Association, 1886, Directory to the iron and steel works of the United States: Philadelphia, 205 p.

American Iron and Steel Association, 1888, Directory to the iron and steel works of the United States: Philadelphia, 231 p.

American Iron and Steel Association, 1890, Directory to the iron and steel works of the United States, 10th ed.: Philadelphia, 248 p.

American Iron and Steel Association, 1892, Directory to the iron and steel works of the United States, 11th ed.: Philadelphia, American Iron and Steel Association, 282 p.

American Iron and Steel Association, 1894, Directory to the iron and steel works of the United States, 12th ed.: Philadelphia, American Iron and Steel Association, 292 p.

American Iron and Steel Association, 1896, Directory to the iron and steel works of the United States, 13th ed.: Philadelphia, American Iron and Steel Association, 320 p.

American Iron and Steel Association, 1901, Directory to the iron and steel works of the United States, 15th ed.: Philadelphia, American Iron and Steel Association, 428 p.

American Iron and Steel Association, 1903, Supplement to the directory to the iron and steel works of the United States: Philadelphia, American Iron and Steel Association, 196 p.

American Iron and Steel Association, 1904, Directory to the iron and steel works of the United States, 16th ed.: Philadelphia, American Iron and Steel Association, 468 p.

American Iron and Steel Association, 1908, Directory to the iron and steel works of the United States, 17th ed.: Philadelphia, American Iron and Steel Association, 500 p.

American Iron and Steel Association, 1910, Supplement to the directory to the iron and steel works of the United States: Philadelphia, American Iron and Steel Association, 160 p.

American Iron and Steel Association, 1912, Second supplement to the 1908 edition of the directory to the iron and steel works of the United States Philadelphia, American Iron and Steel Association, 80 p.

American Iron and Steel Institution, 1916, Directory of the iron and steel works of the United States, 18th ed.: Philadelphia, American Iron and Steel Institution, 437 p.

Anderson Companies, 2012, Dyer Quarry, Inc., accessed November 2, 2012, at *http://www.dyerquarry.com/*.

Ball, C.M., 1895, The magnetic separation of iron-ores: The Engineering Magazine, vol. 9, no. 5, p. 908-915.

Ball, C.M., 1896, The magnetic separation of iron-ores: Transactions of the American Institute of Mining Engineers, vol. 25, p. 533-555.

Ball, J.L., and Crider, J.G., 2000, The deformation of the Jacksonwald Syncline, Pennsylvania [ABS]: Geological Society of America, Abstracts with Programs, vol. 32, no. 1, p. 4.

Barnes, J.H., 1997, Directory of the nonfuel-mineral producers in Pennsylvania: Pennsylvania Geological Survey, 4th ser., Open-File Report 97-04, 295 p.

Barnes, J. H., compiler, 2011, Directory of the nonfuel-mineral producers in Pennsylvania: Pennsylvania Geological Survey, 4th ser., Open-File Report OFMR 11–01.1, 184 p., accessed March 9, 2015, at http://www.dcnr.state.pa.us/cs/groups/public/documents/document/dcnr_009940.pdf.

BIBLIOGRAPHY

Barnes, J. H., and Smith, R. C., II, 2001, The nonfuel mineral resources of Pennsylvania: Pa. Geological Survey, 4th ser., Educational Series 12, 38 p.

Barnett, G.R., 1922, Commonwealth of Pennsylvania] vs Birdsboro Stone Co.: The Dauphin County Reports, Harrisburg, Pa., p. 135-145.

Barton, B.S., 1805, [Zeolite in basalt from Reading, Pennsylvania]: The Philadelphia Medical and Physical Journal, vol. 2, no. 1, p. 178-179.

Bascom, Florence, and Stose, G.W., 1938, Geology and mineral resources of the Honeybrook and Phoenixville quadrangles, Pennsylvania: U.S. Geological Survey Bulletin 891, 145 p.

Bastin, E.S., 1911, Graphite: U.S. Geological Survey Mineral Resources of the United States Calendar Year 1909, part 2, p. 830.

Basu, Debabrata, 1974, Genesis of the Grace Mine magnetite deposit, Morgantown, Berks County, southeastern Pennsylvania: Doctoral dissertation, Lehigh University, Bethlehem, Pa.

Basu, Debabrata, and Sclar, C.B., 1976, Genesis of the Grace Mine magnetite deposit, Morgantown, Berks County [ABS]: Geological Society of America Abstracts with Programs, vol. 8, no. 6, p. 769-770.

Beard, Robert, 2008, Hay Creek; prehnite and zeolites; specimens south of Birdsboro, Pennsylvania: Rock and Gem, vol. 38, no. 11, p. 56-60.

Bever, J.E., and Liddicoat, W.K., 1954, Investigation of the Triassic iron ores of Pennsylvania: typewritten manuscript, 91 p.

Behre, C.H., 1933, Slate in Pennsylvania: Pennsylvania Geological Survey, 4th ser., Mineral Resources Report 16, 400 p.

Berkey, C.P., and others, 1933, Mineral deposits of New Jersey and eastern Pennsylvania: 16th International Geological Congress, Guidebook 8, Excursion A-8, 54 p.

Berkheiser, S.W., Jr., 1984, Fetid barite occurrences, western Berks County, Pennsylvania: Pennsylvania Geological Survey, 4th ser., Mineral Resource Report 84, 43 p.

Berkheiser, S.W., Jr., 1985, High-purity silica occurrences in Pennsylvania: Pennsylvania Geological Survey, 4th ser., Mineral Resources Report 88, 67 p.

Berkheiser, S.W., Jr., 2003, Play ball!: Pennsylvania Geology, vol. 33, no. 1, p. 10-14.

Berkheiser, S.W., Jr., Barnes, J.H., and Smith, R.C. II, 1985, Directory of the nonfuel-mineral producers in Pennsylvania, 4th ed.: Pennsylvania Geological Survey, 4th ser., Information Circular 54, 170 p.

Berkheiser, S.W., Jr., and Smith, R.C. II, 1984, High-alumina clay discovered in Berks County: Pennsylvania Geology, vol. 15, no. 1, p. 2-6.

Bethlehem Steel Company, 1961, Bethlehem Steel Grace Mine at Morgantown, Berks County, Pennsylvania: Bethlehem, Pa., 7 p.

Beutner, E.C., and Charles, E.G., 1985, Large volume loss during cleavage formation, Hamburg Sequence, Pennsylvania: Geology, vol. 13, no. 11, p. 803-805.

Beutner, E.C., Fisher, D.M., and Kirkpatrick, J.L., 1983, Cleavages and related syntectonic fibres in the Hamburg Sequence near Mohrsville, Pa. [ABS]: Geological Society of America Abstracts with Programs, vol. 15, no. 3, p. 175.

Beutner, E.C., Fisher, D.M., and Kirkpatrick, J.L., 1988, Kinematics of deformation at a thrust fault ramp(?) from syntectonic fibers in pressure shadows: Geological Society of America Special Paper 222, p. 77-88.

Biemesderfer, G.K., and Leske, R.H., 1961, Ground water control at Grace Mine: Mining Congress Journal, vol. 47, no. 10, p. 39-44.

Bingham, J.P., 1957, Grace mine: Mining Engineering, vol. 9, no. 1, p. 45-48.

Biographical Publishing Company, 1898, Book of biographies [of the] biographical sketches of leading citizens of Berks county, Pa.: Buffalo, New York, Biographical Publishing Company, 740 p.

Blackwell, P.A.S., 2004, Along the route 100 corridor: Arcadia Publishing, 128 p.

Bliss, E.F., 1913, Glaucophane from eastern Pennsylvania: American Museum of Natural History Bulletin no. 32, p. 517-526.

Bollfras, Susan, 2013, Molltown, PA, accessed January 4, 2013, at *http://www.angelfire.com/pa5/mollpa/molltown.html*.

Braun, D.D., 1996, Surficial geology of the New Tripoli 7.5' quadrangle, Berks, Carbon, Lehigh, and Schuylkill Counties, Pennsylvania: Pennsylvania Geologic Survey, 4th ser., Open-File Report 96-33, 11 p.

Braun, D.D., 1996, Surficial geology of the Allentown West 7.5' quadrangle; Berks and Lehigh counties, Pennsylvania: Pennsylvania Geological Survey, 4th ser., Open-File Report 96-42, 10 p.

Braun, D.D., 1996, Surficial geology of the Hamburg 7.5' quadrangle, Berks County, Pennsylvania: Pennsylvania Geological Survey, 4th ser., Open-File Report 96-39, 8 p.

Braun, D.D., 1996, Surficial geology of the Kutztown 7.5' quadrangle, Berks and Lehigh counties, Pennsylvania: Pennsylvania Geological Survey, 4th ser., Open-File Report 96-40, 7 p.

Braun, D.D., 1996, Surficial geology of the New Ringgold 7.5' quadrangle; Berks, Lehigh, and Schuylkill counties, Pennsylvania: Pennsylvania Geological Survey, 4th ser., Open-File Report 96-32, 10 p.

Braun, D.D., 1996, Surficial geology of the Topton 7.5' quadrangle: Berks and Lehigh Counties, Pennsylvania: Pennsylvania Geological Survey, 4th ser., Open-File Report 96-41, 10 p.

Braun, D.D., and Epstein, J.B., compilers, 1993, Surficial geologic map of part of the Allentown 30'X60' quadrangle, Berks, Bucks, Carbon, Lehigh, Luzerne, Monroe, Northampton, and Schuylkill counties: U.S. Geological Survey Open-File Report OF 93-0723, 15 p.

Brooke, G.C., 1959, Birdsboro: company with a past, built to last: New York, Newcomen Society in North America, 24 p.

Brookmeyer, Bryon, 1978, Pennsylvania's Cornwall-type iron mines: Rocks and Minerals, vol. 53, no. 3, p. 135-139.

Bromery, R.W., Zandle, G.L., and others, 1959, Aeromagnetic map of the Morgantown quadrangle, Berks, Lancaster, and Chester Counties, Pennsylvania: U.S. Geological Survey Geophysical Investigations Map GP–220, scale 1:24,000.

Brown, C.C., 1901, Directory of American cement industries and handbook for cement users: Indianapolis, Municipal Engineering Company, 645 p.

Brown, F.M., 1969, Amity first in Berks, 250th Anniversary 1719-1969: unpaginated.

Broughton, Paul, 1966, Collecting the minerals of southeastern Pennsylvania: Rocks and Minerals, vol. 21, no. 3, p. 169-172.

Brown, A.P., and Ehrenfield, Frederick, 1913, Minerals of Pennsylvania: Pennsylvania Geological Survey, 3rd ser., Report 9, 160 p.

Brunner, D.B., and Smith, E.F., 1883, Some minerals from Berks County, Pa: American Chemical Journal, vol. 5, no. 4, p. 279-280.

Brush, G.J., and Dana, J.D., 1882, Appendixes to the fifth edition of Dana's Mineralogy: New York, John Wiley and Sons, variously paginated.

Buckwalter, T.V., Jr., 1953, Origin and field relations of the Pochuck gneiss in Lebanon, Berks, and Lancaster counties, Pennsylvania: Proceedings of the Pennsylvania Academy of Science, vol. 27, p. 114-119.

Buckwalter, T.V., Jr., 1954, Mica prospects; a report on eastern Berks County: Pennsylvania Department of Internal Affairs Monthly Bulletin, vol. 22, no. 2, p. 17-20.

Buckwalter, T.V., Jr., 1956, A post-Hardyston pre-Triassic diabase dike in Berks County, Pennsylvania: Proceedings of the Pennsylvania Academy of Science, vol. 30, p. 170-175.

Buckwalter, T.V., Jr., 1958, Granitization in the Reading Hills, Berks County, Pennsylvania: Proceedings of the Pennsylvania Academy of Science, vol. 32, p. 133-138.

Buckwalter, T.V., Jr., 1959, Geology of the Precambrian rocks and Hardyston Formation of the Boyertown Quadrangle, Pa.: Pennsylvania Geological Survey, 4th ser., Atlas 197, 15 p.

Buckwalter, T.V., Jr., 1960, Some structural aspects of the Reading Hills: Proceedings of the Pennsylvania Academy of Science, vol. 34, p. 109-116.

Buckwalter, T.V., Jr., 1963, Evidence on the origin of the 'pinite' of the Reading Hills, Pennsylvania: Proceedings of the Pennsylvania Academy of Science, vol. 37, p. 160-165.

Burtner, Roger, Weaver, Richard, and Wise, D.U., 1958, Structure and stratigraphy of Kittatinny Ridge at Schuylkill Gap, Pennsylvania: Proceedings of the Pennsylvania Academy of Science, vol. 32, p. 141-145.

Cassidy, C.J., 1958, Political Career of William A. Witman, Sr.: Historical Review of Berks County, acessed January 3, 2016, at *http://www.berkshistory.org/multimedia/articles/political-career-of-william-a-witman-sr/*.

Charleton, D.E.A., 1921, Iron mining in the United States: Engineering and Mining Journal, vol. 111, p. 144-145.

BIBLIOGRAPHY

Chou, I-Ming, and Eugster, H.P., 1977, Solubility of magnetite in supercritical chloride solutions: American Journal of Science, vol. 277, p. 1296-1314.

Claussen, W.E., 1975, An autumn journey beyond Barto: Historical Review of Berks County, vol. 40, no. 2, p. 52-55.

Claussen, W.E., 1971, A History of Colebrookdale Township: 84 p.

Cleaveland, Parker, 1816, An elementary treatise on mineralogy and geology 1st ed.: Boston, 668 p.

Cleaveland, Parker, 1822, An elementary treatise on mineralogy and geology: 2nd ed.: Boston, 818 p.

Cochran, W., 1969, Grace Mine iron ore waste disposal system and estimated costs: U.S. Bureau of Mines Information Circular 8435, 17 p.

Conway, J.A., 1931, Mining and Crushing Methods and Costs at the Monocacy Quarry of the John T. Dyer Quarry Co., Monocacy, Pa: U.S. Department of Commerce, Bureau of Mines, 28 p.

Commonwealth of Pennsylvania, 1877, Annual report of the Secretary of Internal Affairs of the Commonwealth of Pennsylvania: Harrisburg, Pa, 1064 p.

Cooper, F.D., 1975, The mineral industry of Pennsylvania in 1972: Pennsylvania Geological Survey, 4th ser., Information Circular 78, 52 p.

Corbin, J.R., 1923, Gold in Pennsylvania: Pennsylvania Geological Survey, 4th ser., Bulletin 77, 13 p.

Croll, P.C., 1926, Annals of the Oley Valley in Berks County, Pa.: Reading, Pa., Reading Eagle Press, 148 p.

Dana, E.S., 1892, System of Mineralogy, Sixth Edition: New York, John Wiley and Sons, 1067 p.

Dana, E.S., 1904, The system of mineralogy of James Dwight Dana, Sixth edition with appendix 1: New York, John Wiley and Sons, 1207 p.

Dana, J.D., 1844, A system of mineralogy, Second edition: New York, Wiley and Putnam, 633 p.

Dana, J.D., 1850, A system of mineralogy, Third edition: New York, George P. Putnam. 711 p.

Dana, J.D., 1854, A system of mineralogy, Fourth edition: New York, George P. Putnam. 531 p.

Dana, J.D., and Brush, G.J., 1868, A system of mineralogy, Fifth edition: New York, John Wiley and Sons, 827 p.

Daniels, D.L., 1985, Gravimetric character and anomalies in the Gettysburg Basin, Pennsylvania; a preliminary appraisal: U.S. Geological Survey Circular 946, p. 128-132.

Davidson, F.P., 1885, Stilbite [part] 5 of mineralogical notices: American Chemical Journal, vol. 6, p. 414.

Davis, F.A., and Kochersperger, H.L., 1876, Illustrated historical atlas of Berks County: Reading, Pa., A.M. Davis, 142 p.

Davis, Harold, 1960, New Pennsylvania caves: Speleothems, vol. 8, no. 4.

Day, D.T., 1904, The stone industry in 1903 *in* U.S. Geological Survey Mineral Resources of the United States Calendar Year 1903, 1204 p.

Deasy, G.F., and Greiss, P.R., 1969, Atlas of Pennsylvania's mineral resources, part 2-B, Economics of Pennsylvania's clay and shale production: Pennsylvania Geological Survey, 4th ser., Bulletin M50, variously paginated.

Dewey, F.P., 1891, A preliminary descriptive catalog of the systematic collections in economic geology and mineralogy in the United States Museum: Bulletin of the United States Museum No. 42, 256 p.

D'Invilliers, E.V., 1883, The geology of the South Mountain belt of Berks County: Pennsylvania Geological Survey, 2nd Series, Report D3, vol. 2, part 1, 441 p.

D'Invilliers, E.V., 1886, Report on the iron ore mines and limestone quarries of the Cumberland-Lebanon Valley, PA.: Pennsylvania Geological Survey, 2nd Series, Annual Report 1866, part 4, p. 1409-1567.

D'Invilliers, E.V., 1887, Iron ore mines and limestone quarries of the Great Valley in 1886: Pennsylvania Geological Survey, 2nd. ser., Annual Report for 1886, part 4, p. 1411-1567.

Downs, Sandra, 2000, Pennsylvania's show caves: Rock and Gem, vol. 30, no. 11, p. 80-83.

Drake, A.A., Jr., 1987, Geologic map of the Topton Quadrangle, Lehigh and Berks counties, Pennsylvania: U.S. Geological Survey Geologic Quadrangle Map GQ-1609, 1 sheet.

Drake, A.A., Jr., 1993, Bedrock geologic map of the Allentown West Quadrangle, Lehigh and Berks counties, Pennsylvania: U.S. Geological Survey Geologic Quadrangle Map GQ-1727, 1 sheet.

Drake, A.A., Jr., and Epstein, J.B., 1967, The Martinsburg Formation (Middle and Upper Ordovician) in the Delaware Valley, Pennsylvania-New Jersey: U.S. Geological Survey Bulletin 1244-H, p. 1-16.

Drake, A.A., Jr., Lyttle, P.T., and Lash, G.G., 1978, Ordovician clastic rocks of eastern Pennsylvania: U.S. Geological Survey Professional Paper 1100, p. 50.

Druzba, P.A., 2003, Neversink: Reading's "other mountain": the resort years 1880-1930: Exeter House Books, 192 p.

Mostardi, Michael, and Durant, Joe, eds., 1991, Caves of Berks County, Pennsylvania: Mid-Appalachian Region of the National Speleological Society Bulletin 18, 84 p.

Eaton, H.N., 1912, The geology of South Mountain at the junction of Berks, Lebanon, and Lancaster counties, Pennsylvania: Journal of Geology, vol. 20, p. 331-343.

Eben, C.F., 1996, A brief history of the Grace mine: unpublished report, 1 p.

Eckel, E.C., 1904, Cement-rock deposits of the Lehigh District of Pennsylvania and New Jersey: U.S. Geological Survey Bulletin 225, p. 448-456.

Einaudi, M.T., and Burt, D.M., 1982, Introduction—terminology, classification, and composition of skarn deposits: Economic Geology, vol. 77, no. 4, p. 745-754.

Einaudi, M.T., Meinert, L.D., and Newberry, R.J., 1981, Skarn deposits: Economic Geology, 75th Anniversary Volume, p. 317-391.

Eugster, H.P., and Chow, I-Ming, 1979, A model for the deposition of Cornwall- type magnetite deposits: Economic Geology, vol. 74, p. 763-774.

Eyerman, John, 1889, The mineralogy of Pennsylvania - part 1: Easton, Pa., 54 p.

Eyerman, John, 1911, The mineralogy of Pennsylvania - part 2, chemical analyses: Easton, Pa., 25 p.

Fagan, L. 1860, Map of Berks County Pennsylvania from actual surveys: Philadelphia, H.F. Bridgen, 1 plate.

Fagan, L., 1862, Township map of Berks County, Pennsylvania from actual surveys: Philadelphia, H.F. Bridgens, 32 p.

Faill, R.T., 1987, Tectonics and structural geology of the Central Appalachians in Pennsylvania; recent developments and future prospects [ABS]: Geological Society of America Abstracts with Programs, vol. 19, no. 1, p. 13.

Faill, R.T., 2003, The early Mesozoic Birdsboro central margin basin in Mid-Atlantic region, eastern United States: Geological Society of America Bulletin, vol. 115, no. 4, p. 406-421.

Faill, R.T., 2004, The Birdsboro Basin: Pennsylvania Geology, vol. 34, no. 2, p. 2-11.

Fegley, L.P.G., 1935, The Boyertown ore mines: Historical Sketches, Historical Society of Montgomery County, p. 265-270.

Fenton, C.L., and Fenton, M.A., 1937, Cambrian calcareous algae from Pennsylvania: American Midland Naturalist, vol. 18, no. 3, p. 435-441.

Fisher, P.D., 1957, Ohnmact cave: National Speleological Society, Netherworld News, vol. 5, no. 10, p. 114.

Fisher, P.D., 1957, Ohnmact cave: Speleo Digest, p. 48-49.

Foord, Gene, 2000, Loellingite or safflorite from Huff's Church, PA?: Friends of Mineralogy Pennsylvania Chapter Newsletter, vol. 28, no. 2, p. 4.

Fox, C.T., 1916, Mount Pleasant furnace located; recent discoveries at Barto: *Reading Eagle*, May 21, 1916, p. 20.

Fox, C.T., 1925, Reading and Berks County Pennsylvania, vol. 1: New York, Lewis Publishing Company, 427 p.

Frankhouser, E.M., Sr., 1962, The Morgantown mystery: Historical Review of Berks County, vol. 27, no. 4, p. 102-105.

Frazer, Persifor, 1880, Geology of Lancaster County: Pennsylvania Geological Survey, 2nd Ser., Report CCC, 350 p.

BIBLIOGRAPHY

Frazer, D.M., 1937, Basic rocks in the eastern Pennsylvania Highlands: American Geophysical Union Transactions, vol. 18, p. 249-254.

Frazer, D.M., and Gertz, A.J., 1939, Notes on Hardyston quartzite: Pennsylvania Academy of Science Proceedings, vol. 13, p. 94-97.

Frear, William, 1913, Pennsylvania limestone and lime supplies: Pennsylvania State College Agricultural Experiment Station Bulletin No. 127, 106 p.

Frear, William, and Erb, E.S., 1913, Lime resources of Pennsylvania: Annual Report of the Pennsylvania State College for the year, 1911-1912, p. 272-440.

Frear, William, and Holben, F.J., 1916, Limestone resources of Pennsylvania, Supplementary report: Annual Report of the Pennsylvania State College for the year 1914-1915, p. 306-412

Friends of Mineralogy, 2011, Okenite - a new mineral for Pennsylvania: Friends of Mineralogy, Pennsylvania Chapter Newsletter, vol. 39, no. 1, p. 1.

Friends of the Robesonia Furnace, Inc., 2015, History, accessed December 26, 2015, at *http://www.robesoniafurnace.org/history.html*.

Froelich, A.J., and Gottfried, David, 1999, Early Mesozoic; igneous and contact metamorphic rocks: *in* Shultz, C.H., ed., The geology of Pennsylvania: Pennsylvania Geological Survey, 4th ser., and Pittsburgh Geological Society, Special Publication 1, p. 202-209.

Ganis, G.R., 1972, Geology of the Ordovician clastic rocks of the Bethel quadrangle, Pennsylvania, M.S. thesis, Lehigh University, 31 p.

Ganis, G.R., and Wise, D.U., 2008, Taconic events in Pennsylvania; datable phases of a approximately 20 m.y. orogeny: American Journal of Science, vol. 308, no. 2, p. 167-183.

Gartner, Randy, 2015, The great Robesonia cinder bank: Friends of the Robesonia Furnace, Inc. Newsletter, vol. 35, no. 1, p. 1.

Gault, H.R., 1950, Metallic mineral reserves of Pennsylvania: Symposium on Mineral Resources in Pennsylvania, Proceedings of the Pennsylvania Academy of Science, vol. 24, p. 208-214.

Gault, H.R., and Hamilton, C.L., 1953, Partial log of a deep well, Freidensburg (Oley P. O.), Berks County, Pennsylvania: Proceedings of the Pennsylvania Academy of Science, vol. 27, p. 146-153.

Gedde, R.W., 1965, Geophysical investigation of a magnetite deposit, Chester County, Pennsylvania: M.S. thesis, Pennsylvania State University, University Park, 59 p.

Geil, John, and Shrope, W.B., 1857, Map of the City of Reading from actual surveys: Philadelphia,

Genth, F.A., 1875, Preliminary report on the mineralogy of Pennsylvania, with an appendix on the hydrocarbon compounds by S.P. Sadtler: Pennsylvania Geological Survey, 2nd ser., Report B, 206 p.

Genth, F.A., 1889, Mineral Collection *in* Catalog of the Geological Museum, part 3: Pennsylvania Geological Survey, 2nd ser., Report OOO, p. 93-97.

Genth, F.A., 1886, Contributions to mineralogy: Proceedings of the American Philosophical Society, vol. 22, p. 30-47.

Genth, F.A., Jr., 1882, Notes on damourite from Berks County, Pennsylvania: Proceedings of the Academy of Natural Sciences of Philadelphia, vol. 34, p. 47-48.

Geyer, A.R., 1977, Building stones of Pennsylvania's capital area: Pennsylvania Geological Survey, 4th ser., Environmental Geology Report 5, 47 p.

Geyer, A.R., Buckwalter, T.V., Jr., McLaughlin, D.B., and Gray, Carlyle, 1963, Geology and mineral resources of the Womelsdorf Quadrangle: Pennsylvania Geologic Survey, 4th ser., Altas 177c, 96 p.

Geyer, A.R., Smith, R.C., II, and Barnes, J.H., 1976, Mineral collecting in Pennsylvania, 4th ed.: Pennsylvania Geological Survey, 4th ser., General Geology Report 33, 260 p.

Glaser, Leah, 2005, Hopewell Furnace National Historic Site Administrative History: National Park Service, 341 p.

Glen-Gery Corporation, 2014, History of Glen-Gery Brick accessed December 26, 2014, at *http://www.glengery.com/about-us/history*.

Gordon, S.G., 1918, Proceedings of societies the Philadelphia Mineralogical Society: American Mineralogist, vol. 3, no. 8, p. 163-164.

Gordon, S.G., 1920, Two American occurrences of epidesmine: American Mineralogist, vol. 5, no. 9, p. 167.

Gordon, S.G., 1922, The mineralogy of Pennsylvania: Academy of Natural Sciences of Philadelphia Special Publication No.1, 255 p.

Gottfried, David, Froelich, A.J., and Grossman, J.N., 1991, Geochemical data for Jurassic diabase associated with early Mesozoic basins in the Eastern United States; western Newark Basin, Pennsylvania and New Jersey: U.S. Geological Survey Source Open-File Report OF 91-322-D, 27 p.

Graham, D.A., 1993, Cannon maker for the Revolution: Thomas Rutter III (1732-1795) of Eastern Pennsylvania ; a biographical sketch which includes a brief history of Colebrookdale Furnace, Berks County, Pennsylvania, and a brief history of Warwick Furnace, Chester County, Pennsylvania: privately published, 64 p.

Graham, D.A., 1996, Rutter, Thomas I (c1660-1730) of Germantown, Pennsylvania and the birth of the Pennsylvania iron industry: privately published, 75 p.

Graham, D.A., 2010, Colebrook Dale furnace (1720-1770) and Pine Forge (1720-1844) Berks County, Pennsylvania: Pennsylvania's first blast furnace and refinery forge, a historical sketch of the Pine Forge historical area: privately published, 123 p.

Grant, R.W., 1974b, Pennsylvania Minerals (136): Mineralogical Society of Pennsylvania Keystone Newsletter, vol. 23 no. 4, p. 6.

Gray, Carlyle, 1950, A structural problem near Evansville, Penna.: Proceedings of the Pennsylvania Academy of Science, vol. 24, p. 170-175.

Gray, Carlyle, 1951, Preliminary report on certain limestone and dolomites of Berks County, Pennsylvania: Pennsylvania Geological Survey, 4th ser., Progress Report 136, 85 p.

Gray, Carlyle, 1952, The high calcium limestones of the Annville belt in Lebanon and Berks counties, Pennsylvania: Pennsylvania Geological Survey, 4th ser., Progress Report 140, 17 p.

Gray, Carlyle, 1952, The nature of the base of the Martinsburg Formation: Proceedings of the Pennsylvania Academy of Science, vol. 26, p. 86-92.

Gray, Carlyle, 1959, Nappe structures in Pennsylvania: Geological Society of America Bulletin, vol. 70, no. 12, part 2, p. 1611.

Gray, Carlyle, 1999, Metallic mineral deposits-Cornwall-type iron deposits in Shultz, C.H., ed., The geology of Pennsylvania: Pennsylvania Geological Survey, 4th ser., and Pittsburgh Geological Society, Special Publication 1, p. 567-573.

Gregory, G.E., 1970, Mineral locations in Pennsylvania: Gems and Minerals, no. 397, p. 24-27; 49.

Groth, H.A., The Birdsboro trap-rock quarries: Mineral Collector, vol. 10, no. 11, p. 159-160.

Guiseppe, A.C., 1995, Examination of intrafan variations within a Triassic carbonate clast basal conglomerate; Reading, Pennsylvania [ABS] Geological Society of America Abstracts with Programs, vol. 27, no. 1, p. 50-51.

Hamburg Bicentennial Committee, 1976, Bicentennial history of Hamburg Borough: Hamburg, Pennsylvania.

Harden, J.H., 1886, Early mining operations in Berks and Chester Counties: Engineer's Club of Philadelphia Proceedings, vol. 6, no. 1, p. 23-35.

Harris, A.G., and Repetski, J.E., 1982, Conodonts revise the Lower-Middle Ordovician boundary and timing of miogeoclinal events in the East-Central Appalachian Basin [ABS]: Geological Society of America Abstracts with Programs, vol. 14, no. 5, p. 261.

Hartman, E.L., 1976, Temple Borough, History of Temple, Pennsylvania, Bicentennial Issue, 1776-1976: Temple Borough Committee, 49 p.

Hawes, G.W., 1884, Report on the building stones of the United States and statistics of the quarry industry for 1880: Department of the Interior, Census Office, variously paginated.

Hawkes, H.E., Jr., 1945, Magnetic anomaly near Bechtelsville, Pa: U.S. Geological Survey Preliminary Report, 2 p.

Hawkes, H.E., Wedow, Helmuth, and Balsley, J.R., 1953, Geologic investigation of the Boyertown magnetite deposits in Pennsylvania :U.S. Geological Survey Bulletin 995-D, p. 135-148.

Heizmann, L.J., 1961, From Lyceum to museum scientific societies in Reading and Berks County: Historical Review of Berks County, vol. 26, no. 4, p. 106-114.

BIBLIOGRAPHY

Heizmann, L.J., 1971, The Pagoda: Historical Review of Berks County, vol. 36, no. 2, p. 51-52.

Hermelin, S.G., 1783, Report about the mines in the United States of America, 1783: Johnson, Amandus, translator, 1931, Philadelphia, The John Morton Memorial Museum, 76 p.

Hersey, J.B., 1942, Gravity investigation of central-eastern Pennsylvania: Geological Society of America Bulletin, vol. 55, no. 4, p. 417-444.

Hexamer, E., and Son, 1872-1911, Insurance maps of the City of Philadelphia, Volumes 1-37.

Hice, R.R., 1911, The mineral production of Pennsylvania, 1911: Pennsylvania Geological Survey, 3rd. ser., Report 8, 139 p.

Hice, R.R., 1915, The mineral production of Pennsylvania for the year 1913: Pennsylvania Geological Survey, 3rd ser., Report 11, 108 p.

Hickok, W.O., IV, 1939, The minerals of Pennsylvania, the iron ores of Pennsylvania: Pennsylvania Geological Survey, 4th ser., Bulletin M18-B, 21 p.

Hiester, J.P., and others, 1851, Report on the topography and diseases of Berks County: Transactions of the Medical Society of the State of Pennsylvania, vol. 1, p. 83-96.

Historical Committee of the Blandon Bicentennial Committee, 1976, The story of Blandon, Pa., and Maidencreek Township: Kutztown Publishing Company, Kutztown, Pa., 108 p.

Historical Committee of the Kutztown Centennial Association, 1915, The centennial history of Kutztown Pennsylvania: Kutztown Publishing Company, Kutztown, Pa., 247 p.

Hobson, J.P., Jr., 1957, Lower Ordovician (Beekmantown) succession in Berks County, Pennsylvania: Bulletin of the American Association of Petroleum Geologists, vol. 41, no. 12, p. 2710-2722.

Hobson, J.P., Jr., 1963, Stratigraphy of the Beekmantown Group in southeastern Pennsylvania: Pennsylvania Geol. Survey, 4th ser., Bulletin G37, 331 p.

Hoefert, C.M., Burris, Lea, Yenchik, John, Black, L.E., and Friehauf, K.C., 2003, Mappin' without diggin'; an application of magnetometry techniques in Rittenhouse Gap mining district in Berks County, Pennsylvania [ABS]: Geological Society of America, Abstracts with Programs, vol. 35, no. 3, p. 94.

Hoffman, B.A., 1976, Forge and foundry an historic review of Birdsboro Berks County, Pennsylvania: self published, 60 p.

Hoover, K.V., Saylor, T.E., Lapham, D.M., and Tyrell, M.E., 1971, Properties and uses of Pennsylvania shales and clays, southeastern Pennsylvania: Pennsylvania Geological Survey, 4th ser., Mineral Resources Report M63, 329 p.

Hoskinson, W.S., and Brunner, D.B., 1885, [Stilbite, part 6 of] mineralogical notes: American Chemical Journal, vol. 6, no. 6, p. 414.

Howell, B.F., 1943, Burrows of Skolithos and Planolites in the Cambrian Hardyston sandstone at Reading, Pennsylvania: Wagner Free Institute of Science, vol. 3, 33 p.

Hopkins, T.C., 1897, The building materials of Pennsylvania I - brownstones: Appendix to the Annual Report of Pennsylvania State College, 121 p.

Hopkins, T.C., 1898, Clay and Clay Industries of Pennsylvania Part 2 Clays of southeastern Pennsylvania: Appendix to the Annual Report of Pennsylvania State College for 1898-99, 76 p.

Hopkins, T.C., 1900, Clay and Clay Industries of Pennsylvania Part 3 Clays of the Great Valley and South Mountain areas: Appendix to the Annual Report of Pennsylvania State College for 1899, 45 p.

Huff's Union Church, 2000, The History of Huff's Union Church 1760 - 2000, acessed November 23, 2015 at *http://www.huffschurch.com/History2.html*.

Hunt, T.S., 1876, A new ore of copper and its metallurgy: Transactions of the American Institute of Mining Engineers, vol. 4, p. 325-328.

Ibach, E.W., 1976, The hub of the Tulpehocken: Womelsdorf, Pa., Boyer Printing Co., 653 p.

Inners, J.D., and Ferguson, W.B., 1996, French Creek State Park, Berks and Chester Counties; Piedmont Rocks and Hopewell Furnace: Pennsylvania Geological Survey, 4th ser., Park Guide 6, unpaginated.

BIBLIOGRAPHY

Jambor, J.L., 1976, New occurrences of the hybrid sulphide tochilinite: Geological Survey of Canada Paper 76-1B, Report of activities, part B, p. 65-69.

J.L. Floyd and Company, 1911, Genealogical and biographical annals of Northumberland County Pennsylvania: Chicago, J.L. Floyd and Company, 988 p.

Jonas, A.I., 1917, Pre-Cambrian and Triassic diabase in eastern Pennsylvania: Bulletin of the American Museum of Natural History, vol. 37, p. 173-181.

Jones, Eliot, 1914, The anthracite coal combination in the United States: Harvard University Press, 261 p.

Kaufman, Alvin, 1956, The mineral industry of Pennsylvania in 1953: Pennsylvania Geological Survey, 4th ser., Information Circular 8, 26 p.

Kemp. J.F., 1902, The geologic relations and distribution of platinum and associated metals: U.S. Geological Survey Bulletin 193, 95 p.

Kempton Centennial Committee, undated, Centennial history of Kempton and Albany Townships 1874-1974, unpaginated.

Kerr, J.R., 1965, The Mineral Industry of Pennsylvania: Pennsylvania Geological Survey, 4th ser., Information Circular 53, 42 p.

Koch, K.A., 1949, Weiser Lake-The mine hole: Historical Review of Berks County, vol. 14, no. 2, p. 43-44.

Kochel, R.C., 1975, Morphology, structure, and origin of two blockfields and associated deposits, northern Berks county, Pennsylvania: Senior Thesis, Franklin and Marshall College, Lancaster, Pa.

Kochel, R.C., 1976, Morphology, origin, and structural control of several blockfields and associated deposits, northern Berks County, Pennsylvania (River of Rocks, Devils Potato Patch, and Blue Rocks) [ABS]: Geological Society of America Abstracts with Programs, vol. 8, no. 4, p. 486-487.

Kuhlman, Robert, 1989, The Jacksonwald Syncline; a useful historical geology field site [ABS]: Geological Society of America Abstracts with Programs, vol. 21, no. 2, p. 28.

Kuhlman, Robert, 1990, The Jacksonwald Syncline; a Mesozoic Ridge and Valley fold: Northeastern Geology, vol. 12, no. 1-2, p. 69-72.

Kulp, R.R., and Hughes, R.O., III, 1999, The existence of olivine in several exposures of Pennsylvania York Haven type diabase [ABS]: Geological Society of America Abstracts with Programs, vol. 31, no. 7, p. 165.

Kulp, R.R., and Hughes, R.O., III, 2001, Xenoliths of the Triassic Passaic Formation in the Monocacy Hill diabase intrusion, Amity Township, Berks County, Pennsylvania [ABS]: Geological Society of America, Abstracts with Programs, vol. 33, no. 6, p. 375.

Kurjack, D.C., 1954, Hopewell Village National Historic Site Pennsylvania: Washington, D.C., National Park Service Historical Handbook Series No. 8.

Kurtz, P.S., 1999, The story behind two Bieler photographs: Mogantown-Caernarvon Historical Society, The Local Historian, vol. 28, no. 2, p. 1-2.

Kunz, G.F., 1885, Precious stones: U.S. Geological Survey Mineral Resources of the United States 1883 and 1884, p. 723-771.

Lapham, D.M., and Geyer, A.R., 1959, Mineral collecting in Pennsylvania, 1st ed: Pennsylvania Geological Survey, 4th ser., Bulletin G33, 74 p.

Lapham, D.M., and Geyer, A.R., 1965, Mineral collecting in Pennsylvania, 2nd ed.: Pennsylvania Geological Survey, 4th ser., Bulletin G33, 148 p.

Lapham, D.M., and Geyer, A.R., 1969, Mineral collecting in Pennsylvania 3rd ed.: Pennsylvania Geological Survey, 4th ser., Bulletin G33, 164 p.

Lash, G.G., 1975, The structure and stratigraphy of the Pen Argyl Member of the Martinsburg Formation in Lehigh and Berks counties, Pennsylvania: U.S. Geological Survey Open-File Report 78-391, 225 p.

Lash, G.G., 1978, The structure and stratigraphy of the Pen Argyl Member of the Martinsburg Formation in Lehigh and Berks counties, Pennsylvania: U.S. Geological Survey Open-File Report 78-391, 225 p.

Lash, G.G., 1980, The Eckville Fault; a major Alleghenian upthrust in the Central Appalachian Great Valley [ABS]: Pennsylvania Academy of Science Program and abstracts, vol. 38, no. 2, p. 12.

BIBLIOGRAPHY

Lash, G.G., 1984, Density-modified grain-flow deposits from an early Paleozoic passive margin: Journal of Sedimentary Petrology, vol. 54, no. 2, p. 557-562.

Lash, G.G., 1984, The Richmond and Greenwich slices of the Hamburg Klippe in eastern Pennsylvania: U.S. Geological Survey Professional Paper 1312, 40 p.

Lash, G.G., 1985, Accretion-related deformation of an ancient (early Paleozoic) trench-fill deposit, central Appalachian Orogen: Geological Society of America Bulletin, vol. 96, no. 9, p. 1167-1178.

Lash, G.G., 1985, Geologic map of the Kutztown quadrangle, Berks and Lehigh Counties, Pennsylvania: U.S. Geological Survey Geologic Quadrangle Map GQ-1577, 1 pl.

Lash, G.G., 1986, Anatomy of an early Paleozoic subduction complex in the Central Appalachian Orogen: Sedimentary Geology, vol. 51, no. 1-2, p. 75-95.

Lash, G.G., 1986, Sedimentology of channelized turbidite deposits in an ancient (early Paleozoic) subduction complex, Central Appalachians: Geological Society of America Bulletin, vol. 97, no. 6, p. 703-710.

Lash, G.G., 1986, Structural and sedimentologic characteristics of the Greenwich slice of the Hamburg Klippe; an ancient subduction complex in eastern Pennsylvania: Geological Society of America Centennial field guide.

Lash, G.G., 1987a, Geologic map of the Hamburg quadrangle, Schuylkill and Berks Counties, Pennsylvania: U.S. Geological Survey Geologic Quadrangle 1637.

Lash, G.G., 1987b, Diverse melanges of an ancient subduction complex: Geology, vol. 15, no. 7, p. 652-655.

Lash, G.G., 1989, Documentation and significance of progressive microfabric changes in Middle Ordovician trench mudstones: Geological Society of America Bulletin, vol. 101, no. 10, p. 1268-1279.

Leidy, T.W., and Shenton, D.R., 1958, Titan in Berks, Edison's experiments in iron concentration: Historical Society of Berks County, vol. 23, no. 4, p. 104-110.

Leighton, Henry, 1941, Clay and shale resources in Pennsylvania: Pennsylvania Geological Survey, 4th ser., Bulletin M23, 245 p.

Lesley, J.P., 1864, Professor J.P. Lesleys description of Jones Warwick and other iron ore locations on the "Warwick Reserve" in the counties of Berks and Chester July 23rd 1864: unpublished manuscript on file at Hopewell Furnace National Historical Site.

Lesley, J.P., 1892, A summary description of the Geology of Pennsylvania, volume 1. Pennsylvania Geological Survey, 2nd ser., 719 p.

Lesley, J.P., Hall, C.E., Prime, Frederick, Jr., and Chance, H.M., 1883, The geology of Lehigh and Northampton counties: Pennsylvania Geological Survey, 2nd ser., Report D3, 283 p.

Lesley, J.P., and others, 1883, The geology of Chester County: Pennsylvania Geological Survey, 2nd ser., Report C4, 394 p.

Lininger, J.L, 1986, A trip through the Reading Prong: Friends of Mineralogy Pennsylvania Chapter field guide, unpaginated.

Long, L.E, and Kulp, J.L., 1962, Isotopic age study of the metamorphic history of the Manhattan and Reading Prongs: Geological Society of America Bulletin, vol. 73, no. 8, p. 969-995.

Longwill, S.M., and Wood, C.R., 1965, Ground-water resources of the Brunswick Formation in Montgomery and Berks Counties, Pennsylvania: Pennsylvania Geological Survey, 4th ser., Ground Water Report W22, 59 p.

Love, G.E.W., and Kauffman, M.E., 1969, Structural interpretation of the geology near Morgantown, southeastern Pennsylvania: Proceedings of the Pennsylvania Academy of Science, vol. 43, p. 177-179.

Lynch, M.N., 1995, The Pagoda: Historical Review of Berks County, vol. 60, no. 2, p. 64-65.

Lyttle, P.T., 1979, Tectonic history of the Shochary Ridge region, eastern Pennsylvania [ABS]: Geological Society of America, Abstracts with Programs, vol. 11, no. 1, p. 22, 43.

Lyttle, P.T., and Drake, A.A., Jr., 1979, Tectonic history of Shochary Ridge in Pennsylvania: U.S. Geological Survey Professional Paper 1150, p. 61-62.

Lyttle, P.T., Drake, A .A., Jr., Wright, T.O., and Stephens, G.C., 1979, Regional implications of the stratigraphy and structure of Shochary Ridge, Berks and Lehigh counties, Pennsylvania; discussion and reply: American Journal of Science, vol. 279, no. 6, p. 721-732.

MacLachlan, D.B., 1979, Geology and mineral resources of the Temple and Fleetwood quadrangles, Berks County, Pennsylvania: Pennsylvania Geological Survey, 4th ser., Atlas 187ab, 71 p.

MacLachlan, D.B., 1992, Geology and mineral resources of the Reading and Birdsboro quadrangles, Berks County, Pennsylvania: Pennsylvania Geological Survey, 4th ser., Atlas A187cd.

MacLachlan, D.B., Buckwalter, T.V., Jr., and McLaughlin, D.B., 1975, Geology and mineral resources of the Sinking Spring Quadrangle, Berks and Lancaster counties, Pennsylvania: Pennsylvania Geological Survey, 4th ser., Atlas 177d, 228 p.

Martin Stone Quarries, Inc., 1984, An outstanding mineral-producer: Pennsylvania Geology, vol. 15, no. 5, p. 2-4.

Maslowski, Andy, 1987, Gold in Pennsylvania: Rock and Gem, vol. 17, no. 8, p. 8-9.

Mast, C.Z., and Simpson, R.E., 1942, Annals of the Conestoga Valley in Lancaster, Berks, and Chester Counties Pennsylvania: self-published, 689 p.

McCauley, J.F., 1961, Uranium in Pennsylvania: Pennsylvania Topographic and Geological Survey, 4th ser., Mineral Resource Report 43, 71 p.

McCreath, A.S., 1875, Report of progress in the laboratory of the survey at Harrisburg: Pennsylvania: Pennsylvania Topographic and Geological Survey, 2nd ser., Report M, 105 p.

McCreath, A.S., 1879, Second report of progress in the laboratory of the survey at Harrisburg: Pennsylvania: Pennsylvania Topographic and Geological Survey, 2nd ser., Report MM, 438 p.

MacLachlan, D.B., 1979, Geology and mineral resources of the Temple and Fleetwood quadrangles, Berks County, Pennsylvania: Pennsylvania Geological Survey, 4th ser., Atlas 187ab, 71 p.

MacLachlan, D.B., 1983; revised, 1992: Geology and mineral resources of the Reading and Birdsboro quadrangles, Berks County, Pennsylvania Pennsylvania Geological Survey, 4th ser., Atlas 187cd, 1 pl.

MacLachlan, D.B., Buckwalter, T.V., and McLaughlin, D.B., 1975, Geology and mineral resources of the Sinking Spring quadrangle, Berks and Lancaster Counties, Pennsylvania: Pennsylvania Geological Survey, 4th ser., Atlas 177d, 228 p.

Meiser, G.M., IX, and Meiser, G.J., 1982, The passing scene, vol. 1: Historical Society of Berks County, 273 p.

Metz, Robert, 1998, Nematode trails from the Late Triassic of Pennsylvania: Ichnos, vol. 5, no. 4, p. 303-308.

Metz, Robert, 1999, Nonmarine trace fossils from the Upper Triassic Passaic Formation, Douglassville, Pennsylvania [ABS]: Geological Society of America, Abstracts with Programs, vol. 31, no. 2, p. 57.

Metz, Robert, 1999, Scratch circles; a new specimen from a lake-margin deposit of the Passaic Formation (Upper Triassic), Douglassville, Pennsylvania: Northeastern Geology and Environmental Sciences, vol. 21, no. 3, p. 179-180.

Miller, B.L., 1911, The mineral pigments of Pennsylvania: Pennsylvania Geological Survey, 3rd ser., Report 4, 101 p.

Miller, B.L., 1912a, Graphite deposits of Pennsylvania: Pennsylvania Geological Survey, 3rd ser., Report 6, 147 p.

Miller, B.L., 1912b, The geology of the graphite deposits of Pennsylvania: Economic Geology, vol. 7, no. 8, p. 762-777.

Miller, B.L., 1926, An unusual case of limestone decomposition: Proceedings of the Pennsylvania Academy of Science, vol.1, p. 77-78.

Miller, B.L., 1925, Limestones of Pennsylvania: Pennsylvania Geological Survey, 4th ser., Bulletin M7, 368 p.

Miller, B.L., 1926, The origin and utilization of the Cambro-Ordovician limestones of Pennsylvania: Proceedings of the Pennsylvania Academy of Science, vol. 1, p. 89-99.

Miller, B.L., 1932, The Lehigh Portland cement district, Pennsylvania in Berkey, C.P., and others, Mineral deposits of New Jersey and eastern Pennsylvania: Guidebook, 16th International Geological Congress United States 1933, Guidebook 8, Excursion A-8, p. 30-3es: U.S. Geological Survey Bulletin 370, p. 145-163.

Miller, B.L., 1933, Mineral pigments of Pennsylvania in Wilson, Hewett, Iron oxide mineral pigments of the United States: U.S. Bureau of Mines Bulletin 370, p.

Miller, B.L., 1934, Limestones of Pennsylvania: Pennsylvania Geological Survey, 4th ser., Bulletin M20, 729 p.

Miller, B.L., 1939, Economic Geology in Miller, B.L., Fraser, D.M., and Miller, R.L., 1939, Northampton County, Pennsylvania; Pennsylvania Geological Survey, 4th ser., Bulletin C48, p. 311-408.

Miller, B.L., 1944, Specific data on the so-called "Reading overthrust": Geological Society of America Bulletin, vol. 55, no. 12, p. 211-254.

BIBLIOGRAPHY

Mills, J.W., and Eyrich, H.T., 1966, The role of unconformities in the localization of epigenetic mineral deposits in the United States and Canada: Geology and the Bulletin of the Society of Economic Geologists, vol. 61, no. 7, p. 1232-1257.

Mine and Quarry News Bureau, 1897, The mine, quarry and mineralogical record of the United States, Canada and Mexico: Chicago, 702 p.

Montgomery, Arthur, 1969, Mineralogy of Pennsylvania, 1922-1965: Academy of Natural Science of Philadelphia Special Publication 9, 104 p.

Montgomery, Arthur, 1970, Pennsylvania Minerals (95): Mineralogical Society of Pennsylvania Keystone Newsletter vol. 19, no. 11, p. 9-10.

Montgomery, Arthur, 1971, Pennsylvania Minerals (108): Mineralogical Society of Pennsylvania Keystone Newsletter vol. 20, no. 12, p. 10.

Montgomery, Arthur, 1972a, Pennsylvania Minerals (109): Mineralogical Society of Pennsylvania Keystone Newsletter vol. 21, no. 1, p. 9-10.

Montgomery, Arthur, 1972b, Pennsylvania Minerals (110): Mineralogical Society of Pennsylvania Keystone Newsletter vol. 21, no. 2, p. 7-8.

Montgomery, Arthur, 1972c, Pennsylvania Minerals (111): Mineralogical Society of Pennsylvania Keystone Newsletter vol. 21 no. 3, p. 9-10.

Montgomery, Arthur, 1972d, Pennsylvania Minerals (112): Mineralogical Society of Pennsylvania Keystone Newsletter vol. 21 no. 4, p. 7-8.

Montgomery, Arthur, 1974, Pennsylvania Minerals (142): Mineralogical Society of Pennsylvania Keystone Newsletter, vol .23, no. 10, p. 4-5.

Montgomery, M.L., 1884, Early furnaces and forges of Berks County, Pennsylvania: Pennsylvania Magazine of History and Biography, vol. 8, no. 1, p. 56-81.

Montgomery, M.L., 1886, History of Berks County in Pennsylvania: Philadelphia, Everts, Peck and Richard, 1418 p.

Montgomery, M.L., 1909, Historical and biographical annals of Berks County, Pennsylvania: Chicago, J.H. Beers and Company, two volumes.

Moseley, J.R., 1950, The Ordovician-Silurian contact near Kempton, Pennsylvania: Proceedings of the Pennsylvania Academy of Science, vol. 24, p. 176-187.

Moss, J.H., 1976, Periglacial origin of extensive lobate colluvial deposits on the south flank of Blue Mountain near Shartlesville and Strausstown, Berks County, Pennsylvania: Proceedings of the Pennsylvania Academy of Science, vol. 50, no. 1, p. 42-44.

Mostardi, Michael, and Durant, Joe, 1991, Caves of Berks County Pennsylvania: Mid-Appalachian Region of the National Speleological Society Bulletin 18, 84. p.

Mt. Penn Association of Sciences, 1898, Mt. Penn Association of Sciences: Mineral Collector, vol. 5, no. 6, p. 82.

Mt. Penn Association of Sciences, 1899, Mt. Penn Association of Sciences: Mineral Collector, 1899, vol. 6, no. 8, p. 139.

Mt. Penn Association of Sciences, 1899, Mt. Penn Association of Sciences: Mineral Collector, 1899, vol. 5, no. 11, p. 177.

Mt. Penn Association of Sciences, 1900, Mt. Penn Association of Sciences: Mineral Collector, 1900, vol. 7, no. 9, p. 145

Mt. Penn Association of Sciences, 1902, Mt. Penn Association of Sciences: Mineral Collector, 1902, vol. 9, no. 8, p. 122-123.

Mt. Penn Association of Sciences, 1903, Mt. Penn Association of Sciences: Mineral Collector, 1903, vol. 10, no. 7, p. 112.

Myers, P.B, Jr., 1969, Development of the Hamburg klippe in the Bernville-Strausstown area, Pennsylvania [ABS]: Geological Society of America Abstracts with Programs. Part 4, p. 55-56.

Myers, R.E., 1940, The Hardyston jasper of the Reading Hills in Pennsylvania: Rocks and Minerals, vol. 15, no. 8, p. 219-225.

Natural History Museum, 2014, Mengle, accessed December 31, 2014, at *http://www.nhm.ac.uk/research-curation/library/archives/catalogue/dserve.exe?dsqServer=placid&dsqIni= Dserve.ini&dsqApp=Archive&dsqDb=Persons&dsqSearch=Code=='PX7982'&dsqCmd=Show.tcl*.

Newhouse, W.H., 1933, Mineral zoning in the New Jersey- Pennsylvania-Virginia Triassic area: Economic Geology, vol. 28, p. 613-633.

Noecker, T.C., Noecker, M.L., Smith, R.E., and Smith, I.A., 1965, Two centuries of progress: Shoemakersville, Pennsylvania bicentennial observance: 1765-1965: Shoemakersville Press for Bicentennial Committee, unpaginated.

Olsen, P.E., and Schlische, R.W., 1989, Newark Basin, Pennsylvania and New Jersey; Stop 5.2; Douglasville Quarry of Pottstown Traprock Company [modified]: in Olsen, P.E., Schlische, R.W., and Gore, Pamela J.W., eds., Sedimentation and basin analysis in siliciclastic rock sequences; Volume 2, Tectonic, depositional, and paleoecological history of early Mesozoic rift basins, eastern North America: American Geophysical Union, Field trips for the 28th International Geological Congress, p. 78-79.

Olsen, P.E., 1989, Newark Basin, Pennsylvania and New Jersey; Stop 5.3; Douglasville footprint quarries [modified]: in Olsen, P.E., Schlische, R.W., and Gore, Pamela J.W., eds., Sedimentation and basin analysis in siliciclastic rock sequences; Volume 2, Tectonic, depositional, and paleoecological history of early Mesozoic rift basins, eastern North America: American Geophysical Union, Field trips for the 28th International Geological Congress, p. 79-80.

Olsen, P.E., and Schlische, R.W., 1989b, Newark Basin, Pennsylvania and New Jersey; Stop 5.4; Border fault at Boyertown, PA [modified]: in Olsen, P.E., Schlische, R.W., and Gore, Pamela J.W., eds., Sedimentation and basin analysis in siliciclastic rock sequences; Volume 2, Tectonic, depositional, and paleoecological history of early Mesozoic rift basins, eastern North America: american Geophysical Union, Field trips for the 28th International Geological Congress, p. 80-81.

Omlor, Jean, Miller, P.L., Kutz, W.C., Rudolph, Everett and Weitzel, Stella,1976, History of the Township of Spring - 125th Anniversary: 100 p.

O'Neill, B.J., Jr., 1964, Directory of the mineral industry in Pennsylvania: PAGS, 4th ser., Information Circular 54.

O'Neill, B.J., Jr., 1964, Limestones and dolomites of Pennsylvania in Atlas of Pennsylvania's mineral resources-part 1: Pennsylvania Geological Survey, 4th ser., Bulletin M50, 40 p.

O'Neill, B.J., Jr., 1965, Directory of the mineral industry in Pennsylvania: Pennsylvania Geological Survey, 4th ser., Information Circular 54.

O'Neill, B.J., Jr., 1977, Directory of the mineral industry in Pennsylvania, 3rd ed.: Pennsylvania Geological Survey, 4th ser., Information Circular 54.

O'Neill, B.J., Jr., 1976, Our mineral heritage; Hopewell Village: Pennsylvania Geology, vol. 7, no. 3, p. 2-5.

O'Neill, B.J., Jr., and others, 1965, Properties and uses of Pennsylvania shales and clays: Pennsylvania Geological Survey, 4th ser., Bulletin M51, 448 p.

Orner, R.J., and Friehauf, K.C., 2000, Geology of the Rittenhouse Gap iron district, Berks County, Pennsylvania [ABS]: Geological Society of America, Abstracts with Programs, vol. 32, no. 7, p. 83.

Orth, R.L.T., 2009, Colonial iron masters seek their fortune in our area: Berks-Mont News, accessed January 2, 2016, at *http://www.berksmontnews.com/article/BM/20090902/OPINION03/309029965*.

Overly, Benjamin, 1973, Edward and George Brooke and the Birdsboro iron industry: Historical Review of Berks County, vol. 38, no. 2, p. 48-51; 78-80.

Pennsyvania Department of Environmental Protection, 2010, Pennsylvania active mining permits: acesssed March 9, 2014, at https://www.dep.state.pa.us/dep/deputate/minres/bmr/2010%20Active%20Industrial%20Minerals%20O,perators%2012-1-2010.pdf.

Pennsylvania Department of Labor and Industry, 1914, First industrial directory of Pennsylvania 1913: Harrisburg, Pa., 778 p.

Pennsylvania Department of Labor and Industry, 1916, Second industrial directory of Pennsylvania 1916: Harrisburg, Pa., 1795 p.

Pennsylvania Department of Labor and Industry, 1920, Third industrial directory of Pennsylvania 1919: Harrisburg, Pa., 1212 p.

Pennsylvania Department of Labor and Industry, 1922, Fourth industrial directory of the Commonwealth of Pennsylvania: Harrisburg, Pa., 1409 p.

Pennsylvania Geological Survey, 1883, Atlas to D3: Pennsylvania Topographic and Geologic Survey, 2nd ser., 22 sheets.

Pennsylvania Geological Survey, 1885, Pennsylvania grand atlas division IV south and great valley topographical maps part I: Pennsylvania Topographic and Geologic Survey, 2nd ser., 43 sheets.

Pennsylvania Geological Survey, 1978, Uranium near Oley, Berks County: Pennsylvania Geology, vol. 9, no. 4, p. 29-31.

BIBLIOGRAPHY

Phillips, E.H., 1877, Levans review directory of Reading and Berks County for 1877-78: Reading, Pa., Levan and Company, 277 p.

Pierce, Patricia, 1950, Iron and copper mining in Caernarnon Township: Historical Review of Berks County, vol. 16, no. 1. p. 12.

Pierotti, Gregory, Mathur, Ryan, Smith, R.C. II, and Barra, Fernando, undated, Ages for the Antietam Reservoir, Eastern Pennsylvania, a story of open system behavior Re-Os isotopes in molybdenite, accessed June 17, 2011 at *http://jcsites.juniata.edu/faculty/mathur/molyposter2.pdf*.

Pit and Quarry, 1932, Handbook and directory of cement, gypsum, lime, sand, gravel, and crushed-stone plants, 24th ed.: Chicago, Complete Service Publishing Company.

Pit and Quarry, 1937, Handbook and directory of cement, gypsum, lime, sand, gravel, and crushed-stone plants, 29th ed.: Chicago, Complete Service Publishing Company.

Pit and Quarry, 1952, Handbook and directory of the nonmetallic minerals industry, 45th ed.: Chicago, Complete Service Publishing Company.

Plank, D.H., 1904, Historical sketch of Morgantown: Transactions of the Historical Society of Berks County, vol. 1, unpaginated.

Platt, Franklin, 1875, Report on the Warwick and Hopewell iron mines in Chester County and the Jones mine in Berks County: unpublished manuscript on file at the Hagley Museum, Wilmington, Del., 21 p.

Platt, L.B.; Loring, R.B.; Stephens, G.C., 1969, Taconic events in the Hamburg 15' Quadrangle, Pennsylvania [ABS]: Geological Society of America Abstracts with Programs. Part 1, p. 48-49.

Poor's Manual Company, 1916, Poor's Manual of Industrials, 7th ed.: New York, 3106 p.

Popovich, D.E., 1965, Distribution of certain elements in the major rock units at the Cornwall and Morgantown mines: Master's Thesis, Pennsylvania State University, University Park, Pa., 73 p.

Potter, Noel, Jr., and Moss, J.H., 1968, Origin of the Blue Rocks block field and adjacent deposits, Berks County, Pennsylvania: Geological Society of America Bulletin, vol. 79, no. 2, p. 255-262.

Prime, Frederick, Jr., 1875, Report of progress on the brown hematite ore ranges of Lehigh County: Pennsylvania Geological Survey, 2nd Ser., Report D, 73 p.

Prime, Frederick, Jr., 1878, The brown hematie deposits of the siluro-cambrian limestones of Lehigh County with a description of the mines lying between Emaus, Alburtis, and Fogelsville: Pennsylvania Geological Survey, 2nd Ser., Report D2, 99 p.

Prime, Frederick, Jr., and others, 1883, The geology of Lehigh and Northampton Counties: Pennsylvania Geological Survey, 2nd ser., Report D3, vol. 1, 283 p.

Prouty, C.E., 1959, The Annville, Myerstown, and Hershey Formations of Pennsylvania: Pennsylvania Geological Survey, 4th ser., Bulletin G31, 47 p.

Putnam, B.T., 1886, Notes on the iron ores of Pennsylvania, *in* Pumpelly, Raphael, Report of the mining industries of the United States: Tenth Census, Department of the Interior Census Office, p. 179-221.

Pumpelly, Raphael, 1886, Report of the mining industries of the United States: Tenth Census, Department of the Interior Census Office, variously paginated.

Rand, T.D., 1898, The Birdsboro trap quarries: Proceedings of the Academy of Natural Sciences of Philadelphia, vol. 50, p. 10.

Raring, A.M., and Ganis, Robert, 1973, An occurrence of conodont-bearing allochthonous carbonates in the Martinsburg Formation, Berks County, Pennsylvania [ABS]: Geological Society of America Abstracts with Programs, vol. 5, no. 2, p. 210.

Reading Public Museum, 2014, History, accessed April 4, 2014, at *http://www.readingpublicmuseum.org/about_us/history.php*.

Reed, J.C., 1976, Annotated bibliography of minerals new to the Pennsylvania list 1965-1974: Mineralogical Society of Pennsylvania, 83 p.

Repetski, J.E., 1979, Paleontological studies of the Ordovician in Pennsylvania [ABS]: Geological Society of America Abstracts with Programs, vol. 11, no. 1, p. 50.

Repetski, J.E., 1984, Conodonts from Spitzenberg: Field Conference of Pennsylvania Geologists, Guidebook for the Annual Field Conference of Pennsylvania Geologists, vol. 49, p. 94-101.

Robarts, J.O.K., 1904, The Raudenbush iron ore mine: Transactions of the Historical Society of Berks County, vol. 1, p. 2-5.

Roberts, Elwood, ed., 1904, Biographical annals of Montgomery County, Pennsylvania: T. S. Benham and Company and the Lewis Publishing Company, 1556 p.

Robinson, G.R., Jr., 1985, Magnetite skarn deposits of the Cornwall (Pennsylvania) type; a potential cobalt, gold, and silver resource: U.S. Geological Survey Circular 946, p. 126-128.

Robinson, G.R., Jr., 1988, Base and precious metals associated with diabase in the Newark, Gettysburg, and Culpeper Basins of the eastern United States-a review: U.S. Geological Survey Bulletin 1776, p. 303-320.

Robinson, G.R., Jr., 1998, Base- and precious-metal mineralization associated with igneous and thermally altered rocks in the Newark, Gettysburg, and Culpeper early Mesozoic basins of New Jersey, Pennsylvania, and Virginia in Manspeizer, Warren, ed., Triassic-Jurassic rifting continental breakup and the origin of the Atlantic Ocean and passive margins: Developments in Geotectonics, vol. 22, part B, p. 621-648.

Robinson, G.R., Jr., and Sears, C.M., 1988, Inventory of metal mines and occurrences associated with the early Mesozoic basins of the eastern United States-summary tables: U.S. Geological Survey Bulletin 1776, p. 265-303.

Robinson, Ryan, 2007, Titanic town, accessed November 3, 2015, at *http://lancasteronline.com/news/titanic-town/article_a3936160-5fd1-5711-8f07-5840b5d612ef.html*.

Robinson, S.A., 1825, A catalog of American minerals with their localities: Boston, Cummings, Hillard, and Co., 316 p.

Rogers, H.D., 1839, Third annual report of the geological survey of the state of Pennsylvania: Harrisburg, Pa., 119 p.

Rogers, H.D., 1840, Fourth annual report of the geological survey of the state of Pennsylvania: Harrisburg, Pa., 252 p.

Rogers, H.D., 1841, Fifth annual report on the geological exploration of the Commonwealth of Pennsylvania: Harrisburg, Pa., 179 p.

Rogers, H.D., 1858, The geology of Pennsylvania, a government survey: Philadelphia, vol. 1, 586 p.; vol. 2, 1046 p.

Rolling Rock Building Stone, Inc., 2010, About us, accessed April 30, 2010, at *http://www.rollrock.com/about-us/*.

Rose, A.W., 1970, Atlas of Pennsylvania's mineral resources-part 3, metal mines and occurrences in Pennsylvania: Pennsylvania Geological Survey, 4th ser., Mineral Resources Report 50, 14 p.

Rose, A.W., Herrick, D.C., and Deines, Peter, 1985, An oxygen and sulfer isotope study of skarn-type magnetite deposits of the Cornwall type, southeastern Pennsylvania: Economic Geology and the Bulletin of the Society of Economic Geologists, vol. 80, p. 418-443.

Rothwell, R.P., ed., 1899: The mineral industry, its statistics, technology and trade in the united states and other countries to the end of 1899, vol. 8, 980 p.

Rukavina, N.A., 1968, Grain orientation vs. texture and internal structure in Martinsburg turbidites near Hamburg, Pennsylvania: Geological Society of America Special Paper 101, p. 185.

Rupp, I.D., 1844, History of the Counties of Berks and Lebanon: Lancaster, Pa., G. Hills, 519 p.

Russell, G.C., Jr., 1941, Hydrothermal alteration of a dike at Seisholtzville, Pennsylvania: Proceedings of the Pennsylvania Academy of Science vol. 15, p. 89-93.

Sadtler, Benjamin, Jr., 1883, Minerals from Fritz Island, Pa.: American Chemical Journal, vol. 4, p. 356-357.

Sanders, R.H., 1883, The slate region in Prime, Frederick, Jr., Geology of Lehigh and Northampton Counties: Pennsylvania Geological Survey, 2nd ser., Report D3, part 1, p. 83-148.

Savoy, Lauret, Harris, A.G., and Repetski, J.E., 1981, Paleogeographic implications of the Lower/Middle Ordovician boundary, northern Great Valley, eastern Pennsylvania to southeastern New York [ABS]: Geological Society of America Abstracts with Programs, vol. 13, no. 3, p. 174.

Schaffer, W.I., and Weimer, A.B., 1916, Campbell v. Manatawny Bessemer Ore Co. appellant: Pennsylvania Superior Court Reporter, vol. 62, p. 57-61.

Scharnberger, C.K., 1993, Relationship of seismicity to geologic structure in Berks County, Pennsylvania [ABS]: Seismological Research Letters, vol. 64, no. 3-4, p. 263.

Schlische, R.W., and Withjack, M.O., 2005, The early Mesozoic Birdsboro central Atlantic margin basin in the mid-Atlantic region, eastern United States: Discussion: Geological Society of America Bulletin, vol. 116, p. 823–832.

BIBLIOGRAPHY

Schlische, R.W., and Olsen, P.E., 1988, Structural evolution of the Newark basin *in* Husch, J.M., and Hozic, M.J., eds., Geology of the Newark Basin: Field guide and proceedings of the fifth annual meeting of the Geological Association of New Jersey, Rider College, Lawrenceville, New Jersey, p. 43-65.

Schoenfeld, J., and Smith, E.F., 1884, Brucite: American Chemical Journal, vol. 5, p. 279.

Schoepf, J.D., 1788, Travels in the Confederation 1783-1784: Morrison, A.J., translator and editor, 1911, Baltimore, Lord Baltimore Press, 426 p.

Schultz, G.W., 1936, Major General Daniel Udree: Historical Review of Berks County, vol. 1, no. 3, p. 66-70.

Schuylkill Copper Mining Company, Reading, Pennsylvania, 1864, [Prospectus]: New York, Clayton and Medole, 8 p.

Senior, L.A., and Sloto, R.A., 2006, Arsenic, boron, and fluoride concentrations in ground water in and near diabase intrusions, Newark Basin, southeastern Pennsylvania: U.S. Geological Survey Scientific Investigations Report 2006-5261, 105 p.

Sevon, W.D., 1975, Sandstone saprolite, roundstone diamicton and the Harrisburg peneplain in eastern Pennsylvania [ABS]: Geological Society of America Abstracts with Programs, vol. 7, no. 1, p. 118.

Seybert, Adam, 1808, A catalogue of some American minerals which are found in different parts of the United States: Philadelphia Medical Museum, v. 5, p. 152-159, 256-268.

Shaler, N.S., 1883, Descriptions of quarries and quarry regions *in* Report on the building stone of the United States and statistics of the quarry industry for 1880: Washington, D.C., Department of the Interior Census Office, 410 p.

Shaw, J.B., 1930, The Ceramic Industries of Pennsylvania: Pennsylvania State College Bulletin 7, 206 p.

Silliman, Benjamin, Jr., 1886, Description of a double-muffle furnace designed for the reduction of hydrous silicates containing copper, etc., like the so-called "clay ore" of the Jones mine in Pennsylvania: Engineering and Mining Journal, vol. 22, p. 264-266.

Silvestri, S.M.M., Schlische, R.W., and Olsen, P.E., 1991, Analysis of deformed tetrapod footprints from the Jacksonwald Syncline of the Newark Basin; implications for the Triassic-Jurassic extinctions [ABS]: Geological Society of America Abstracts with Programs, vol. 23, no. 1, p. 127.

Simpson, D.R., 1969, Hornfels associated with Triassic diabase [ABS]: Geological Society of America Abstracts with Programs, vol. 20, part 1, p. 55-56.

Simpson, E.L., Dilliard, K.A., Rowell, B.F., and Higgins, D., 2002, The fluvial-to-marine transition within the post-rift Lower Cambrian Hardyston Formation, eastern Pennsylvania, USA: Geology, vol. 147, no. 1-2, p. 127-142.

Sims, S.J., 1968, The Grace Mine magnetite deposit, Berks County, Pennsylvania *in* Ridge, J.D., ed., Ore deposits of the United States, 1933-1967, vol. 1: New York, The American Institute of Mining, Metallurgical, and Petroleum Engineers, p. 108-124.

Stephens, G.C., 1969, Stratigraphy and structure of a portion of the basal Silurian clastics of eastern Pennsylvania: M.S. thesis, George Washington University, 50 p.

Sloto, R.A., 2009, Mines and Minerals of Chester County, Pennsylvania: privately published, 469 p.

Sloto, R.A., 2011, Trace metals related to historical iron smelting at Hopewell Furnace National Historic Site, Berks and Chester Counties, Pennsylvania: U.S. Geological Survey Fact Sheet 2011-3101, 2 p.

Sloto, R.A., and Helmke, M.F., 2011, An innovative method for nondestructive analysis of cast iron artifacts at Hopewell Furnace National Historic Site, Pennsylvania: Park Science, vol. 27, no. 3. p. 50-53.

Sloto, R.A., and Reif, A.G., 2011, Distribution of trace metals at Hopewell Furnace National Historic Site, Berks and Chester Counties, Pennsylvania: U.S. Geological Survey Scientific Investigations Report 2011-5014, 38 p.

Smeltzer, B.L., 1960, Shouts Bridge Cave, Berks County: Nittany Grotto News, vol. 9, no. 2, p. 29-30.

Smeltzer, B.L., 1964, Host Cave (Berks County): York Grotto Newsletter, vol. 5, no. 8, p. 106-108.

Smeltzer, B.L., 1979, Schofer Cave: York Grotto Newsletter, vol. 16, no. 2, p. 26-29.

Smeltzer, B.L., 1985, Crystal Cave, Berks County, Pennsylvania: York Grotto Newsletter, vol. 20, no. 3, p. 45-51.

Smith, E.F., 1885, Mineralogical notes 5. stilbite: American Chemical Journal, vol. 6, p. 414.

Smith, E.F., 1910, Some Berks County minerals: Proceedings of the Academy of Natural Sciences of Philadelphia, v. 62, part 3, p. 538-540.

Smith, R.C., II, 1973, Geochemistry of Triassic diabase from southeastern Pennsylvania: PhD dissertation, Pennsylvania State University, 240 p.

Smith, R.C., II, 1974, Fetid barite from Berks County, Pennsylvania: Pennsylvania Geology, vol. 5, no. 6, p. 4-7.

Smith, R.C., II, 1975, New molybdenite occurrence in Berks County: Pennsylvania Geology, vol. 6, no. 6, p. 16.

Smith, R.C., II, 1977a, Zinc and lead occurrences in Pennsylvania: Pennsylvania Geological Survey, 4th ser., Mineral resources report 72, 318 p.

Smith, R.C., II, 1977b, Mystery prospects, Berks County: Pennsylvania Geology, vol. 8, no. 3, p. 15-16.

Smith, R.C., II, 1978, The mineralogy of Pennsylvania, 1966-1975: Friends of Mineralogy Pennsylvania Chapter Special Publication No. 1, 304 p.

Smith, R.C., II, 2003, Late Neoproterozoic felsite (602 + or - 2 Ma) and associated metadiabase dikes in the Reading Prong, Pennsylvania, and rifting of Laurenti: Northeastern Geology and Environmental Sciences, vol. 25, no. 3, p. 175-185.

Smith, R.C., II, Berkheiser, S.W., Jr., and Hoff, D.T., 1988, Locations and analyses of selected early Mesozoic copper occurrences in Pennsylvania: U.S. Geological Survey Bulletin 1776, p. 320-332.

Smith, Sanderson, 1855, On some new localities of minerals: American Association for the Advancement of Science Proceedings, vol. 9, p. 188-190.

Smitheringale, W.G., and Jensen, M.L., 1963, Sulfur isotopic composition of the Triassic igneous rocks of eastern United States: Geochimica et Cosmochimica Acta, vol. 27, p. 1183-1207.

Snyder, D.H., 2000, The hidden green diamond: privately published, 74 p.

Speece, J.H., 1979, Dragon Cave; America's oldest cave reference (Berks County, Pa.): Journal of Spelean History, vol. 13, no. 1-2, p. 4-5.

Spencer, A.C., 1907, Magnetite deposits of the Cornwall types in Berks and Lebanon counties, Pennsylvania: U.S. Geological Survey Bulletin 315-D, p. 185-189.

Spencer, A.C., 1908, Magnetite deposits of the Cornwall type in Pennsylvania: U.S. Geological Survey Bulletin 359, 102 p.

Srogi, LeeAnn, and Lynde, Nicole, 2004, Layering in diabase of the Jurassic Morgantown sheet; preliminary results and a working hypothesis [ABS]: Geological Society of America Abstracts with Programs, vol. 36, no. 2, p. 71.

Srogi, LeeAnn, Lutz, Tim, Dickson, L.D., Pollock, Meagan, Gimson, Kim, and Lynde, Nicole, 2010, Magmatic layering and intrusive plumbing in the Jurassic Morgantown sheet, central Atlantic magmatic province: Geological Society of America Field Guide 16, p. 51-67.

Stephens, G.C., and Wright, T.O., 1967, Spitzenberg Hill, Berks Co., Pa.; Triassic or Ordovician? [ABS]: Geological Society of America Abstracts with Programs, vol. 8, no. 2, p. 276-277.

Stephens, G.C., Wright, T.O., and Platt, L.B., 1982, Geology of the Middle Ordovician Martinsburg Formation and related rocks in Pennsylvania: Field Conference of Pennsylvania Geologists, Guidebook for the Annual Field Conference of Pennsylvania Geologists, vol. 47, 87 p.

Stitzer, Susan, 2013, History of Fleetwood, Pennsylvania, accessed January 28, 2013, at http://www.fleetwoodpa.org/fleetwood_history.htm.

Stoddard, J.C., and Callen, A.C., 1910, Ocher deposits of eastern Pennsylvania: U. S. Geological Survey Bulletin 430-G, p. 424-439.

Stone, R.W., 1932, Building stones of Pennsylvania: Pennsylvania Geological Survey, 4th ser., Bulletin M15, 316 p.

Stone, R.W., 1932, Pennsylvania Caves: Pennsylvania Geological Survey, 4th ser., Bulletin G3, 120 p.

Stone, R.W., 1939, The minerals of Pennsylvania non-metallic minerals: Pennsylvania Geological Survey, 4th ser., Bulletin M18-C, 49 p.

Stose, G.W., and Jonas, A.I., 1935, Highlands near Reading, Pennsylvania, an erosion remnant of a great overthrust sheet: Geological Society of America Bulletin, vol. 46, no. 5, p. 757-779.

Stuck, R.J., Vanderslice, J.E., and Hozik, M.J., 1988, Paleomagnetism of igneous rocks in the Jacksonwald Syncline, Pennsylvania [ABS]: Geological Society of America Abstracts with Programs, vol. 20, no. 1, p. 74.

BIBLIOGRAPHY

Szajna, M.J., and Hartline, B.W., 1995, New vertebrate footprint assemblages from the Rhaetian interval of the Passaic Formation in Berks County, Pennsylvania [ABS]: Geological Society of America Abstracts with Programs, vol. 27, no. 1, p. 86.

Szajna, M.J., and Hartline, B.W., 1996, A new vertebrate footprint locality from the Late Triassic Passaic Formation near Birdsboro, Pennsylvania in LeTourneau, P.M., and Olsen, P.E., eds., The great rift valleys of Pangea in eastern North America; Volume 2, Sedimentology, stratigraphy, and paleontology: New York, Columbia University Press, p. 264-272.

Szajna, M.J., and Hartline, B.W., 2000, An update of reptile footprint discoveries in the Late Triassic (Rhaetian) section of the Newark rift basin, Pennsylvania [ABS]: Geological Society of America Abstracts with Programs, vol. 32, no. 1, p. 77.

Szajna, M.J., and Hartline, B.W., 2003, A new vertebrate footprint locality from the late Triassic Passaic Formation near Birdsboro, Pennsylvania, in Letourneau, P.M., and Olsen, P.E., The great rift valleys of Pangea in eastern North America, vol. 2: Columbia University Press, New York, p. 264-272.

Tanzola, A.P., 1995, Structural and petrological analysis of the Precambrian gneiss at Chapel Hill, Reading Prong, eastern Pennsylvania [ABS]: Geological Society of America Abstracts with Programs, vol. 27, no. 1, p. 86-87.

Thomas, C.A., 1946, The Dyer quarry zeolites: Rocks and Minerals, part 1, vol. 21, no. 3, p. 142-143; part 2, vol. 21, no. 4, p. 205-206; part 3, vol. 21, no. 5, p. 278-279.

Thomas, C.A., 1948, Dyer's apophyllite: Rocks and Minerals, vol. 23, no. 11-12, p. 914-915.

Thomas, C.A., 1950, Mineral-logicals--from A to Z: Rocks and Minerals, vol. 25, no. 7-8, p. 356-361.

Thomas, C.A., 1961a, Pennsylvania fluorescents [part 5]: Mineralogical Society of Pennsylvania Keystone Newsletter, v. 10, no. 4, p. 7-8.

Thomas, C.A., 1961b, Pennsylvania fluorescents [part 6]: Mineralogical Society of Pennsylvania Keystone Newsletter, v. 10, no. 5, p. 7-8.

Thomson, R.D., 1957, The mineral industry of Pennsylvania in 1954: Pennsylvania Geological Survey, 4th ser., Information Circular 9, 34 p.

Thomson, R.D., 1958, The mineral industry of Pennsylvania in 1957: Pennsylvania Geological Survey, 4th ser., Information Circular 11, 40 p.

Tsusue, Akio, 1964, Mineral aspects of the Grace magnetite deposit, Pennsylvania: Pennsylvania Geological Survey, 4th ser., Bulletin M49, 10 p.

Wagner, A.E., Balthaser, F.W., and Hoch, D.K., 1913, The story of Berks County, Pennsylvania: Reading, Eagle Book and Job Press, 254 p.

Walker, G.W., Osterwald, F.W., and Adams, J.W., 1963, Geology of uranium-bearing veins in the conterminous United States: U. S. Geological Survey Professional Paper 455 A-F, 120 p.

Walker, J.E., 1966, Hopewell Village, the dynamics of a nineteenth century iron-making community: Philadelphia, University of Pennsylvania Press, 526 p.

Washington Township Anniversary Fest Committee, 2004, Worth remembering profiles of a community 2002: ???

Washington Township Historical Committee, 2014, Making a living, accessed December 30, 2014 at *http://greg.quuxuum.org/twp/work/work.html*.

Weiler, Warren, 1976, Stones useful or useless: Morgantown-Caernarvon Historical Society, The Local Historian, vol. 5, no. 1.

Wetherill, C.M., 1853, Chemical examination of two minerals from the neighborhood of Reading and on the occurrence of gold in Pennsylvania: American Philosophical Society Transactions, no. 10, p. 345-351.

Wetherill, C.M., 1854, Chemical notices [on the occurrence of gold near Reading]: Academy of Natural Science Proceedings, vol. 7, p. 233-234.

Wharton, H.M., 2002, Recollections of the discovery in 1949 of the Grace mine iron deposit at Morgantown, Berks Co., PA: Friends of Mineralogy, Pennsylvania Chapter Newsletter, vol. 30, no. 4, p. 3-8.

Wharton, H.M., 2003, The 1949 discovery of the Grace mine iron deposit: Pennsylvania Geology, vol. 33, no. 1, p. 2-9.

Wheatley, C.M., 1882, Some new Pennsylvania mineral localities: Proceedings of the Academy of Natural Science of Philadelphia, vol. 34, p. 36.

Wherry, E.T., 1908a, The Newark copper deposits of southeastern Pennsylvania: Economic Geology, vol. 3, p. 726-738.

Wherry, E.T., 1908b, Notes on copper mining in the American colonies: Mining and Metallurgical Section, Journal of the Franklin Institute, vol. 166, p. 309-314.

Wherry, E.T., 1910, Contributions to the mineralogy of the Newark Group in Pennsylvania: Transactions Wagner Free Institute of Science, vol. 7, p.7-27.

Wherry, E.T., 1913, North border relations of the Triassic in Pennsylvania: Proceedings of the Academy of Natural Science of Philadelphia, vol. 65, p. 114-125.

Wherry, E.T., 1916, Notes on the geology near Reading, Pennsylvania: Journal of the Washington Academy of Sciences, vol. 6, p. 23.

Wherry, E.T., 1923, The blue rocks in Greenwich Township: Berks County Historical Society Transactions, vol. 3, p. 204-208.

Wherry, E.T., and Shannon, E.V., 1922, Crocidolite from eastern Pennsylvania: Journal of the Washington Academy of Sciences, vol. 12, no. 10, p. 242-244.

Whitcomb, Lawrence, 1942, Spitzenberg conglomerate as a Triasssic outlier in Pennsylvania: Geological Society of America Bulletin vol. 53, p. 755-764.

Whitcomb, Lawrence, and Engel, J.A., 1934, The probable Triassic age of the Spitzenberg conglomerate, Berks County, Pennsylvania: Proceedings of the Pennsylvania Academy of Science, vol. 8, p. 37-43.

Wikipedia contributors, 2014, Catasauqua and Fogelsville Railroad: Wikipedia, the free encyclopedia, accessed on March 22, 2014.

Wilkens, Hans, 1955, Notes on the geology of the Reading Hills; the Hardyston quartzite in Berks County, Pennsylvania: Proceedings of the Pennsylvania Academy of Science, vol. 29, p. 193-194.

Wilkens, Hans, 1957, An erosion surface in the Reading Hills, Berks County, Pennsylvania: Proceedings of the Pennsylvania Academy of Science, vol. 31, p. 98.

Wilkens, Hans, 1957, Some terrace deposits in the Schuylkill Valley, Berks County, Pennsylvania: Proceedings of the Pennsylvania Academy of Science, vol. 31, p. 99.

Willard, Bradford, 1955, Cambrian contacts in eastern Pennsylvania: Geological Society of America Bulletin, vol. 66, no. 7, p. 819-833.

Willard, Bradford, McLaughlin, D.B., and Ryan, J.D., 1947, Triassic of the Delaware Valley: Geological Society of America Bulletin, vol. 58, no. 12, part 2, p. 1240-1241.

Willis, Bailey, 1886, Report on certain magnetites in eastern Pennsylvania in Pumpelly, Raphael, Report of the mining industries of the United States exclusive of the precious metals, with special investigations into the iron resources of the Republic and into the Cretaceous coals of the northwest: Washington, Bureau of the Census, Final report on the tenth census, 1880, vol. 15, p. 223-234.

Wilmer Atkinson Company, 1914, Farm and business directory of Berks County Pennsylvania: Philadelphia, 240 p.

Wilson, Hewett, 1933, Iron oxide mineral pigments of the United States: U.S. Bureau of Mines Bulletin 370, 198 p.

Witte, W.K., Kent, D.V., and Olsen, P.E., 1991, Magnetostratigraphy and paleomagnetic poles from Late Triassic-earliest Jurassic strata of the Newark Basin: Geological Society of America Bulletin, vol. 103, no. 12, p. 1648-1662.

Wittits, Thomas, and Dissinger, Charlotte, 1990, Leesport Furnace in Goda, B.R., ed., Sesquicentennial history of Leesport, 1840-1990: Leesport Borough, 140 p.

Wood, C.R., 1980, Groundwater resources of the Gettysburg and Hammer Creek Formations, southeastern Pennsylvania: Pennsylvania Geological Survey, 4th ser., Water Resources Report 49, 87 p.

Wood, C.R., and MacLachlan, D.B., 1978, Geology and groundwater resources of northern Berks County, Pennsylvania: Pennsylvania Geological Survey, 4th ser., Water Resources Report 44, 91 p.

Wood, G.H., and Kehn, T.M., 1968, Geologic map of the Pine Grove quadrangle Schuylkill, Lebanon, and Berks Counties, Pennsylvania: U.S. Geological Survey Map Map GQ-691, 1 pl.

Wood, G.H., and Kehn, T.M., 1968, Geologic map of the Swatara Hill quadrangle Schuylkill and Berks Counties Pennsylvania: U.S. Geological Survey Map GQ-689, 1 pl.

BIBLIOGRAPHY

Wright, T.O., and Caldemeyer, R.D., 1976, Invertebrate fossils from Shochary Ridge, Berks and Lehigh counties, Pennsylvania: Proceedings of the Pennsylvania Academy of Science, vol. 51, no. 2, p. 153-156.

Wright, T.O., and Kreps, L.F., 1979, Provenance of the Hamburg Klippe and Shochary Ridge, eastern Pennsylvania: Proceedings of the Pennsylvania Academy of Science, vol. 53, no. 1, p. 98-100.

Wright, T.O., and Stephens, G.C., 1976, Structure, age and depositional environment of Shochary Ridge, Martinsburg Formation, Berks and Lehigh counties, Pennsylvania [ABS]: Geological Society of America Abstracts with Programs, vol. 8, no. 2, p. 305.

Wright, T.O., and Stephens, G.C., 1978, Regional implications of the stratigraphy and structure of Shochary Ridge, Berks and Lehigh counties, Pennsylvania: American Journal of Science, vol. 278, no. 7, p. 1000-1017.

Wright, T.O., and Stephens, G.C., 1979, Reply [to Regional implications of the stratigraphy and structure of Shochary Ridge, Berks and Lehigh counties, Pennsylvania]: American Journal of Science, vol. 279, no. 6, p. 728-732.

APPENDIX 1

INDEX MAP

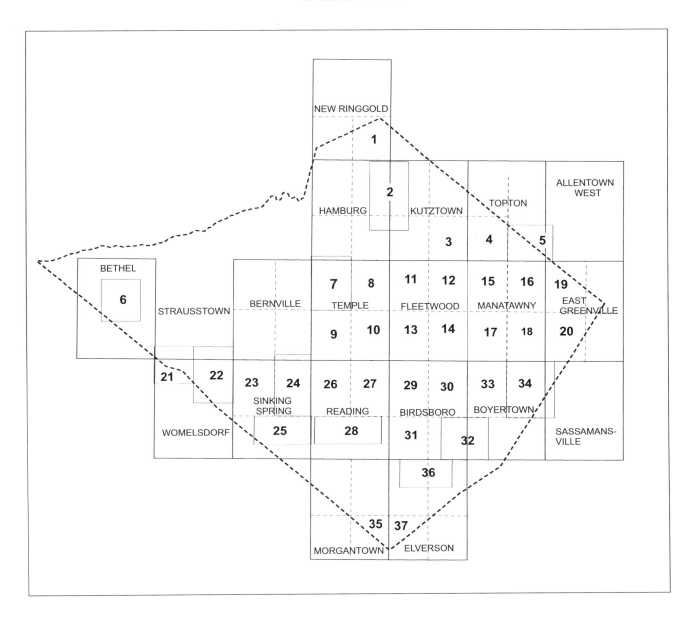

EXPLANATION

READING   QUADRANGLE NAME

5   FIGURE NUMBER

APPENDIX 1

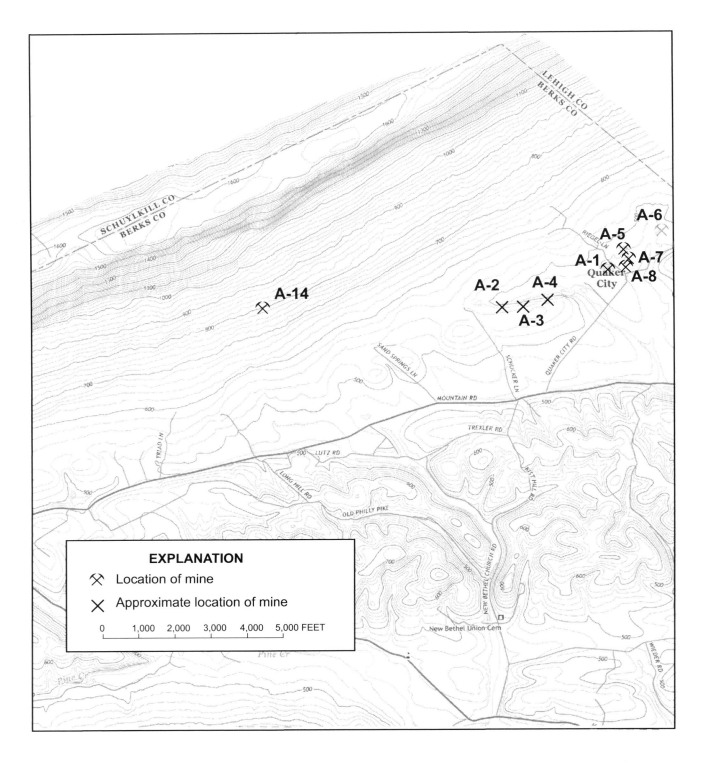

Figure 1. Mines on the southeastern quarter of the U.S. Geological Survey New Ringgold 7.5-minute topographic quadrangle map.

APPENDIX 1

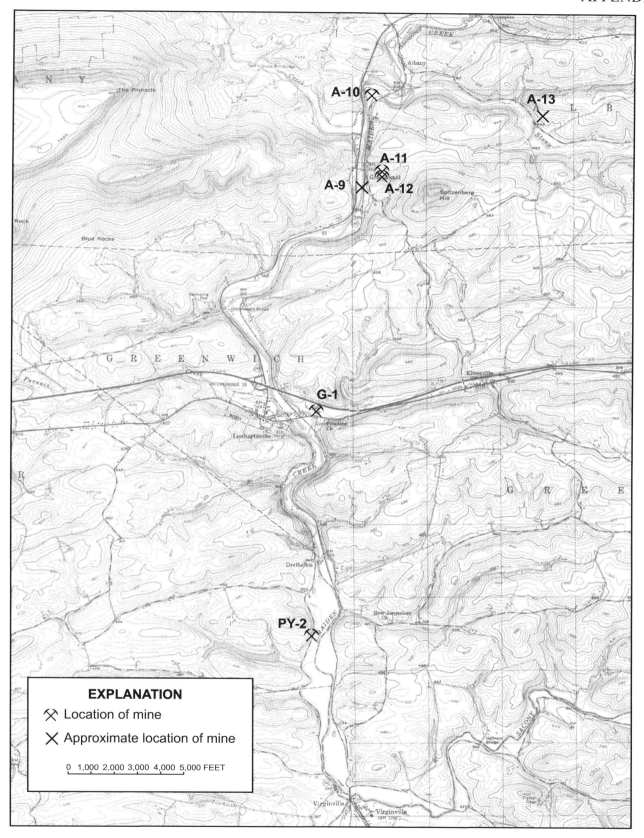

Figure 2. Mines and on part of the eastern part of the U.S. Geological Survey Hamburg and western part of the Kutztown 7.5-minute topographic quadrangle maps.

APPENDIX 1

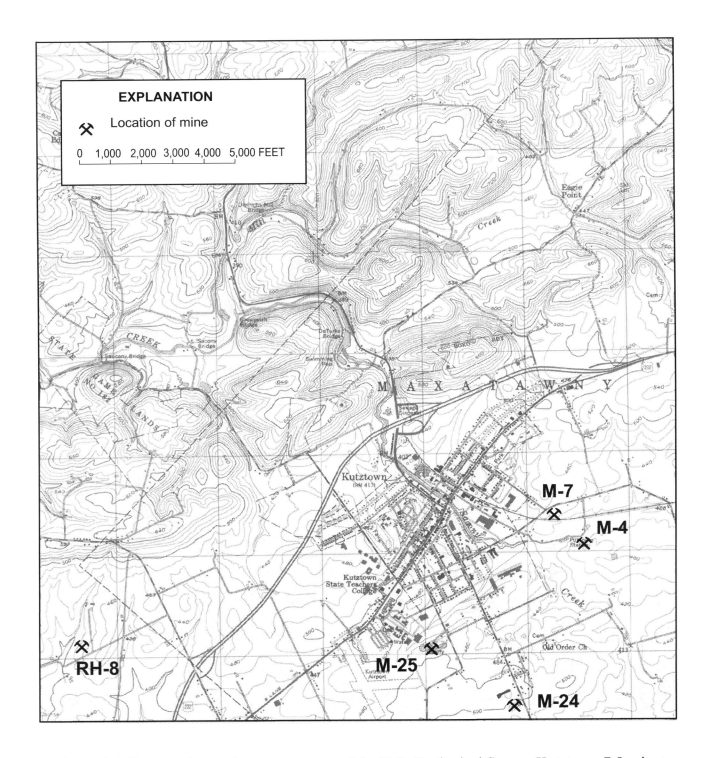

Figure 3. Mines on the southeastern quarter of the U.S. Geological Survey Kutztown 7.5-minute topographic quadrangle map.

APPENDIX 1

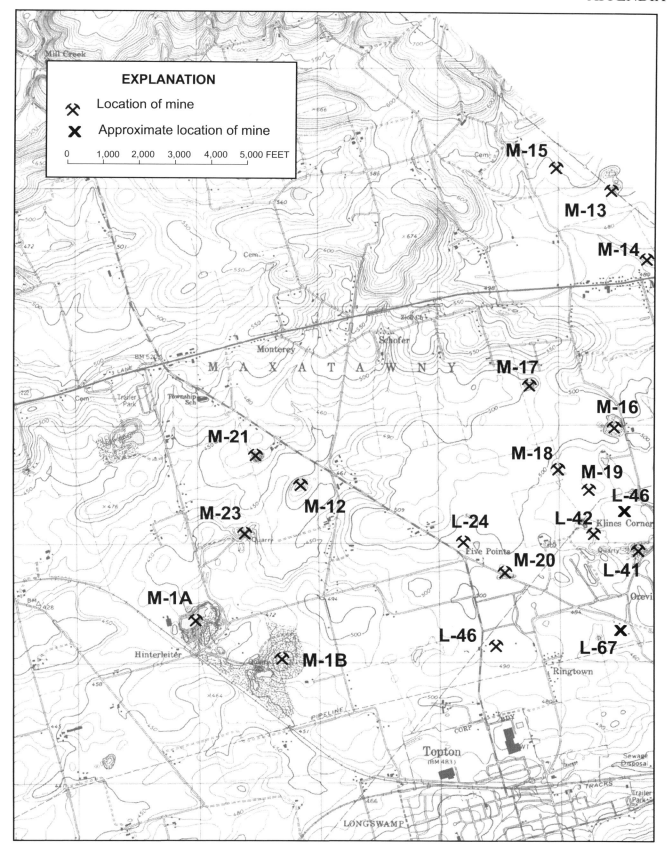

Figure 4. Mines on the southwestern quarter of the U.S. Geological Survey Topton 7.5-minute topographic quadrangle map.

APPENDIX 1

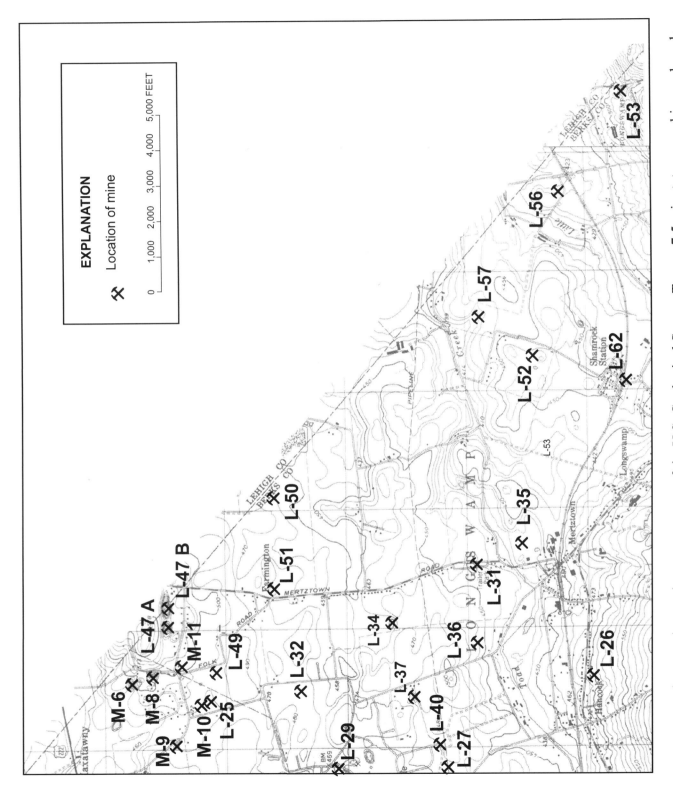

Figure 5. Mines on the southeastern quarter of the U.S. Geological Survey Topton 7.5 minute topographic quadrangle map and part of the southwestern quarter of the U.S. Geological Survey Allentown West 7.5 minute topographic quadrangle map.

APPENDIX 1

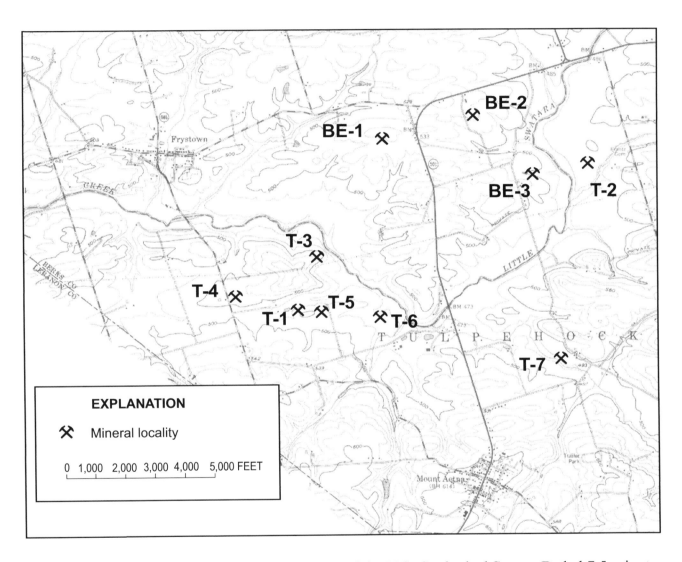

Figure 6. Mineral localities on the central part of the U.S. Geological Survey Bethel 7.5-minute topographic quadrangle map.

APPENDIX 1

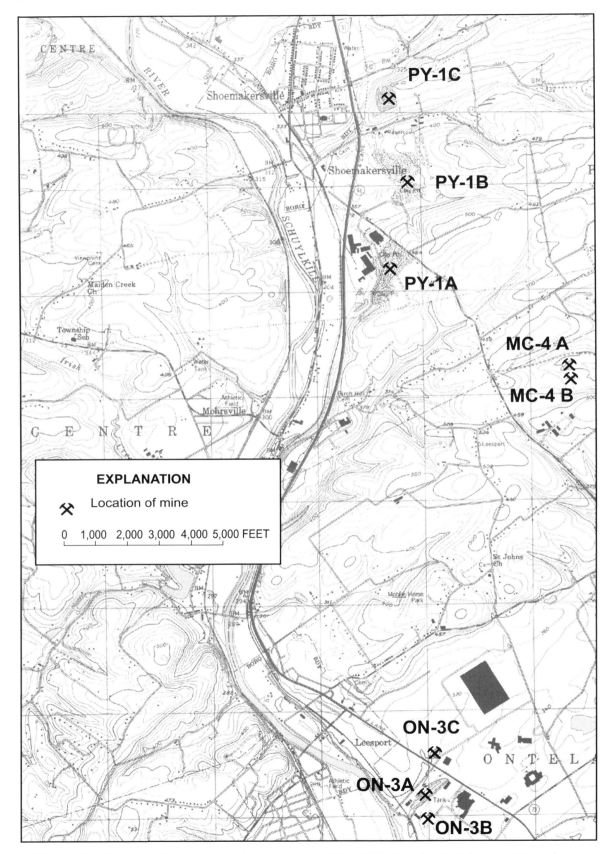

Figure 7. Mines on the northwestern quarter of the U.S. Geological Survey Temple and southwest part of the Hamburg 7.5-minute topographic quadrangle maps.

APPENDIX 1

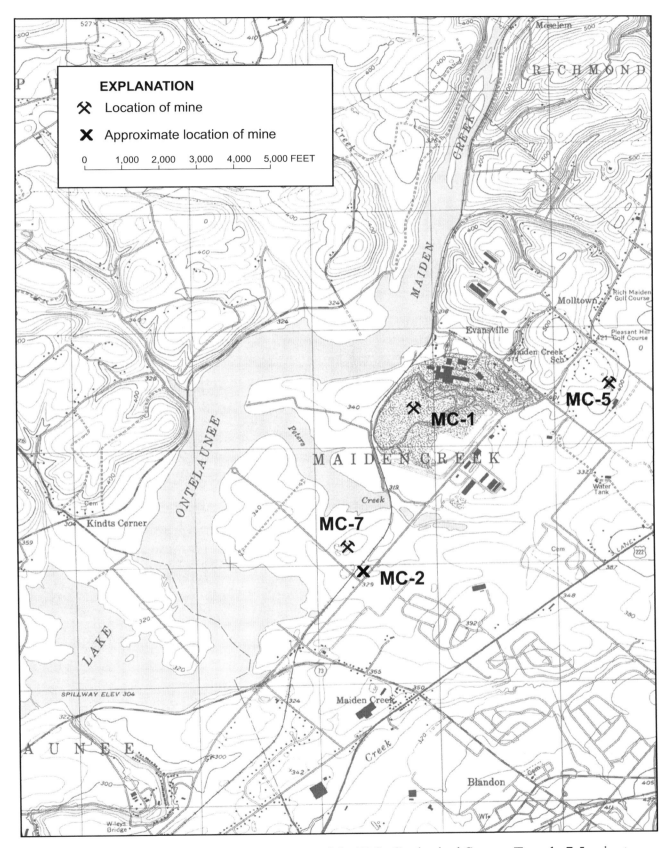

Figure 8. Mines on the northeastern quarter of the U.S. Geological Survey Temple 7.5-minute topographic quadrangle map.

APPENDIX 1

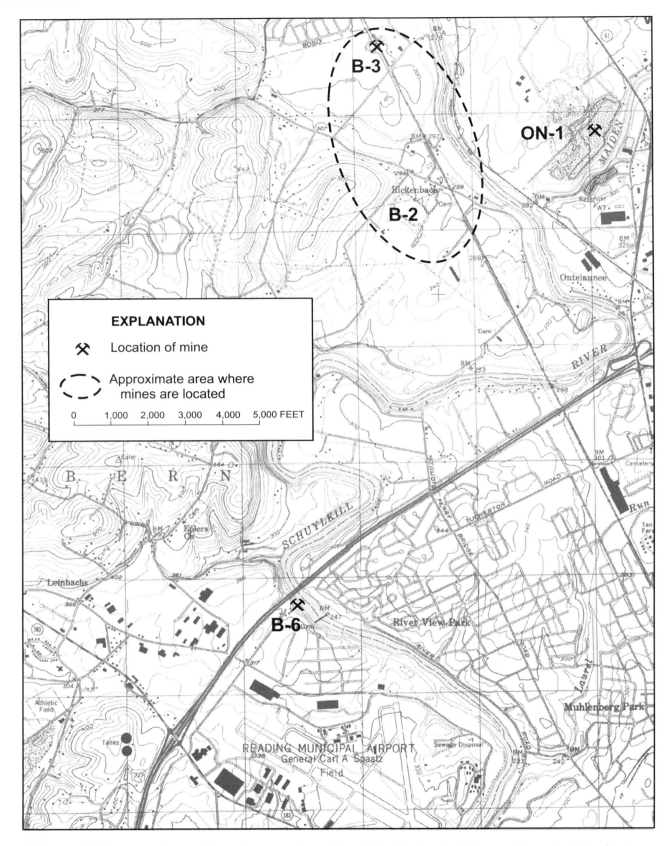

Figure 9. Mines on the southwestern quarter of the U.S. Geological Survey Temple 7.5-minute topographic quadrangle map.

APPENDIX 1

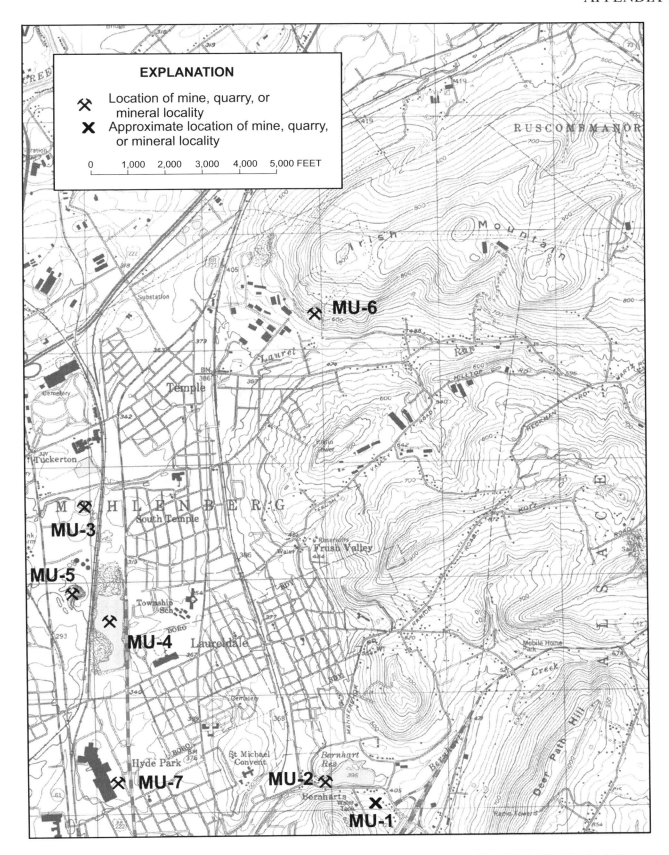

Figure 10. Mines and mineral localities on the southeastern quarter of the U.S. Geological Survey Temple 7.5-minute topographic quadrangle map.

The Mines and Minerals of Berks County

APPENDIX 1

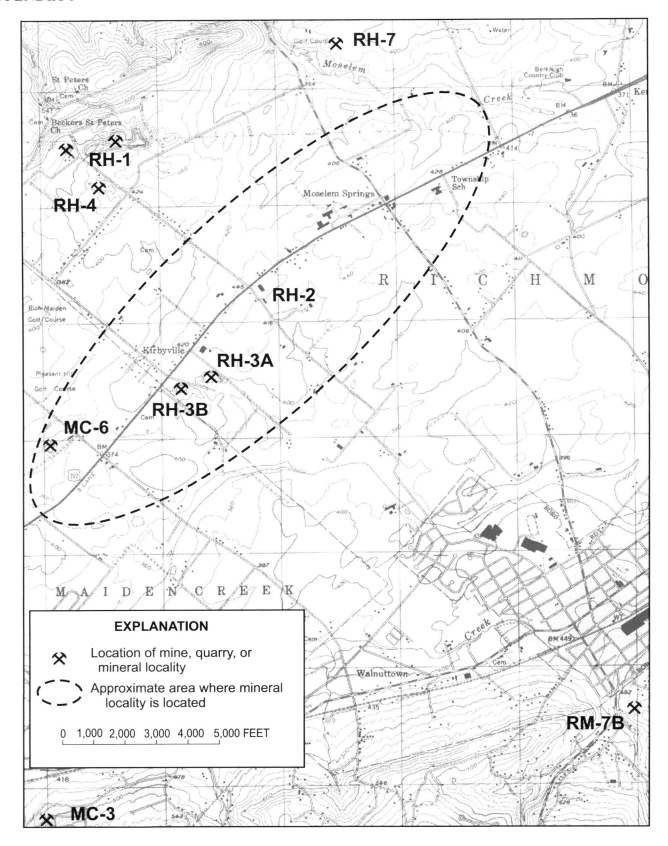

Figure 11. Mines and mineral localities on the northwestern quarter of the U.S. Geological Survey Fleetwood 7.5-minute topographic quadrangle map.

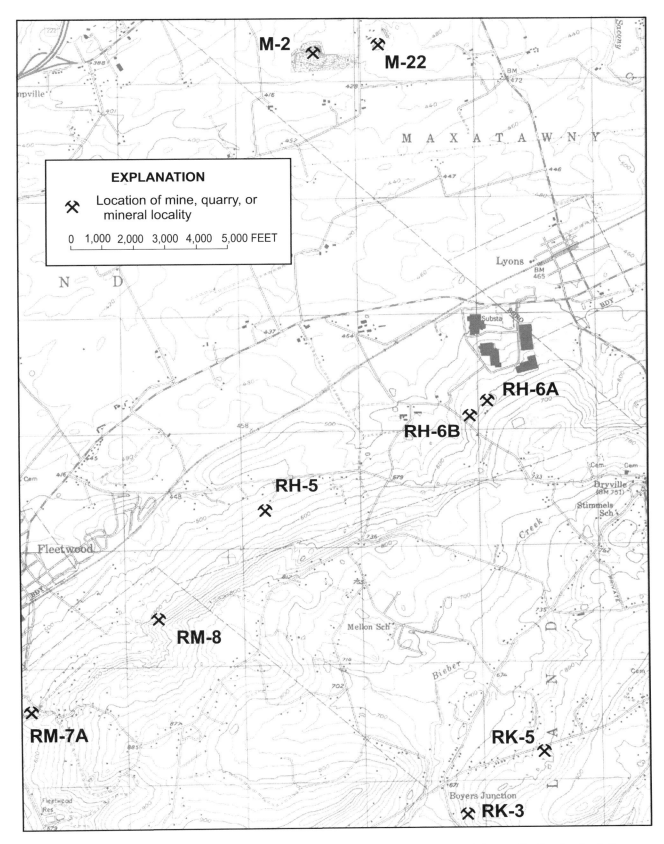

Figure 12. Mines and mineral localities on the northeastern quarter of the U.S. Geological Survey Fleetwood 7.5-minute topographic quadrangle map.

APPENDIX 1

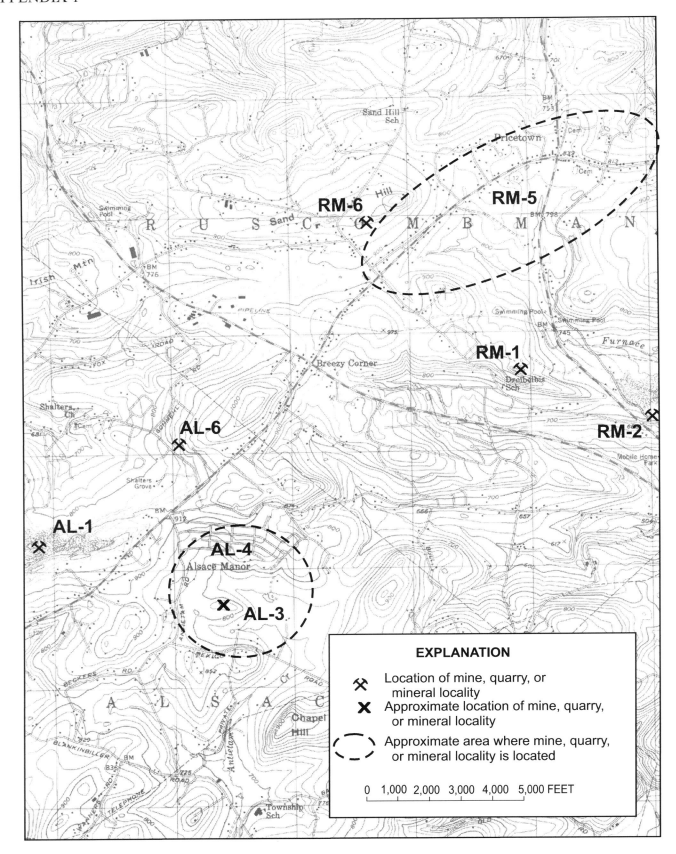

Figure 13. Mines and mineral localities on the southwestern quarter of the U.S. Geological Survey Fleetwood 7.5-minute topographic quadrangle map.

APPENDIX 1

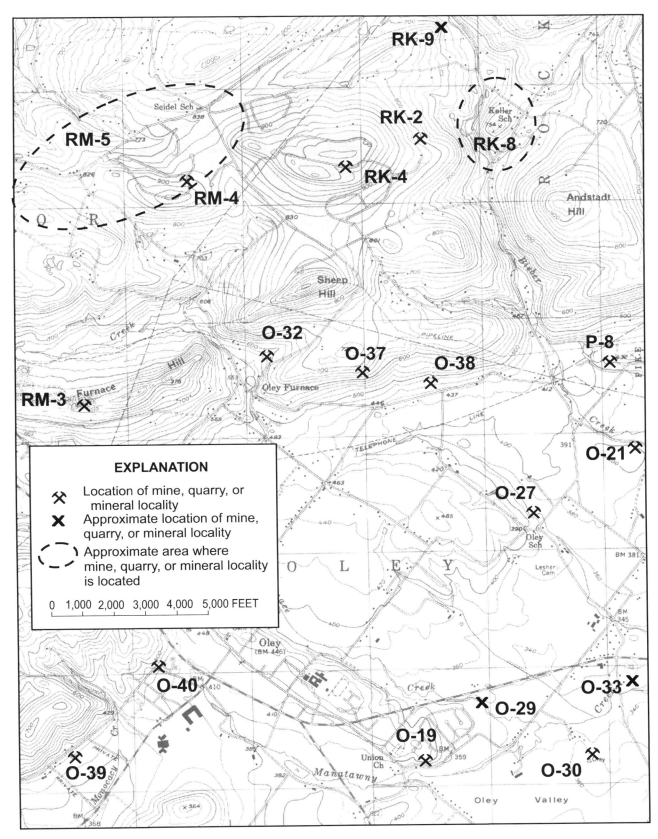

Figure 14. Mines and mineral localities on the southeastern quarter of the U.S. Geological Survey Fleetwood 7.5-minute topographic quadrangle map.

APPENDIX 1

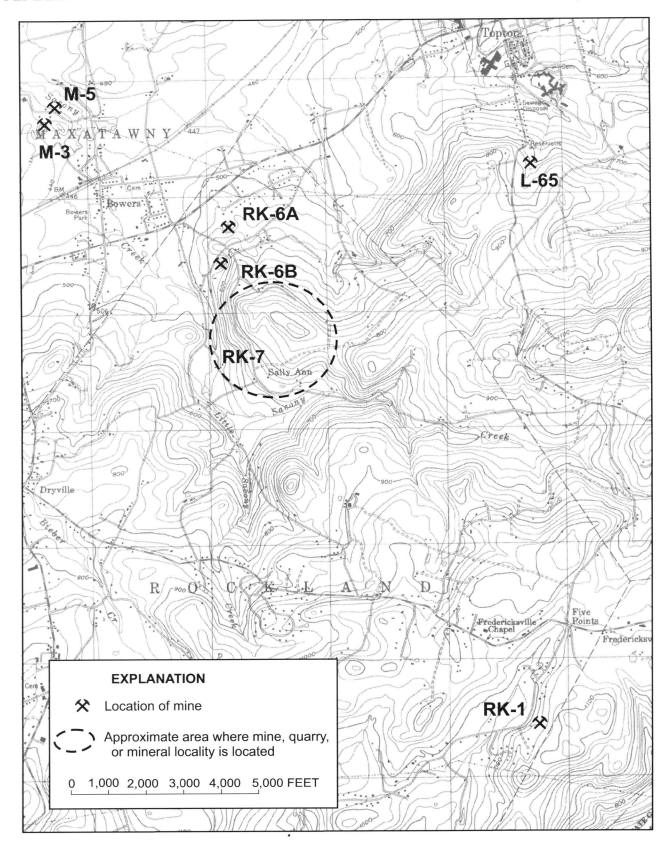

Figure 15. Mines and mineral localities on the northwestern quarter of the U.S. Geological Survey Manatawny 7.5-minute topographic quadrangle map.

APPENDIX 1

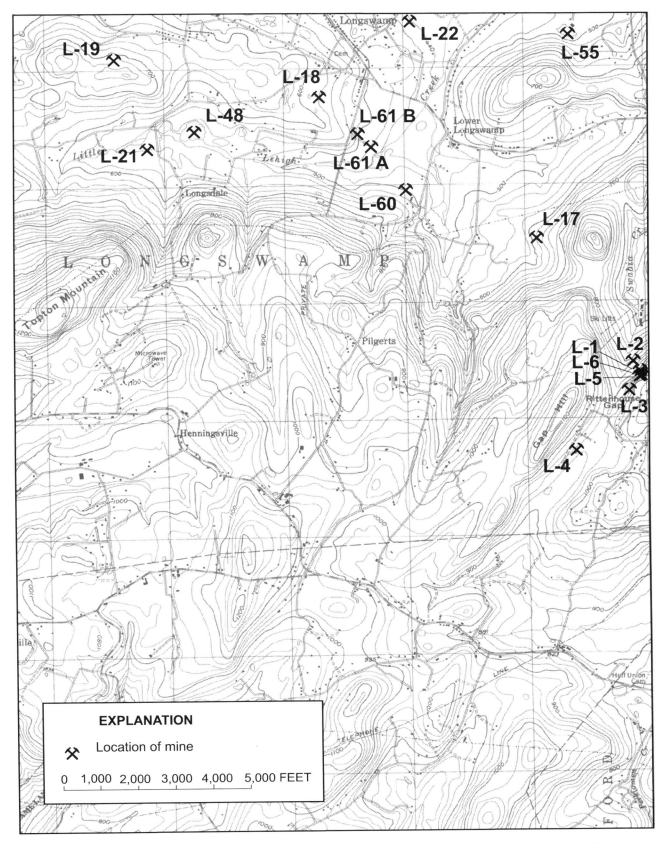

Figure 16. Mines on the northeastern quarter of the U.S. Geological Survey Manatawny 7.5-minute topographic quadrangle map.

The Mines and Minerals of Berks County

APPENDIX 1

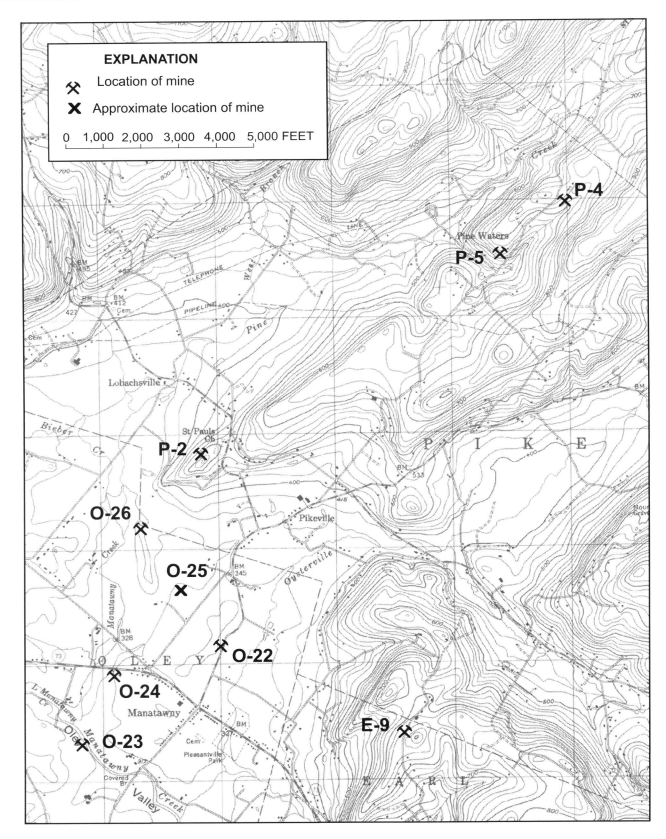

Figure 17. Mines on the southwestern quarter of the U.S. Geological Survey Manatawny 7.5-minute topographic quadrangle map.

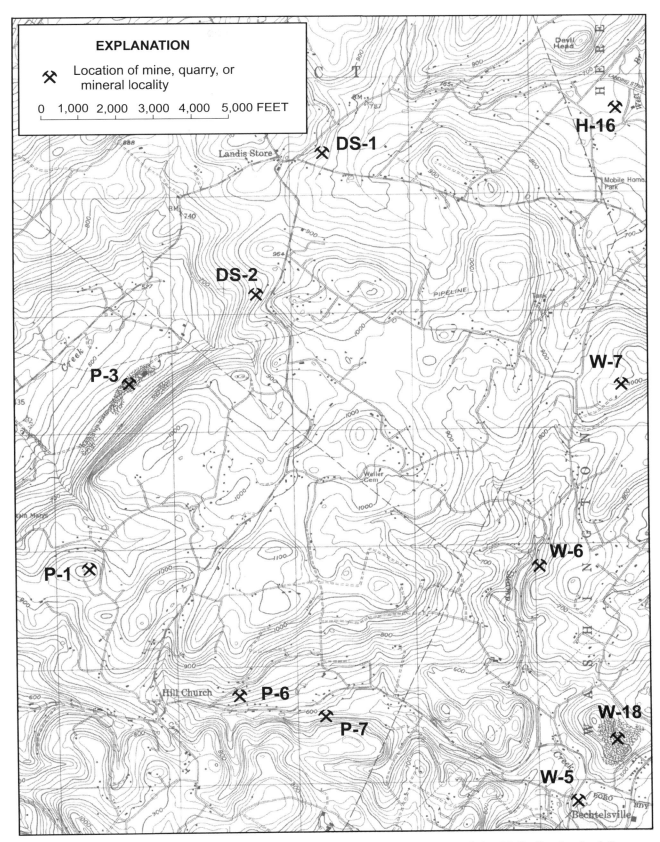

Figure 18. Mines and mineral localities on the southeastern quarter of the U.S. Geological Survey Manatawny 7.5-minute topographic quadrangle map.

APPENDIX 1

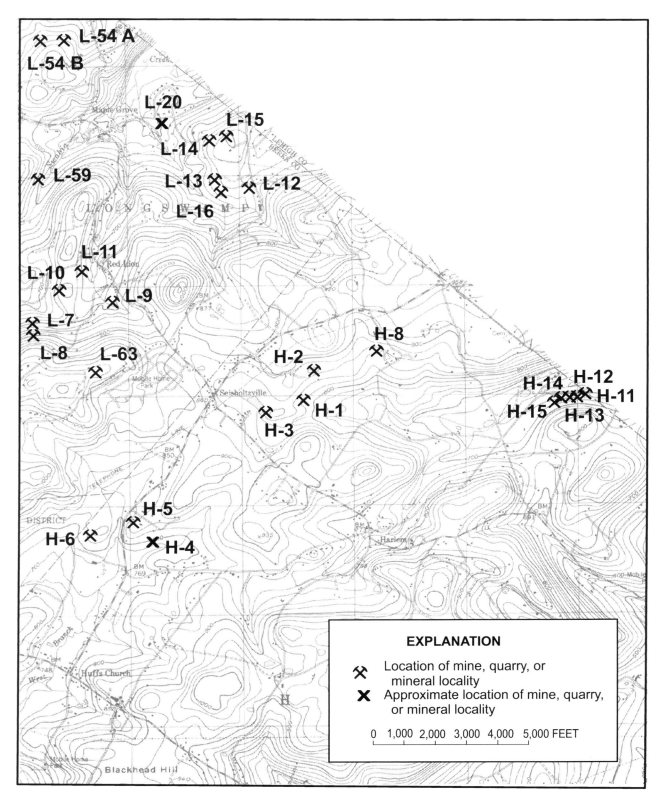

Figure 19. Mines and mineral localities on the northwestern part of the U.S. Geological Survey East Greenville 7.5-minute topographic quadrangle map.

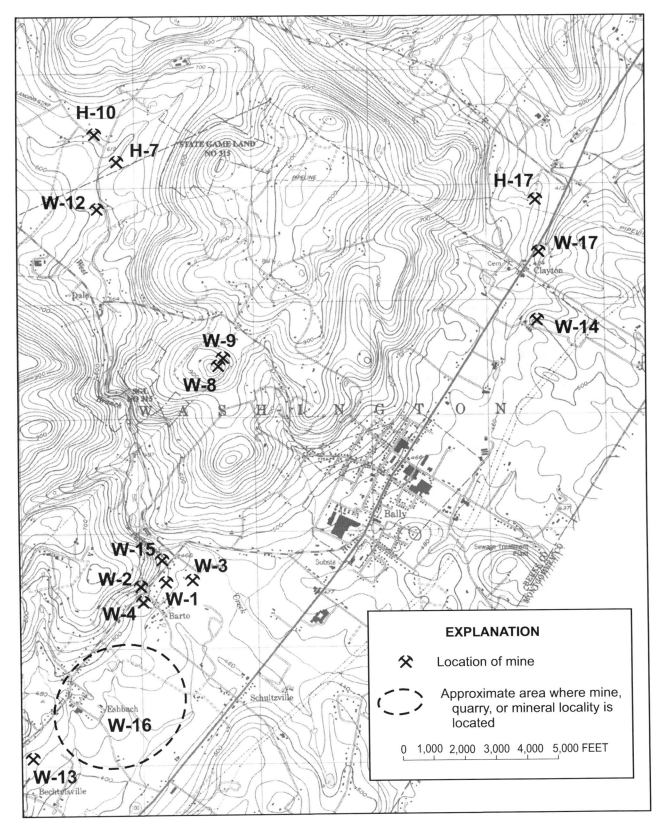

Figure 20. Mines on the southwestern part of the U.S. Geological Survey East Greenville 7.5-minute topographic quadrangle map.

APPENDIX 1

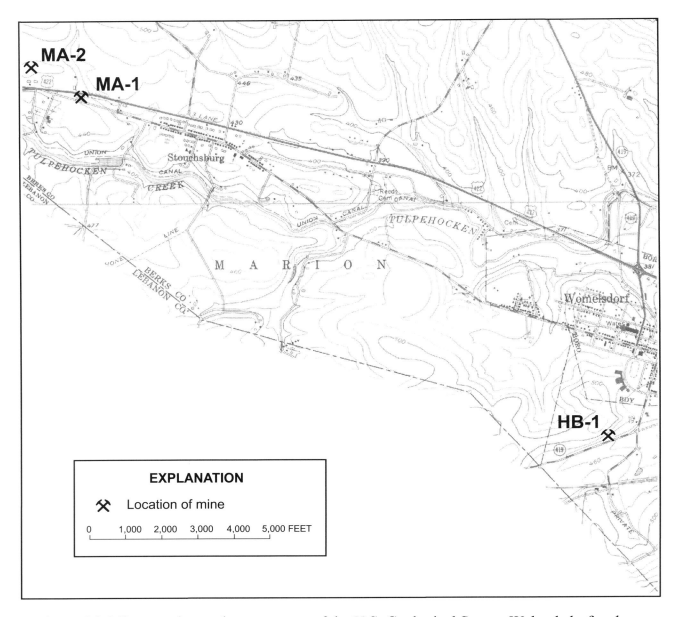

Figure 21. Mines on the northwestern part of the U.S. Geological Survey Wolmelsdorf and southwestern part of the Strausstown 7.5-minute topographic quadrangle maps.

APPENDIX 1

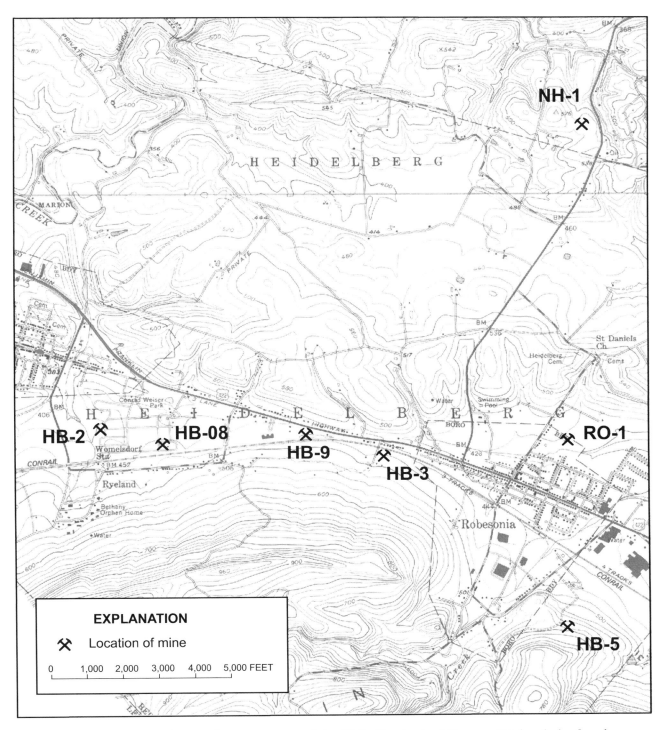

Figure 22. Mines on the northeastern part of the U.S. Geological Survey Wolmelsdorf and southeastern part of the Strausstown 7.5-minute topographic quadrangle maps.

APPENDIX 1

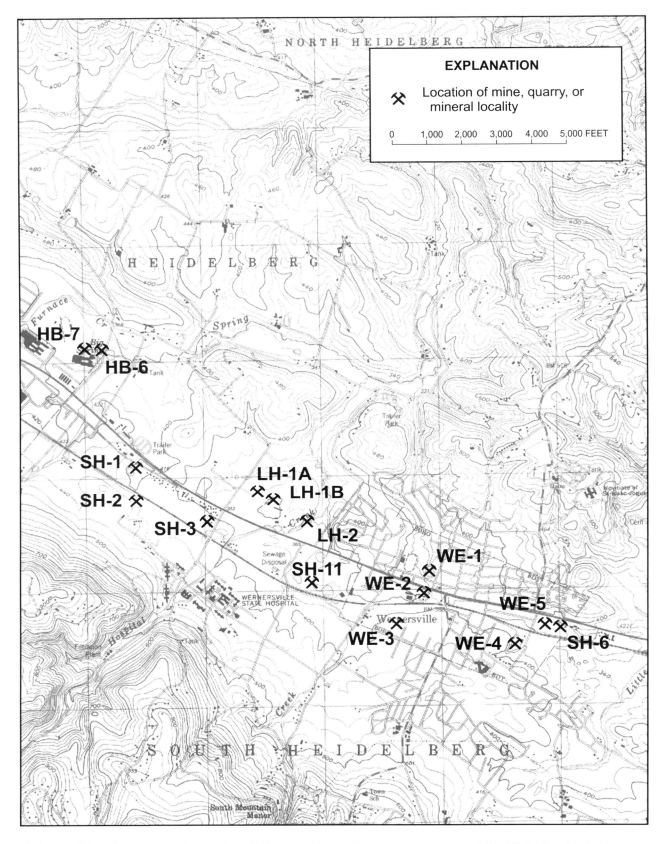

Figure 23. Mines and mineral localities on the northwestern quarter of the U.S. Geological Survey Sinking Spring 7.5-minute topographic quadrangle map.

APPENDIX 1

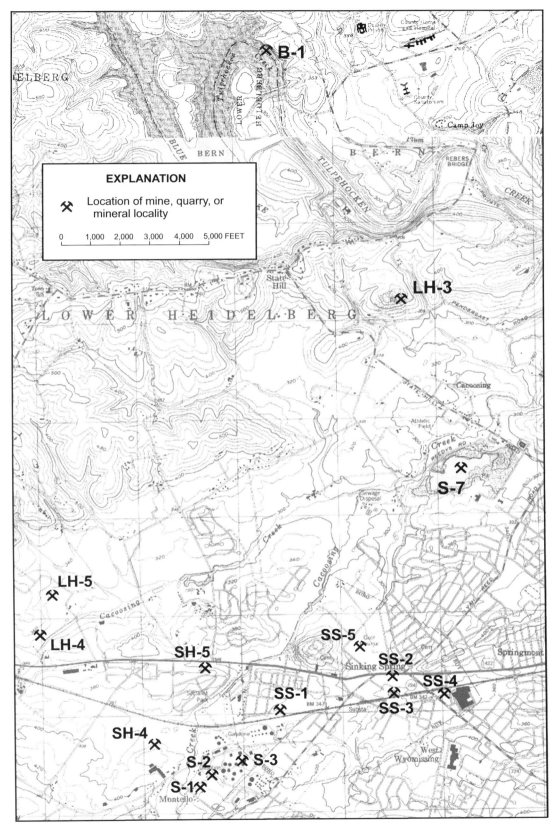

Figure 24. Mines and mineral localities on the northeastern quarter of the U.S. Geological Survey Sinking Spring and southeastern part of the Bernville 7.5-minute topographic quadrangle maps.

APPENDIX 1

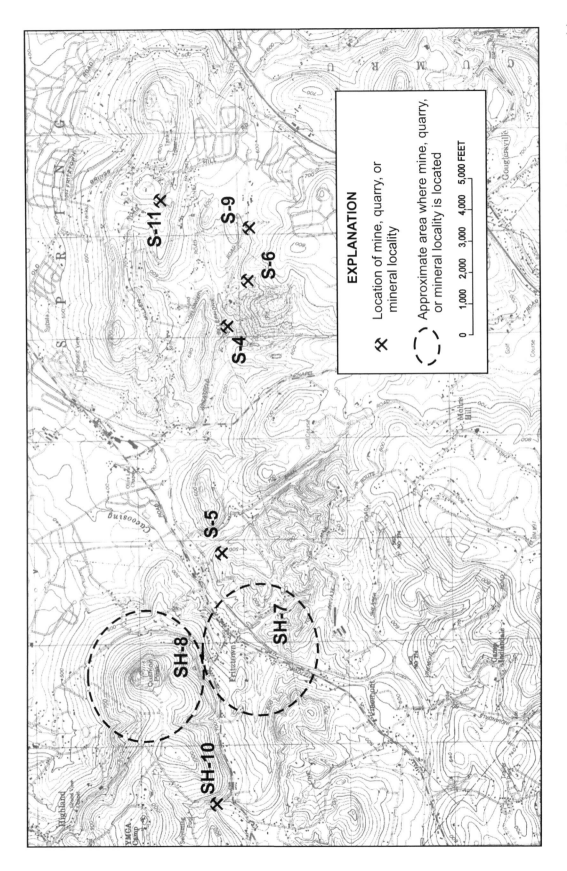

Figure 25. Mines and mineral localities on the southern part of the U.S. Geological Survey Sinking Spring 7.5-minute topographic quadrangle map.

APPENDIX 1

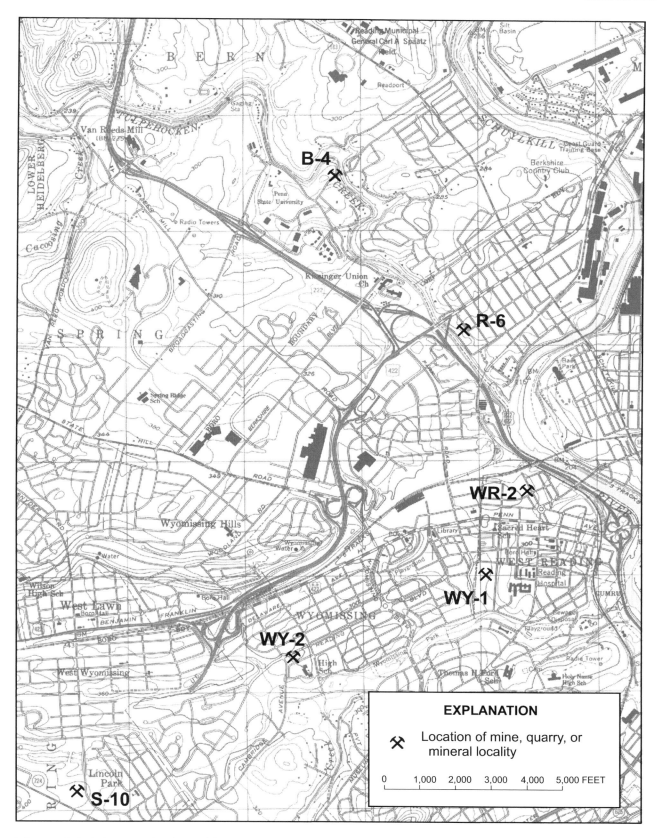

Figure 26. Mines and mineral localities on the northwestern quarter of the U.S. Geological Survey Reading 7.5-minute topographic quadrangle map.

APPENDIX 1

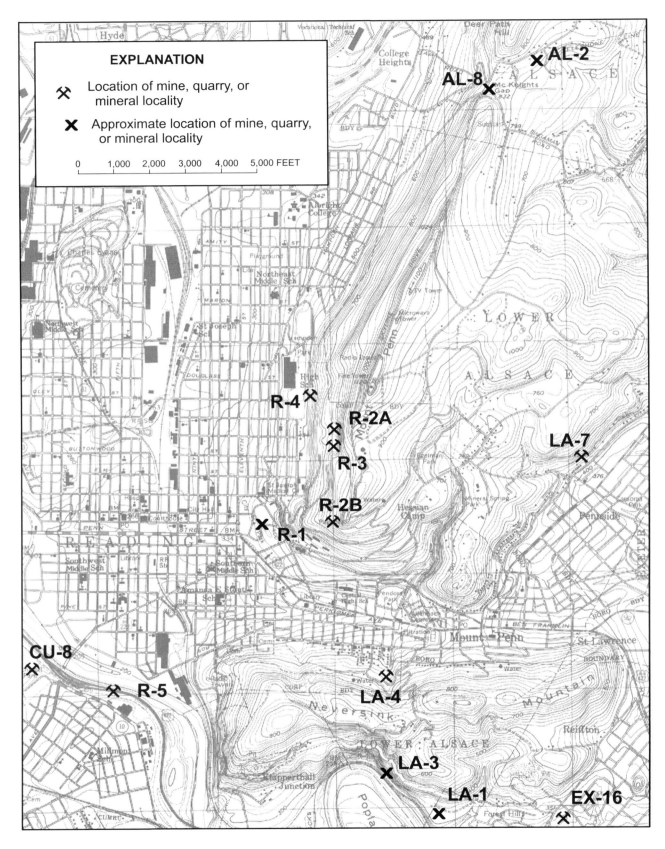

Figure 27. Mines and mineral localities on the northeastern quarter of the U.S. Geological Survey Reading 7.5-minute topographic quadrangle map.

APPENDIX 1

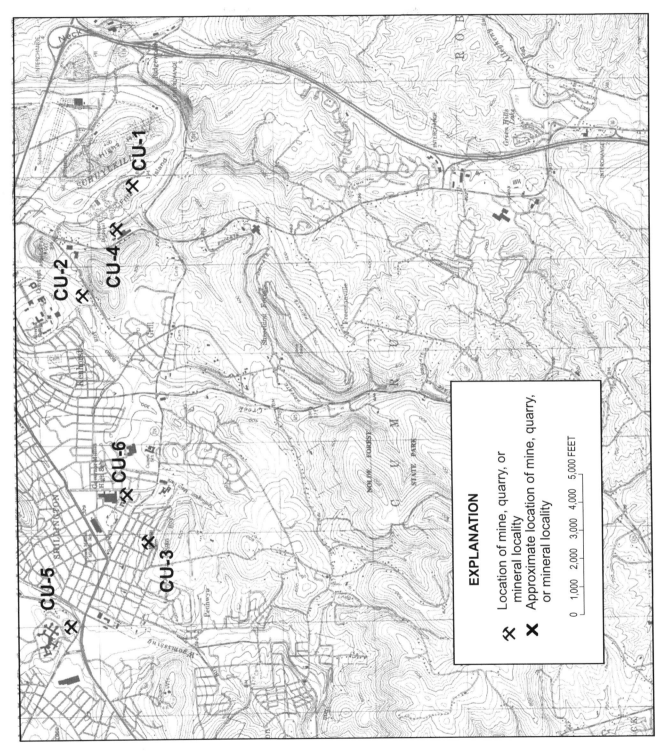

Figure 28. Mines on the southern part of the U.S. Geological Survey Reading 7.5-minute topographic quadrangle map.

APPENDIX 1

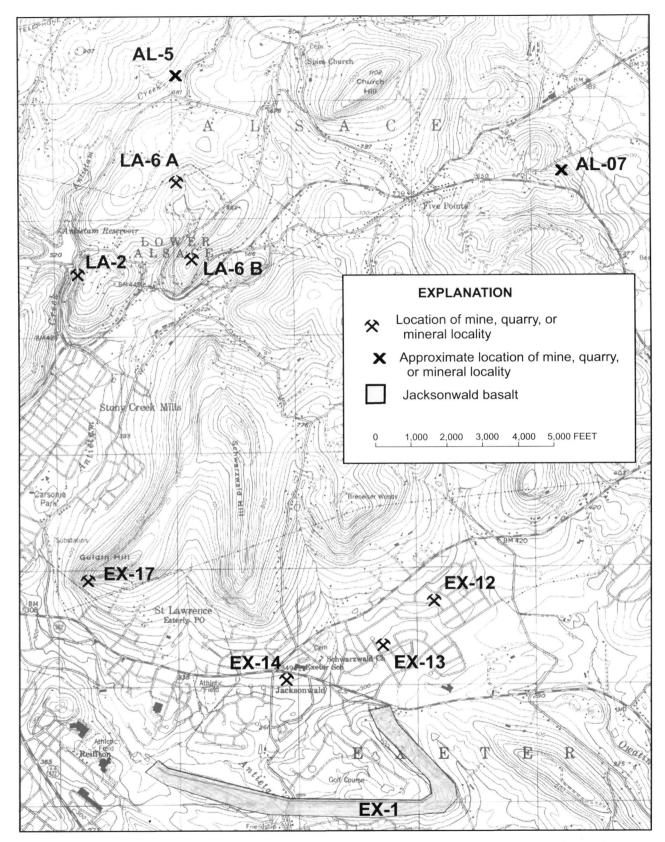

Figure 29. Mines and mineral localities on the northwestern quarter of the U.S. Geological Survey Birdsboro 7.5-minute topographic quadrangle map.

APPENDIX 1

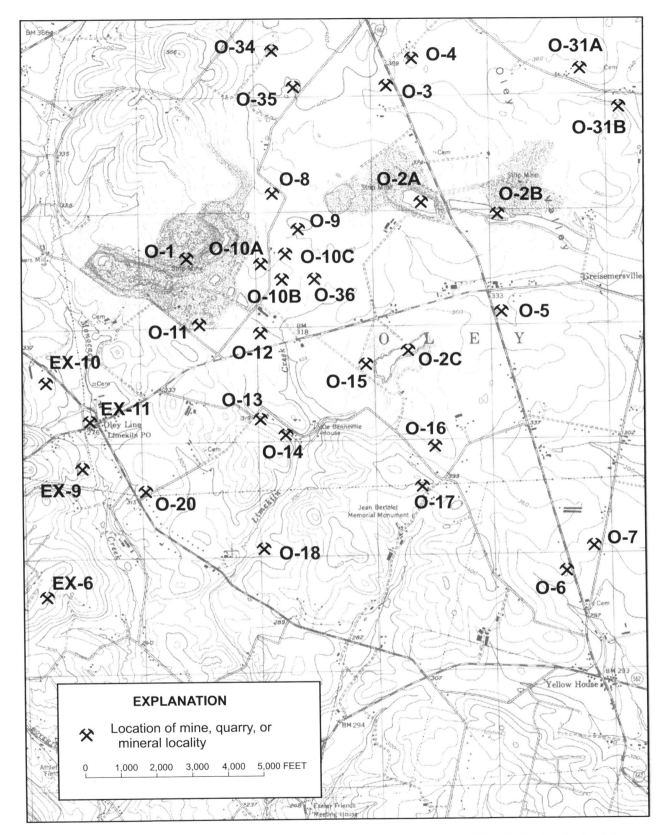

Figure 30. Mines and mineral localities on the northeastern quarter of the U.S. Geological Survey Birdsboro 7.5-minute topographic quadrangle map.

APPENDIX 1

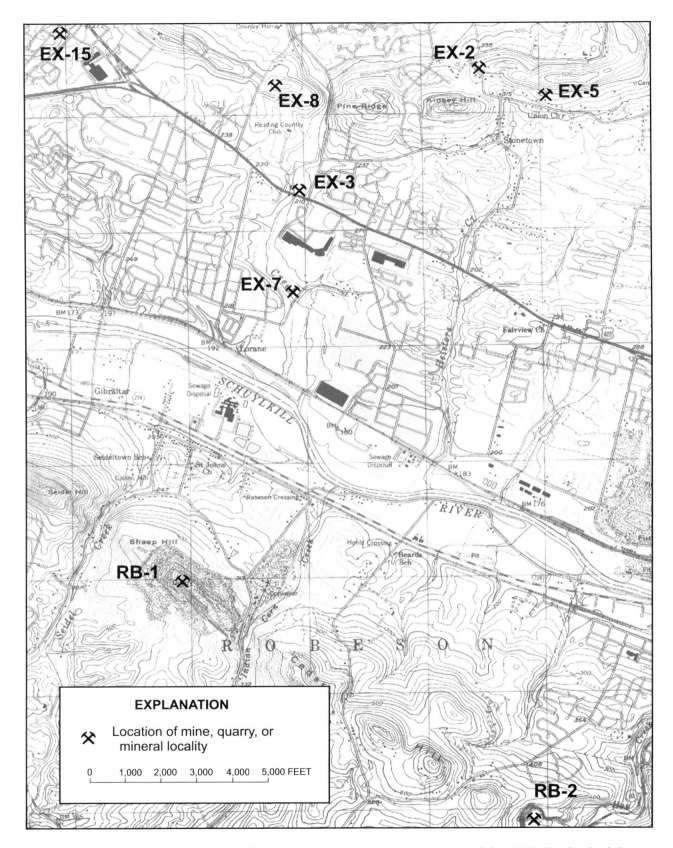

Figure 31. Mines and mineral localities on the southwestern quarter of the U.S. Geological Survey Birdsboro 7.5-minute topographic quadrangle map.

APPENDIX 1

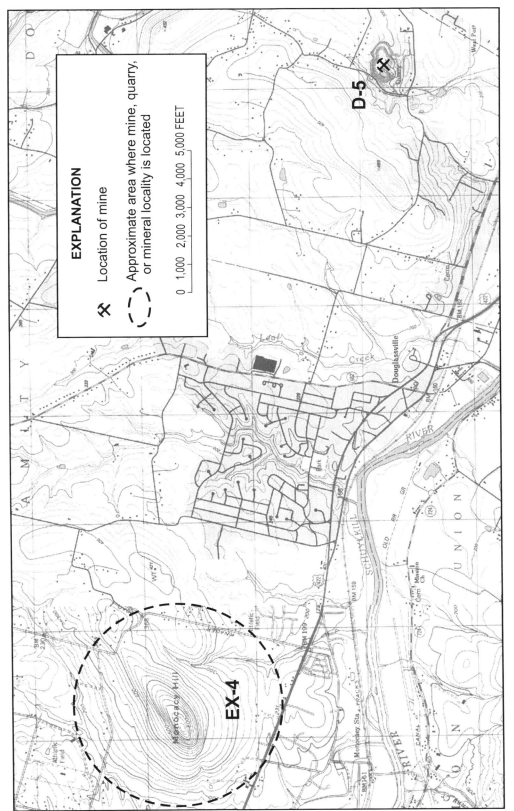

Figure 32. Mines and mineral localities on the southeastern part of the Birdsboro and southwestern part of the Boyertown U.S. Geological Survey 7.5-minute topographic quadrangle maps.

APPENDIX 1

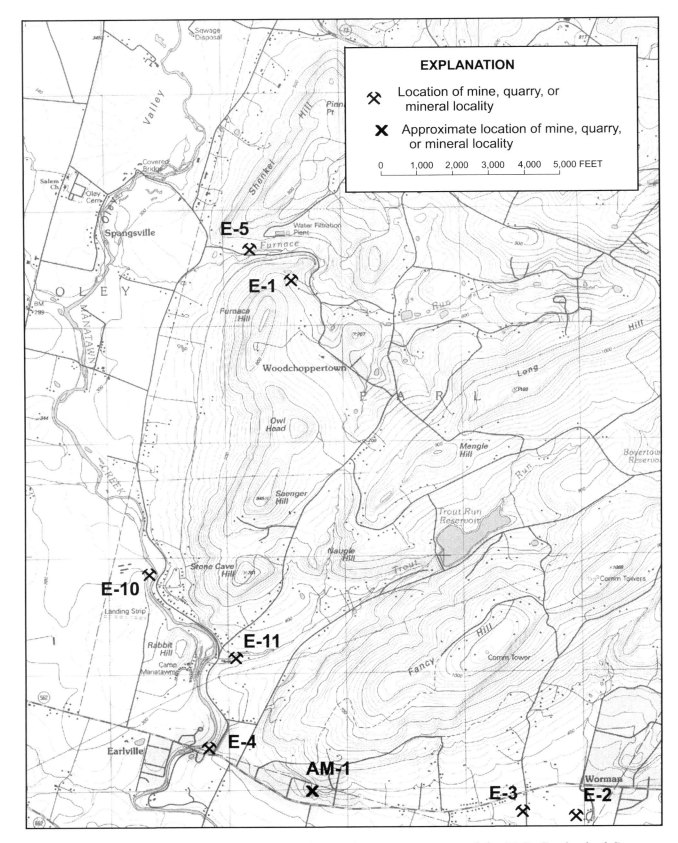

Figure 33. Mines and mineral localities on the northwestern quarter of the U.S. Geological Survey Boyertown 7.5-minute topographic quadrangle map.

APPENDIX 1

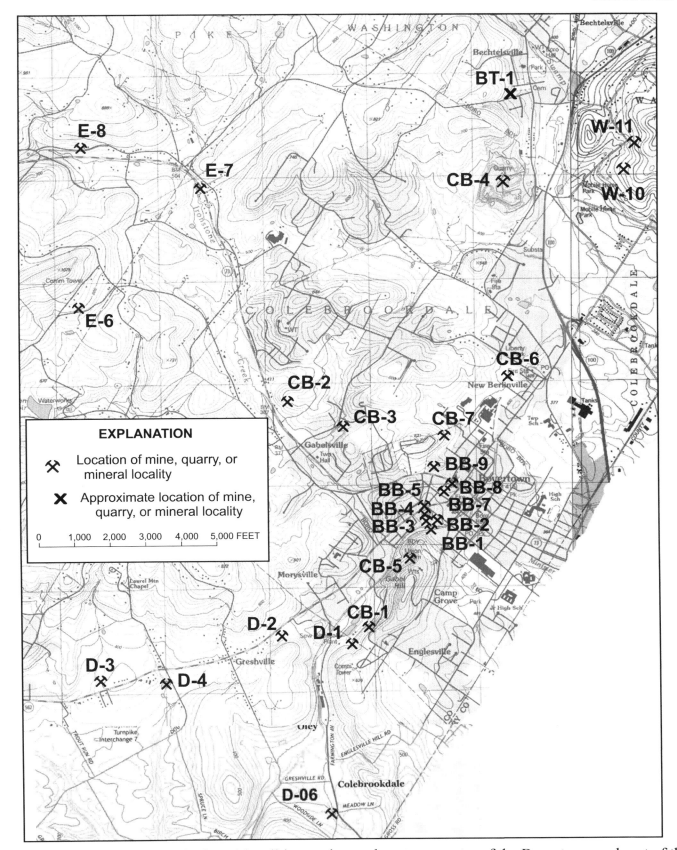

Figure 34. Mines and mineral localities on the northeastern quarter of the Boyertown and part of the northwestern Sassamansville U.S. Geological Survey 7.5-minute topographic quadrangle maps.

APPENDIX 1

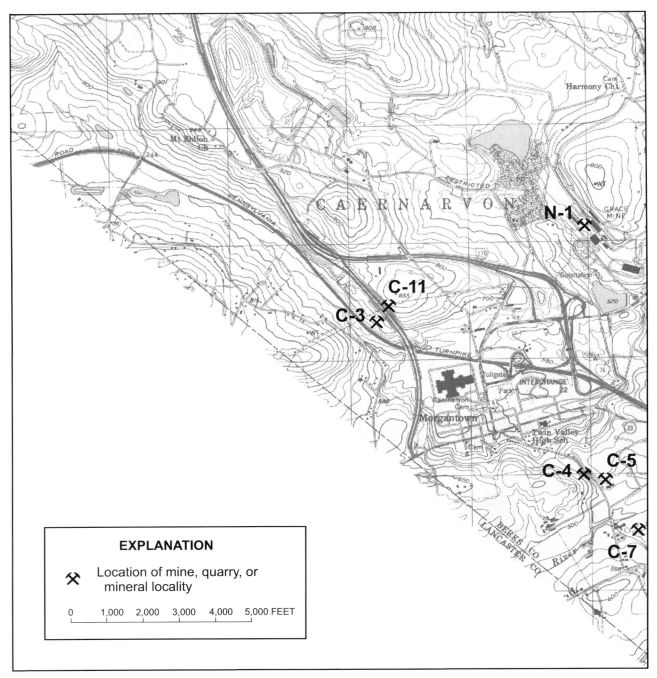

Figure 35. Mines and mineral localities on the southeastern quarter of the U.S. Geological Survey Morgantown 7.5-minute topographic quadrangle map.

APPENDIX 1

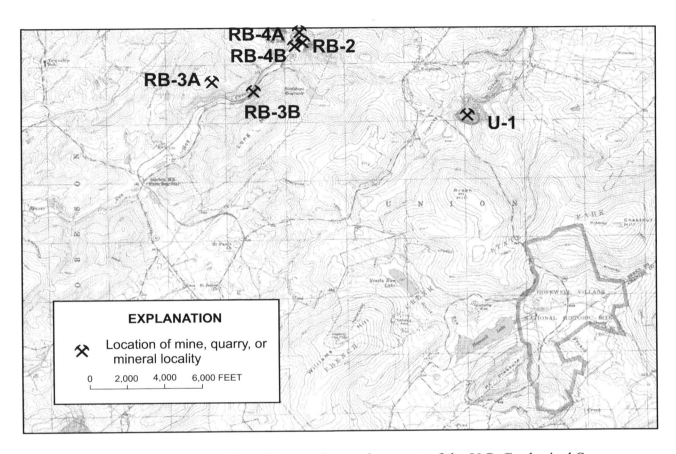

Figure 36. Mines and mineral localities on the northern part of the U.S. Geological Survey Elverson 7.5-minute topographic quadrangle map.

APPENDIX 1

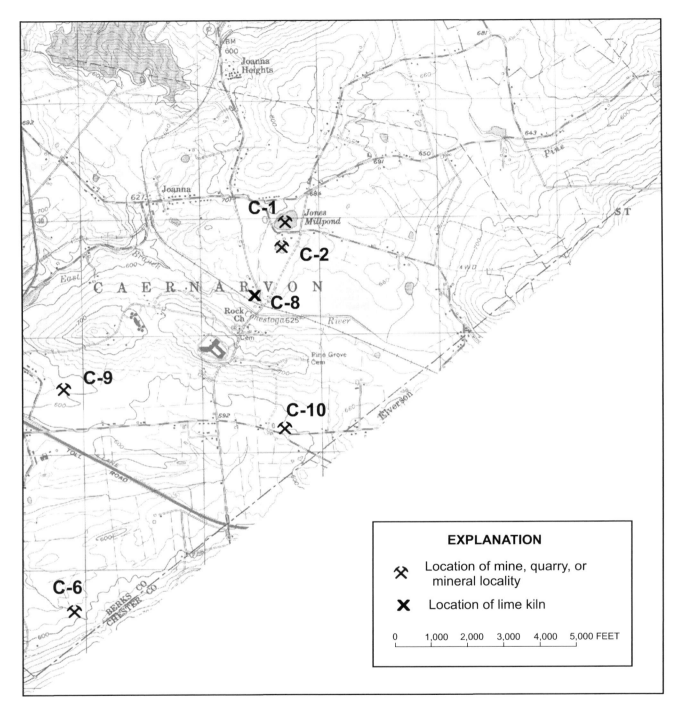

Figure 37. Mines and localities on the southwestern quarter of the U.S. Geological Survey Elverson 7.5-minute topographic quadrangle map.

APPENDIX 2

Mine Name (Principle mine name in parenthesis)	Township	Page
A&J Aggregates (Temple quarry)	Alsace	19
Abandoned quarry at Jacksonwald	Exeter	124
Adams quarry (McQuade Brothers quarry)	West Reading	356
Ahrens & Tobias quarry (Whitman quarry)	Reading	275
Ahrens quarry	Reading	279
Ajax Cement Co (Molltown Cement Co)	Maiden Creek	193
Albany quarry	Albany	14
Allentown Iron Company mine	Longswamp	168
Allentown Portland Cement Co (Evansville quarry)	Maiden Creek	196
Allentown Portland Cement Co (Kirbyville quarries)	Richmond	286
Allentown Portland Cement Co (Maidencreek quarry)	Maiden Creek	202
Allentown Portland Cement Co quarry (Lehigh Cement Co)	Oley	254
Almshouse quarry	Cumru	106
American Mining Company (Jones mine)	Caernarvon	69
Anderson quarry (Dyer Gibraltar quarry)	Robeson	291
Angstadt quarry	Maxatawny	211
Antietam Lake (Antietam Reservoir locality)	Lower Alsace	187
Antietam Reservoir locality	Lower Alsace	187
Atlas Mineral Products quarry (Albany quarry)	Albany	14
Atlas Ocher Mine	Longswamp	185
B. Frank Ruth and Company quarry	Albany	15
Bailey quarry	Maxatawny	214
Baldy Street quarry (Keystone quarry)	Maxatawny	210
Barr quarry	Washington	352
Barr, Joseph G. quarry (Essig quarry)	Bern	26
Barto mine	Washington	346
Barto Stone & Cement Block Co (Barr quarry)	Washington	352
Bechel, David (Trexler mica mine)	Alsace	18
Bechtel farm (Boyertown Graphite Co mine)	Colebrookdale	89
Bechtel, Jesse quarry (Greshville quarry)	Douglass	108
Bechtelsville (Martin) quarry	Colebrookdale	85
Becker quarry	Muhlenburg	224
Becker, Bauer & Company quarry	Reading	278
Beidler (Muhlenberg) mine	Wyomissing	358
Beitler mine	Rockland	307
Benfield, D. (Gilberg mine)	Washington	350
Benfield, H.B. (Dresher mine)	Longswamp	176
Benjamin Moore & Company (Fritch & Brother's mines)	Longswamp	181
Benson quarry	Reading	278
Berck quarry (Keystone quarry)	Maxatawny	210
Bergs farm locality	Albany	18
Berks and Chester Mining Company (Jones mine)	Caernarvon	66
Berks Cast Stone Company quarry (South Temple quarry)	Muhlenburg	221
Berks County China Clay Works (Long mines)	Longswamp	177
Berks Development Company	Earl	116
Berks Products Co Ontelaunee quarry	Ontelauee	265
Berks Products Company quarry (South Temple quarry)	Muhlenburg	221
Berks Products Corp (Fehr and O'Rourke quarry)	Cumru	103
Berks Products Corp quarry	Maxatawny	211
Berks Silica Sand Corporation (Temple quarry)	Alsace	19

APPENDIX 2

Mine Name (Principle mine name in parenthesis)	Township	Page
Berkshire Furnace mine	Lower Heidelberg	191
Bernhart's Dam locality	Muhlenburg	216
Bernville area	Berneville	28
Berthou mine	Washington	347
Bertolet & Hunter quarry (Temple Sand Company quarry)	Muhlenburg	224
Bertolet, D.F. quarry	Oley	261
Bertolet, Israel quarry	Oley	261
Bertolet, J.G. quarry	Oley	260
Bertolet, L.J. quarry	Oley	259
Bessemer Iron Company (S. Bittenbender mine)	Hereford	136
Bethlehem Iron Company (Dunkle mines)	Longswamp	160
Bethlehem Iron Company (Frederick mines)	Longswamp	182
Bethlehem Iron Company (Gap mine)	Longswamp	156
Bethlehem Iron Company (J.B. Gehman mine)	Hereford	139
Bethlehem Iron Company (Miller farm mines)	Longswamp	161
Bethlehem Iron Company (Smoyer mine)	Longswamp	162
Bethlehem Iron Company (Weiler mine)	Longswamp	160
Bethlehem Iron Company Gardner Station mine	Longswamp	161
Bethlehem Steel Company (Grace mine)	New Morgan Borough	227
Bieber mine	Rockland	309
Bieber quarry (Hinterleiter quarry)	Maxatawny	208
Bieber quarry (Keystone quarry)	Maxatawny	210
Big Bed Slate Company (Centennial quarry)	Albany	12
Big Dam Quarry	Lower Alsace	187
Big Spring quarry	Heidelberg	128
Binder mine (Warwick mine)	Boyertown	46
Birdsboro Materials quarry (Dyer Trap Rock quarry)	Robeson	297
Birdsboro quarry (Dyer Gibraltar quarry)	Robeson	291
Birdsboro Stone Company quarry (Dyer Monocacy quarry)	Union	339
Bishop mine	Exeter	122
Bishop's Mill Locality	Exeter	120
Bittenbender, Christian mine	Hereford	138
Bittenbender, Samuel mine	Hereford	133
Bittenbender, William mine	Hereford	133
Blandon area wavellite locality	Ruscombmanor	317
Blue Bell Lime Company	Muhlenburg	224
Blue quartzite quarry	Ruscombmanor	317
Bohn farm fetid barite locality	Tulpehocken	336
Bomegratz's farm quartz locality	Ruscombmanor	317
Bowers quarry	Maxatawny	211
Boyer quarry	Earl	117
Boyertown Graphite Company mine	Colebrookdale	89
Boyertown Iron Mines, Inc. (Boyertown mines)	Boyertown	56
Boyertown mines	Boyertown	36
Boyertown Mining Company (Boyertown mines)	Boyertown	55
Boyertown Ore Company (Boyertown mines)	Boyertown	55
Breneman quarry	Spring	324
Brinkley & Zinn quarry (Wernersville Lime & Stone Co quarry)	Wernersville	354
Brook, George quarry	Union	344
Brook's quarry	Muhlenburg	216

Mine Name (Principle mine name in parenthesis)	Township	Page
Brower mine	Colebrookdale	87
Brown quarry	Reading	278
Brownback Mine (Charles Heffner mines)	Richmond	291
Brumbach, Isaac quary	Oley	259
Brumbach, Percival mine	Rockland	308
Brunner farm quarry	Caernarvon	83
Brusstar & Brother quarry (Humphrey sandstone quarry)	Robeson	304
Burkholder farm fetid barite locality	Tulpehocken	335
Buttonwood & Fourteenth Streets quarry	Reading	278
Byler mine	Caernarvon	78
Byler quarry	Caernarvon	84
C.K. Williams & Co. (Neversink Mt sienna mine)	Lower Alsace	190
C.K. Williams & Company ocher mine	Richmond	282
California mine	Boyertown	41
Centennial quarry	Albany	12
Centennial Slate Company (Centennial quarry)	Albany	12
Chalins & Company (J.B. Gehman mine)	Hereford	139
Charles Brensinger farm graphite prospect	Longswamp	186
Christman, Jacob quarry	Hereford	141
Christman, James & Lewis quarry	Hereford	141
Clemmerer quarry	Hereford	143
Cline & Weiler (Trollingers quarry)	Hereford	143
Clingan quarry (DyerGibraltar quarry)	Robeson	291
Clymer Iron Company (Clymer mines)	Ruscombmanor	313
Clymer Iron Company (Charles Heffner mines)	Richmond	290
Clymer Iron Company (Fritch & Brother's mines)	Longswamp	181
Clymer Iron Company (Hunter mine)	Oley	262
Clymer Iron Company (Old Tunnel mine)	Ruscombmanor	314
Clymer Iron Company (Udree mine)	Ruscombmanor	312
Clymer Iron Company (Weaver mine)	Oley	262
Clymer Iron Company (Wheatfield mine)	Spring	324
Clymer Iron Company quarry (Bowers quarry)	Maxatawny	211
Clymer mines	Ruscombmanor	313
Colerain Iron Company (Gilberg mine)	Washington	350
Colerain Iron Company (J.B. Gehman mine)	Hereford	139
Collier's flagstone quarry	Perry	270
Columbia Graphite Co (Boyertown Graphite Co mine)	Colebrookdale	89
Conrad's slope	Longswamp	156
Crane Iron Company (C. Bittenbender mine)	Hereford	138
Crane Iron Company (Frederick mines)	Longswamp	182
Crane Iron Company (Lichenwallner mine)	Longswamp	176
Crane Iron Company (Miller farm mines)	Longswamp	161
Crane Iron Company (S. Bittenbender mine)	Hereford	133
Crane iron Company (Wetzel mlne)	Longswamp	160
Crane Iron Company mine	Maiden Creek	202
Crouse and Plank (P.W. Plank & Company)	Caernarvon	83
Cushion Mountain iron mines	South Heidelberg	322
Cushion Mountain locality	South Heidelberg	322
Dale mine	Hereford	140
Darman, Edwin (Amos Fisher mine)	Longswamp	170

APPENDIX 2

Mine Name (Principle mine name in parenthesis)	Township	Page
Davidheiser, David quarry	Earl	116
Davidheiser, J. quarry	Douglass	109
Davis, Charles quarry	Reading	279
Deckert quarry	Sinking Spring Borough	318
Deisher quarry	Oley	260
Deppen quarry No. 1	Wernersville	352
Deppen quarry No. 2	Wernersville	352
Deppen. Samuel quarry	Heidelberg	128
DeTurk, Jonas quarry	Exeter	125
DeTurk, L. quarry (Lehigh Cement Co)	Oley	255
Devil's Lake (Deppen quarry No. 1)	Wernersville	352
Deysher mine	Washington	347
Dielh quarry	Washington	351
Dietrick quarry (Laureldale quarry)	Muhlenburg	219
Dishong quarry	Marion	207
Dotterer mine	Earl	114
Douglassville quarry (Pottstown Trap Rock quarry)	Douglass	111
Dr. Funk's fish dam graphite Mine	Colebrookdale	89
Dragon Cave	Richmond	281
Dreibelbis quarry (Laureldale quarry)	Muhlenburg	219
Dresher mine	Longswamp	176
Drexel quarry	West Reading	357
Dunkle mines	Longswamp	160
Dyer Gibraltar quarry	Robeson	291
Dyer Monocacy quarry	Union	336
Dyer Trap Rock (Birdsboro Materials) quarry	Robeson	297
E & G Brooke Iron Company (Barto mine)	Washington	346
E & G Brooke Iron Company (Big Dam Quarry)	Lower Alsace	187
E & G Brooke Iron Company (Byler mine)	Caernarvon	79
E & G Brooke Iron Company (Jones mine)	Caernarvon	71
East Penn Graphite Company mine	Longswamp	177
Easterly area	St. Lawrence	333
Eastern Industries quarry	Oley	250
Eastern Industries quarry (Hinterleiter quarry)	Maxatawny	208
Eastern Lime Company (A.G. Smith quarry)	Maxatawny	212
Eastern Lime Company quarry (Hinterleiter quarry)	Maxatawny	208
Eastern Lime Corp (Eastern Industries quarry)	Oley	250
Eastern Steel Compnay (Boyertown mines)	Boyertown	56
Eban, Christian quarry	Reading	279
Eckenrode & Schlott (Ruth quarry No. 1)	Spring	322
Eckert & Brother (Wheatfield mine)	Spring	324
Eckert & Brother (Fritz Island mine)	Cumru	92
Eckert & Brother (Mount Penn mines)	Reading	273
Eckert & Brother (Schitter mine)	Ruscombmanor	315
Eckert mine (Moatz & Schrader mine)	Longswamp	166
Eckert open cut	Boyertown	41
Eckert slope (Eckert open cut)	Boyertown	41
Edison mine (L. Gilbert mine)	Washington	348
Eline mine	Washington	350
Elm & Rose Streets quarry	Reading	278

APPENDIX 2

Mine Name (Principle mine name in parenthesis)	Township	Page
Emaus Furnace mine (Dresher mine)	Longswamp	176
Epler & Leinback quarrry (Epler quarries)	Bern	28
Epler quarries	Bern	27
Eppler & Rischville sandstone quarry (Shonour's quarry)	Centre	85
Eshbach area	Washington	353
Essig quarry	Bern	26
Esterly mine	Exeter	122
Eureka mine	Spring	331
Eureka Ore Company (Eureka mine)	Spring	331
Evans quarry	Sinking Spring Borough	318
Evansville quarry	Maiden Creek	196
Fairgrounds mine (Mount Penn mines)	Reading	273
Fegley & Walbert mine	Longswamp	172
Fegley mine	Washington	351
Fegley, Issac (Charles Heffner mines)	Richmond	290
Fegley, John mine	Longswamp	174
Fegley, Jones and Gable (S. Bittenbender mine)+C715	Hereford	134
Fehr & O'Rourke quarry	Cumru	102
Fenstermacher mine	Longswamp	168
Filbert, Hiester ocher mine	Marion	207
Finley mine	Longswamp	162
Fischer prospect	Lower Alsace	190
Fisher mine (Leesport Iron Company mine)	Richmond	289
Fisher quarry (Bowers quarry)	Maxatawny	211
Fisher, Amos mine	Longswamp	170
Fisher, Daniel (Ruth mine)	Spring	329
Fisher, Daniel mine	Rockland	310
Fisher, J. ore holes (Daniel Fisher mine)	Rockland	310
Fisher, J.G. quarry	Oley	259
Fleetwood area	Richmond	286
Fleetwood Iron Company mine	Longswamp	170
Flint Hill locality	Rockland	310
Focht quarry	Albany	17
Focht, G.W. Stone Company quarry	Muhlenburg	222
Fox, Jacob farm locality	Ruscombmanor	317
Frank Thomson (S. Bittenbender mine)	Hereford	134
Franklin DeLong farm graphite prospect	Longswamp	186
Frantz farm fetid barite locality (McLain farm fetid barite locality)	Tulpehocken	335
Frederick mines	Longswamp	182
Freyermouth Quarry	Earl	116
Frill quarry (McQuade Brothers quarry)	West Reading	355
Fritch & Brother Klines Corner mine	Longswamp	168
Fritch & Brother's mines	Longswamp	181
Fritz Island mine	Cumru	92
Fritztown area	South Heidelberg	322
Fritztown area	Spring	333
Frystown fetid barite locality	Tulpehocken	334
Frystown fetid barite nodule locality	Bethel	28
Fulton quarry (McQuade Brothers quarry)	West Reading	356
G.W. Focht Stone Company quarry	Muhlenburg	222

APPENDIX 2

Mine Name (Principle mine name in parenthesis)	Township	Page
Gabel & Jones (Gabel mine)	Boyertown	50
Gabel mine	Boyertown	50
Gabel quarry	Washington	352
Gabel, Jones & Gabel mine (Gabel mine)	Boyertown	50
Gabelsville Graphite Works (Boyertown Graphite Co mine)	Colebrookdale	89
Gabelsville riebeckite locality	Earl	119
Gap mine	Longswamp	156
Gay, Thomas (Eline mine)	Washington	350
Gehman, J.B. heir's tract	Hereford	139
Gehman, J.B. mine	Hereford	139
Gehret quarry (G.W. Focht Stone Company quarry)	Muhlenburg	222
Geigley, John quarry	Union	344
Geist quarry (Miller quarry)	Bechtelsville	23
Gery's Gap mine (Gap mine)	Longswamp	156
Gibble farm fetid barite locality	Tulpehocken	336
Gibraltar quarry (Dyer Gibraltar quarry)	Robeson	291
Gickerville quarry (Dyer Gibraltar quarry)	Robeson	291
Gilberg, Jacob (J.B. Gehman mine)	Hereford	139
Gilberg, Jacob (S. Bittenbender mine)	Hereford	134
Gilberg, Jacob mine	Washington	350
Gilbert mine	Washington	349
Gilbert, L. (Edison) mine	Washington	347
Gilt flagstone quarry	Albany	17
Ginkinger mine	Longswamp	157
Glen-Gery quarry	Perry	268
Glen-Gery quarry	Lower Heidelberg	192
Goss, William quarry (Hottenstein quarry)	Maxatawny	210
Gottschall's farm locality	Alsace	20
Gottschall's mine	Alsace	20
Goul quarry No. 1	South Heidelberg	320
Goul quarry No. 2	South Heidelberg	320
Goul quarry No. 3	South Heidelberg	320
Grace mine	New Morgan Borough	227
Green Hill area	Rockland	309
Greenawalt quarry	Albany	17
Greenwich Manufacturing Company quarry	Greenwich	125
Gregory graphite prospect	Hereford	145
Greis & Company(C. Bittenbender mine)	Hereford	139
Greis & Wendling (BIC Gardner Station mine)	Longswamp	161
Greis, Gregory (Sparr mine)	Washington	350
Greiss & Hartzel (Wetzel mlne)	Longswamp	160
Gresh and Bechtel quarry (Greshville quarry)	Douglass	108
Gresh, Levi & David (Greshville quarry)	Douglass	108
Greshville quarry	Douglass	108
Grett Ore Company (Crane Iron Company mine)	Maiden Creek	203
Griesemer slate quarry	Oley	261
Griesermer, Ezra quarry	Oley	259
Grill mine	Spring	331
Grim, Seth quarry	Oley	257
Gring's quarry	Reading	277

APPENDIX 2

Mine Name (Principle mine name in parenthesis)	Township	Page
Gring's sand quarry	Spring	324
Grossler quarry	Muhlenberg	224
Grove Brothers mine (Landis mine)	Washington	345
Grubb & Geigley quarry	Union	344
Grubb Brother's quarry	Union	344
Grubb, Benjamin quarry	Union	344
Gudlin, William quarry	Wyomissing	358
Guiterman, Henry (J.B. Gehman mine)	Hereford	139
Guldin Hill sandstone quarry	Exeter	122
Guldin, Peter quarry	Oley	256
Guldin, S.P. quarry	Oley	256
H&K Group (Pottstown Trap Rock quarry)	Douglass	111
H&K Group quarry (Dyer Trap Rock/Birdsboro Materials quarry)	Robeson	297
Hagy & Rhoades mine (California mine)	Boyertown	41
Hagy pit (Eckert open cut)	Boyertown	41
Hain mine	South Heidelberg	322
Hain mine (Schitter mine)	Ruscombmanor	315
Haines & Kibblehouse (Pottstown Trap Rock quarry)	Douglass	111
Haines farm (Pricetown Area)	Ruscombmanor	316
Haines mine (Schitter mine)	Ruscombmanor	315
Hamlin, Danner (Deckert quarry)	Sinking Spring Borough	318
Hancock mud dam deposit	Longswamp	181
Hartman, Levi quarry	Oley	256
Hartman, Valentine mine	Alsace	20
Hartman, W. farm locality	Muhlenburg	216
Hartzel & Shimer (J.B. Gehman heir's tract)	Hereford	139
Hartzell (S. Bittenbender mine)	Hereford	136
Hartzog's Mill (Bishop's Mill Locality)	Exeter	120
Harvey, William (S. Bittenbender mine)	Hereford	133
Hassler & Lerch (Henry Miller quarry(	Wernersville	354
Hay Creek prehenite locality	Robeson	304
Heffner, Charles mines	Richmond	290
Heinly, Faust & Bros (Centennial quarry)	Albany	12
Heins quarry (Oberholtzer quarry)	Washington	351
Hemerly quarry	Albany	12
Henry Erwin and Sons ocher mine	Longswamp	183
Herb & Reitenauer (Oberholtzer quarry)	Washington	351
Herbein, Levi quarry	Oley	259
Hertzel & Swoyer mine	Oley	264
Hertzler quarry	Caernarvon	81
Hertzog mine	Longswamp	174
Hess & Ziegenfus (C. Bittenbender mine)	Hereford	139
Hess quarry (Hinterleiter quarry)	Maxatawny	208
Hess, George (J.B. Gehman heir's tract)	Hereford	139
Highway Materials, Inc. (Temple quarry)	Alsace	19
Hill Church riebeckite locality 5	Pike	271
Hill Church riebeckite locality 6	Pike	271
Hill quarry (Fehr and O'Rourke quarry)	Cumru	103
Hine quarry	Oley	258
Hinterleiter Crossing quarry (Hinterleiter quarry)	Maxatawny	208
Hinterleiter quarry	Maxatawny	208

APPENDIX 2

Mine Name (Principle mine name in parenthesis)	Township	Page
Hobart & Nauman (Sparr mine)	Washington	350
Hoffman prospect	District	107
Hospital Creek quarry	Lower Heidelberg	191
Hottenstein quarry	Maxatawny	210
Houck quarry	Oley	256
Huber farm fetid barite occurrence	Bethel	29
Huber quarry	Reading	278
Huff Church locality	Hereford	140
Hull, Benjamin quarries	Lower Heidelberg	191
Humphrey sandstone quarry	Robeson	304
Humphrey Stone Co (Humphrey sandstone quarry)	Robeson	304
Hunter mine	Oley	262
Hunter mine (Moslem mines)	Richmond	288
Huyett quarry	South Heidelberg	321
Iron mine number 7	Maxatawny	215
Iron mine number 8	Maxatawny	215
Iron mine number 12	Maxatawny	215
Iron mine number 13	Maxatawny	215
Iron mine number 14	Maxatawny	215
Iron mine number 15	Maxatawny	215
Iron mine number 17	Longswamp	173
Iron mine number 18	Longswamp	173
Iron mine number 19	Longswamp	173
Iron mine number 20	Longswamp	173
Iron mine number 22	Maxatawny	216
Iron mine number 23	Maxatawny	216
Iron mine number 24	Maxatawny	216
Iron mine number 27	Longswamp	170
Iron mine number 30	Longswamp	170
Iron mine number 31	Longswamp	171
Iron mine number 32	Longswamp	171
Iron mine number 33	Longswamp	171
Iron mine number 34	Longswamp	171
Iron mine number 35	Longswamp	171
Iron mine number 37	Longswamp	173
Iron mine number 38	Longswamp	174
Iron mine number 40	Longswamp	171
Iron mine number 41	Longswamp	171
Iron mine number 42	Longswamp	171
Iron mine number 43	Maxatawny	216
Iron mine number 46	Longswamp	171
Iron mine number 58	Maxatawny	215
Iron mine number 9	Maxatawny	215
Island mine (Fritz Island mine)	Cumru	92
J. Wilbur Company quarry	Albany	13
Jacksonwald occurrence	Exeter	119
Johnson quarry (Laureldale quarry(	Muhlenburg	219
Jones Good Luck (Jones mine)	Caernarvon	62
Jones mine	Caernarvon	62
Jones, Griffith (Hunter mine)	Oley	262

APPENDIX 2

Mine Name (Principle mine name in parenthesis)	Township	Page
Kauffman & Company (Seitzinger mine)	Wyomissing	357
Kauffman & Company mine (Moslem mines)	Richmond	289
Kauffman, F.V. quarry north	Oley	261
Kauffman, S. quarry	Exeter	123
Kaufman & Spang mine	Earl	111
Keehn quarry	Reading	278
Keely quarry	Douglass	109
Keim, John quarries	Pike	271
Kemmerer quarry	Oley	258
Kepple & Reber (Grill mine)	Spring	331
Kestner quarry	Reading	278
Keystone Ocher Co mine (C.K. Williams & Company ocher mine)	Richmond	282
Keystone quarry	Maxatawny	210
King quarry (Dishong quarry)	Marion	207
Kinney mine	Caernarvon	78
Kinsey Hill locality	Exeter	120
Kinzi's Mill (Bishop's Mill Locality)	Exeter	120
Kirbyville quarries	Richmond	286
Kirbyville-Moselem Springs quartz crystal locality	Richmond	290
Kirschmann quarry (Mount Penn quarries)	Reading	277
Klapp quarry	Reading	279
Klein, D.K. mine	Longswamp	166
Kline Umber Prospect	Bethel	29
Kline, Peter mine	Longswamp	162
Kline, Simon quarry	Reading	278
Klines Corner mine	Longswamp	170
Knabb, Albert quarry	Exeter	124
Knabb, David mine	Lower Alsace	190
Knabb, Levi quarry	Oley	258
Knabb, William slate quarry (Griesemer slate quarry)	Oley	261
Knauers sandstone quarry	Brecknock	59
Knorr (Witmoyer & Brother quarry)	Wernersville	353
Kohler quarry	Maxatawny	212
Kohler Road quarry	Maxatawny	214
Kostenbader quarry (Mount Penn quarries)	Reading	277
Kunkle quarry	Albany	17
Kurtz, Elvin farm fetid barite locality	Tulpehocken	336
Kurztown Crushed Stone & Lime Co quarry (Keystone quarry)	Maxatawny	210
Kutz, Lewis (Boyertown Graphite Co mine)	Colebrookdale	89
Kutz, Sel quarry (Hinterleiter quarry)	Maxatawny	208
Kutz-Landis farm fetid barite locality	Tulpehocken	336
Kutztown quarry (Hinterleiter quarry)	Maxatawny	208
Kutztown quarry (Keystone quarry)	Maxatawny	211
Lance (Neversink Mt sienna mine)	Lower Alsace	190
Landis mine	Washington	345
Landis, H.N. mine (Dotterer mine)	Earl	114
Laros, Jesse mine No. 4	Longswamp	174
Laros, Jesse mine No.3	Longswamp	174
Laureldale quarry	Muhlenberg	216
Lebrandt & McDowell (Moslem mines)	Richmond	289

The Mines and Minerals of Berks County

APPENDIX 2

Mine Name (Principle mine name in parenthesis)	Township	Page
Lee quarry and farm Locality	Oley	261
Leesport Iron Company mine	Richmond	286
Leesport Iron Company quarries	Ontelauee	267
Legler flagstone quarry	Albany	17
Lehigh Cement Company (Richmond No. 5 quarry)	Richmond	291
Lehigh Cement Company quarry	Oley	254
Lehigh Cement LLC (Evansville quarry)	Maiden Creek	201
Leiby flagstone quarry	Perry	270
Leinbach's Hill locality (McQuade Brothers quarry)	West Reading	355
Leinbach's Hill quarry (McQuade Brothers quarry)	West Reading	355
Lenhartsville quarry (Greenwich Manufacturing Co quarry)	Greenwich	125
Lesher, Jacob (Rock mine)	Longswamp	153
Lewis brothers mine (California mine)	Boyertown	41
Lewis mine (Phoenix middle slope)	Boyertown	43
Lewis tract mines (Phoenix upper slope)	Boyertown	44
Lewis, E. (L. Gilbert mine)	Washington	348
Lewis, Samuel mine	Maxatawny	215
Lichenwallner mine	Longswamp	176
Limestone quarry No. 2	South Heidelberg	321
Little Oley riebeckite locality	Douglass	111
Litzenberger mine	Longswamp	176
Livingood quarry (Greshville quarry)	Douglass	108
Livingood, J.S. (Ruth mine)	Spring	329
Lobachsville mines	Pike	270
Long mines	Longswamp	177
Long's quarry	Reading	277
Ludwig & Mauger (Hunter mine)	Oley	262
Ludwig quarry	Spring	322
Lux and Relser quarry	Union	344
Lyken-Byler mine (Byler mine)	Caernarvon	78
Lykens quarry	Union	344
Lyons area	Maxatawny	216
Maidencreek Portland Cement Co	Maiden Creek	195
Maidencreek quarry	Maiden Creek	201
Maidencreek Station quarry	Ontelauee	267
Mammouth quarry	Albany	12
Manatawny Bessemer Ore Co (Kauffman & Spang mine)	Earl	111
Manatawny Mining Co (Kauffman & Spang mine)	Earl	111
Manwiller mine	Oley	263
Marcasite locality	Centre	84
Marquart quarry (L.J. Bertolet quarry)	Oley	259
Marshall, John quarry	Heidelberg	127
Martin quarry (Gabel quarry)	Washington	353
Martin Stone Quarries, Inc. quarry (Gabel quarry)	Washington	353
Mast, Christian quarry	Caernarvon	83
Mast, Daniel quarry	Caernarvon	80
Mast, Stephen quarry (Daniel Mast quarry)	Caernarvon	81
Mayberry (Rock mine)	Longswamp	153
McCellan quarry (Hottenstein quarry)	Maxatawny	210
McKnight's Gap	Alsace	21

APPENDIX 2

Mine Name (Principle mine name in parenthesis)	Township	Page
McLain farm fetid barite locality	Tulpehocken	335
McManus (Wheatfield mine)	Spring	324
McQuade Brothers quarry	West Reading	355
Meckley mine (Old Mickley mine)	Longswamp	162
Mengle's quarry	Earl	117
Mensch quarry (Pottstown Trap Rock quarry)	Douglass	110
Meredith, J.M. quarry	Maiden Creek	201
Merkel mine	Longswamp	172
Merkel quarries	Richmond	291
Mica prospect No. 1	Earl	117
Mica prospect No. 2	District	107
Mica prospect No. 3	Pike	271
Michael Haak (Neversink Mountain sienna mine)	Lower Alsace	190
Miller farm fetid barite occurrence	Bethel	29
Miller farm mines	Longswamp	161
Miller mine (Mount Penn mines)	Reading	273
Miller quarry	Bechtelsville	22
Miller, Charles mine	Maxatawny	215
Miller, Henry quarry	Wernersville	353
Moatz & Schrader mine	Longswamp	166
Moll & Gery tract (Gap mine)	Longswamp	156
Moll & Spinner (Gap mine)	Longswamp	156
Moll & Worst (C. Bittenbender mine)	Hereford	139
Molltown Cement Co	Maiden Creek	193
Monocacy Hill	Exeter	120
Monocacy Iron Company (Ruth mine)	Spring	329
Monocacy quarry (Dyer Monocacy quarry)	Union	336
Montello Brick Company quarry	Spring	332
Moore farm fetid barite locality	Tulpehocken	335
Moore, William quarry	Heidelberg	127
Morganroth & Hiteman (J.B. Gehman mine)	Hereford	139
Morgantown roadcut scapolite occurrence	Caernarvon	84
Moselem Iron Company (Moslem mines)	Richmond	289
Moselem mines	Richmond	286
Moselem Mining Company (Moslem mines)	Richmond	289
Moser, James (BIC Gardner Station mine)	Longswamp	161
Moss Street quarry	Reading	278
Mount Penn Iron mines	Reading	273
Mount Penn quarries	Reading	277
Mount Penn Sand & Stone Works quarry (Mount Penn quarries)	Reading	277
Mount Penn white spot	Reading	275
Mt. Etna fetid barite locality	Tulpehocken	334
Muhlenberg (Beidler) mine	Wyomissing	358
Nathan Haas mine	Longswamp	176
National Gypsum Co (Evansville quarry)	Maiden Creek	201
Nester, Harvey (Bechtelsville quarry)	Colebrookdale	85
Neversink Mountain sienna mine	Lower Alsace	189
New Berlinville clay mine	Colebrookdale	88
New Jerusalem mines	Rockland	310
Nine (Hine) quarry	Oley	258

The Mines and Minerals of Berks County

APPENDIX 2

Mine Name (Principle mine name in parenthesis)	Township	Page
Nolan's Cave	Muhlenberg	221
Noll's mine	Richmond	290
Northampton Iron Company (Amos Fisher mine)	Longswamp	170
Northampton iron Company (Wetzel mine)	Longswamp	160
Oakley & Richards mine (Mount Penn mines)	Reading	273
Oberholtzer quarry	Washington	351
Ohlinger dam (Antietam Reservoir locality)	Lower Alsace	187
Ohlinger Mill Locality (Stony Creek Mills locality)	Lower Alsace	190
Ohnmacht cave (Essig quarry)	Bern	26
Olafson mine	Hereford	139
Old Mickley mine	Longswamp	162
Old Millert mine	Rockland	309
Old quarry No. 1	Lower Heidelberg	192
Old quarry No. 2	Lower Heidelberg	192
Old quarry No. 3 (Breneman quarry)	Spring	324
Old tunnel mine	Ruscombmanor	314
Oley (Friedensburg) area	Oley	264
Oley No. 1 quarry (Lehigh Portland Cement Co)	Oley	254
Oley No. 2 quarry (Lehigh Portland Cement Co)	Oley	254
Oley uranium occurrence	Oley	264
Oley Valley Electric Railway cut riebeckite locality	Earl	119
Oley Valley quarry (Eastern Industries quarry)	Oley	250
Oley West quarry (Lehigh Cement Co)	Oley	254
Ontelaunee quarry	Ontelauee	265
Opposite Fritz Island locality	Cumru	102
Oswold quarry	Albany	12
Peacock quarry (Rickenbach Station quarries)	Bern	27
Peipher quarry	Sinking Spring Borough	318
Penn Limestone quarry (Hinterleiter quarry)	Maxatawny	208
Pennsylvania Copper Company (Jones mine)	Caernarvon	63
Pennsylvania Uranium Mining Co Mine	Earl	117
Philadelphia & Reading Coal & Iron Co (S. Bittenbender mine)	Hereford	134
Philadelphia Slate Mantle Company quarry	Albany	11
Phoenix Iron Company (California mine)	Boyertown	41
Phoenix Iron Company (Jones mine)	Caernarvon	71
Phoenix iron Company (Mount Penn mines)	Reading	273
Phoenix Iron Company (Phoenix middle slope)	Boyertown	43
Phoenix Iron Company (Phoenix upper slope)	Boyertown	44
Phoenix Iron Company (Radenbush mine)	Cumru	102
Phoenix Iron Company (Stauffer mine)	Washington	350
Phoenix lower slope (California mine)	Boyertown	41
Phoenix middle slope	Boyertown	43
Phoenix upper slope	Boyertown	43
Pine Iron Works (Hunter mine)	Oley	262
Pittsburg quarry	Albany	12
Pittsburg Slate Company (Pittsburg quarry)	Albany	13
Plank, David quarry (Hertzler quarry)	Caernarvon	81
Plank, John quarry	Caernarvon	82
Plank, P.W. & Company lime kilns	Caernarvon	83
Pleasant Hill Road quarry	Maiden Creek	201

APPENDIX 2

Mine Name (Principle mine name in parenthesis)	Township	Page
Potteiger & Weitzel (Ruth quarry No. 2)	Spring	324
Pottstown Iron Compant (Hunter mine)	Oley	262
Pottstown Iron Company (Barto mine)	Washington	346
Pottstown Trap Rock quarry	Douglass	109
Price's quarry	Centre	85
Pricetown Area	Ruscombmanor	316
Pupek quarry (Dyer Trap Rock/Birdsboro Materials quarry)	Robeson	297
Quaker City Slate Company quarry	Albany	11
Radenbush mine	Cumru	100
Railroad cut opposite Polar Neck	Lower Alsace	189
Railroad cuts south of Boyertown	Colebrookdale	89
Rapp quarry	Earl	117
Raub mine	Spring	330
Rauch mine	Hereford	140
Raudenbush quarry	Oley	259
Reading Cement Company quarry	Maiden Creek	193
Reading Crushed Stone Co quarry (Rickenbach Station quarries)	Bern	27
Reading Iron Company (Deysher mine)	Washington	347
Reading Iron Company quarry (Angstadt quarry)	Maxatawny	212
Reading Iron Works (Fritz Island mine)	Cumru	92
Reading locality of Dana (1854) (Valentine Hartman mine)	Alsace	20
Reading Quarry Company quarry (Laureldale quarry)	Muhlenburg	216
Reading Railroad quarry	Bern	28
Reading Railroad quarry (G.W. Focht Stone Company quarry)	Muhlenburg	222
Reading Sand & Stone Company quarry (Mount Penn quarries)	Reading	277
Reber shaft (Raub mine)	Spring	331
Red Bridge quartz crystal locality	Bern	28
Red Oxide iron ore mine (Dotterer mine)	Earl	114
Reed (Wheatfield mine)	Spring	324
Reed quarry	Heidelberg	128
Reeves, Buck & Company (Radenbush mine)	Cumru	102
Refractory Sand Co (Gring's sand quarry)	Spring	324
Refractory Sand Co (Sand Hill sand quarries)	Ruscombmanor	317
Reichard-Coulston, Inc.(Henry Erwin& Sons ocher mine)	Longswamp	183
Reiffe quarry	Oley	261
Reifsnyder quarry (McQuade Brothers quarry)	West Reading	356
Reifton quarry	Exeter	125
Reily, William (East Penn Graphite Co)	Longswamp	177
Reily, William (Long mines)	Longswamp	177
Reinhardt (Frederick mines)	Longswamp	182
Reitnaur & Strohl mine (Atlas Ocher mine)	Longswamp	185
Reitnaur mine	Hereford	141
Rhoades & Grim mine	Colebrookdale	87
Rhoades, J. sandstone quarry	Amity	22
Rhoads farm (Pricetown Area)	Ruscombmanor	316
Rhoads Mining Company	Boyertown	54
Richmond quarry No. 1 (Evansville quarry)	Maiden Creek	196
Richmond quarry No. 5	Richmond	291
Rickenbach Station quarries	Bern	27
Rickenbach, J.B.K. quarry (Rickenbach Station quarries)	Bern	27

APPENDIX 2

Mine Name (Principle mine name in parenthesis)	Township	Page
Rittenhouse cut (Rock mine)	Longswamp	153
Rittenhouse Gap District	Longswamp	149
Rittenhouse Iron Company of Pennsylvania	Longswamp	150
Ritter, Benjamin quarry	Exeter	124
Robeson quarry (Dyer Gibraltar quarry)	Robeson	291
Rock mine	Longswamp	153
Rohrback mine	Pike	270
Rolling Rock building stone quarry	Pike	271
Romig, Simon (Boyertown Graphite Co mine)	Colebrookdale	89
Romig, Simon (Tatham mine)	Longswamp	182
Roth, Henry quarry	Hereford	141
Rothermel mine (Leesport Iron Company mine)	Richmond	288
Rothermel quarry (Temple Sand Company quarry)	Muhlenburg	224
Route 222 Roadcut Locality	Cumru	105
Rowe, William (Deysher mine)	Washington	347
Rowe, William (Landis mine)	Washington	345
Rush quarries	Washington	352
Rush's ore pit	Hereford	140
Ruth & Foreman quarries (Ruth quarry No. 1)	Spring	322
Ruth & Foreman quarries (Ruth quarry No. 2)	Spring	322
Ruth mine	Spring	329
Ruth quarry No. 1	Spring	322
Ruth quarry No. 2	Spring	322
Ruth, John sand quarry (Sheetz sand quarry)	Heidelberg	129
Rutter's iron mine (Boyertown mines)	Boyertown	36
Ryland Road quarry	Heidelberg	129
Sadler farm fetid barite occurrence	Bethel	29
Salem Church ocher pit locality	Alsace	21
Sand Hill sand quarries	Ruscombmanor	316
Schaeffer, E. quarry	Oley	257
Schall quarry	Washington	351
Schankweiler, Henry (C. Bittenbender mine)	Hereford	139
Schantz quarry	Hereford	142
Schearer. Reuben quarry	Oley	260
Schitter mine	Ruscombmanor	314
Schlegel's farm locality	Cumru	105
Schmeck's farm graphite locality	Longswamp	186
Schollenberger quarry	Oley	260
Schrading mine	Rockland	310
Schroeder farm (Pricetown Area)	Ruscombmanor	316
Schroeder, Weiss & Brensinger (Dunkle mines)	Longswamp	160
Schuler, John (Boyertown Graphite Co mine)	Colebrookdale	89
Schuylkill Copper Mining Company mine	Cumru	105
Schuylkill Stone Company quarry (Dyer Monocacy quarry)	Union	336
Schuylkill Valley Clay Manufacturing Co (Glen-Gery quarry)	Perry	268
Schuylkill Valley Stone Company (Dyer Monocacy quarry)	Union	336
Schuylkilll Valley Sand Company	Tilden	333
Schwartz (Wheatfield mine)	Spring	324
Schweinbinz & Company (J.B. Gehman mine)	Hereford	139
Schweiter & Kutz mine	Ruscombmanor	315

APPENDIX 2

Mine Name (Principle mine name in parenthesis)	Township	Page
Schweitzer farm sandstone quarry	Brecknock	59
Schweyer & Leiss mine	Maxatawny	215
See, George sand quarry (Sheetz sand quarry)	Heidelberg	129
Seitzinger mine	Wyomissing	357
Seltzer, Daniel quarry	South Heidelberg	321
Seltzer, John sand quarry (Sheetz sand quarry)	Heidelberg	129
Seltzer, Mary sand quarry (Sheetz sand quarry)	Heidelberg	129
Seminary shaft	Boyertown	55
Shaeffer's new mine	Maiden Creek	206
Shaeffer's old mine	Maiden Creek	206
Shalters quarry (Laureldale quarry(	Muhlenburg	219
Shanesville riebeckite locality	Earl	119
Shankweller (Fritch & Brother mine)	Longswamp	168
Sharadin quarry (Berks Products quarry)	Maxatawny	210
Shaub, Jonas quarry	Hereford	141
Sheetz sand quarry	Heidelberg	129
Shillington Road quarry	Sinking Spring Borough	319
Shiloh Hills locality (Route 222 Roadcut Locality)	Cumru	105
Shimer, F.S. (J.B. Gehman mine)	Hereford	139
Shoemakersville Clay Works (Glen-Gery quarry)	Perry	268
Shonour's quarry	Centre	85
Siesholtzville granite quarry	Hereford	143
Simons (Jones mine)	Caernarvon	69
Sinking Springs area	Sinking Spring Borough	319
Six Penny quarry (Dyer Monocacy quarry)	Union	336
Smaltz Road quarry	Marion	207
Smith & Morganroth (J.B. Gehman heir's tract)	Hereford	139
Smith, A.G. quarry	Maxatawny	212
Smith, Peter farm graphite prospect	District	107
Smith, S. mine	Maxatawny	215
Smoyer mine	Longswamp	162
Sneider, P. quarry	Oley	258
Snyder, D. quarry	Exeter	123
Snyder, John quarry	Oley	259
Snydersville malachite occurrence	Exeter	122
Solomon Boyer & Company mine	Longswamp	174
South Temple quarry	Muhlenburg	221
Southern Coal and Iron Company	Longswamp	150
Spang's iron mine (Kauffman & Spang's mine)	Earl	111
Sparr mine	Washington	350
Standard Slate Company (Centennial quarry)	Albany	12
Star Clay Company	Longswamp	178
Stauffer mine	Washington	350
Stauffer quarry	St. Lawrence	333
Stauffer quarry (Mount Penn quarries)	Reading	277
Stauffer's quarries	Colebrookdale	86
Steel Ore Company (Gabel mine)	Boyertown	50
Steel Ore Company (S. Bittenbender mine)	Hereford	134
Stein, Henry mine	Longswamp	175
Sternburg quarry	Reading	279

APPENDIX 2

Mine Name (Principle mine name in parenthesis)	Township	Page
Stitzel quarry	Reading	278
Stoltzfus, John quarry	Caernarvon	82
Stoltzfus, Mast quarry (Styer quarry)	Caernarvon	80
Stoltzfus, Walter (John Stoltzfus quarry)	Caernarvon	83
Stonersville magnetite occurrence	Exeter	121
Stony Creek Mills locality	Lower Alsace	190
Storb Crushed Stone quarry (Pottstown Trap Rock quarry)	Douglass	110
Stouchsburg quarry (Dishong quarry)	Marion	207
Stowe Trap Rock quarry (Pottstown Trap Rock quarry)	Douglass	110
Styer quarry	Caernarvon	79
Sweitzer quarry (Angstadt quarry)	Maxatawny	212
Sweitzer quarry (Bowers quarry)	Maxatawny	211
Sweyer, Allen (Klines Corner mine)	Longswamp	170
Talley mine	Oley	264
Tatham mine	Longswamp	182
Temple Crushed Stone Company (Temple quarry)	Alsace	19
Temple Iron Company (Charles Heffner mines)	Richmond	290
Temple Iron Company (Ziegler mine)	Longswamp	172
Temple quarry	Alsace	19
Temple Sand Company quarry	Muhlenburg	223
Temple Silica Sand Co quarry (Temple Sand Co quarry)	Muhlenburg	224
Temple Slag Company (Temple quarry)	Alsace	19
Thalheimer's sand hole	Reading	278
Theirwechler mine	Sinking Spring Borough	319
Thomas Iron Company (Gap mine)	Longswamp	156
Thomas Iron Company (Henry Stein mine)	Longswamp	175
Thomas Iron Company (Litzenberger mine)	Longswamp	176
Thomas Iron Company (Long mines)	Longswamp	177
Thomas Iron Company (Old Mickley mine)	Longswamp	162
Thomas Iron Company (Rock mine)	Longswamp	154
Thomas Iron Company (Tunnel mine)	Longswamp	154
Thomas Iron Company (Weiler mine)	Longswamp	160
Thomas Iron Company Gardner Station mines	Longswamp	162
Thomas Iron Company Klines Corner mine	Longswamp	166
Thomas Iron Company mine	Longswamp	160
Thomas Iron Company Mine (Trexler mine)	Longswamp	160
Tilli quarry (Pottstown Trap Rock quarry)	Douglass	110
Topton Furnace quarry (Hinterleiter quarry)	Maxatawny	208
Trap Rock quarry (Dyer Trap Rock/Birdsboro Materials quarry)	Robeson	297
Trexler mica mine	Alsace	19
Trexler mine	Longswamp	160
Trexler, D.L mine	Longswamp	166
Trexler, E.H. mine (D.L. Trexler mine)	Longswamp	166
Trexler, E.H. quarry	Longswamp	186
Trexler, Edwin (D.L. Trexler mine)	Longswamp	166
Trexler, Jonas (D.L. Trexler mine)	Longswamp	166
Trexler, Reuben (Rock mine)	Longswamp	153
Trollingers quarry	Hereford	142
Trout quarry (Fehr and O'Rourke quarry)	Cumru	103
Tulpehocken Stone Company quarry	North Heidelberg	250

Mine Name (Principle mine name in parenthesis)	Township	Page
Tunnel mine	Longswamp	154
Twelfth Street quarry	Reading	278
Tyson, Cornelius quarry	Exeter	124
U.S. Route 222 Roadcut Locality	Cumru	105
Udree mine	Ruscombmanor	312
Unnamed slate mine	Albany	17
Unnamed slate quarry 1	Albany	13
Unnamed slate quarry 2	Albany	13
Vindex Portland Cement Co quarry (Reading Cement Co quarry)	Maiden Creek	194
Virginville area	Richmond	290
W.M. Kauffman & Company (Fenstermacher mine)	Longswamp	170
Wade mine	Maiden Creek	206
Wagenhorst mine	Longswamp	174
Wainright and Son (Boyertown Graphite Co mine)	Colebrookdale	89
Walbert mine (Fegley & Walbert mine)	Longswamp	172
Walker granite quarry	Longswamp	185
Wanner, John (Westley's quarry)	Centre	85
Warner quarry (Dyer Gibraltar quarry)	Robeson	291
Warwick Iron Company (Manwiller mine)	Oley	263
Warwick Iron Company (S. Bittenbender mine)	Hereford	136
Warwick Iron Company (Warwick mine)	Boyertown	46
Warwick mine	Boyertown	46
Washington & Pear Streets quarry	Reading	278
Waste Management (Dyer Gibraltar quarry)	Robeson	291
Weaver & Company (Beitler mine)	Rockland	307
Weaver mine	Oley	262
Weaver, Col. J. quarry	Oley	257
Weaver, William & Sons (Boyertown Graphite Co mine)	Colebrookdale	89
Weidman farm shale prospect	Perry	270
Weidner, William quarry	Oley	260
Weiler mine	Longswamp	160
Weiser Lake (Seitzinger mine)	Wyomissing	357
Weist School locality	Alsace	21
Wells & Hoffman quarry	Union	344
Wenrich, Albert quarry	Robesonia Borough	305
Wenrich, W. quarry	Heidelberg	128
Wernersville Lime & Stone Co quarry (Keystone quarry)	Maxatawny	211
Wernersville Lime & Stone Company quarry	Wernersville	354
Wescoe mine	Longswamp	174
West Reading mine hole (Seitzinger mine)	Wyomissing	357
Westley's quarry	Centre	84
Wetzel mine	Longswamp	160
Wheatfield mine	Spring	324
Wheatfield Mining Company (Wheatfield mine)	Spring	325
Whitman quarry	Reading	275
Whitman, D. Harvey quarry	Union	344
Wilman quarry	Oley	260
Wilson, B. quarry (Christian Mast quarry)	Caernarvon	83
Witmoyer & Brother quarry	Wernersville	353
Wren mine	Douglass	109

APPENDIX 2

Mine Name (Principle mine name in parenthesis)	Township	Page
Wren, Major mine (Boyertown Graphite Co mine)	Colebrookdale	89
Yale quarry	Oley	261
Yoder, David quarry	Oley	260
Yoder, S. mine	Pike	270
Young & Dowdy (Barto mine)	Washington	347
Zeigler mine	Longswamp	172
Ziegenfus, Nathan (Solomon Boyer & Company mine)	Longswamp	174
Ziemer farm sandstone quarry	Brecknock	59
Zion Church locality (Valentine Hartman mine)	Alsace	20

APPENDIX 3

Map Number	Name of Mine	Appendex 1 Figure Number
A-1	Quaker City Slate Company quarry	1
A-2	Hemerly quarry	1
A-3	Mammouth quarry	1
A-4	Oswold quarry	1
A-5	Centennial quarry	1
A-6	Pittsburg quarry	1
A-7	Unnamed slate quarry 1	1
A-8	Unnamed slate quarry 2	1
A-9	J. Wilbur Company quarry	2
A-10	Albany quarry	2
A-11	B. Frank Ruth and Company quarry	2
A-12	Focht quarry	2
A-13	Unnamed slate mine	2
A-14	Kunkle quarry	1
AL-1	Temple quarry	13
AL-2	Trexler mica mine	27
AL-3	Gottschall's mine	13
AL-4	Gottschall's farm locality	13
AL-5	Hartman, Valentine mine	29
AL-6	Salem Church ocher pit locality	13
AL-7	Weist School locality	29
AL-8	McKnight's Gap	27
AM-1	Rhoades, J. sandstone quarry	33
B-1	Essig quarry	24
B-2	Rickenbach Station quarries	9
B-3	Epler quarries	9
B-4	Red Bridge quartz crystal locality	26
B-6	Reading Railroad quarry	9
BB-1	Gabel mine	34
BB-2	Warwick mine	34
BB-3	Rhoads Mining Company	34
BB-4	California mine	34
BB-5	Eckert open cut	34
BB-7	Phoenix middle slope	34
BB-8	Phoenix upper slope	34
BB-9	Seminary shaft	34
BE-1	Huber farm fetid barite occurrence	6
BE-2	Miller farm fetid barite occurrence	6
BE-3	Sadler farm fetid barite occurrence	6
BT-1	Miller quarry	34
C-1	Jones mine	37
C-2	Kinney mine	37
C-3	Byler mine	35
C-4	Styer quarry	35
C-5	Mast, Daniel quarry	35

APPENDIX 3

Map Number	Name of Mine	Appendex 1 Figure Number
C-6	Hertzler quarry	37
C-7	Plank, John quarry	35
C-8	Plank, P.W. & Company lime kilns	37
C-9	Stoltzfus, John quarry	35
C-10	Mast, Christian quarry	37
C-11	Morgantown roadcut scapolite occurrence	35
CB-1	Brower mine	34
CB-2	Boyertown Graphite Company mine	34
CB-3	Dr. Funk's fish dam graphite Mine	34
CB-4	Bechtelsville (Martin) quarry	34
CB-5	Rhoades & Grim mine	34
CB-6	New Berlinville clay mine	34
CB-7	Stauffer's quarries	34
CU-1	Fritz Island mine	28
CU-2	Radenbush mine	28
CU-3	Opposite Fritz Island locality	28
CU-4	Schuylkill Copper Mining Company mine	28
CU-5	Fehr & O'Rourke quarry	28
CU-6	Almshouse quarry	28
CU-8	Schlegel's farm locality	27
D-1	Wren mine	34
D-2	Greshville quarry	34
D-3	Davidheiser, J. quarry	34
D-4	Keely quarry	34
D-5	Pottstown Trap Rock quarry	32
D-6	Little Oley riebeckite locality	34
DS-1	Mica prospect No. 2	18
DS-2	Hoffman prospect	18
E-1	Kaufman & Spang mine	33
E-2	Freyermouth Quarry	33
E-3	Rapp quarry	33
E-4	Boyer quarry	33
E-5	Mica prospect No. 1	33
E-6	Gabelsville riebeckite locality	34
E-7	Shanesville riebeckite locality	34
E-8	Oley Valley Electric Railway cut riebeckite locality	34
E-9	Dotterer mine	17
E-10	Davidheiser, David quarry	33
E-11	Mengle's quarry	33
EX-1	Jacksonwald occurrence	29
EX-2	Kinsey Hill locality	31
EX-3	Bishop's Mill Locality	31
EX-4	Monocacy Hill	32
EX-5	Snydersville malachite occurrence	31
EX-6	Stonersville magnetite occurrence	30

APPENDIX 3

Map Number	Name of Mine	Appendex 1 Figure Number
EX-7	Esterly mine	31
EX-8	Bishop mine	31
EX-9	Kauffman, S. quarry	30
EX-10	Knabb, Albert quarry	30
EX-11	Snyder, D. quarry	30
EX-12	Ritter, Benjamin quarry	29
EX-13	Tyson, Cornelius quarry	29
EX-14	Abandoned quarry at Jacksonwald	29
EX-15	Reifton quarry	31
EX-16	DeTurk, Jonas quarry	27
EX-17	Guldin Hill sandstone quarry	29
G-1	Greenwich Manufacturing Company quarry	2
H-1	Bittenbender, William mine	19
H-2	Bittenbender, Samuel mine	19
H-3	Bittenbender, Christian mine	19
H-4	Siesholtzville granite quarry	19
H-5	Olafson mine	19
H-6	Rauch mine	19
H-7	Dale mine	20
H-8	Gehman, J.B. mine	19
H-10	Rush's ore pit	20
H-11	Christman, James & Lewis quarry	19
H-12	Christman, Jacob quarry	19
H-13	Roth, Henry quarry	19
H-14	Shaub, Jonas quarry	19
H-15	Schantz quarry	19
H-16	Trollingers quarry	18
H-17	Clemmerer quarry	20
HB-1	Moore, William quarry	21
HB-2	Marshall, John quarry	22
HB-3	Deppen. Samuel quarry	22
HB-5	Wenrich, W. quarry	22
HB-6	Reed quarry	23
HB-7	Big Spring quarry	23
HB-8	Ryland Road quarry	22
HB-9	Sheetz sand quarry	22
L-1	Rock mine	16
L-2	Tunnel mine	16
L-3	Conrad's slope	16
L-4	Gap mine	16
L-5	Ginkinger mine	16
L-6	Thomas Iron Company mine	16
L-7	Trexler mine	19
L-8	Weiler mine	19
L-9	Wetzel mIne	19

APPENDIX 3

Map Number	Name of Mine	Appendex 1 Figure Number
L-10	Dunkle mines	19
L-11	Miller farm mines	19
L-12	Bethlehem Iron Company Gardner Station mine	19
L-13	Thomas Iron Company Gardner Station mines	19
L-14	Finley mine	19
L-15	Old Mickley mine	19
L-16	Smoyer mine	19
L-17	Frederick mines	16
L-18	Fritch & Brother's mines	16
L-19	Tatham mine	16
L-20	Kline, Peter mine	19
L-21	Hancock mud dam deposit	16
L-22	Laros, Jesse mine No.3	16
L-23	Allentown Iron Company mine	5
L-24	Thomas Iron Company Klines Corner mine	4
L-25	Merkel mine	5
L-26	Iron mine number 35	5
L-27	Fisher, Amos mine	5
L-29	Iron mine number 40	4
L-31	Trexler, E.H. quarry	5
L-32	Iron mine number 27	5
L-34	Iron mine number 30	5
L-35	Iron mine number 31	5
L-36	Iron mine number 32	5
L-37	Iron mine number 33	5
L-38	Iron mine number 34	5
L-40	Iron mine number 38	5
L-41	Iron mine number 41	4
L-42	Iron mine number 42	4
L-45	Iron mine number 46	4
L-46	Trexler, D.L mine	4
L-47	Zeigler mine	5
L-48	East Penn Graphite Company mine	16
L-48	Long mines	16
L-48	Star Clay Company	16
L-49	Iron mine number 17	5
L-50	Iron mine number 20	5
L-51	Fegley & Walbert mine	5
L-52	Laros, Jesse mine No. 4	5
L-53	Hertzog mine	5
L-54	Wescoe mine	19
L-55	Wagenhorst mine	16
L-56	Stein, Henry mine	5
L-57	Dresher mine	5
L-58	Fegley, John mine	5

APPENDIX 3

Map Number	Name of Mine	Appendex 1 Figure Number
L-59	Solomon Boyer & Company mine	19
L-60	Litzenberger mine	16
L-61	Lichenwallner mine	16
L-62	Nathan Haas mine	5
L-63	Walker granite quarry	19
L-64	Iron mine number 18	5
L-65	Henry Erwin and Sons ocher mine	15
L-66	Iron mine number 19	5
L-67	Iron mine number 37	5
LA-1	Big Dam Quarry	27
LA-2	Antietam Reservoir locality	29
LA-3	Railroad cut opposite Polar Neck	27
LA-4	Neversink Mountain sienna mine	27
LA-6	Knabb, David mine	29
LA-7	Fischer prospect	27
LH-1	Hull, Benjamin quarries	23
LH-2	Hospital Creek quarry	23
LH-3	Glen-Gery quarry	24
LH-4	Old quarry No. 1	24
LH-5	Old quarry No. 2	24
M-1	Hinterleiter quarry	4
M-2	Berks Products Corp quarry	12
M-3	Bowers quarry	15
M-4	Hottenstein quarry	3
M-5	Angstadt quarry	15
M-6	Iron mine number 9	5
M-7	Kohler Road quarry	3
M-8	Iron mine number 12	5
M-9	Iron mine number 13	5
M-10	Iron mine number 14	5
M-11	Iron mine number 15	5
M-12	Bailey quarry	4
M-13	Iron mine number 7	4
M-14	Iron mine number 8	4
M-15	Smith, S. mine	4
M-16	Lewis, Samuel mine	4
M-17	Iron mine number 22	4
M-18	Iron mine number 23	4
M-19	Iron mine number 24	4
M-20	Iron mine number 43	4
M-21	Miller, Charles mine	4
M-22	Iron mine number 58	12
M-23	Smith, A.G. quarry	4
M-24	Kohler quarry	3
M-25	Keystone quarry	3

APPENDIX 3

Map Number	Name of Mine	Appendex 1 Figure Number
MA-1	Dishong quarry	21
MA-2	Smaltz Road quarry	21
MC-1	Evansville quarry	8
MC-2	Meredith, J.M. quarry	8
MC-3	Shaeffer's new mine	11
MC-4	Crane Iron Company mine	7
MC-5	Reading Cement Company quarry	8
MC-6	Pleasant Hill Road quarry	11
MC-7	Maidencreek quarry	8
MU-1	Hartman, W. farm locality	10
MU-2	Bernhart's Dam locality	10
MU-3	Laureldale quarry	10
MU-4	South Temple quarry	10
MU-5	Focht, G.W. Stone Company quarry	10
MU-5	G.W. Focht Stone Company quarry	10
MU-6	Temple Sand Company quarry	10
MU-7	Grossler quarry	10
N-1	Grace mine	35
NH-1	Tulpehocken Stone Company quarry	22
O-1	Eastern Industries quarry	30
O-2	Lehigh Cement Company quarry	30
O-2A	Oley West quarry (Lehigh Cement Co)	30
O-2B	Oley No. 2 quarry (Lehigh Portland Cement Co)	30
O-2C	Oley No. 1 quarry (Lehigh Portland Cement Co)	30
O-3	Houck quarry	30
O-4	Guldin, Peter quarry	30
O-5	Hartman, Levi quarry	30
O-6	Guldin, S.P. quarry	30
O-7	Schaeffer, E. quarry	30
O-8	Weaver, Col. J. quarry	30
O-9	Grim, Seth quarry	30
O-10	Kemmerer quarry	30
O-11	Sneider, P. quarry	30
O-12	Hine quarry	30
O-13	Knabb, Levi quarry	30
O-14	Raudenbush quarry	30
O-15	Griesermer, Ezra quarry	30
O-16	Fisher, J.G. quarry	30
O-17	Herbein, Levi quarry	30
O-18	Bertolet, L.J. quarry	30
O-19	Yale quarry	19
O-20	Snyder, John quarry	30
O-21	Brumbach, Isaac quary	14
O-22	Yoder, David quarry	17
O-23	Schearer. Reuben quarry	17

APPENDIX 3

Map Number	Name of Mine	Appendex 1 Figure Number
O-24	Wilman quarry	17
O-25	Schollenberger quarry	17
O-26	Weidner, William quarry	17
O-27	Deisher quarry	14
O-29	Bertolet, J.G. quarry	14
O-30	Bertolet, D.F. quarry	14
O-31	Kauffman, F.V. quarry north	30
O-32	Bertolet, Israel quarry	14
O-33	Lee quarry and farm Locality	14
O-34	Hunter mine	30
O-35	Weaver mine	30
O-36	Manwiller mine	30
O-37	Hertzel & Swoyer mine	14
O-38	Talley mine	14
O-39	Reiffe quarry	14
O-40	Oley uranium occurrence	14
ON-1	Ontelaunee quarry	9
ON-2	Maidencreek Station quarry	10
ON-3	Leesport Iron Company quarries	7
P-1	Mica prospect No. 3	18
P-2	Lobachsville mines	17
P-4	Rohrback mine	17
P-5	Yoder, S. mine	17
P-6	Hill Church riebeckite locality 5	18
P-7	Hill Church riebeckite locality 6	18
P-8	Keim, John quarries	14
PY-1	Glen-Gery quarry	7
PY-2	Leiby flagstone quarry	2
R-1	Mount Penn Iron mines	27
R-2	Mount Penn white spot	27
R-3	Whitman quarry	27
R-4	Mount Penn quarries	27
R-5	Long's quarry	27
R-6	Gring's quarry	26
RB-1	Dyer Gibraltar quarry	31
RB-2	Dyer Trap Rock (Birdsboro Materials) quarry	31
RB-3	Humphrey sandstone quarry	36
RB-4	Hay Creek prehenite locality	36
RH-1	Moselem mines	11
RH-2	Kirbyville-Moselem Springs quartz crystal locality	11
RH-3	Kirbyville quarries	11
RH-4	Leesport Iron Company mine	11
RH-5	C.K. Williams & Company ocher mine	12
RH-6	Heffner, Charles mines	12
RH-7	Merkel quarries	11

APPENDIX 3

Map Number	Name of Mine	Appendex 1 Figure Number
RH-8	Richmond quarry No. 5	3
RK-1	Beitler mine	15
RK-2	Brumbach, Percival mine	14
RK-3	Bieber mine	12
RK-4	Fisher, Daniel mine	14
RK-5	New Jerusalem mines	12
RK-6	Schrading mine	15
RK-7	Flint Hill locality	15
RK-8	Green Hill area	14
RK-9	Old Millert mine	14
RM-1	Udree mine	13
RM-2	Clymer mines	13
RM-3	Old tunnel mine	14
RM-4	Schitter mine	14
RM-5	Pricetown Area	13
RM-6	Sand Hill sand quarries	13
RM-7	Schweiter & Kutz mine	11, 12
RM-8	Blue quartzite quarry	12
RO-1	Wenrich, Albert quarry	22
S-1	Ludwig quarry	24
S-2	Ruth quarry No. 1	24
S-3	Ruth quarry No. 2	24
S-4	Wheatfield mine	25
S-5	Ruth mine	25
S-6	Raub mine	25
S-7	Breneman quarry	24
S-8	Fritztown area	25
S-9	Grill mine	25
S-10	Eureka mine	26
S-11	Gring's sand quarry	25
SH-1	Goul quarry No. 1	23
SH-2	Goul quarry No. 2	23
SH-3	Goul quarry No. 3	23
SH-4	Seltzer, Daniel quarry	24
SH-5	Huyett quarry	24
SH-6	Limestone quarry No. 2	23
SH-7	Fritztown area	25
SH-8	Cushion Mountain locality	25
SH-10	Cushion Mountain iron mines	25
SH-11	Hain mine	23
SS-1	Evans quarry	24
SS-2	Peipher quarry	24
SS-3	Deckert quarry	24
SS-4	Shillington Road quarry	24
SS-5	Theirwechler mine	24

APPENDIX 3

Map Number	Name of Mine	Appendex 1 Figure Number
T-1	McLain farm fetid barite locality	6
T-2	Burkholder farm fetid barite locality	6
T-3	Moore farm fetid barite locality	6
T-4	Gibble farm fetid barite locality	6
T-5	Kurtz, Elvin farm fetid barite locality	6
T-6	Kutz-Landis farm fetid barite locality	6
T-7	Bohn farm fetid barite locality	6
U-1	Dyer Monocacy quarry	36
W-1	Landis mine	20
W-2	Barto mine	20
W-3	Berthou mine	20
W-4	Deysher mine	20
W-5	Stauffer mine	18
W-6	Gilbert, L. (Edison) mine	18
W-7	Gilberg, Jacob mine	18
W-8	Sparr mine	20
W-9	Eline mine	20
W-10	Gilbert mine	34
W-11	Fegley mine	34
W-12	Schall quarry	20
W-13	Oberholtzer quarry	20
W-14	Dielh quarry	20
W-15	Barr quarry	20
W-16	Eshbach area	20
W-17	Rush quarries	20
W-18	Gabel quarry	18
WE-1	Deppen quarry No. 1	23
WE-2	Deppen quarry No. 2	23
WE-3	Witmoyer & Brother quarry	23
WE-4	Wernersville Lime & Stone Company quarry	23
WE-4	Miller, Henry quarry	23
WR-2	Drexel quarry	26
WY-1	Seitzinger mine	26
WY-2	Muhlenberg (Beidler) mine	26

Made in the USA
Middletown, DE
11 June 2023

32057250R00263